Nitrides with Nonpolar Surfaces

Edited by
Tanya Paskova

Related Titles

Neumark, G. F., Kuskovsky, I. L., Jiang, H. (eds.)

Wide Bandgap Light Emitting Materials and Devices

2007

ISBN: 978-3-527-40331-8

Piprek, J. (ed.)

Nitride Semiconductor Devices: Principles and Simulation

2007

ISBN: 978-3-527-40667-8

Sze, S. M., Ng, K. K.

Physics of Semiconductor Devices

2006

ISBN: 978-0-471-14323-9

Adachi, S.

Properties of Group-IV, III-V and II-VI Semiconductors

2005

ISBN: 978-0-470-09032-9

Ruterana, P., Albrecht, M., Neugebauer, J. (eds.)

Nitride Semiconductors

Handbook on Materials and Devices

2003

ISBN: 978-3-527-40387-5

Nitrides with Nonpolar Surfaces

Growth, Properties, and Devices

Edited by
Tanya Paskova

WILEY-VCH Verlag GmbH & Co. KGaA

The Editor

Dr. Tanya Paskova
Kyma Technologies
Raleigh, NC, USA
E-mail: paskova@kymatech.com

Cover

Schematic representation of nitride crystal structure (B. Haskell, Santa Barbara, USA)

Library of Congress Card No.: applied for

British Library Cataloguing-in-Publication Data
A catalogue record for this book is available from the British Library.

Bibliographic information published by the Deutsche Nationalbibliothek
Die Deutsche Nationalbibliothek lists this publication in the Deutsche Nationalbibliografie; detailed bibliographic data are available on the Internet at <http://dnb.d-nb.de>.

Composition Laserwords Private Ltd., Chennai, India

Printing Strauss GmbH, Darmstadt

Bookbinding Litges & Dopf GmbH, Heppenheim

Printed in the Federal Republic of Germany
Printed on acid-free paper

ISBN: 978-3-527-40768-2

Contents

Nitrides with Nonpolar Surfaces. Growth, Properties, and Devices. Edited by Tanya Paskova

ISBN: 978-3-527-40768-2

Preface

Starting with the pioneering work of Paul Maruska and Jacques Pankove on GaN in the late 1960s and early 1970s [1, 2], and the subsequent major breakthroughs by Amano and Akasaki in the 1980s [3, 4] and then by Nakamura in the early to mid-1990s [5–9], nearly all work on GaN-based materials and devices has been on *c*-plane (0001) oriented layers. Since the wurtzite structure is noncentrosymmetric (space group $P6_3mc$; point group $6mm$), it is polar with its spontaneous polarization parallel to the *c*-axis. Through the mid-to-late 1990s, there was a broadening realization of the importance of piezoelectricity and polarization in group III nitrides [10–12]. It is clear now that polarization-related effects are a dominant feature in common *c*-axis GaN-based heterostructures.

At the same time that Nakamura was making his major breakthroughs in GaN-based materials and devices, Vanderbilt was developing the quantum theory of polarization [13, 14]. In this landmark work, Vanderbilt showed that polarization in a solid is a bulk property and can be determined quantum mechanically with knowledge of the phase of the valence electron wavefunctions. Subsequently, Bernardini, Fiorintini, and Vanderbilt calculated the spontaneous polarization and piezoelectric constants for GaN, AlN, and InN [15] – this work is the standard used today for these important physical properties.

It is now well established that in *c*-plane GaN-based heterostructures fixed charges form at heterointerfaces due to discontinuities in spontaneous polarization and piezoelectric polarization between constituent layers. For electron devices, such as now commonly on AlGaN/GaN heterostructures, the polarization discontinuity at the AlGaN/GaN interface gives rise to a significant positive fixed sheet charge, which may be screened by free electrons in a two-dimensional electron gas (2DEG) when the AlGaN layer has suitable thickness. For commonly used $Al_{0.30}Ga_{0.70}N$/GaN high-electron mobility transistor structures, the electron concentration in the 2DEG is on the order of 10^{13} cm^{-2}. For electron devices, polarization is largely seen as beneficial as it gives rise to 2DEG sheet concentrations nearly an order of magnitude larger than what is realizable in conventional III–V systems.

In contrast, for light emitting devices, for example, those based on GaN barriers and InGaN quantum wells, the fixed sheet charges at the

Nitrides with Nonpolar Surfaces. Growth, Properties, and Devices. Edited by Tanya Paskova

ISBN: 978-3-527-40768-2

heterointerfaces cause large electric fields in the quantum wells (fields on the order of 1 MV cm^{-1}). These internal electric fields cause separation of electron and hole wave function (quantum confined Stark effect), reduced radiative recombination rate, and possibly reduced efficiency. Other features of *c*-plane GaN-based LEDs include a significant blue shift in the electroluminescence with increasing drive current.

In 2000, Waltereit, Brandt, and coworkers in Klaus Ploog's group at the Paul Drude Institute in Berlin reported the first planar nonpolar $(1\bar{1}00)$ *m*-plane films on (100) oriented γ-$LiAlO_2$ substrates [16]. In their breakthrough paper in *Nature*, Waltereit demonstrated the absence of internal electric fields in *m*-plane GaN quantum wells with AlGaN barriers. Waltereit's work motivated a worldwide research effort in nonpolar and semipolar nitrides.

This volume presents major developments in nonpolar nitrides over the last five to six years. Extended defects, namely threading dislocations, basal plane stacking faults, and associated partial threading dislocations, have been a major obstacle to the development of nonpolar and semipolar nitrides, as described by Liliental–Weber and Kröger. Haskell, Fini, and Nakamura write about efforts to reduce extended defect densities via lateral epitaxial overgrowth (LEO) in HVPE. Grzegory and coworkers present the work from the Polish Academy of Sciences to produce nearly defect-free nonpolar bulk GaN substrates. The reduced in-plane symmetry of nonpolar orientations, in comparison to the usual *c*-plane GaN orientations, facilitates polarized light detection and emission as described by Grahn. The early efforts on nonpolar LEDs are described by Chakbraborty, Chichibu, and Mishra. Funato, Kawakami, Narukawa, and Mukai present the Kyoto University and Nichia's work on semipolar LEDs on inclined semipolar facets in *c*-plane LEO GaN structures and on bulk semipolar substrates – the latter representing the first reports of high-performance semipolar emitters [17].

Since this volume was prepared, the progress in nonpolar and semipolar nitrides has been remarkable. In 2006, the Mitsubishi Chemical Corporation developed low defect density nonpolar and semipolar substrates by wafering thick *c*-plane GaN boules produced by HVPE. The groups at Rohm and at UCSB have produced nonpolar and semipolar LEDs with performance that rivals state-of-the-art *c*-plane devices [18–21]. The Rohm and UCSB groups demonstrated the first nonpolar laser diodes and the first AlGaN clad-free laser diodes [22]. The promise of nonpolar light emitters has now been realized – in the coming years, we will learn whether nonpolar structures will be the ultimate solution for nitride emitters and for a range of electron devices.

References

1. Maruska, H.P. and Tietjen, J.J. (1969) *Applied Physics Letters*, **15**, 327.
2. Maruska, H.P., Stevenson, D.A., and Pankove, J.I. (1973) *Applied Physics Letters*, **22**, 303.
3. Amano, H., Sawaki, N., Akasaki, I., and Toyoda, Y. (1986) *Applied Physics Letters*, **48**, 333.
4. Amano, H., Kito, M., Hiramatsu, K.I., and Akasaki, I. (1989) *Japanese*

Journal of Applied Physics, **28**, L2112.
5. Nakamura, S. (1991) *Japanese Journal of Applied Physics*, **30**, L1705.
6. Nakamura, S., Mukai, T., and Senoh, M. (1991) *Japanese Journal of Applied Physics*, **30**, L1998.
7. Nakamura, S., Iwasa, N., Senoh, M., and Mukai, T. (1992) *Japanese Journal of Applied Physics*, **31**, 1258.
8. Nakamura, S., Mukai, T., and Senoh, M. (1994) *Applied Physics Letters*, **64**, 1687.
9. Nakamura, S., Senoh, M., Nagahama, S. *et al.* (1996) *Japanese Journal of Applied Physics*, **35**, L74.
10. Bykhovski, A., Gelmont, B., and Shur, M. (1993) *Journal of Applied Physics*, **74**, 6734.
11. Im, J.S., Kollmer, H., Off, J. *et al.* (1998) *Physical Review*, **B57**, R9435.
12. Takeuchi, T., Wetzel, C., Yamaguchi, S. *et al.* (1998) *Applied Physics Letters*, **73**, 1691.
13. Kingsmith, R.D. and Vanderbilt, D. (1993) *Physical Review*, **B47**, 1651.
14. Vanderbilt, D. and Kingsmith, R.D. (1993) *Physical Review*, **B48**, 4442.
15. Bernardini, F., Fiorentini, V., and Vanderbilt, D. (1997) *Physical Review*, **B56**, 10024.
16. Waltereit, P., Brandt, O., Trampert, A. *et al.* (2000) *Nature*, **406**, 865.
17. Funato, M., Ueda, M., Kawakami, Y. *et al.* (2006) *Japanese Journal of Applied Physics*, **45**, L659.
18. Okamoto, K., Ohta, H., Nakagawa, D. *et al.* (2006) *Japanese Journal of Applied Physics*, **45**, L1197.
19. Schmidt, M.C., Kim, K.C., Sato, H. *et al.* (2007) *Japanese Journal of Applied Physics*, **46**, L126.
20. Schmidt, M.C., Kim, K.C., Farrell, R.M. *et al.* (2007) *Japanese Journal of Applied Physics*, **46**, L190.
21. Okamoto, K., Ohta, H., Chichibu, S.F. *et al.* (2007) *Japanese Journal of Applied Physics*, **46**, L187.
22. Feezell, D.F., Schmidt, M.C., Farrell, R.M. *et al.* (2007) *Japanese Journal of Applied Physics*, **46**, L284.

James S. Speck and Steven P. DenBaars
Materials Department
University of California
Santa Barbara, CA 93106

List of Contributors

Isamu Akasaki
Meijo University
1-501 Shiogamaguchi
Tempaku-ku
Nagoya 468-8502
Japan

Hiroshi Amano
Meijo University
1-501 Shiogamaguchi
Tempaku-ku
Nagoya 468-8502
Japan

Michał Boćkowski
Institute of High Pressure Physics
Polish Academy of Sciences
ul. Sokołowska 29/37
01-142 Warsaw
Poland

Oliver Brandt
Paul-Drude-Institut für
Festkörperelektronik
Hausvogteiplatz 5-7
10117 Berlin
Germany

Arpan Chakraborty
Cree Santa Barbara Technology Center
340 Storke Road
Goleta, CA 93117
USA

Shigefusa Chichibu
Institute of Multidisciplinary Research
for Advanced Materials
Tohoku University
2-1-1 Katahira
Aoba
Sendai 980-8577
Japan

Vanya Darakchieva
Department of Physics,
Chemistry and Biology
Linköping University
581 83 Linköping
Sweden

Bruno Daudin
CEA-Grenoble
Département de Recherche
Fondamentale sur la Matière
Condensée, SP2M
17 rue des Martyrs
38054 Grenoble cedex 9
France

Steven P. DenBaars
Materials Department
University of California
Santa Barbara, CA 93106
USA

Nitrides with Nonpolar Surfaces. Growth, Properties, and Devices. Edited by Tanya Paskova

ISBN: 978-3-527-40768-2

Paul T. Fini
Inlustra Technologies LLC
5385 Hollister Avenue
Santa Barbara, CA 93111
USA

Sebastien Founta
CEA-Grenoble
Département de Recherche
Fondamentale sur la Matière
Condensée, SP2M
17 rue des Martyrs
38054 Grenoble cedex 9
France

Mitsuru Funato
Department of Electronic
Science and Engineering
Kyoto University
Kyoto 615-8510
Japan

Bruno Gayral
CEA-Grenoble
Département de Recherche
Fondamentale sur la Matière
Condensée, SP2M
17 rue des Martyrs
38054 Grenoble cedex 9
France

Holger T. Grahn
Paul-Drude-Institut für
Festkörperelektronik
Hausvogteiplatz 5-7
10117 Berlin
Germany

Izabella Grzegory
Institute of High Pressure Physics
Polish Academy of Sciences
ul. Sokołowska 29/37
01-142 Warsaw
Poland

Benjamin A. Haskell
Inlustra Technologies LLC
5385 Hollister Avenue
Santa Barbara, CA 93111
USA

Daisuke Iida
Meijo University
1-501 Shiogamaguchi
Tempaku-ku
Nagoya 468-8502
Japan

Masataka Imura
Meijo University
1-501 Shiogamaguchi
Tempaku-ku
Nagoya 468-8502
Japan

Motoaki Iwaya
Meijo University
1-501 Shiogamaguchi
Tempaku-ku
Nagoya 468-8502
Japan

Satoshi Kamiyama
Meijo University
1-501 Shiogamaguchi
Tempaku-ku
Nagoya 468-8502
Japan

Yoichi Kawakami
Department of Electronic
Science and Engineering
Kyoto University
Kyoto 615-8510
Japan

Takeshi Kawashima
Meijo University
1-501 Shiogamaguchi
Tempaku-ku
Nagoya 468-8502
Japan

Roland Kröger
Institute of Solid State Physics
University of Bremen
Otto-Hahn Allee NW1
28359 Bremen
Germany

Zuzanna Liliental-Weber
Lawrence Berkeley National Laboratory
Materials Science Department
1 Cycolotron Road
Berkeley, CA 94720
USA

Bolesław Łucznik
Institute of High Pressure Physics
Polish Academy of Sciences
ul. Sokołowska 29/37
01-142 Warsaw
Poland

Umesh Mishra
ECE Department
University of California
Santa Barbara, CA 93106-9560
USA

Bo Monemar
Department of Physics and Measurement Technology
Linköping University
581 83 Linköping
Sweden

Takashi Mukai
Nichia Corporation
Nitride Semiconductor Research Laboratory
Anan
Tokushima 774-8601
Japan

Shuji Nakamura
Materials Department
University of California
Santa Barbara, CA 93106-5050
USA

Yukio Narukawa
Nichia Corporation
Nitride Semiconductor Research Laboratory
Anan
Tokushima 774-8601
Japan

Plamen P. Paskov
Department of Physics, Chemistry and Biology
Linköping University
581 83 Linköping
Sweden

Tanya Paskova
Kyma Technologies, Inc.
8829 Midway West Road
Raleigh, NC 27617
USA

B. Pastuszka
Institute of High Pressure Physics
Polish Academy of Sciences
ul. Sokołowska 29/37
01-142 Warsaw
Poland

Sylwester Porowski
Institute of High Pressure Physics
Polish Academy of Sciences
ul. Sokołowska 29/37
01-142 Warsaw
Poland

Fabian Rol
CEA-Grenoble
Département de Recherche
Fondamentale sur la Matière
Condensée, SP2M
17 rue des Martyrs
38054 Grenoble cedex 9
France

Mathias Schubert
Department of Electrical Engineering
and Center for Materials Research
and Analysis
University of Nebraska-Lincoln
Lincoln, NE 68588-0511
USA

James S. Speck
Materials Department
University of California
Santa Barbara, CA 93106
USA

Jun Suda
Semiconductor Science and
Engineering Laboratory
Department of Electronic
Science and Engineering
Kyoto University Katsura Campus
Kyoto 6158510
Japan

Henryk Teisseyre
Institute of High Pressure Physics
Polish Academy of Sciences
ul. Sokołowska 29/37
01-142 Warsaw
Poland

Dmitri Nikdaj Zakharov
Lawrence Berkeley National Laboratory
Berkeley, CA 94720
USA
at present: Birck Nanotechnology
Center
Purdue University
West Lafayette, IN 47907
USA

Color Plates

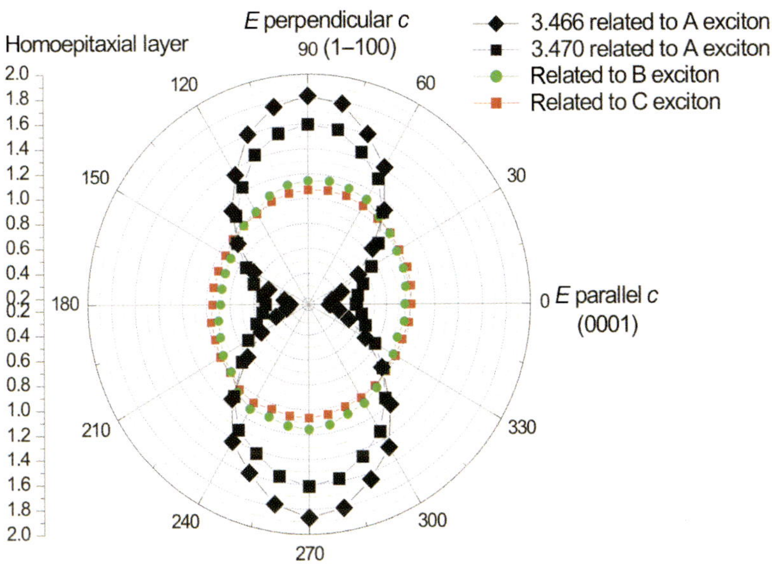

Figure 3.14 (p. 66) Polarization of light emitted from GaN homo-epitaxial layer grown on the nonpolar GaN substrate.

Nitrides with Nonpolar Surfaces. Growth, Properties, and Devices. Edited by Tanya Paskova

ISBN: 978-3-527-40768-2

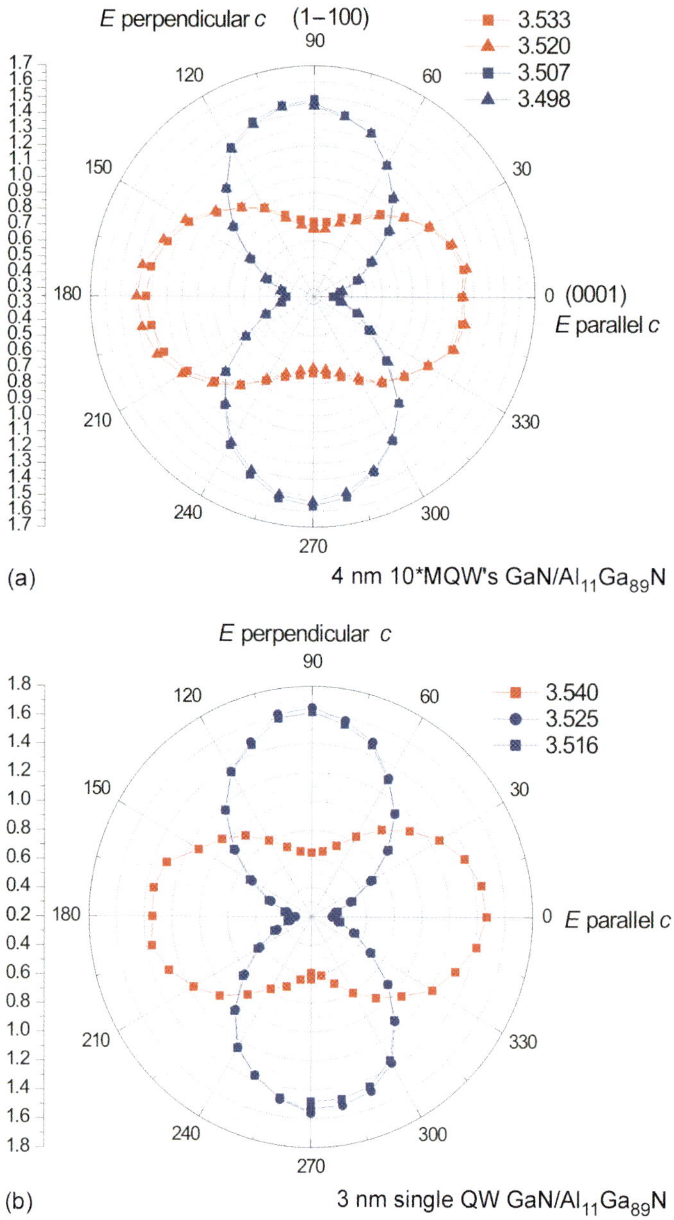

Figure 3.15 (p. 67) Polarization of light emitted from GaN–AlGaN MQW of different thicknesses grown on nonpolar GaN substrates.

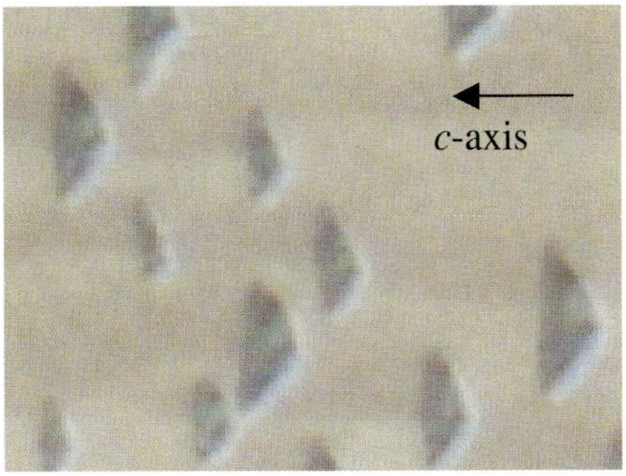

Figure 5.3 (p. 105) Surface pits of the *a*-plane GaN grown at atmospheric pressure on *r*-plane sapphire.

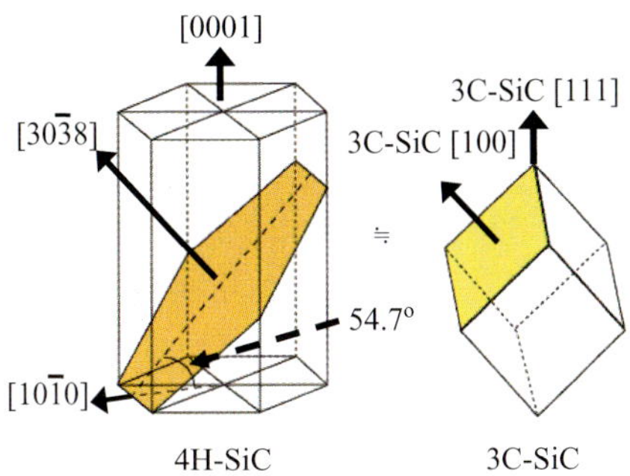

Figure 5.10 (p. 108) Similarity between (30$\bar{3}$8) of 4H-SiC and (100) of 3C-SiC.

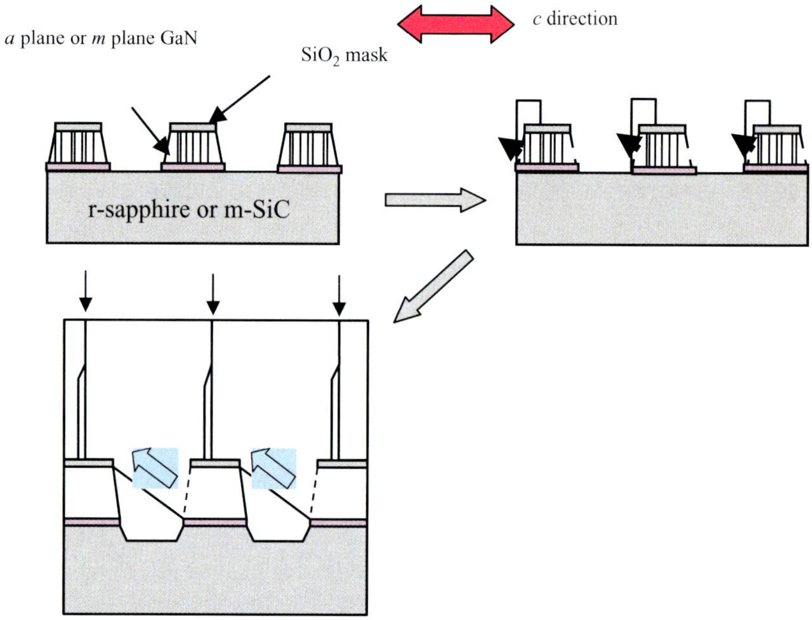

Figure 5.14 (p. 111) Process of sidewall-seeded epitaxial lateral overgrowth of nonpolar GaN grown on *r*-plane sapphire or *m*-plane SiC.

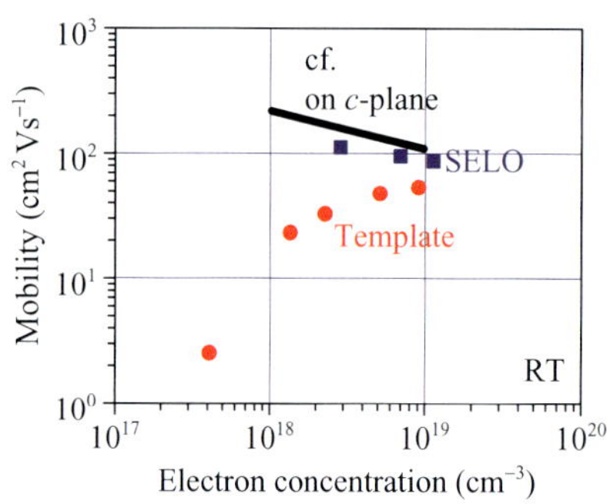

Figure 5.17 (p. 113) Electron mobility at room temperature of *n*-type Si-doped *a*-plane GaN grown on *r*-plane sapphire. Planar GaN grown on AlGaN/AlN intermediate layer is indicated by Template (circles). SELO indicates GaN grown with the SELO process (squares). That of *c*-plane GaN grown on sapphire using a low-temperature buffer layer is also shown for comparison.

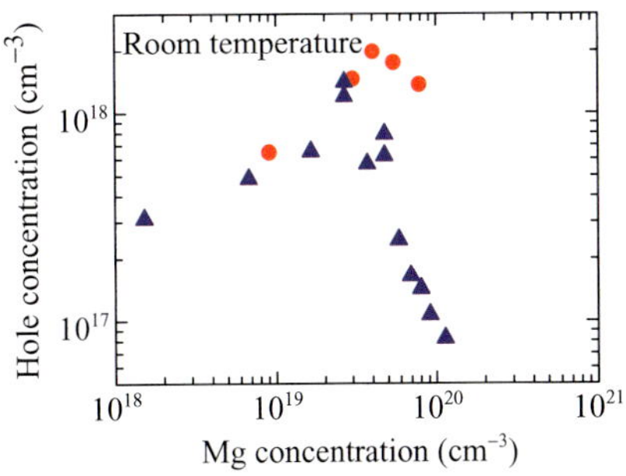

Figure 5.18 (p. 114)
Room-temperature hole concentration of *a*-plane GaN grown on *r*-plane sapphire using AlGaN/AlN intermediate layer (red circle) and *c*-plane GaN grown on *c*-plane sapphire using low-temperature buffer layer (blue triangle).

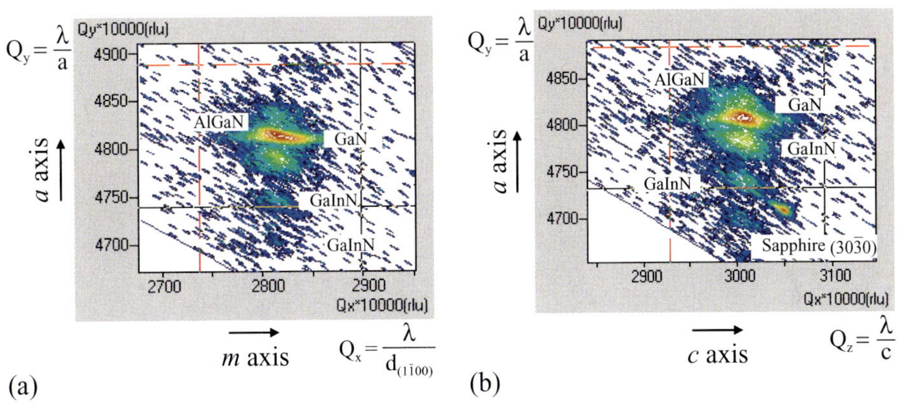

Figure 5.19 (p. 115) RSM of the green LED having GaInN (3 nm)/GaN (12 nm) MQWs. (a) near (20$\bar{2}$0) diffraction, and (b) near (11$\bar{2}$2) diffraction.

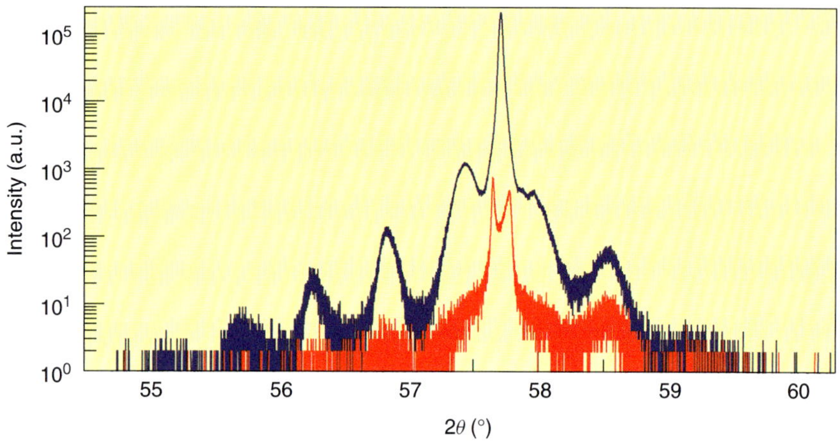

Figure 5.20 (p. 115) $2\theta/\omega$ scan X-ray diffraction profiles from the *a*-plane green LED having GaInN/GaN MQWs. Lower Line: MQWs on *a*-plane GaN template. Upper line: MQWs on SELO-grown GaN.

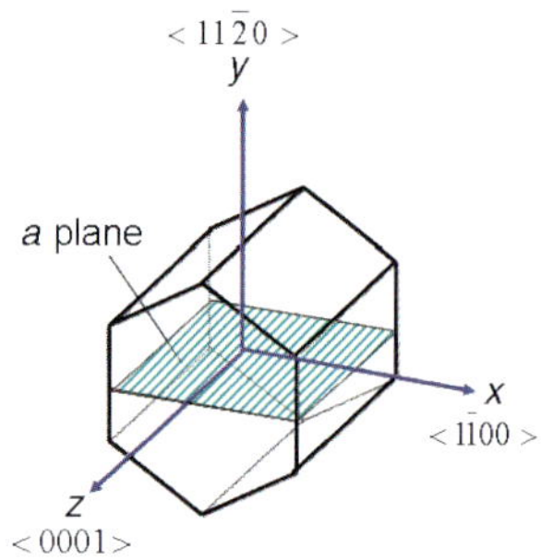

Figure 5.21 (p. 116) Directions of *x*, *y* and *z* axes corresponding to the crystal axes *m*, *a* and *c*.

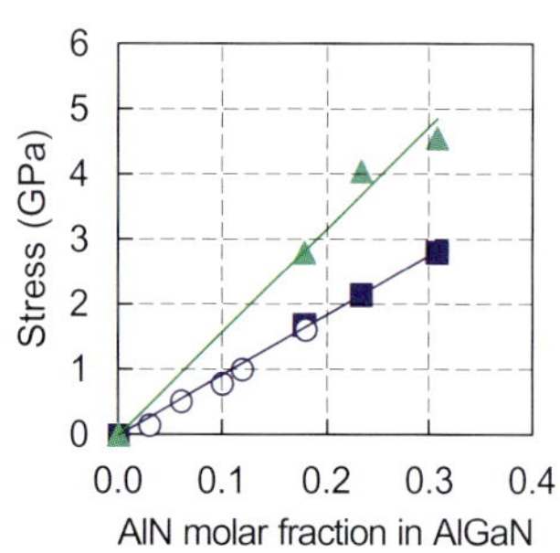

Figure 5.22 (p. 117) Stress in the AlGaN layer coherently grown on GaN. Solid squares: stress along *m* axis in *a* plane. Solid triangle: stress along *c* axis in *a* plane. Open circles: in-plane stress in *c* plane.

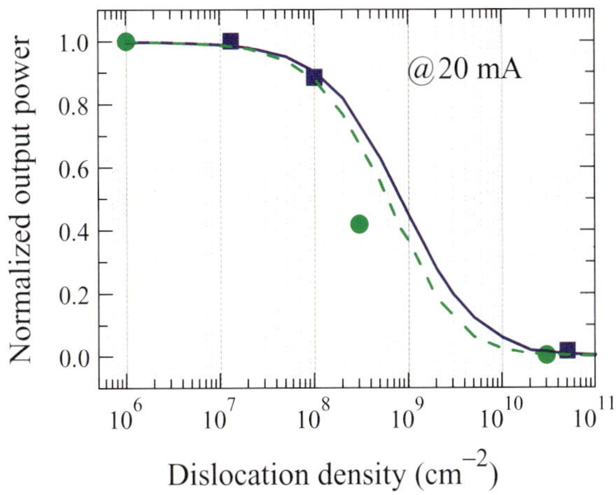

Figure 5.23 (p. 117) Normalized output power of *m*-plane green LEDs (circle) and *a*-plane violet LEDs (square) as a function of dislocation density. Each output power is normalized by the output power of SELO-grown LED. Solid and broken lines show the internal quantum efficiencies of violet and green LEDs as a function of dislocation density.

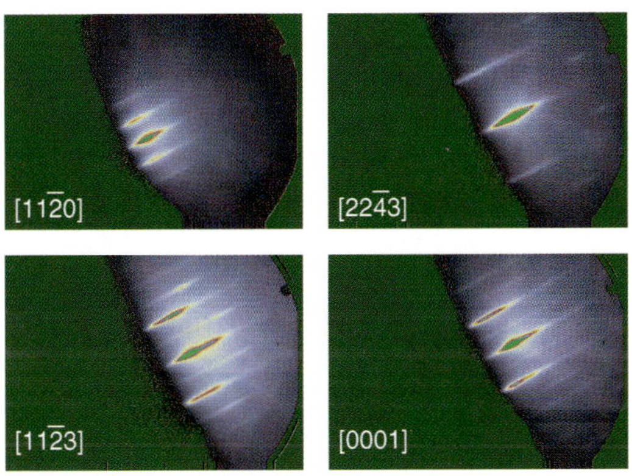

Figure 6.8 (p. 132) RHEED patterns of the GaN(1$\bar{1}$00) film recorded at 100 °C after a two-step deposition of 2 MLs Ga at 580 °C. Note the intense half-order streaks along all azimuths.

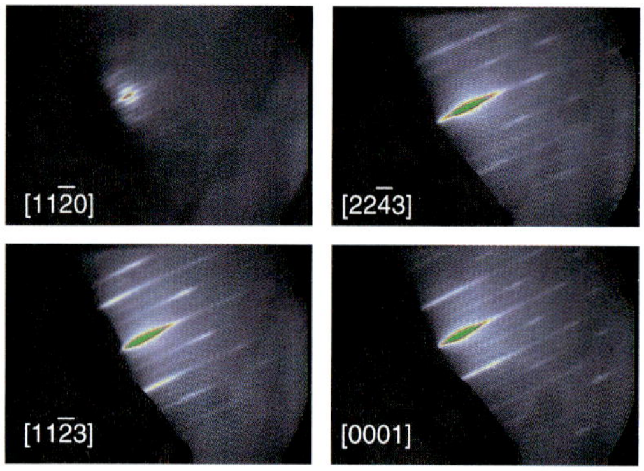

Figure 6.9 (p. 133) RHEED patterns of the GaN(1$\bar{1}$00) film recorded at 100 °C after a three-step deposition of 3 MLs Ga at 580 °C. Note the intense 1/4-order streaks along all azimuths.

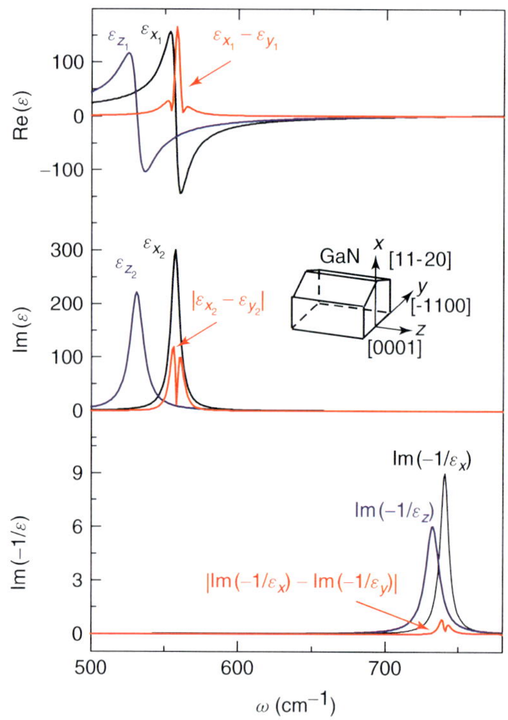

Figure 9.14 (p. 241) Real and imaginary parts of the dielectric function ε and imaginary part of the dielectric loss functions $-1/\varepsilon$ for in- (x) and out-of-plane (z) directions of one representative a-plane GaN film grown on r-plane sapphire surface. For clarity, the functions for directions y are given through the differences between those for directions (x) parallel to the GaN [11$\bar{2}$0] and (y) parallel to the GaN [1$\bar{1}$00] directions.

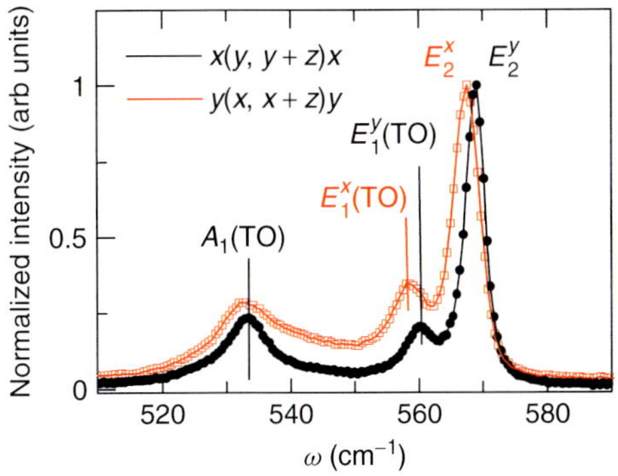

Figure 9.16 (p. 245) Raman spectra in $x(y, y+z)\bar{x}$ and $y(x, x+z)\bar{y}$ scattering geometries from an *a*-plane GaN film on *r*-plane sapphire.

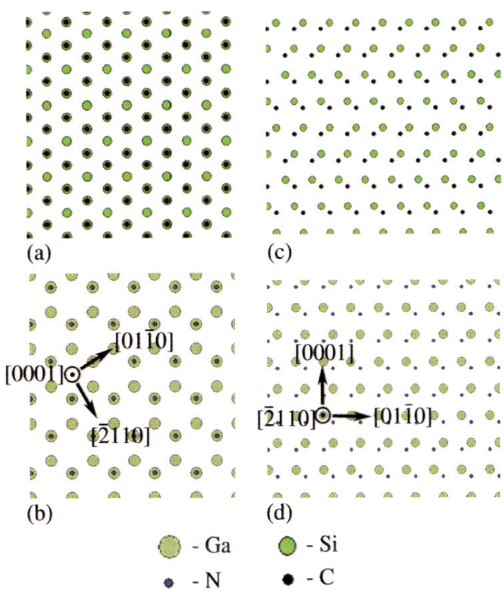

Figure 10.3 (p. 258) Schematic of the atomic arrangement on different crystallographic planes: (a) (0001) 4H-SiC; (b) (0001)GaN; (c) (2110) 4H-SiC; (d) (2110) GaN.

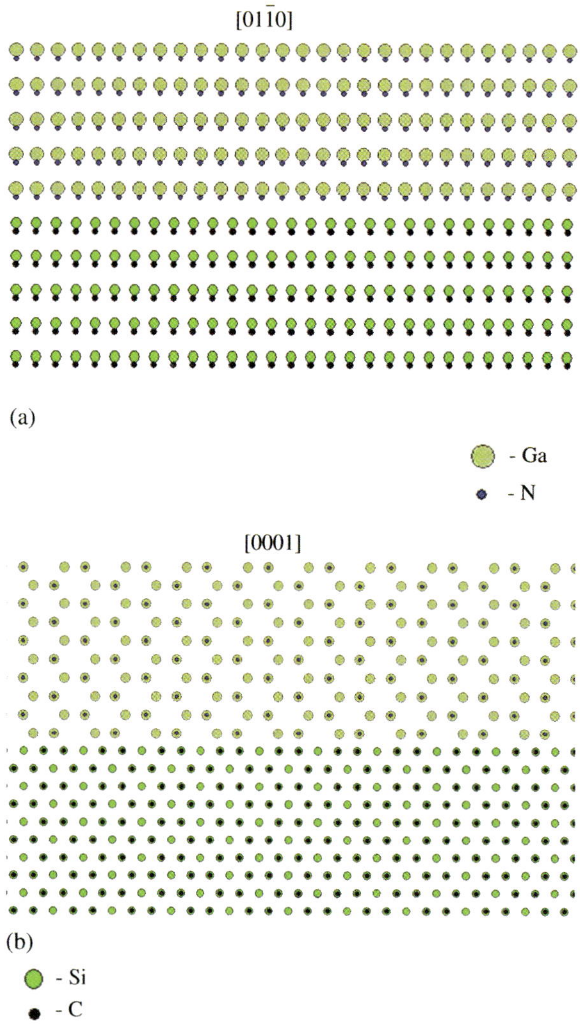

Figure 10.4 (p. 259) Schematic of the atomic arrangement in GaN/4H-SiC cross-sectional samples in (a) $[01\bar{1}0]_{GaN}$ and (b) $[0001]_{GaN}$ orientations.

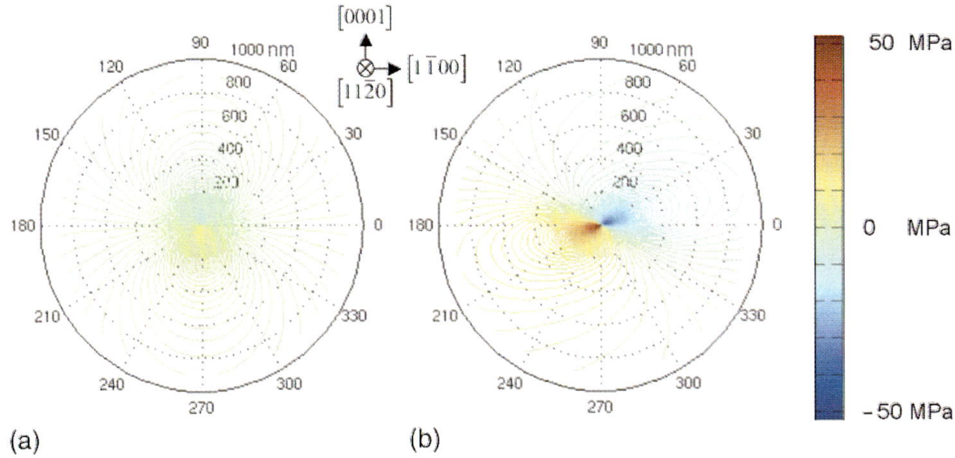

Figure 11.11 (p. 301) Polar plots of the hydrostatic pressure field of a 1/6[2–203] Frank–Shockley partial dislocation in *a*-plane GaN (a) assuming the isotropic case, (b) taking anisotropy into account.

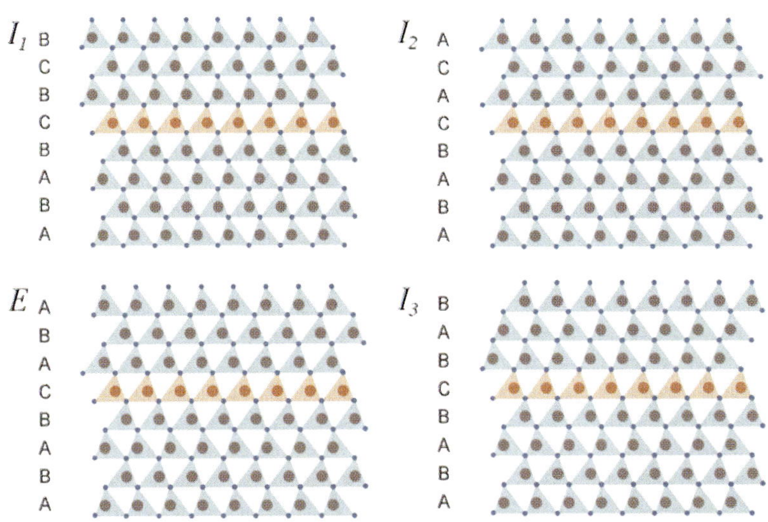

Figure 11.12 (p. 302) Types of intrinsic (*I*) and extrinsic (*E*) BSF types in the wurtzite structure showing the change of stacking sequence of the tetrahedrons drawn around the N-atoms: I_1, I_2, E, and I_3.

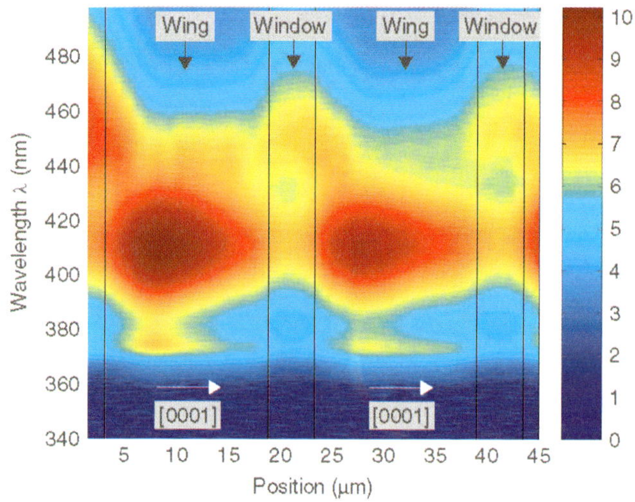

Figure 12.4 (p. 325) Contour plot of the μ-PL intensity on the wing and window regions of the LEO sample as a function of the wavelength (the intensity is plotted in a logarithmic scale).

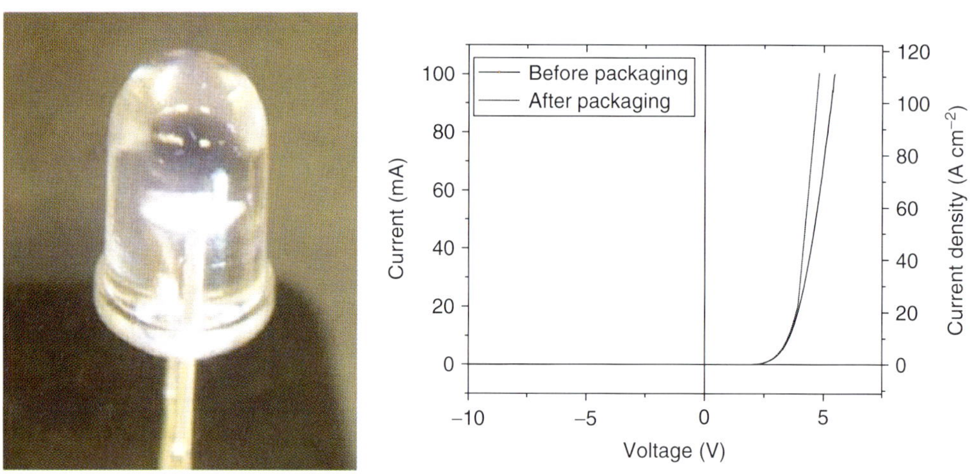

Figure 12.26 (p. 347) An image of the packaged *m*-plane lamp and comparison of the *I–V* characteristics of *m*-plane LED before and after packaging.

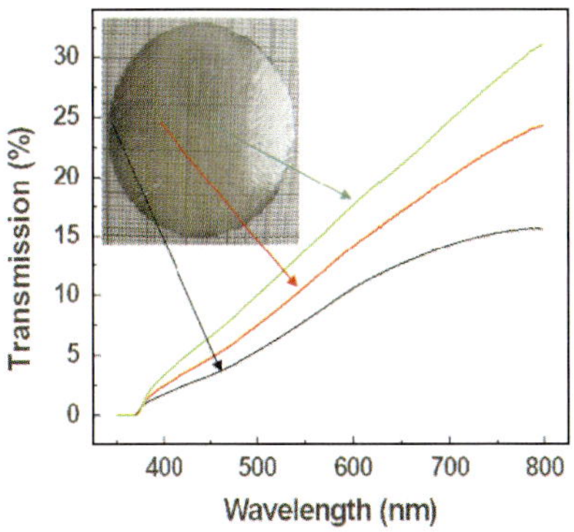

Figure 12.30 (p. 349) Transmission through *m*-plane substrate measured as a function of wavelength.

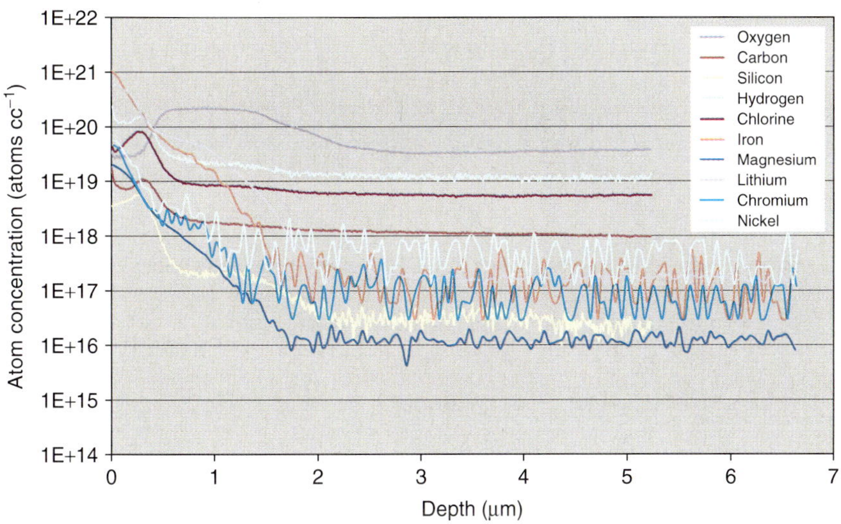

Figure 12.31 (p. 350) SIMS measurement performed at the back of the free-standing *m*-plane GaN substrate (courtesy of Haskell *et al.*).

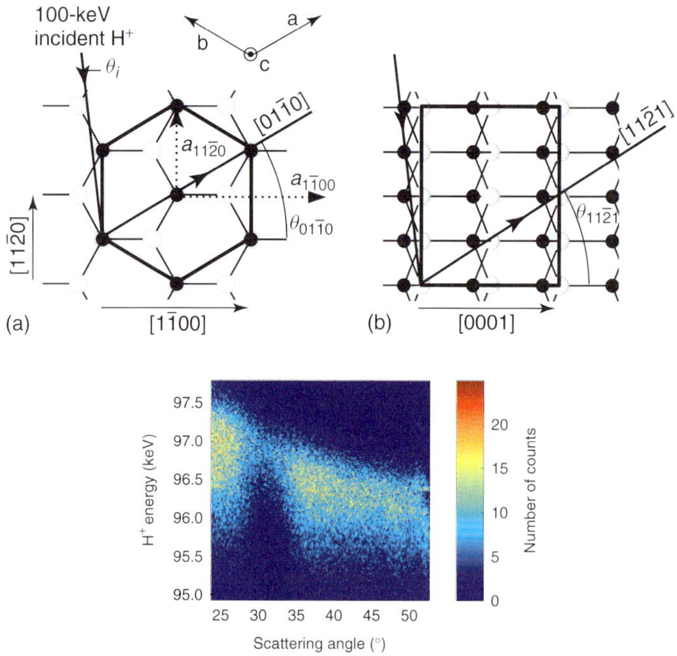

Figure 13.5 (p. 367) Scheme of the scattering geometry in the (a) (0001) and (b) (1-100) planes for medium energy ion scattering, and (c) energy and angular distributions of the backscattered protons for sample C, close to the [01-10] blocking direction. An angle of 30° corresponds to the SiC substrate, which is used as a reference.

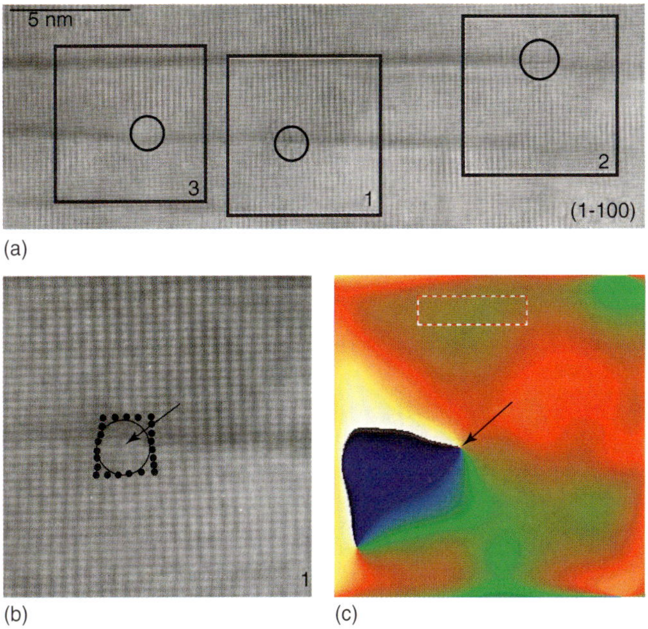

Figure 13.7 (p. 370) HRTEM cross-sectional image taken along the [1-100] zone axis of part of a sample containing 19 *a*-plane GaN QD planes showing two GaN wetting layers in between AlN spacers (a). Circles in the areas marked 1, 2, and 3 indicate the position of misfit dislocations. An enlargement of area 1 is presented in (b), where a Burgers circuit has been drawn around the dislocation (indicated by an arrow). The phase image of (0002) lattice fringes is shown in (c). The selection on the upper part of the image corresponds to the area chosen for phase reference. The 2π-rad variation highlighted by the arrow is characteristic of the presence of a dislocation.

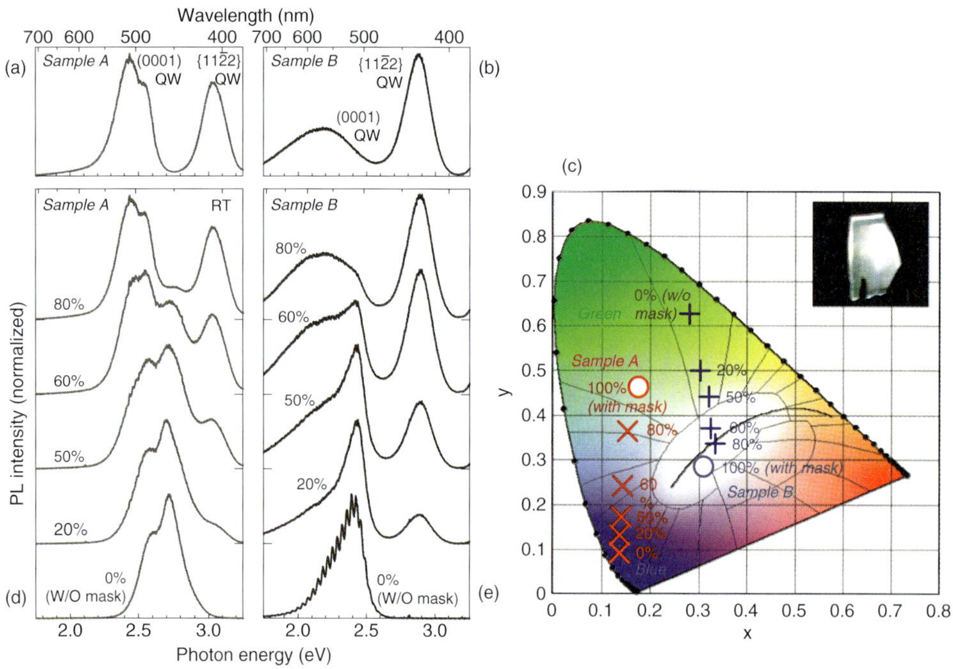

Figure 14.12 (p. 399) PL spectrum of (a) sample **A** and (b) sample **B** acquired at RT. (c) CIE chromaticity diagram where the open circles indicate the emissions from samples **A** (Figure 14.12a) and **B** (Figure 14.12b), while the crosses indicate emissions from samples with designed mask patterns. See text for more details. Variation of PL spectra due to mask geometries for (d) sample **A** and (e) sample **B**.

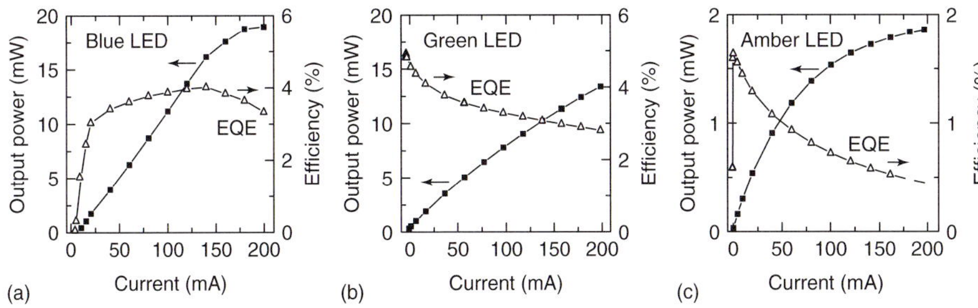

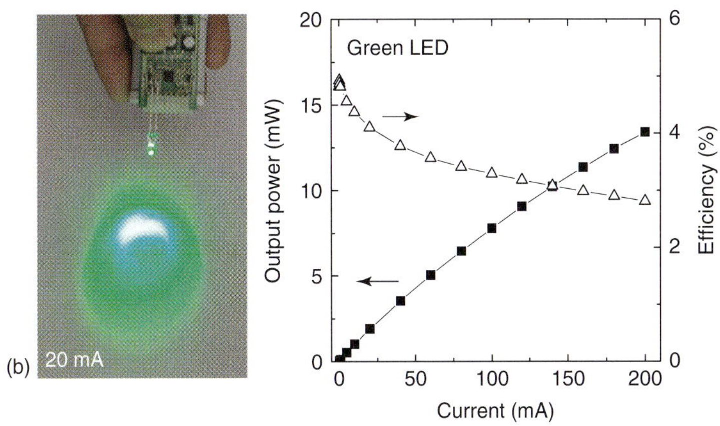

Figure 14.19 (p. 408) Output power and EQE of (a) blue, (b) green, and (c) amber LEDs.

Introduction

Nitrides with Nonpolar Surfaces. Growth, Properties, and Devices. Edited by Tanya Paskova

ISBN: 978-3-527-40768-2

1
Nitride Materials and Devices with Nonpolar Surfaces: Development and Prospects

Tanya Paskova

1.1
Introduction

The commercial fabrication of light emitting diodes (LED) and laser diodes (LD) from the ultraviolet to the amber as well as the development of high-power high electron mobility transistors (HEMT) suitable for numerous telecom applications have been driven by the tremendous improvement in the quality of nitride materials. Despite remarkable achievements during last decade, however, the device performance is strongly hampered by the presence of strong, spontaneous, and piezoelectric polarization fields along the typical growth [0001] direction in the wurtzite nitride materials [1]. As a consequence of these strong internal fields, the quantum wells (QWs), typically used in the active region of optical emitter devices, have a triangle potential profile and the oscillator strength for optical transitions is reduced by several orders of magnitude, depending on the QW thickness [2, 3]. A similar deterioration effect is observed in the nitride-based HEMT devices owing to polarization-generated charge. A typical AlGaN/GaN HEMT built in the [0001] growth direction experiences a channel charge of more than 1×10^{13} cm^{-2} [4], which, in turn, influences the doping flexibility of the two-dimensional electron gas. The latter makes the fabrication of both enhancement- and depletion-mode devices, which are highly needed for the nitride-based logic devices, more difficult. Another complication factor in the nitride wurtzite devices grown in the [0001] direction is related to the limitations of *p*-type magnesium doping levels. The best-reported electrically active levels are only in the range of $(1-2) \times 10^{18}$ cm^{-3} [5]. A potential increase of the carrier concentration would lead to a decrease in the contact resistance, a reduction in the $p-n$ junction turn-on voltage and series resistance, and would allow a production of LEDs and LDs with higher optical output efficiency.

Most of the standard devices today are based on wurtzite nitrides grown in the [0001] direction (*c*-axis) direction (Figure 1.1a) on foreign substrates.

Nitrides with Nonpolar Surfaces. Growth, Properties, and Devices. Edited by Tanya Paskova

ISBN: 978-3-527-40768-2

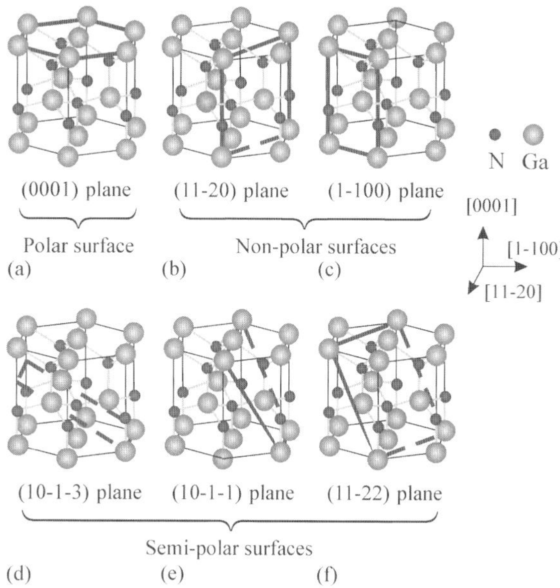

Figure 1.1 Schematic views of (a) the polar (0001) plane (b, c), the nonpolar (11-20) and (1-100) planes; and (d–f) the semipolar (10-1-3), (10-1-1), and (11-22) planes.

By using native AlN or GaN substrates, although offered in limited size and amount, the piezoelectric polarization field, being a result of the lattice-mismatch-induced strain in heteroepitaxial nitride device structures could be reduced. The spontaneous polarization field, however, is an inherent physical property of the wurtzite semiconductors, having a polar-atomic arrangement along the *c* axis (Figure 1.2a) with ionic chemical bonds. A potential remedy to the spontaneous induced charge could be the use of cubic nitrides, having no polar-atomic arrangement in the [001] direction and thus no fixed interfacial or surface polarization charges exist. Unfortunately, the present quality of the cubic nitride materials is unsuitable for device production. Another approach to resolve this issue is to use wurtzite materials that have either no polarization field or a reduced one in the growth direction and across the active device regions, respectively. Schematic views of such wurtzite crystal planes, which are either perpendicular or inclined to the [0001] polar direction, are shown in Figures 1.1b–f. There are two surfaces perpendicular to the *c* axis, which have equal numbers of the Group III and Group V atoms, thus forming the so-called nonpolar surfaces, the (11-20), which is also called *a-plane* (Figure 1.1b) and the (1-100), also called *m-plane* (Figure 1.1c). Both are the most preferable surfaces because they do not possess a spontaneous polarization field [6]. Alternatively, inclined surfaces such as (10-1-3), (10-1-1), and (11-22) are known to have lower polarization fields [6] and are often called *semipolar surfaces* (Figures 1.1d–f). The atomic arrangements of the polar

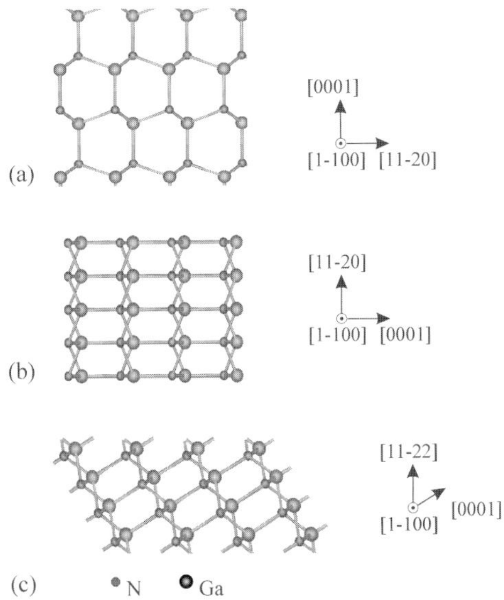

Figure 1.2 Atomic configurations of (a) the polar {0001} planes; (b) the nonpolar {11-20} planes; (c) the semipolar {11-22} planes, viewed along the [1-100] direction. (Courtesy of R. Kroeger.)

and representative nonpolar and semipolar surfaces viewed along the [1-100] direction are shown in Figures 1.2a–c, respectively.

Nitride material with such surfaces, often referred to today and hereafter for seek of simplicity as *nonpolar,* was first produced more than 30 years ago [7]. At that time, the main purpose was to optimize the material quality and heterostructure properties by finding the best possible substrate material and orientation [7–27]. The enormous potential of this type of material, with no built-in electric fields, has been emphasised by the Paul Drude Institute team in 2000. The publications by Waltereit *et al.* [28, 29] are considered to be the beginning of a 'new generation' nonpolar nitride materials. During the last few years, an extensive research activity, focused on growth and characterization of the nonpolar nitride material system has taken place worldwide. Figure 1.3 depicts the publication record in the scientific journals, clearly indicating the strong increase of interest in this area of research. As a result, significant technological achievements have been realized, which paved the way to a better understanding of the specific features, properties, and challenges in both growth and characterization of this material system.

1.2
Historical Survey of Nonpolar Nitride Growth Achievements

Here we present a very brief survey of the most important steps in the growth development of Group III-nitride materials and devices with nonpolar and semipolar surfaces. We follow a chronological order, and try to establish the connections between the steps and their driving forces. We do not intend to

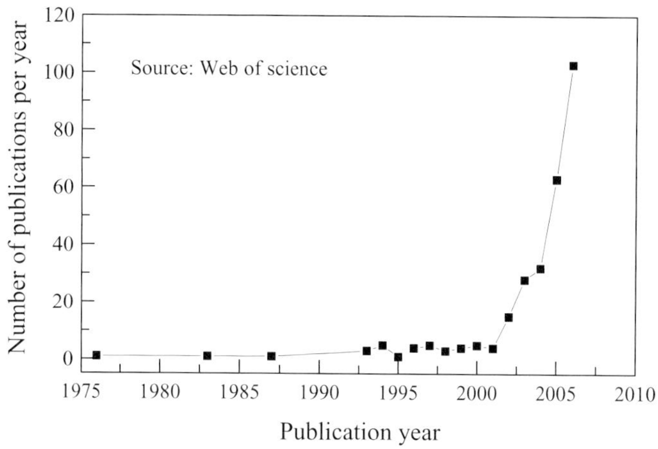

Figure 1.3 The number of publications per year focused on nitrides with nonpolar and semipolar surfaces versus time, showing the strong increase in research activity in this field after 2000.

provide a detailed picture of every step of the nonpolar nitride research. In fact, the full story of nonpolar nitride growth development using different growth techniques and different substrates will be presented in the chapters to follow. This is one of the main purposes of the present book.

1976–1999: initial results The first GaN layer grown with [11-20] orientation was reported in 1976 by Sano and Aoki [7], using hydride vapor phase epitaxy (HVPE), and the first AlN with [11-20] orientation was reported later, in 1983, by Yoshida *et al.* [8], which was grown by molecular beam epitaxy (MBE). Over the years, GaN with different orientations have been achieved by using practically all growth techniques – HVPE [7, 15], metal organic vapor phase epitaxy (MOVPE) [9, 10, 14, 17–19, 22–26] and MBE [8, 11, 12, 20, 21, 27]. For more than 20 years, in the course of optimizing the growth conditions and finding the most promising crystal orientation, with the aim of improving the quality of the material, all heteroepitaxially grown nitride materials with orientations different from the [0001] one have been found to be of very poor crystalline quality. Yet, the material was the basis for the study of epitaxial relationships [13, 26], for establishing the key properties and for identifying the main growth challenges.

2000: the beginning The first nonpolar *m*-plane GaN/AlGaN heterostructures have been produced on $LiAlO_2$ substrates by plasma-assisted MBE [28, 29]. Although the structural quality of this material was not even comparable to that of material grown in the [0001] direction, it was sufficient to experimentally verify that these *m*-plane nitrides were free of electrical polarization. Since then, several research teams have started to evaluate in detail the potential of the nonpolar nitrides and practically all their attempts have been directed toward device applications.

2002: the first planar MOVPE grown *a*-plane GaN layers on *r*-plane sapphire and *a*-plane SiC For more than 20 years, all publications consistently reported that the nitride growth in directions different from the [0001] one was difficult. It was impossible to achieve a smooth morphology despite the growth parameter optimizations. The first GaN layer with planar surface morphology was reported by the University of California Santa Barbara (UCSB) research group, utilizing epitaxial lateral overgrowth (ELOG) technique in the MOVPE growth on *r*-plane sapphire, [30] and later on *a*-plane SiC substrates [31].

2002: the first planar MOVPE grown *a*-plane AlN layers on *a*-plane SiC Although the first realizations of AlN with nonpolar (11-20) surface [8, 22, 23, 25, 26] and (1-100) surface [24] were reported before 2000, a systematic study of nonpolar AlN-based materials, aiming at development of AlN/AlGaN heterostructures for device applications, started at Kyoto University only in 2002 [32, 33]. By exploring both 6H and 4H-SiC substrates, both *a*- and *m*-plane structures have been brought into focus for further research.

2002: the first planar MBE grown *a*-plane multiple quantum well heterostructures The first planar nonpolar GaN/AlGaN [34] and GaN/AlN [35] multiple quantum wells (MQWs) were grown by MBE in Bell Laboratories in 2002. This first experimental realization of device structures based on *a*-plane material on *r*-plane sapphire clearly showed the peak transition energies around 352 and 326 nm in agreement with the transition energies calculated for the respective well thicknesses, assuming a flat-band model, and thus confirming the predicted absence of built-in electrostatic fields.

2003: the first planar HVPE grown *a*-plane GaN layers on *r*-plane sapphire In 2003, the research team at UCSB achieved the first planar *a*-plane GaN grown on *r*-plane sapphire by HVPE for significantly thicker layers, aiming at development of nonpolar quasi-substrates [36, 37]. By also employing the ELOG technique, different SiO_2 stripe orientations have been evaluated to improve both the morphology and structural-defect density.

2003: the first planar MBE grown *a*-plane InN layers on *r*-plane sapphire The first planar nonpolar *a*-plane InN grown on *r*-plane sapphire using an AlN nucleation layer or GaN buffer layer by plasma-assisted MBE was reported by the research group from Cornell University in 2003 [38]. With this, all the nitrides with nonpolar orientations were completed, thus paving the way toward a whole range of nitride-based nonpolar heterostructures and devices.

2003: the first planar nonpolar Schottky diode The first planar nonpolar Schottky diode based on *a*-plane GaN on *r*-plane sapphire was produced by MOVPE at University of South Carolina (USC) [39]. The feasibility of Schottky barriers and ohmic contacts was demonstrated by employing the ELOG technique. This opened up the possibility for fabricating nonpolar electron devices.

2003: the first UV LED based on *a*-plane GaN/InGaN MQWs grown on *r*-plane sapphire The same group at USC also demonstrated the first ultraviolet (UV) LED using nonpolar *a*-plane GaN/AlGaN MQWs grown on *r*-plane sapphire [40]. Even these first experimental prototypes showed peak emission at 363 nm, with intensity almost 30 times stronger than that in the structures grown on *c*-plane sapphire. Furthermore, the emission peak position of the nonpolar MQWs remained completely stable with increasing pump current or optical pump density, whereas the ones with the *c*-plane structures exhibited a blueshift as large as 250 meV.

2004: the first visible LED based on *a*-plane GaN/InGaN MQWs Visible LEDs in the range of 413–425 nm based on *pn*-junction GaN/InGaN MQWs were reported in 2004 by research groups from USC [41] and UCSB [42]. In both approaches, the *r*-plane sapphire substrates were used with the typical optimizations developed for the *c*-plane LEDs devices, including a thick reduced-dislocation density *a*-plane ELOG GaN template fabricated by HVPE [42]. The electroluminescence peak position was found to be independent of the drive current [41], or only initially blueshifted (at low excitation densities) owing to band-tail states and was further saturated under a change of the pump power over six orders of magnitudes [42], indicating absence of the polarization field.

2005: the first planar thick *m*-GaN layers grown by HVPE on $LiAlO_2$ and *m*-plane 6H-SiC While significant progress had been made in the fabrication of *a*-plane GaN thick layers for quasi-substrate applications, that of the (1-100) *m*-plane thick layers lagged behind. Fabrication of the first planar *m*-plane thick GaN layers were achieved by HVPE growth, again using the ELOG approach and employing either $LiAlO_2$ [43, 44] or *m*-plane SiC substrates [44]. The $LiAlO_2$ substrate turned out to be especially beneficial not only because it is cheaper compared to the SiC but also because it serves as a self-separated substrate, thus leaving a freestanding crack-free GaN film.

2005: highly doped *p*-type *m*-GaN layers grown MBE on *m*-plane 6H-SiC The first realization of *p*-type doping of nonpolar *a*-plane grown GaN on *r*-plane sapphire by MBE was reported in 2003 [45]. The demonstration of experimentally achieved high *p*-type doping levels of 7.2×10^{18} cm^{-3}, which is higher than the best value of 1×10^{18} cm^{-3} achieved in the *c*-plane GaN, was reported for *m*-plane GaN grown on *m*-plane 6H-SiC by MBE in 2005, again by the UCSB research group [46]. The mobility parallel to the [11-20] direction was as high as $\sim$11.5 cm^2 V^{-1}s^{-1} (at $p \sim 1 \times 10^{18}$ cm^{-3}), and higher than that parallel to the [0001] axes for the same hole concentration.

2005: the first planar semipolar GaN layers grown by MOVPE Despite the advances in the growth of nonpolar heterostructures and devices, fabrication of high-performance nonpolar optoelectronic devices remains challenging and has not yet been fully optimized. To obtain the best possible optimizations, research of the wurtzite nitride structure grown in alternative orientations has also attracted interest. So-called *semipolar* materials with growth surfaces inclined to the (0001) polar plane, expected to possess reduced built-in electric, field have also been grown. The first planar semipolar GaN layers with (10-1-1) and (10-1-3) surfaces were produced on (100) $MgAl_2O_2$ and (110) $MgAl_2O_2$ (spinel) substrates, respectively, by MOVPE [47] and HVPE [48]. Further optimizations of the growth revealed that a small substrate miscut by 3° in the [011] direction is required to obtain coalesced films with smooth surface morphology.

2005: the first LED based on semipolar GaN/InGaN MQWs Semipolar ⟨11-22⟩-oriented GaN/InGaN MQWs with efficient radiative recombination were first fabricated by means of faceting regrowth technique on patterned GaN template with stripe geometry on *c*-plane sapphire [49]. A comparison of the recombination efficiency of the structures containing stripes with semipolar facets, that were not fully coalesced, with structures with regular stripes revealed the superiority of the former. Alternative approach was undertaken by the group at the UCSB. Semipolar ⟨10-1-1⟩- and ⟨10-1-3⟩-oriented GaN/InGaN LEDs on $MgAl_2O_2$ substrate were produced [50]. Consistent with the expectations, electroluminescence measurements of the semipolar LEDs yielded a reduced blueshift of the peak emission wavelength with increasing drive current in comparison with commercial *c*-plane LEDs.

2005: the first nonpolar quantum dot and nanowire structures grown by MOVPE on *a*-plane SiC Despite the promising future of these materials, the structural quality of heteroepitaxially grown nonpolar nitride layers is still poor. The high density of structural defects such as dislocations in all types of polar, nonpolar, and semipolar layers, as well as stacking faults (SFs) in the nonpolar and semipolar materials and device structures, deteriorates the device performance. Confining the carriers in dislocation-free quantum dots (QDs) or nanowires (NWs) is a possible approach to overcome the impact of structural defects on the radiative recombination yield. The first growth of *a*-plane GaN/AlN QD structure was demonstrated by the Grenoble research group in 2005 [51]. The variation of the emission energy in these structures as compared to that of (0001) GaN QDs was a clear fingerprint of the reduced internal electric field present in the nonpolar nanostructures. Similarly, defect-free GaN NWs aligned along the nonpolar [11-20] direction have been grown by MOVPE on sapphire substrates [52]. Optical excitation studies of such structures demonstrated thresholds for stimulated emission that were substantially lower than previously achieved values for polar-oriented NWs.

2005: bulk nonpolar GaN grown by high-pressure solution growth and HVPE The most desired approach for the growth of high-quality nitride devices by avoiding the electric field due to spontaneous and piezoelectric polarizations would be a homoepitaxial growth of nonpolar device structure on bulk nonpolar nitride substrates. The classical high-pressure solution growth (HPSG) of bulk GaN produces material that is still limited in size and amount, and one way to resolve this drawback is to use a combination of two growth techniques. The Polish research team at Unipress has demonstrated bulk GaN substrates produced by HVPE overgrowth of defect-free thin platelets initially grown by HPSG [53]. This material has high structural quality, but because of the limited lateral sizes of the HPSG grown crystals, the approach has no strong industrial potential at the moment.

2006: bulk nonpolar nitride substrates sliced from HVPE grown boules The most promising approach for producing nonpolar nitride substrates is the real boule growth. For the moment, the only reliable techniques are the HVPE to produce GaN and the sublimation growth to produce AlN boules. Such GaN boules grown in the [0001] direction have been reported by a few groups: ATMI-Cree [54], Kyma Technologies [55], Furukawa Co, and Mitsubishi Chemical Co. Slicing the boules with diameter of 2 in. and height up to 10 mm is possible in any direction of interest and nonpolar bars with lateral sizes of up to $\sim 10 \times 25$ mm are currently achievable. High structural quality, of such material manifested by low dislocation density in the range of 10^5 cm^{-2}, low residual doping in the range of 10^{15}–10^{16} cm^{-3}, and particularly no SFs and strain anisotropy, have been recently demonstrated.

2006: semipolar LEDs grown on bulk nonpolar GaN sliced from boules Homoepitaxial LEDs devices on GaN templates and freestanding substrates, initially grown in nonpolar and semipolar direction by HVPE, have been fabricated [41, 42, 56, 57]. Although the performance of these nonpolar and semipolar nitride-based LEDs showed noticeable improvement, it was partially compromised by the significant density of structural defects such as dislocations and SFs, typical for the growth in these directions, even after being reduced by using the ELOG technique. Homoepitaxial LED and LD devices with nonpolar and semipolar surfaces grown on bulk low-defect density GaN substrates (manufactured by Mitsubishi Chemical Co and Furukawa Co), initially grown in the [0001] direction, are the most recent and promising results achieved by the research teams at UCSB, Rohm Co. [58, 59] and Kyoto University [60]. These devices have shown a performance that is superior to that of all the nitride devices using nonpolar surfaces, up to now. LED devices based on the GaN/InGaN system emitting in blue, green, and amber wavelengths, depending on the In composition, have been demonstrated [60].

2006: the first polarization-sensitive photodetector The first polarization-sensitive photodetector (PSPD) based on *m*-plane GaN [61] and the first very narrow band photodetector [62], combining PSPD with a polarization filter with rotated in-plane crystal orientation, have been proposed very recently, demonstrating a valuable application of the concept for polarization anisotropy.

1.3
Nonpolar Nitrides Today – Key Properties and Challenges

Despite the intensive growth optimizations and significant progress of the crystal quality of the nonpolar and semipolar nitrides, materials, and devices, there is still considerable scope for development. All the material properties appear more complicated and impose more challenges on measurements and analyses. As it will be described in the following, in order to find a way to control and improve the material characteristics, several phenomena have to be evaluated and better understood, as whereas their nature and driving forces have to be clarified.

1.3.1
Morphology

Ever since the investigation of nitride growth on substrates with different orientations began, the morphology of the epitaxial films grown in the [0001] direction has always been found to be superior with respect to any of the others [7, 10, 63]. Only recently, with considerable improvement in the growth techniques, and by employing different nucleation schemes, it has become possible to grow nitrides with planar surfaces different from the (0001) plane [30, 36, 37, 64–74]. Several approaches have been suggested for achieving a planar nonpolar morphology. Among them, the use of the ELOG approach [30, 37, 70], substrates with offset orientations [47, 72–74], and high temperature grown templates [63, 71, 72] are considered the most successful. Neverthless the growth development of nonpolar smooth surfaces continues to lag behind.

The most commonly observed micromorphology of thin nitride layers grown by both the MOVPE and MBE techniques is striated. It is also often called *slate* morphology, dominated by ridges with well-defined orientations, that appeared to be oriented along the ⟨11-20⟩ directions in the *m*-plane GaN films grown on (100) $LiAlO_2$ [67, 69], and along the ⟨0001⟩ directions in the *a*-plane GaN films grown on *r*-plane sapphire [68]. Detailed studies of the substrate morphology revealed that the trench-like patterns follow the corresponding ones of the substrates and hence, it appears that the morphologies of the nonpolar nitrides replicate the substrate morphology [68, 69].

Whereas the morphology of thick *c*-plane GaN films grown by HVPE is typically hillock-type on a large scale macromorphology level with respect to the unit cell and very smooth on micromorphology level, as revealed by scanning electron microscopy (SEM) (Figure 1.4a), the morphology of the thick nonpolar *a*-plane films is stripe-shaped on a large scale with no perfect stripe orientation (Figure 1.4c). In addition, the morphology of both thin and thick nonpolar nitride layers was often reported as decorated by crystallographic triangle pits (Figure 1.5a). They were observed in films grown by different techniques, and have been considered as typical morphological defects under

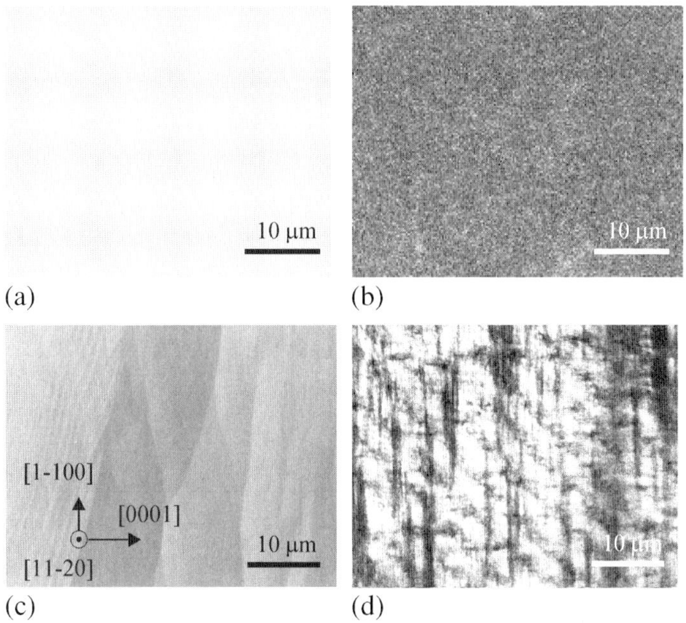

Figure 1.4 SEM (a, c) and panchromatic cathodoluminescence (CL) (b, d) images of top surfaces of polar (a, b), and nonpolar (c, d) *a*-plane GaN grown by HVPE.

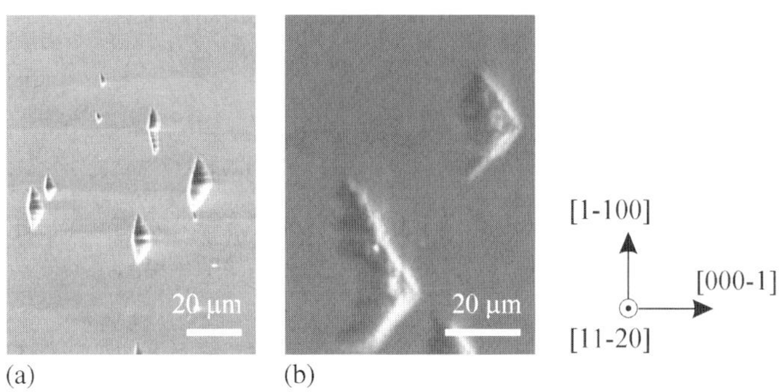

Figure 1.5 SEM images of typical surfaces defects in nonpolar *a*-plane GaN grown by HVPE: (a) triangle surface pits and (b) surface depressions overgrown to a different extent.

nonoptimized growth conditions [9, 75–78]. The pit defects observed in *a*-plane GaN films possess a common orientation with the base perpendicular to the *c*-axis. A detailed investigation performed by Wu *et al.* [77] revealed that the surface pits are indeed parts of three-dimensional hexagonal pyramids lying in the growth plane with the tips pointing to the [000-1] (so-called $-c$) direction. Thus, the facets opposite to the tips are vertical, coinciding with the *c*-plane and forming an N-face wall. The defects are similar to the well-known V-type defects in *c*-plane GaN, where they appear as open hexagons on the film surfaces, being part of inverted pyramids with tips pointing to the film–substrate interface. Triangular-shaped defects are another type, as shown in Figure 1.5b. They have the same orientation with tips along the [0001] direction, but the difference lies in the extent of overgrowth. The base of the defects is not well formed; assuming a similar pyramidal shape, it appears that the overgrowth process was preferentially initiated from the base.

The specific features of the morphology of the nonpolar heteroepitaxial films and heterostructure are directly related to the specific microstructure, including both point and structural defects. They are also related to the electrical and emission properties of the nonpolar nitrides. These relations are discussed in greater detail in several chapters of the book.

1.3.2 Microstructure

The heteroepitaxially grown nitrides have always been characterized by a huge defect density, especially with respect to other material systems such as arsenide, although in the nitrides, they do not have such a crucial detrimental effect on the device performance as in the case of the Group II–VI material-based devices, for example. In fact, the biggest problem for crystal quality development of the nonpolar nitrides is their microstructure. One of the decisive issues for the realization of high-quality GaN films with nonpolar surfaces is the high density of SFs commonly found in such films, in addition to the high density of threading dislocations (TDs). The SFs hamper the applicability of the material to the near ultraviolet spectral range. The SFs and TDs are detrimental for the lifetime of such devices because they act as current shortcut pathways. In several reviews on the microstructure of nonpolar nitride films grown either on *r*-plane sapphire [26, 30], *a*-plane SiC [79, 80], or γ-$LiAlO_2$ [81] substrates, the SFs are reported to have densities in the range of 10^5 cm^{-1} and dislocation densities in the range of 10^{10} cm^{-2}, almost independently from the substrate material and surface polarity. Different types of SFs have been revealed in the nonpolar epitaxial films by transmission electron microscopy (TEM), but the dominating defects were basal plane SFs (Figure 1.6a), being bonded by two partial dislocations (Figure 1.6b) which is in accordance with the theory. Somewhat lower values of the defect densities in the nonpolar grown nitrides can only be achieved if either an SiN_x micromasking [82] or lateral epitaxial overgrowth techniques are used [30, 37, 70]. Since the TD and SFs originate predominantly at the interface between the substrate and the film, by conducting detailed studies of the atomistic structure of the interface, the epitaxial relationships [83, 84] and the character of the misfit dislocations [84] in the nonpolar *a*-plane GaN have been evaluated. However, a further in-depth knowledge of these properties is of critical importance to understand the defect-forming mechanisms in the early stages of the nonpolar

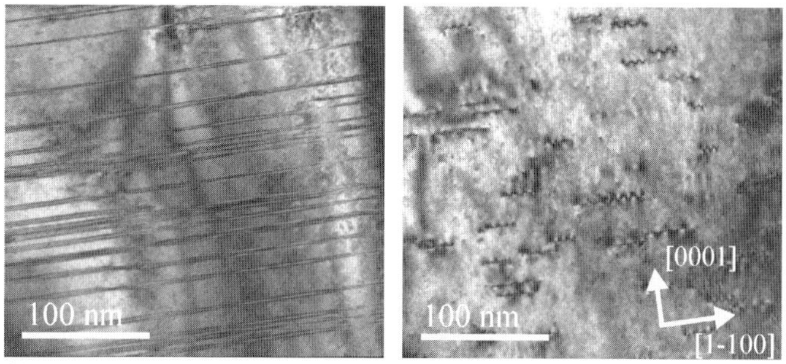

Figure 1.6 Bright-field plan-view TEM images of an *a*-plane GaN film grown on *r*-plane sapphire taken close to the [11-20] zone axis with (a) tilting by 4° toward the 0002 reflection for visualizing basal plane stacking faults; (b) tilting by 5° toward the 1-100 reflection for visualizing dislocations with a nonzero [0001] component.

nitride growth to be able to control and, hopefully, significantly reduce or avoid their formation.

Another interesting observation regarding the distribution of the defects in the nonpolar *a*-plane GaN is that the concentration of all the grown-in defects, such as basal stacking faults (BSFs), partial dislocations, oxygen impurities as well as Ga vacancy complexes show constant dependences with the thickness, contrary to the strong reduction of the defect density by several orders of magnitude over a thickness of ~50 μm, which is observed in GaN grown along the [0001] direction [85].

An alternative for producing nonpolar material with significantly reduced defect density is a material grown in the [0001] direction up to a significant thickness, followed by a slicing of bars and/or wafers with a desired nonpolar orientation. The material produced in this way with both *a*- and *m*-plane orientations has been shown to have no SFs (Figure 1.7a), and the dislocation density to be in the range of 10^5 cm^{-2} (Figure 1.7b) [55, 86].

1.3.3 Strain

The lattice and thermal mismatch between the substrate and the film, characteristic for the heteroepitaxial growth of the nitrides on foreign substrates, results in the presence of strain. This strain has a major effect on all fundamental material properties and device-relevant characteristics, and has been a subject of intense investigation. The strain in the nonpolar nitrides is supposed to be anisotropic owing to the in-plane anisotropies in the lattice matching and the thermal expansion coefficients of both the substrates and the layers along the [0001] and [1-100] directions.

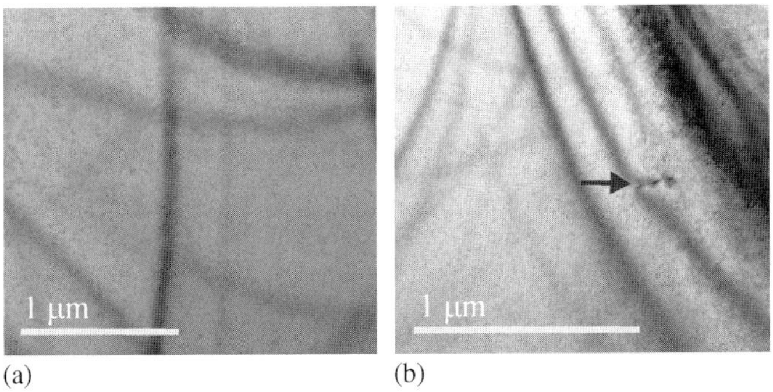

Figure 1.7 Bright-field TEM images taken under similar conditions as for Figure 1.6(a) and 6(b), respectively, for the *a*-plane GaN bar sliced from HVPE grown GaN boule in the [0001] direction.

In linear elasticity theory the stress and strain tensor σ_{ij} and ε_{ij} are related by a complex matrix relation via the tensor of the elastic stiffness constants C_{ij}. Since the *x*, *y*, and *z* axes chosen along the GaN [11-20], [1-100] and [0001] directions, respectively, are parallel to the principal axes, the shear stress and strain components ($i \neq j$) are zero. Also, since the surface is free to expand or contract, the stress component along the growth direction, σ_{xx} or σ_{yy} for the *a* and *m*-planes layers, respectively, vanishes. This leads to slightly simplified relations between the three main strain components [87, 88]:

$$\varepsilon_{yy} = -\frac{C_{11}}{C_{12}}\varepsilon_{xx} - \frac{C_{13}}{C_{12}}\varepsilon_{zz} \quad \text{for } a\text{-plane material} \tag{1}$$

and

$$\varepsilon_{yy} = -\frac{C_{12}}{C_{11}}\varepsilon_{xx} - \frac{C_{13}}{C_{11}}\varepsilon_{zz} \quad \text{for } m\text{-plane material} \tag{2}$$

Figure 1.8a shows the contours of the strain in growth direction ε_{xx} versus the in-plane strain components ε_{yy} and ε_{zz} in the *a*-plane GaN, for elastic stiffness constants reported by Polian *et al.* [89]. The plot clearly demonstrates that the strain in growth direction could have zero value for epitaxial films, which are under biaxial strain. More complicated measurements are needed to independently assess at least two strain components to arrive at conclusions about the strain status of the material. Two approaches have been suggested for strain estimation in nonpolar nitride films by determining the lattice constants of the crystal using high-resolution X-ray diffraction. The first approach [90, 91] uses both plane and edge symmetric geometries and applies the Bragg's equation directly. Using this method, the three strain components can be independently determined, but the accuracy for thicker films is reduced because of the different volumes of the film that are probed. The second approach [92, 93] uses various reflections under symmetric, asymmetric, and

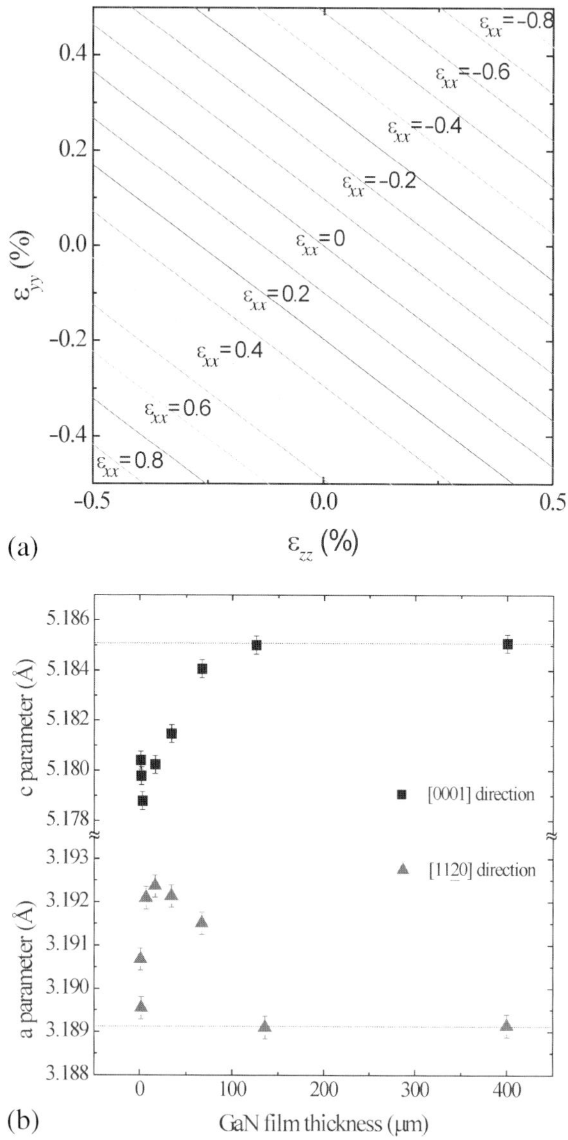

Figure 1.8 (a) Contour plots showing the variation of the strain ε_{xx} versus the in-plane strains ε_{yy} and ε_{zz} in *a*-plane GaN; (b) The lattice parameters versus GaN layer thickness; the dotted lines present the strain-free values for powder material.

skew-symmetric diffraction geometries. The lattice constants are derived by applying Bragg's law to determine the respective lattice plane spacings that are subsequently fitted to the lattice plane spacings of an orthorhombic crystal. This fitting procedure is, however, very sensitive to the selection of reflections since each of them corresponds to different probing depth. Although both approaches have lower accuracy compared to that typically achieved for the *c*-plane materials, they give consistent results for anisotropic type of strain in the nonpolar nitrides. In addition, the experimental results show that whereas for thicker epitaxial layers (5–100 μm) the strain is compressive in the growth plane and tensile in the growth direction, for thin films (<3 μm), the type of in-plane strain in direction perpendicular to the [0001] direction may also be tensile. This complex behavior of the lattice parameters and of the strain components as a function of the thickness (Figure 1.8b) is in contrast with the well-known monotonic decrease of strain with the thickness in the *c*-plane nitrides, and imposes more difficulties for the measurement and data analyses, as well as for further device processing.

Both the anisotropic in-plane character of the strain and the complex behavior versus the layer thickness are confirmed by the behavior and anisotropy of the bending in the two in-plane directions (Figure 1.9). This, in turn, influences other structural characteristics of the epitaxial materials. For instance, the full width at half maximum (FWHM) of the ω rocking curves is found to be strongly dependent of the azimuth angle with respect to the scattering plane (Figure 1.10). This variation was observed to have either M^- or W^- type dependence on the azimuth angle [55, 63, 68]. The range of variation of the FWHM of the ω rocking curves between 0.2 and 0.6° over a 360° azimuth range is found to be almost independent of the growth technique and the nucleation scheme/optimization used. The latter suggests that such an anisotropic behavior is a specific characteristic for the anisotropic nature of the growth in this direction, and although the exact reason is currently

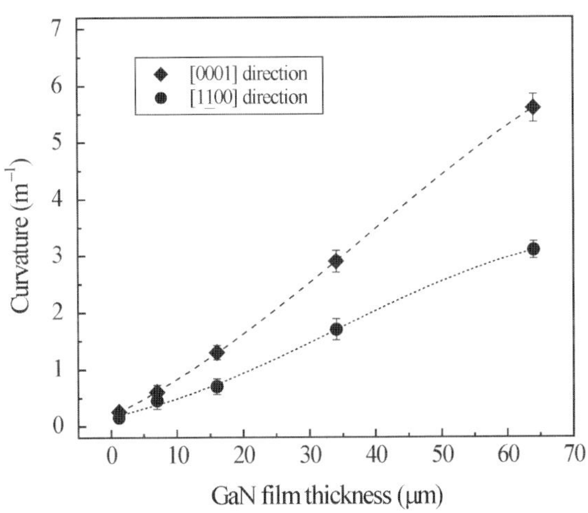

Figure 1.9 Wafer curvature in the two in-plane directions [0001] and [1-100] versus the thickness of *a*-plane GaN films. The dashed lines are guides to the eye.

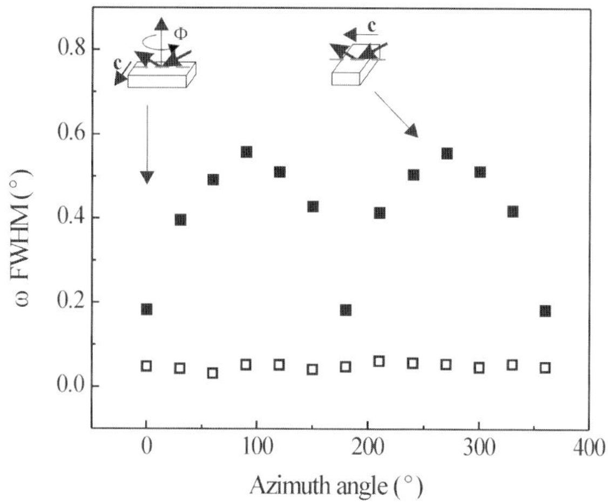

Figure 1.10 Azimuth angle dependence of the FWHM of the 11-20 rocking curves for an heteroepitaxially grown *a*-plane GaN film in the [11–20] direction (full symbols) and for an *a*-plane GaN bar sliced from GaN boule grown in [0001] direction (open symbols).

not well understood, it is most likely related to unavoidable defects with a specific distribution. In contrast, the azimuth dependence of the rocking curve width (Figure 1.10, open symbols) of an *a*-plane oriented GaN bar sliced from a boule grown in the [0001] direction [55] shows an isotropic behavior with a value of about 90–160 arcsec, which is almost an order of magnitude lower than the best values achieved for the heteroepitaxially grown nonpolar material. The latter is indicative of the superior structural quality of the material grown in the [0001] direction and sliced along a nonpolar surface.

1.3.4
Optical Properties

The optical properties of the nonpolar nitrides have been extensively studied [94–106] by photoluminescence (PL), cathodoluminescence (CL), and photoreflectivity (PR). All the studies consistently reported two important features of this type of material that differ significantly from the optical properties of the nitrides grown in the [0001] direction. The first difference is related to the specific microstructure of these materials and is characterized by additional emission bands attributed to structural defects such as SFs. The second difference is related to the anisotropic in-plane character of the optical properties, in agreement with the theoretical predictions.

The emission spectra of *m*-plane [94, 95] and *a*-plane GaN [96–99] films grown on *r*-plane sapphire typically show, in addition to the near-band-edge (NBE) emission band at 356–357.2 nm (the peak position slightly varies in films with different thickness owing to a change of the strain), defect-related bands due to different structural defects that are enhanced/reduced to a different extents in samples grown on different templates. At helium temperature, the PL and the CL spectra of nonpolar GaN films are usually dominated by an emission band at 362–363.4 nm, depending on the film thickness (see Figure 1.11a). In some samples grown on nonoptimized ELOG templates with stripes oriented along the [1-101] direction, an additional well pronounced emission band at ~369.5 nm has been observed [78]. Such an emission could also be noticed in some samples grown without buffers, although this was often only an asymmetry on the lower energy side of the stronger 363-nm emission. A detailed optical study [99] of the emissions in *a*-plane GaN grown on *r*-plane sapphire was performed by PL at different excitation densities and temperatures. The analyses of the PL data, together with previous correlations of emission at 3.42 eV in *c*-plane GaN with BSF structural defects and with BSF of type I_1 in MOVPE grown *a*-plane GaN [101], leads to the suggestion of a BSF origin for these emissions in *a*-plane GaN. Since the BSF could be considered as a very thin zinc-blende (ZB) quantum well, embedded into the wurtzite matrix with type-II band alignment, the 363-nm emission could be treated as recombination between electrons confined in the ZB QW and holes localized in the wurtzite barrier. There are two suggestions for the possible origin of the peak at 369.5 nm – due to BSFs of type I_2 or due to prismatic stacking faults (PSFs). There is no convincing proof for any of these as yet.

Spatially resolved panchromatic CL imaging [78, 98] reveals nonuniform in-plane distribution of the integral emission in nonpolar materials (Figure 1.4d), in strong contrast to the uniform emission intensity distribution in polar materials (Figure 1.4b). The monochromatic CL imaging shows the in-plane distributions of the respective emission bands (Figure 1.11c–e), and allows a correlation of the emissions with particular SF structural defects independently revealed by plan-view TEM [98].

The other well-known emission bands in GaN, in lower energy regions such as donor–acceptor pair emission together with its phonon replicas, as well as the blue–green, yellow, and red emission bands, have also been observed in GaN films with nonpolar orientations and their origin was, in principle, related to the same defects as the respective emissions in polar materials [95, 99].

It should be underlined that the luminescence spectra of low-defect density nonpolar GaN material grown by optimized ELOG templates are dominated by the exciton-related emissions, and the defect-related recombinations have been significantly reduced [70]. Furthermore, for the bulk GaN material with nonpolar surfaces sliced from boules, which have been grown in the *c* direction, the low-defect density is reflected in a high optical quality [86, 100]. Namely, low temperature PL spectra in the NBE region show free-exciton emissions

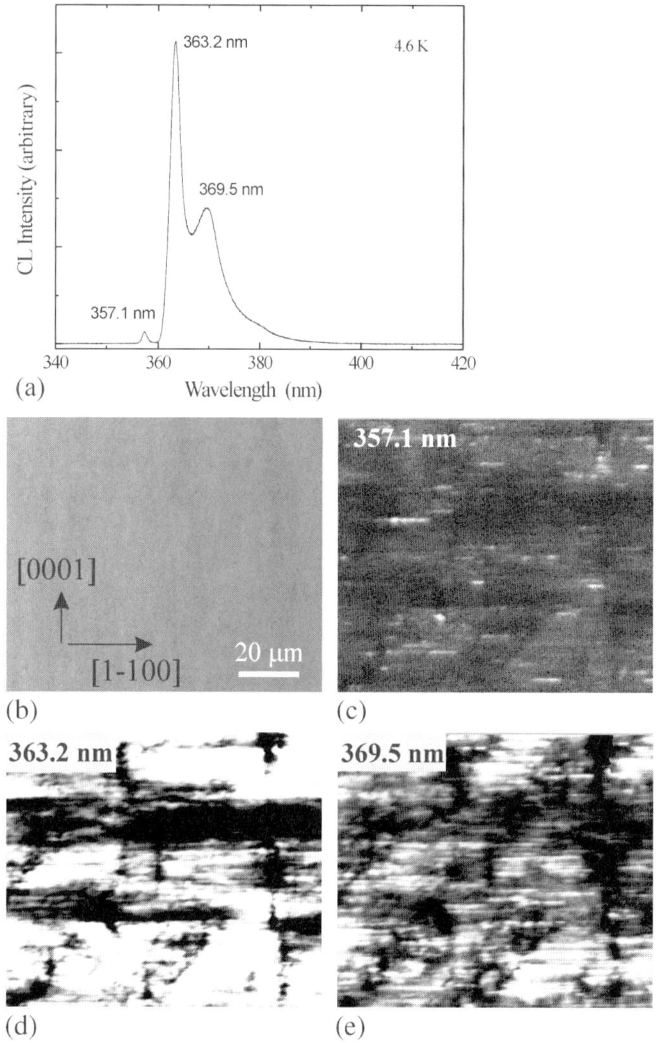

Figure 1.11 CL spectrum (a), secondary electron image of the surface morphology (b) and monochromatic CL images at the wavelength of the peak maxima 357.1 nm (c), 363.2 nm (d), and 369.5 nm (e), as seen in the CL spectrum for an *a*-plane GaN film grown with nonoptimized ELOG stripes oriented along the [1-101] direction.

and very narrow lines of donor-bound excitons, involving the O and Si donors, with FWHM of order of 0.6–0.7 meV. This type of material exhibits superior emission properties in comparison to the heteroepitaxially GaN material grown in nonpolar and semipolar directions.

The origin and the behavior of the emission in GaN with [11-22], [10-1-1], and [10-1-3] semipolar orientations have been less studied [101, 102]. In principle, as expected, the observed emissions are related to the microstructure of the materials and in case of homoepitaxial growth on GaN bulk substrates (sliced from boules) with low-defect density, the emission spectra show the well-known dominating NBE peaks [101].

The emission properties of nonpolar *a*-plane AlN and InN films grown on *r*-plane sapphire are the least explored. Only a few papers present the initial data on this topic. The NBE emissions were confirmed very close to that in polar AlN [103, 104] and InN [38, 105] materials with no detailed studies of the effect of strain on them. Significant improvement of the emission intensity was observed in nonpolar AlN [104] as a result of microstructure improvement in case of ELOG [104] or Pendeo [106] template employment. Similar to the observation for GaN, the AlN bulk material with nonpolar orientations sliced from boules grown in the [0001] direction was found to possess superior emission quality [103].

One of the most important aspects of the films and heterostructures with nonpolar and semipolar surfaces is related to the polarization dependence of their optical properties. The polarization anisotropy has been studied both theoretically and experimentally in nonpolar GaN [87, 88], as well as in InN [105]. The optical polarization anisotropy in wurtzite nitrides originates from their valence band structure, which can be significantly modified by the anisotropic in-plane strain in the films.

1.3.5 Optical Phonons

The optical phonon spectrum is one of the most fundamental characteristics of the crystals. It reflects the specific features of the interatomic interactions and gives very comprehensive and detailed information about the thermal and optical properties involving the efficiency of the optoelectronic devices. The vibrational properties in all the nitride systems have been investigated in detail over the years by Raman scattering (RS) spectroscopy. Recent studies of nonpolar *a*-plane GaN by RS confirmed the finding in the *c*-plane GaN [107, 108]. However, in some cases there is a lack of agreement between the values of some phonon deformation potentials and strain-free phonon-mode positions in GaN and AlN, as determined theoretically and by employing RS spectroscopy. The nonpolar materials allow an access to the complete set of phonons by infrared spectroscopic ellipsometry (IRSE). This provides an alternative tool to study the vibrational properties and to establish very important and useful fundamental parameters of the nitrides.

The anisotropic strain in the nonpolar nitride materials films heteroepitaxially grown in nonpolar directions inevitably leads to more complex behavior of the optical phonons. The theory predicts that some of the phonons in the nitrides split under anisotropic strain. Recent reports on *a*-plane nonpolar GaN films grown on *r*-plane sapphire, employing both RS and IRSE techniques to study the phonons, and XRD to study the strain experimentally, confirm the theoretical predictions [91, 109]. The determination of the A_1(TO) and E_1(LO) phonon deformation potentials by IRSE allowed a comparison with the RS experimental values and the theoretical values, and in some cases called for advanced theoretical models.

The vibrational properties of nonpolar AlN and InN films have not been explored in detail. Only a few studies of the RS phonon spectra presented a confirmation for the expected crystallographic orientations of *a*-plane AlN films and the quality of the material [32].

1.3.6
Electrical Properties

Practically all intentionally nondoped nonpolar heteroepitaxially grown nitride materials possess *n*-type conductivity. The typical free carrier concentrations were reported, in the range of 10^{17}–10^{18} cm^{-3} for *a*-plane GaN grown by MOVPE [110] and HVPE [63], and in the range of 10^{18} cm^{-3} for *a*-plane InN grown by MBE on *r*-plane sapphire [38]. Somewhat better values of the free carrier concentration in the range of 10^{17} cm^{-3} were reported for *a*-plane GaN grown on *r*-plane sapphire by employing the ELOG technique [110]. In contrast to the *c*-plane GaN and InN films grown on *c*-plane sapphire, the apparent improvement of electrical properties of the nitride materials grown in nonpolar directions with increasing film thickness has not been observed [38]. The latter is most likely related to the constant depth profiles of the point native and the impurity defects in the nonpolar materials [85].

Although the *p*-type conduction in nonpolar materials was achieved later, the first *p*-type doping of nonpolar *a*-plane grown GaN doped with Mg was demonstrated by MBE growth [45]. Moreover, shortly after that, experimentally achieved concentrations of 7.2×10^{18} cm^{-3} in *m*-plane GaN grown by MBE on *m*-plane SiC [46] exceeded the highest *p*-type doping levels of 1×10^{18} cm^{-3}, achieved in the *c*-plane GaN. Similar effects of enhanced *p*-type conductivity have been observed later in *a*-plane GaN films grown by MOVPE on *r*-plane sapphire [111] and semipolar (10-1-1) GaN films grown by MOVPE on (100)-plane $MgAl_2O_4$ substrates [112].

Theoretical calculations predict significant anisotropy in effective hole masses for the different valence bands of GaN, though the various calculations differ in the predicted magnitude of the anisotropy in each band [113–115]. Direct measurements of hole masses are, however, very difficult and most reported experimental values are indeed inferred from luminescence, magneto-optical studies of exciton luminescence, infrared reflectance and transmittance studies of polar materials, which are often isotropically averaged and thus not informative about the anisotropy. Directionally dependent Hall-effect measurements in *m*-plane GaN films grown on *m*-plane SiC substrates revealed strong anisotropy for both electrons and holes [46], and thus directly confirmed the theoretical predictions and expectations. The mobility parallel to the [11-20] direction was measured to be as high as $\sim 11.5\ \mathrm{cm^2\ Vs^{-1}}$ (at $p \sim 1.8 \times 10^{18}\ \mathrm{cm^{-3}}$), which was higher than that parallel to the [0001] axes for the same hole concentration [46]. Since the conduction band curvature of GaN shows no significant anisotropy and is insensitive to the strain, the anisotropy in electron mobility was attributed most likely to anisotropic scattering rates which are influenced by the presence of high-density SFs and consequently the formation of band-edge discontinuities in heteroepitaxial polar and semipolar nitrides. The exact mechanisms of the influence of SFs on the scattering processes and anisotropic mobility are currently not well understood. Recent reports on directional dependent transport measurements of *p*-type *m*-plane GaN films grown on low-defect density *m*-plane GaN substrates sliced from boules (grown in *c* direction) revealed no significant anisotropy in in-plane mobility [116]. This contradiction to the theoretical predictions was tentatively explained by the absence of significant anisotropy in the curvature of the highest energy and most densely populated valence band and/or by two equal but opposite anisotropic channels. Clearly, the topic requires further detailed investigation and a better understanding of the scattering mechanisms in nonpolar nitride materials.

1.4 Nonpolar and Semipolar Nitride-based Devices Today

Nearly two decades of research and development of the growth of *c*-plane polar nitrides have opened the way for the commercial production of nitride-based lasers, LEDs, and HEMTs. Much of this experience can be applied to the growth and fabrication of nitride-based devices with *a*, *m* and semipolar surfaces. In fact, several groups have now produced devices based on nonpolar and semipolar nitrides. Even the first optoelectronic devices on nonpolar and semipolar nitrides show significant reduction in the shift of emission wavelength in accordance with the theoretical predictions. This 10-fold or even higher reduction in the shift, compared with that in the polar devices, indicates absence of polarization-induced charges in the nonpolar devices. However, several challenges, especially with regard to the control of SFs, have not been experienced in the polar nitride materials and these remain to be overcome.

The realization of first nonpolar *a*-plane InGaN/GaN emitters by the UCSB group is based on the employment of the well developed for polar nitrides ELOG approach [42]. The reduced defect density resulted in a drive-current-independent electroluminescence peak position at about 412 nm and output power as high as 1.5 mW at a driving current of 250 mA. In addition, a small linewidth broadening with increasing drive current indicated a low device heating. The same growth approach applied to *m*-plane device gave similar results of output power of 2.95 mW at 300 mA for LED emitting at 450 nm with no power saturation at high drive currents [43]. Blue–green–amber LED emitters have also been demonstrated. Even better device performance was demonstrated very recently for *m*-plane LDs, operating under pulsed and continuous conditions, in case of homoepitaxial growth on bulk nonpolar GaN substrates sliced from boules grown in the *c*-direction.

The polarization-free interfaces in the nonpolar nitride heterostructures can also have a significant impact on HEMT electronic devices. A polar AlGaN/GaN HEMT has a typical channel charge of more than 1×10^{13} cm^{-2}, being completely polarization generated. Elimination of the quantum-confined Stark effect in the nonpolar HEMT devices will bring in the need for silicon doping to form the two-dimensional electron gas, which is the approach used in GaAs- and InP-based devices. This will allow the doping level to be independent of the polarization and strain-related effects. The elimination of the quantum-confined Stark effect is also expected to help the fabrication of both enhancement and depletion-mode devices, which can significantly simplify the nitride-based logic devices.

The nonpolar and semipolar nitride devices also have the potential to use the polarization light emission. This property is especially beneficial for backlighting liquid crystal displays (LCD) and projectors. The polarized light sources do not require polarizing filters and thus the screens can be made thinner, lighter, and more energy efficient.

1.5
Prospects in the Development of Nonpolar Nitrides and Devices

The future development of the nitride materials and devices with nonpolar and semipolar surfaces inevitably goes in several directions, namely, further optimizations of the material quality, better understanding of the effects, phenomena, and mechanisms typical for this material system, further optimizations of the device performance, and expansion of the device capabilities.

The first step for improvement in the quality of the nonpolar nitride materials is aimed at decreasing the dislocation and SF densities. This is indeed the key toward the desired substantial improvement of the device performance. There are two possible ways to reach this goal: the first way is a heteroepitaxial device growth in nonpolar direction with the need for further optimization or of new approaches for producing films with lower defect density; the second is a homoepitaxial device growth on nitride substrates grown in the *c* direction with the need to develop and enlarge the boule growth, which will enable slicing of high-quality nonpolar bulk substrates with any orientation of interest.

The thorough understanding of the specific properties of this type of material is of critical importance toward disclosing their full potential. As already mentioned, there are strong research efforts worldwide to investigate and develop high-quality nitride materials with nonpolar and semipolar surfaces.

At present time, several issues remain to be clarified, but the present high activity indicates that many of the questions about nonpolar materials will be answered and the material quality will be better controlled in the near future.

The optoelectronic devices based on nonpolar and semipolar nitrides will be further developed aiming higher efficiency at different emission wavelengths. The accent will be on development of the nonpolar and semipolar LDs, emitting in the blue and green regions with low threshold current density. The applicability of the nonpolar and semipolar nitride material for devices emitting up to red wavelengths remains to be established. The contribution of low-dimensional QD and NW structures to low threshold lasing in nonpolar nitride cavities and the development of electrically pumped nanostructure lasers are currently under discussion.

1.6
Summary

In this book, the present stage of development of the nonpolar nitrides is reviewed in detail, with emphasis on the three main topics, namely, crystal growth challenges, key properties and prospects for devices.

We aim to cover all nonpolar nitride materials, as well as all the growth techniques, presently under development. Several chapters are focused on the specific features of the growth approaches, as used for the different types of material. The bulk of the discussion in each chapter is related to the physical properties of the nitride material obtained by the different techniques. In particular, the type, generation, evolution, and density of the defects, which are different from the well-known microstructure in conventionally grown nitrides in the polar [0001] direction and are very abundant in the heteroepitaxial nitride nonpolar materials, are discussed in detail. The optical and vibrational properties, being a complex function of the nonuniform strain anisotropy in these materials, are also addressed in detail in several chapters. The progress made in heterostructures and QWs based on the AlGaN/GaN and the InGaN/GaN systems is also partly reviewed, in relation to the present development of nonpolar and semipolar devices. An outlook of the application areas of the differently grown materials is presented in most of the chapters, together with the capabilities and limitations of the respective growth approaches used.

We feel that sharing the present knowledge on the key issues of nonpolar nitrides from successfully working experts in the field would be very useful for further developments in the field.

Acknowledgments

The author is indebted to P.P. Paskov and R. Kroeger for their vital contributions to this work. The author would also like to thank V. Darakchieva,

F. Tuomisto, C. Roder, S. Figge, R. Schifano, E. Valcheva, J. Birch, A. Kasic, S. Tungasmita, B. Arnaudov, E. Preble, A. Hanser, K. Evans, D. Hommel, and B. Monemar for their valuable discussions and collaboration in the course of the work on this topic.

References

1. Bernardini, F., Fiorentini, V., and Vanderbilt, D. (1997) *Physical Review B*, **56**, R10024.
2. Fiorentini, V., Bernardini, F., Della Sala, F. *et al.* (1999) *Physical Review B*, **60**, 8849.
3. Monemar, B. and Pozina, G. (2000) *Progress in Quantum Electronics*, **24**, 239.
4. Morkoc, H. and Liu, L. (2003) in *Nitride Semiconductors*, (eds P. Ruterana, M. Albrecht, and J. Neugebauer), Wiley-VCH, Berlin, p. 557.
5. McLaurin, M., Mates, T.E., Wu, F., and Speck, J.S. (2006) *Journal of Applied Physics*, **100**, 063707.
6. Nye, J.F. (1985) *Physical Properties of Crystals*, Oxford University Press, New York.
7. Sano, M. and Aoki, M. (1976) *Japanese Journal of Applied Physics*, **15**, 1943.
8. Yoshida, S., Misawa, S., and Gonda, S. (1983) *Applied Physics Letters*, **42**, 427.
9. Sasaki, T. and Zembutsu, S. (1987) *Journal of Applied Physics*, **61**, 2533.
10. Sun, C.J. and Razeghi, M. (1993) *Applied Physics Letters*, **63**, 973.
11. Eddy, C.R., Moustakas, T.D., and Scanton, J. (1993) *Journal of Applied Physics*, **73**, 448.
12. Lei, T., Ludwig, K.F. Jr, and Moustakas, T.D. (1993) *Journal of Applied Physics*, **74**, 4430.
13. Kung, P., Sun, C.J., Saxler, A. *et al.* (1994) *Journal of Applied Physics*, **75**, 4515.
14. Sun, C.J., Kung, P., Saxler, A. *et al.* (1994) *Journal of Applied Physics*, **76**, 236.
15. Hwang, J.S., Kuznetsov, A.V., Lee, S.S. *et al.* (1994) *Journal of Crystal Growth*, **142**, 5.
16. Melton, W.A. and Pankove, J.I. (1997) *Journal of Crystal Growth*, **178**, 168.
17. Hellman, E.S., Liliental-Weber, Z., and Buchann, D.N.E. (1997) *MRS Internet Journal of Nitride Semiconductor Research*, **2**, 30.
18. Xu, K., Xu, J., Deng, P. *et al.* (1998) *Journal of Crystal Growth*, **193**, 127.
19. Xu, K., Xu, J. Deng, P. *et al.* (1999) *Physica Status Solidi A*, **176**, 589.
20. Iwata, K., Asahi, H., Asami, K. *et al.* (1997) *Japanese Journal of Applied Physics*, **36**, L661.
21. Tripathy, S., Soni, R.K., Asahi, H. *et al.* (1999) *Journal of Applied Physics*, **85**, 8386.
22. Masu, K., Nakamura, Y., Yamazaki, T. *et al.* (1995) *Japanese Journal of Applied Physics*, **34**, L760.
23. Kaya, K., Kanno, Y., Takahashi, H. *et al.* (1996) *Japanese Journal of Applied Physics*, **35**, 2782.
24. Stemmer, S., Pirouz, P., Ikuhara, Y., and Davis, R.F. (1996) *Physical Review Letters*, **77**, 1797.
25. Dovidenko, K., Oktyabrsky, S., Narayan, J., and Razeghi, M. (1996) *Journal of Applied Physics*, **79**, 2439.
26. Dovidenko, K., Oktyabrsky, S., and Narayan, J. (1997) *Journal of Applied Physics*, **82**, 4296.
27. Waltereit, P., Brandt, O., and Ploog, K.H. (1999) *Applied Physics Letters*, **75**, 2029.
28. Waltereit, P., Brandt, O., Ramsteiner, M. *et al.* (2000) *Physica Status Solidi A*, **180**, 133.
29. Waltereit, P., Brandt, O., Trampert, A. *et al.* (2000) *Nature*, **406**, 865.
30. Craven, M.D., Lim, S.H., Wu, F. *et al.* (2002) *Applied Physics Letters*, **81**, 1201.

31. Craven, M.D., Chakraborty, A., Imer, B. *et al.* (2003) *Physica Status Solidi C*, **0**, 2132.
32. Onojima, N., Suda, J., and Matsunami, H. (2002) *Japanese Journal of Applied Physics*, **41**, L1348.
33. Onojima, N., Suda, J., Kimoto, T., and Matsunami, H. (2003) *Applied Physics Letters*, **83**, 5208.
34. Ng, H.M. (2002) *Applied Physics Letters*, **80**, 4369.
35. Ng, H.M., Bell, A., Ponce, F.A., and Chu, S.N.G. (2003) *Applied Physics Letters*, **83**, 653.
36. Haskell, B.A., Wu, F., Matsuda, S. *et al.* (2003) *Applied Physics Letters*, **83**, 1554.
37. Haskell, B.A., Wu, F., Craven, M.D. *et al.* (2003) *Applied Physics Letters*, **83**, 644.
38. Lu, H., Schaff, W.J., Eastman, L. *et al.* (2003) *Applied Physics Letters*, **83**, 1136.
39. Adivarahan, V., Chen, C., Yang, J. *et al.* (2003) *Japanese Journal of Applied Physics*, **42**, L1136.
40. Chen, C.Q., Adivarahan, V., Yang, J.W. *et al.* (2003) *Japanese Journal of Applied Physics*, **42**, L1039.
41. Chitnis, A., Chen, C., Adivarahan, V. *et al.* (2004) *Applied Physics Letters*, **84**, 3663.
42. Chakraborty, A., Haskell, B.A., Keller, S. *et al.* (2004) *Applied Physics Letters*, **85**, 5143.
43. Haskell, B.A., Baker, T.J., McLaurin, M.B. *et al.* (2005) *Journal of Electronic Materials*, **34**, 357.
44. Haskell, B.A., Baker, T.J., McLaurin, M.B. *et al.* (2005) *Applied Physics Letters*, **86**, 111917.
45. Armitage, R., Yang, Q., and Weber, E.R. (2003) *MRS Internet Journal of Nitride Semiconductor Research*, **8**, 6.
46. McLaurin, M., Mates, T.E., and Speck, J.S. (2005) *Applied Physics Letters*, **86**, 262104.
47. Kaeding, J.F., Iza, M., Sato, H. *et al.* (2006) *Japanese Journal of Applied Physics*, **45**, L536.
48. Baker, T.J., Haskell, B.A., Wu, F. *et al.* (2005) *Japanese Journal of Applied Physics*, **44**, L920.
49. Nishizuka, K., Funato, M., Kawakami, Y. *et al.* (2004) *Applied Physics Letters*, **85**, 3122.
50. Chakraborty, A., Baker, T.J., Haskell, B.A. *et al.* (2005) *Japanese Journal of Applied Physics*, **44**, L945.
51. Founta, S., Rol, F., Bellet-Amalric, E. *et al.* (2005) *Applied Physics Letters*, **86**, 171901.
52. Gradecak, S., Qian, F., Li, Y. *et al.* (2005) *Applied Physics Letters*, **87**, 173111.
53. Teisseyre, H., Skierbiszewski, C., Lucznik, B. *et al.* (2005) *Applied Physics Letters*, **86**, 162112.
54. Vaudo, R.P., Xu, X., Loria, C. *et al.* (2002) *Physica Status Solidi A*, **194**, 494.
55. Paskova, T., Kroeger, R., Figge, S. *et al.* (2006) *Applied Physics Letters*, **89**, 051914.
56. Chakraborty, A., Haskell, B.A., Keller, S. *et al.* (2005) *Japanese Journal of Applied Physics*, **44**, L173.
57. Koyama, T., Onuma, T., Masui, H. *et al.* (2006) *Applied Physics Letters*, **89**, 091906.
58. Schmidt, M.C., Kim, K.-C., Farrell, R.M. *et al.* (2007) *Japanese Journal of Applied Physics*, **46**, L190.
59. Okamoto, K., Ohta, H., Chichibu, S.F. *et al.* (2007) *Japanese Journal of Applied Physics*, **46**, 4787.
60. Funato, M., Ueda, M., Kawakami, Y. *et al.* (2006) *Japanese Journal of Applied Physics*, **45**, L659.
61. Rivera, C., Pau, J.L., Munoz, E. *et al.* (2006) *Applied Physics Letters*, **88**, 213507.
62. Ghosh, S., Rivera, C., Pau, J.L. *et al.* (2007) *Applied Physics Letters*, **90**, 091110.
63. Paskova, T., Paskov, P.P., Valcheva, E. *et al.* (2004) *Physica Status Solidi A*, **201**, 2265.
64. Kusakabe, K. and Ohkawa, K. (2005) *Japanese Journal of Applied Physics*, **44**, 7931.
65. Craven, M.D., Lim, S.H., Wu, F. *et al.* (2002) *Applied Physics Letters*, **81**, 469.
66. Imer, B., Wu, F., Craven, M.D. *et al.* (2006) *Japanese Journal of Applied Physics*, **45**, 8644.

67. Haskell, B.A., Chakraborty, A., Wu, F. *et al.* (2005) *Journal of Electronic Materials*, **34**, 357.
68. Wang, H., Chen, C., Gong, Z. *et al.* (2004) *Applied Physics Letters*, **84**, 499.
69. Sun, Y.J., Brandt, O., and Ploog, K.H. (2003) *Journal of Vacuum Science and Technology B*, **21**, 1350.
70. Bougrioua, Z., Laägt, M., Vennegues, P. *et al.* (2007) *Physica Status Solidi A*, **204**, 282.
71. Ni, X., Fu, Y., Moon, Y.T. *et al.* (2006) *Journal of Crystal Growth*, **290**, 166.
72. Ikamura, M., Hoshino, A., Nakano, K. *et al.* (2005) *Japanese Journal of Applied Physics*, **44**, 7418.
73. Matsuoka, T. and Hagiwara, E. (2001) *Physica Status Solidi A*, **188**, 485.
74. Waltereit, P., Brandt, O., Ramsteiner, M. *et al.* (2000) *Journal of Crystal Growth*, **218**, 218.
75. Craven, M.D., Wu, F., Chakraborty, A. *et al.* (2004) *Applied Physics Letters*, **84**, 1281.
76. Sun, H.W., Yang, J.W., Chen, C.Q. *et al.* (2003) *Applied Physics Letters*, **83**, 2599.
77. Wu, F., Craven, M.D., Lim, S.-H., and Speck, J.S. (2003) *Journal of Applied Physics*, **94**, 942.
78. Paskova, T., Kroeger, R., Paskov, P.P. *et al.* (2006) *Proceedings of SPIE Vol. 6121, Gallium Nitride Materials and Devices*, (eds C.W. Litton, J.G. Grode, H. Markoc, and A. Madhukar), p. 47.
79. Zakharov, D.N., Liliental-Weber, Z., Wagner, B. *et al.* (2005) *Physical Review B*, **71**, 235334.
80. Horita, M., Suda, J., and Kimoto, T. (2006) *Applied Physics Letters*, **89**, 112117.
81. Liu, T.Y., Trampert, A., Sun, Y.J. *et al.* (2004) *Philosophical Magazine Letters*, **84**, 435.
82. Chakraborty, A., Kim, K.C., Wu, F. *et al.* (2006) *Applied Physics Letters*, **89**, 041903.
83. Vennéguès, P. and Bougrioua, Z. (2006) *Applied Physics Letters*, **89**, 111915.
84. Kroeger, R., Paskova, T., Figge, S. *et al.* (2007) *Applied Physics Letters*, **90**, 081918.
85. Tuomisto, F., Paskova, T., Kröger, R. *et al.* (2007) *Applied Physics Letters*, **90**, 121915.
86. Paskova, T., Kroeger, R., Hommel, D. *et al.* (2007) *Physica Status Solidi C*, **4**, 2536.
87. Ghosh, S., Misra, P., Grahn, H.T. *et al.* (2005) *Journal of Applied Physics*, **98**, 026105.
88. Ghosh, S., Waltereit, P., Brandt, O. *et al.* (2002) *Physical Review B*, **65**, 075202.
89. Polian, A., Grimsditch, M., and Grzegory, I. (1996) *Journal of Applied Physics*, **79**, 3343.
90. Paskova, T., Darakchieva, V., Paskov, P.P. *et al.* (2005) *Journal of Crystal Growth*, **281**, 55.
91. Darakchieva, V., Paskova, T., Schubert, M. *et al.* (2007) *Physical Review B*, **75**, 195217.
92. Roder, C., Einfeldt, S., Figge, S. *et al.* (2006) *Journal of Applied Physics*, **100**, 103511.
93. Roder, C., Einfeldt, S., Figge, S. *et al.* (2006) *Physica Status Solidi A*, **203**, 1672.
94. Sun, Y.J., Brandt, O., Jahn, U. *et al.* (2002) *Journal of Applied Physics*, **92**, 5714.
95. Reshchikov, M.A., Teke, A., Maruska, H.P. *et al.* (2004) *Materials Research Society Symposium Proceedings*, **798**, Y5.53.1.
96. Chen, Z., Lu, D.C., Liu, X. *et al.* (2003) *Journal of Applied Physics*, **93**, 316.
97. Yu, H.B., Chen, H., Li, D.S. *et al.* (2004) *Journal of Crystal Growth*, **263**, 94.
98. Liu, R., Bell, A., Ponce, F.A. *et al.* (2005) *Applied Physics Letters*, **86**, 021908.
99. Paskov, P.P., Schifano, R., Monemar, B. *et al.* (2005) *Journal of Applied Physics*, **98**, 093519.
100. Paskova, T., Paskov, P.P., Darakchieva, V. *et al.* (2007) *Materials Research Society Symposium Proceedings*, **955E**, I3.4.

101. Ueda, M., Kojima, K., Funato, M. *et al.* (2006) *Applied Physics Letters*, **89**, 211907.
102. Gil, B. (2007) *Applied Physics Letters*, **90**, 121903.
103. Kuokstis, E., Zhang, J., Fareed, Q. *et al.* (2002) *Applied Physics Letters*, **81**, 2755.
104. Okada, N., Kato, N., sato, S. *et al.* (2007) *Journal of Crystal Growth*, **300**, 141.
105. Bhattachariyya, J., Ghosh, S., Gokhale, M.R. *et al.* (2006) *Applied Physics Letters*, **89**, 151910.
106. Bishop, S.M., Park, J.-S., Gu, J. *et al.* (2007) *Journal of Crystal Growth*, **300**, 83.
107. Gao, H., Yan, F., Li, J. *et al.* (2006) *Physica Status Solidi A*, **203**, 3788.
108. Yan, F., Gao, H., Zhang, H. *et al.* (2007) *Journal of Applied Physics*, **101**, 023506.
109. Darakchieva, V., Paskova, T., Schubert, M. *et al.* (2007) *Journal of Crystal Growth*, **300**, 233.
110. Amano, H. (2006) *Presented at the 1st International Symposium on Nitride Growth*, June 4–6, 2006, Linkoping, Sweden.
111. Tsuchiya, Y., Okadome, Y., Honshio, A. *et al.* (2005) *Japanese Journal of Applied Physics*, **44**, L1516.
112. Kaeding, J.F., Asamizu, H., Sato, H. *et al.* (2006) *Applied Physics Letters*, **89**, 202104.
113. Santic, B. (2003) *Semiconductor Science and Technology*, **18**, 219.
114. Vurgaftman, I. and Meyer, J.R. (2003) *Journal of Applied Physics*, **94**, 3675.
115. Rodina, A.V., Dietrich, M., Goeldner, A. *et al.* (2001) *Physical Review B*, **64**, 115204.
116. McLaurin, M. and Speck, J.S. (2007) *Physica Status Solidi (RRL)*, **3**, 110.

PART I
Growth

Nitrides with Nonpolar Surfaces. Growth, Properties, and Devices. Edited by Tanya Paskova

ISBN: 978-3-527-40768-2

2
Growth of Planar and Reduced-defect Density Nonpolar GaN Films by Hydride Vapor Phase Epitaxy

Benjamin A. Haskell, Paul T. Fini, and Shuji Nakamura

2.1
Introduction

Gallium nitride and its alloys with indium and aluminum nitride have attracted significant attention in recent years, owing to the successful development of visible and ultraviolet light emitting diodes (LEDs) [1], blue/violet laser diodes [2], and high-power electronic devices [3] based on the system comprising these materials. Revolutions in lighting, display technology, data storage, and power switching are underway as a result of the unique optical, electronic, and structural properties of this (Al, In, Ga)N semiconductor system.

Gallium nitride is most stable in the wurtzite structure that belongs to space group $P6_3mc$. The wurtzite structure, which is depicted in Figure 2.1, is described by a hexagonal coordinate system having three equivalent basal-plane lattice vectors, $\mathbf{a}_1$, $\mathbf{a}_2$, and $\mathbf{a}_3$, and a unique $\mathbf{c}$ translation vector. In three-index notation ($\mathbf{a}_1$, $\mathbf{a}_2$, $\mathbf{c}$), gallium atoms are located at (0,0,0,) and (1/3, 2/3, 1/2), while the nitrogen atoms sit on an equivalent sublattice at (0, 0, u) and (2/3, 1/3, $1/2 + u$), where $u = 0.377$. Referring to Figure 2.1, low-index planes in the wurtzite structure include $\{11\bar{2}0\}$ a planes, $\{1\bar{1}00\}$ m planes, as well as the (0001) $+c$ plane (also referred to as the 'Ga face') and the $(000\bar{1})$ $-c$ plane (also referred to as the 'N face').

Historically, almost all III-nitride research has focused on epitaxial growth and properties of c-plane-oriented thin films and devices. While significant breakthroughs have been recently achieved with c-plane-oriented (Al, In, Ga)N films and devices, these heterostructures and optoelectronic devices are strongly impacted by spontaneous and piezoelectric polarization effects. Since the wurtzite III-nitrides are polar crystals, discontinuities in the net polarization in heterostructures give rise to fixed space charges. In c-plane quantum wells, the polarization-related charges at the hetero-interfaces lead to large internal electric fields in the quantum wells – these fields

Nitrides with Nonpolar Surfaces. Growth, Properties, and Devices. Edited by Tanya Paskova

ISBN: 978-3-527-40768-2

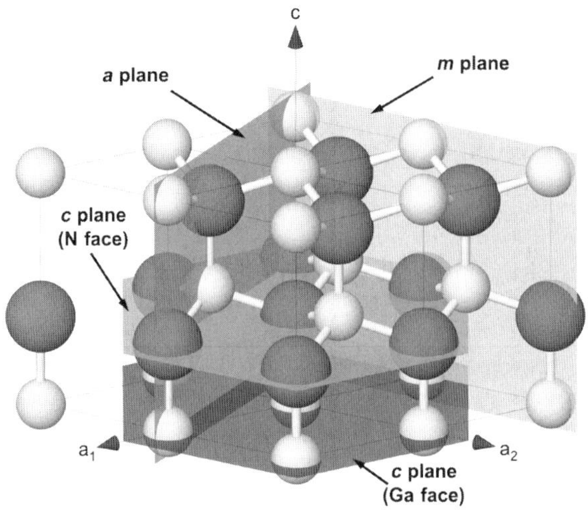

Figure 2.1 Scale representation of the GaN crystal structure in hexagonal coordinates. Ga atoms are shaded lightly and N atoms are shaded dark.

are the basis for the quantum-confined Stark effect (QCSE) [4], in which spatial separation electron and hole wave functions occur [5]. The radiative recombination lifetime in the films increases, raising the probability of nonradiative recombination and decreasing achievable internal quantum efficiency.

The polarization-related effects in wurtzite heterostructures can be entirely avoided by growing devices on alternative orientations of GaN crystals, such as the $\{1\bar{1}00\}$ *m* planes or the $\{11\bar{2}0\}$ *a* planes, in which the polar *c*-axis is parallel to the free surface and any planar hetero-interfaces. Since polarization fields exist entirely within the plane of the device structure, heterostructures grown on these planes exhibit flat-band conditions under zero bias.

Unfortunately, attempts over 30 years of GaN epitaxy research repeatedly indicated that the nonpolar planes were 'unstable' [6–13]. One study after another yielded nonpolar films with surfaces too rough and/or faceted for device growth. In 2000, Waltereit *et al.* first demonstrated planar *m*-plane GaN growth via molecular beam epitaxy (MBE) [5]. This demonstration was followed by metalorganic chemical vapor deposition (MOCVD) growth of planar *a*-plane GaN films in 2002 by Craven *et al.* [14].

While considerable progress was made in thin-film growth of nonpolar GaN from 2000–2002, thick-film or bulk growth of nonpolar orientations continued to be elusive until late 2002. The performance of nonpolar GaN-based devices would be limited by the lack of low-defect density film and substrate options. This chapter describes the progress achieved in thick-film nonpolar GaN growth via hydride vapor phase epitaxy (HVPE) toward the goal of producing low-defect density nonpolar GaN thick-films and substrates.

2.2 Planar *a*-plane GaN Growth

While heteroepitaxial growth of planar *a*-plane GaN films presents considerable challenges, substrate selection is fortunately not among them. Previous efforts, while failing to achieve smooth surface morphology, had demonstrated that *r*-plane sapphire substrates provide a good lattice match and yield uniformly oriented *a*-plane GaN films. Figure 2.2 shows the sapphire crystal structure and its theoretical lattice match to *a*-plane GaN. The sapphire *r* plane contains a rectangular grid of oxygen ions with an additional O^{2-} ion within

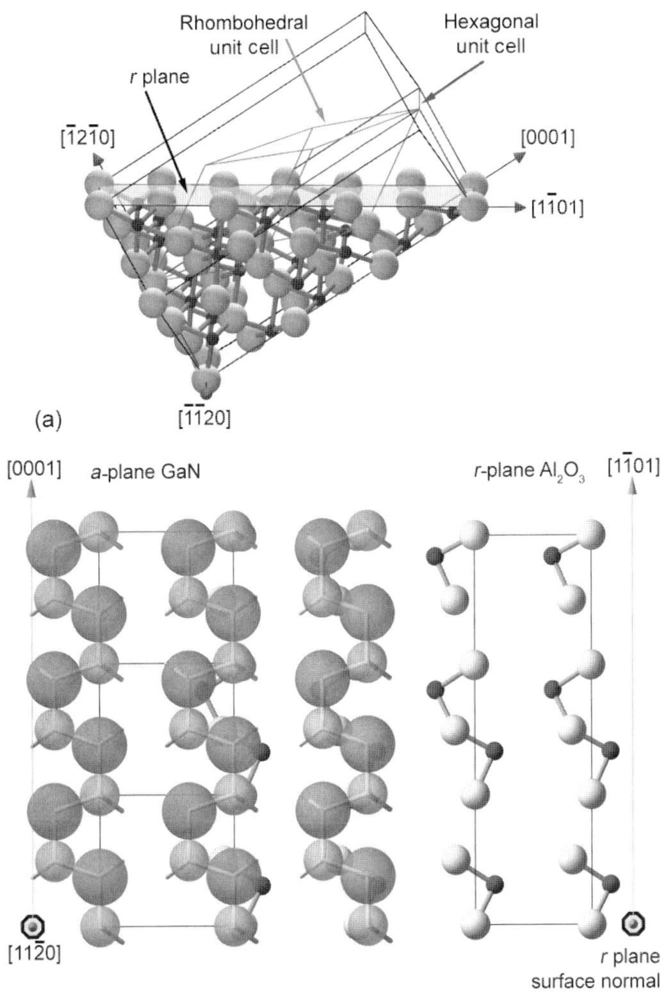

Figure 2.2 (a) Three-dimensional ball and stick representation of the sapphire *r* plane. (b) Atomic arrangement of the sapphire *r* plane with the GaN *a* plane overlaid.

each rectangle, as shown in Figure 2.2b. One edge of the grid is the [11$\bar{2}$0] direction, while the other is the [1$\bar{1}$01], which lies parallel to the projection of the *c*-axis on the *r*-plane surface. The lattice mismatch between *r*-plane Al_2O_3 and *a*-plane GaN is ~1% along the GaN *c* direction and ~16% along the GaN *m* direction [15].

The planar *a*-plane GaN films described below were grown by HVPE directly on *r*-plane sapphire substrates without the use of any low-temperature buffer or nucleation layers. Typical V/III ratios ranged from 9 to 40 with nitrogen and hydrogen mixtures serving as carrier gases. Typical growth rates ranged from 15 to 95 $\mu m\ h^{-1}$ at substrate temperatures between 1040 and 1070 °C [16].

Figure 2.3a shows a Nomarski optical contrast micrograph of a representative HVPE-grown *a*-plane GaN film. Its surface was characterized by long-range 'flow patterns' that had peak-to-valley heights of 100–500 nm over 75–500 μm lateral extents, as measured by profilometry. These low-angle surface features scattered light minimally; the films were specular and optically transparent. Other observed long-range surface features included faint ridges or scales of similar heights, but with varying directions over the wafer surface. No correlation was found between these features and the structural properties of the films. In contrast to previously reported *a*-plane GaN surfaces having facets inclined 30–50° from the surface normal [8, 13, 17], the angular variation of these surface features was 0.2–0.8° and could not be attributed to faceting. Subsurface cracks oriented nearly perpendicular to the GaN *c*-axis were observed in this film, one of which is detailed in the cross-sectional scanning electron micrograph (SEM) image in Figure 2.3b. These internal cracks, which are similar to those observed in *c*-plane GaN films [18], generally did not reach the free surface during growth. These cracks resulted from plastic relief of tensile strain that may have been a consequence of grain coalescence [19]. The cracks subsequently healed via lateral overgrowth to reduce surface energy.

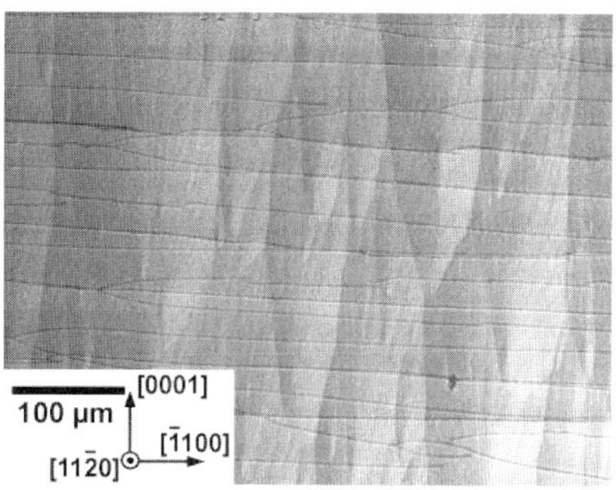

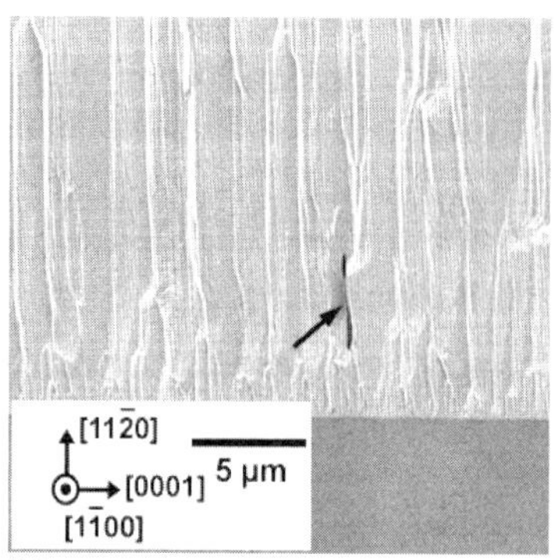

Figure 2.3 (a) Nomarski optical contrast micrograph of an *a*-plane GaN film. (b) Cross-sectional SEM image revealing two internal cracks in an *a*-plane GaN film.

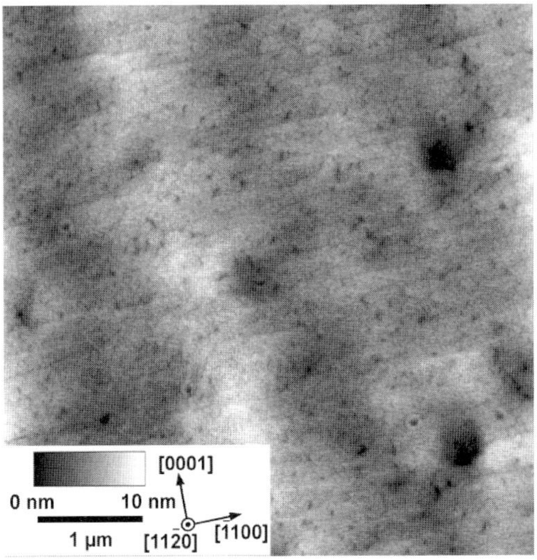

Figure 2.4 A 5 × 5 μm AFM image of a characteristic *a*-plane GaN surface.

Figure 2.4 shows a representative atomic force microscopy (AFM) scan of an *a*-plane GaN film. Local root mean square (RMS) roughness over 2 × 2 μm sampling areas was typically 0.5–0.8 nm. The RMS roughness over 100–400 μm^2 areas remained below 2 nm. The surface was dominated by a high density of 3–7-nm-deep pits with densities ranging from 2×10^9 to 9×10^9 cm^{-2}. These pits likely decorated intersections of threading dislocation with the free surface. Additionally, ~1-nm-high lines with a density of $\sim 7 \times 10^4$ cm^{-1} were apparent in the image. In contrast to the long-range surface features, these steps were oriented roughly perpendicular to the GaN *c*-axis regardless of HVPE flow conditions. These lines have been correlated with the presence of basal-plane stacking faults in the films.

Structural characteristics of the planar *a*-plane GaN films were evaluated by X-ray diffraction (XRD) and plan-view transmission electron microscopy (TEM). XRD was performed using Cu $K\alpha$ radiation in a Philips materials research diffractometer (MRD) Pro four-circle X-ray diffractometer operating in receiving slit mode. $2\theta - \omega$ scans of the *a*-GaN films, such as that shown in Figure 2.5, exhibited peaks that were indexed to the *r*-plane sapphire 1$\bar{1}$02, 2$\bar{2}$04, and 3$\bar{3}$06 reflections, and the 11$\bar{2}$0 GaN reflection. No GaN 0002 reflection was observed, demonstrating that within the detection limits of this technique, the films were uniformly *a*-plane-oriented.

ω rocking curves, shown in Figure 2.6a, were measured on the on-axis 11$\bar{2}$0 reflection for geometries in which the GaN [1$\bar{1}$00] and [0001] directions were in a coplanar geometry. Though full widths at half maximum (FWHM) as small as 660 arcsec have been observed, typical FWHM for the 11$\bar{2}$0 reflection in these geometries were generally comparable 1040–1250 arcsec. The on-axis

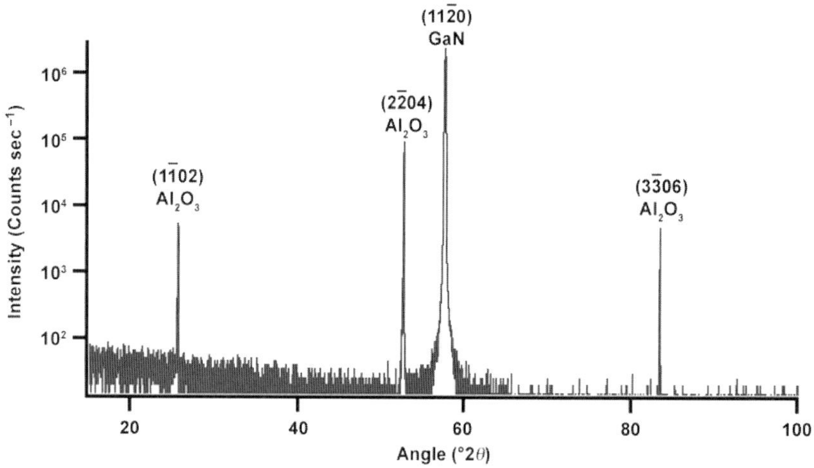

Figure 2.5 $2\theta-\omega$ scan for a planar *a*-plane GaN film.

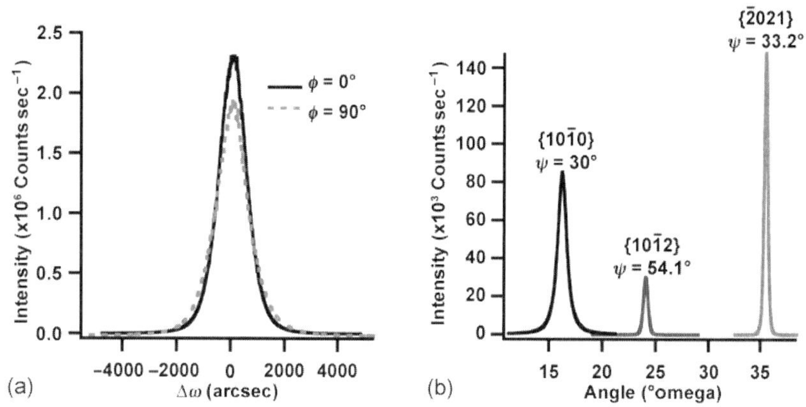

Figure 2.6 (a) On-axis ω rocking curves for a planar *a*-plane GaN film. (b) Off-axis skew geometry rocking curves.

peak widths were comparable to those observed for planar MOCVD-grown *a*-plane GaN films [14]. However, in contrast to MOCVD-grown *a*-plane GaN films [14], the peak width of the on-axis reflection measured with the *c*-axis within the scattering plane ($\phi = 90^\circ$) was comparable to the peak width with the *c*-axis normal to the scattering plane, whereas a substantial asymmetry is commonly observed in MOCVD-grown *a*-plane films. The asymmetric mosaic of the MOCVD-grown films was correlated with growth condition variation; however, no such correlation has been identified for the HVPE-grown films.

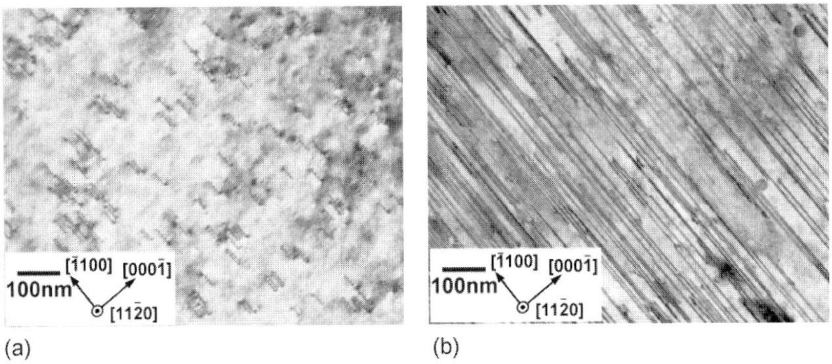

Figure 2.7 Plan-view TEM images as observed with (a) $\mathbf{g} = 0002$ to reveal dislocations having a nonzero [0001] component, and (b) $\mathbf{g} = 1\bar{1}00$ to observe stacking faults parallel to the (0001) plane.

Rocking curves were measured for several off-axis reflections, as shown in Figure 2.6b. The 30° off-axis $10\bar{1}0$ reflections were measured by tilting the samples relative to the scattering plane (skew geometry), yielding a FWHM ranging from 2900 to 3700 arcsec from sample to sample. Typical FWHM for the 33° off-axis $\bar{2}021$ and 54° off-axis $10\bar{1}2$ ranged from 1150 to 1950 arcsec and 1330 to 2660 arcsec, respectively. Minimal variation was observed in the off-axis peak widths with the in-plane rotation of the crystal.

Figure 2.7 shows plan-view transmission electron micrographs of an *a*-plane GaN film. Figure 2.7a was imaged under the $\mathbf{g} = 0002$ diffraction condition, revealing threading dislocations having a Burgers vector component parallel to the GaN [0001] direction. Thus, these are edge-component dislocations. The *c*-component dislocation density ranged from 9×10^9 cm^{-2} to 2×10^{10} cm^{-2} in these samples. The TEM image in Figure 2.7b, taken under the $\mathbf{g} = 1\bar{1}00$ diffraction condition, shows a stacking fault density of $\sim 4 \times 10^5$ cm^{-1}. These basal-plane stacking faults are likely related to the presence of exposed nitrogen-face (000$\bar{1}$) surfaces during the early stages of growth. Additional imaging with varying sample tilt in the $\mathbf{g} = 1\bar{1}00$ diffraction condition revealed partial dislocations of $\sim 7 \times 10^9$ cm^{-2} having Burgers vectors $\mathbf{b} = \frac{1}{3}\langle 1\bar{1}00 \rangle$.

2.3 Lateral Epitaxial Overgrowth of *a*-plane GaN Films

While the surfaces of the planar *a*-plane GaN films described above were smooth, the TEM images shown in Figure 2.7 and X-ray diffraction data shown in Figure 2.6 indicated that their microstructural quality needed improvement. Threading dislocations in *c*-plane GaN have been found to act as nonradiative recombination centers, carrier scattering centers, and possibly fast diffusion pathways, limiting device efficiency, speed, and reliability. Less is known about the influence of stacking faults on device performance. However, it was clear that both the threading dislocation and stacking fault densities in these *a*-plane GaN films needed to be reduced to be suitable for use as templates for optoelectronic devices.

To improve the microstructural quality of *a*-plane GaN films, we developed an HVPE-based variant of lateral epitaxial overgrowth (LEO) for nonpolar GaN films [20]. LEO *a*-plane GaN films were grown either directly on *r*-plane Al_2O_3 substrates or on *a*-plane GaN template layers grown by either MOCVD or HVPE. The masks for the LEO process were prepared by applying conventional photolithographic processing and wet etching to ~130-nm-thick SiO_2 layers deposited by plasma-enhanced chemical vapor deposition. A variety of mask designs were investigated, including nonparallel stripes arranged in a 'wagon-wheel' pattern, arrays of circular apertures, parallel stripes oriented along the $\langle 0001\rangle$ direction, parallel stripes oriented along the $\langle 1\bar{1}00\rangle$ direction, and parallel stripes oriented along higher-index directions. The LEO growth process was carried out under similar HVPE growth conditions to those used for the *a*-plane GaN films described above. It was found that masks consisting of parallel stripes oriented along the $\langle 1\bar{1}00\rangle$ direction were most effective at improving microstructural quality of the GaN films, and will be the focus of the discussion that follows.

Figure 2.8a shows a schematic representation of the $\langle 1\bar{1}00\rangle$ stripe geometry that was used for the samples to be discussed in this section. Stripes oriented along this direction exhibited vertical $\{0001\}$ sidewalls, one of which was Ga-terminated and the other N-terminated. Interrupted growths have shown that the (0001) Ga-face wing advances roughly six times as rapidly as the $(000\bar{1})$ N-face wing. This ratio indicates that the relative growth rate of the $(000\bar{1})$ wing is measurably greater than in MOCVD LEO of *a*-plane GaN, in which the ratio of Ga- to N-face growth is roughly 10 [21]. One benefit of the large difference in lateral growth rates between the $\{0001\}$ faces is that the coalescence front was offset toward the N-face side of the window region, yielding broad wing regions uninterrupted by defective coalescence fronts. These stripes are shown in greater detail in the cross-sectional SEM images of Figure 2.8b and c. The inclined cross section in Figure 2.8b of nearly coalesced 5/35 stripes demonstrates the vertical $\{0001\}$ sidewalls that are prevalent for $\langle 1\bar{1}00\rangle$-oriented stripes throughout lateral growth and immediately preceding coalescence. Figure 2.8c shows a cross-sectional view of coalesced 5/15 stripes. Only contrast variation at the film-template interface due to SEM charging effects allows the window and wing regions to be distinguished.

Figure 2.9a shows a Nomarski optical contrast micrograph of a ~ 12-μm-thick coalesced LEO film formed with $\langle 1\bar{1}00\rangle$-oriented 2/8 stripes. The faint 'fish scale' features on the surface demonstrated that the film's surface is in focus, while the refractive index contrast from the SiO_2 allowed the out-of-focus mask pattern to be observed. Similar to the directly grown *a*-plane GaN films described above, these surface textures contained vertical displacements

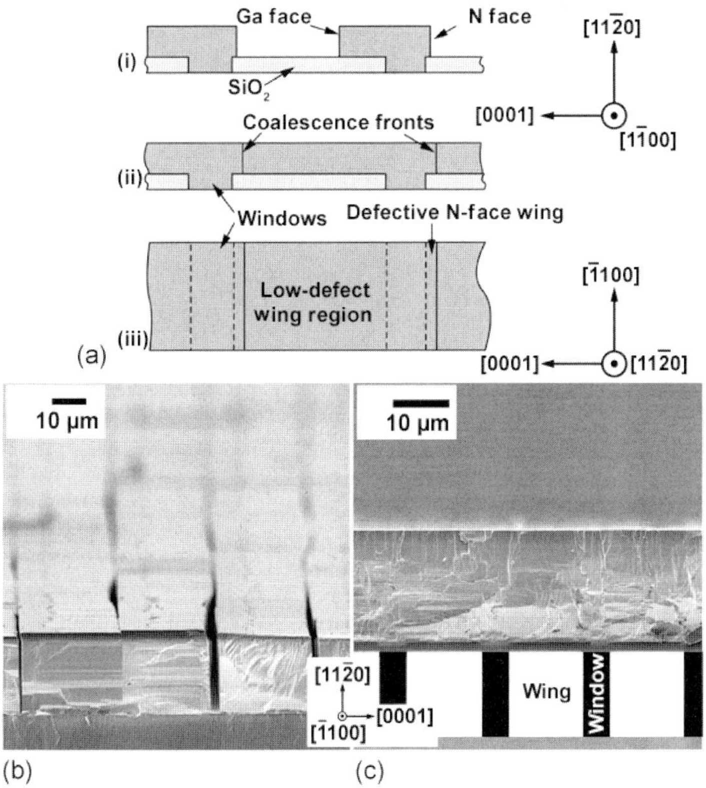

Figure 2.8 (a) Schematics of LEO growth using mask stripes/openings oriented along the GaN ⟨1$\bar{1}$00⟩ direction. (i) Cross section of uncoalesced stripes. (ii) Cross section of coalesced stripes showing the offset coalescence front. (iii) Plan view of coalesced stripes identifying the large low-defect wing region. (b) Scanning electron micrograph of uncoalesced 5/35 stripes. (c) SEM image of coalesced *a*-plane LEO GaN.

on the order of ~200 nm over lateral extents of hundreds of micrometers, and thus are not believed to be related to faceting.

AFM was performed to compare the surface quality in the window and wing regions of the *a*-plane LEO films. Figure 2.9b shows a 10 × 10 μm AFM topograph of two coalesced stripes. The window region appeared as the darker band of pitted material with the coalescence front roughly 1 μm to the left of the window. The Ga-face wing, apparent on the left side of the image, had superior surface quality and was nearly featureless. The wing regions exhibited average pit densities of less than 3×10^6 cm^{-2}, compared to ~10^9 cm in the window regions. The RMS roughness of the wing regions was less than 0.9 nm (11.9 nm peak-to-valley height), compared to 1.3 nm (20.8 nm peak-to-valley) in the window regions. The faint lines parallel to the ⟨1$\bar{1}$00⟩ direction in the window region were not present in the Ga-face wing region.

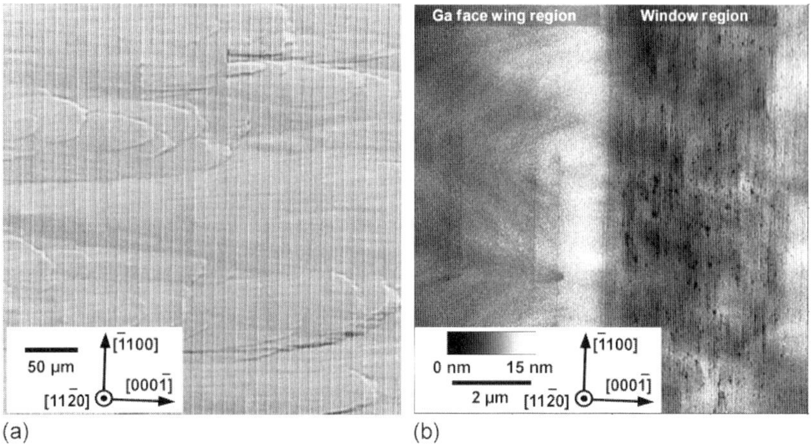

Figure 2.9 (a) Nomarski optical contrast micrograph of a coalesced LEO *a*-plane GaN film. (b) A 10 × 10 μm AFM image of coalesced ⟨1̄100⟩-oriented GaN stripes.

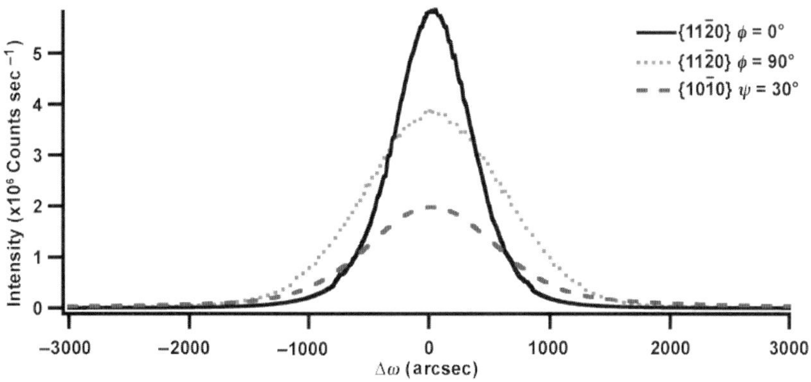

Figure 2.10 X-ray rocking curves for a representative *a*-plane GaN LEO film grown using ⟨1̄100⟩-oriented stripes.

The structural quality of the *a*-plane LEO films was characterized by X-ray diffraction and TEM. Figure 2.10 shows X-ray rocking curve details for a characteristic *a*-plane GaN LEO film grown with ⟨1̄100⟩-oriented stripes. Rocking curves of the 112̄0 GaN reflection taken perpendicular to the LEO stripe direction of coalesced films were single-peaked, indicating a lack of wing tilt within the structural mosaic width of the coalesced films. In isolated cases, wing tilt of less than 0.2° was observed for interrupted growths with uncoalesced stripes. Narrowing of both on- and off-axis reflections was observed in the LEO films compared to planar *a*-plane GaN films grown directly on *r*-plane sapphire [16]. Typical FWHM for the 112̄0 and 101̄0 reflections were

750 and 1250 arcsec, respectively. The non-LEO a-plane GaN films described above exhibited comparable on-axis rocking curve widths for all orientations of the c-axis relative to the X-ray scattering plane. In contrast, the FWHM of LEO a-plane GaN films when the GaN c-axis was within the scattering plane ($\phi = 90°$) was up to five times larger than when the c-axis was perpendicular to the scattering plane ($\phi = 0$). This peak width variation was independent of stripe orientation and was rather a consequence of the mosaic of the crystal.

Figure 2.11 shows plan view and cross-sectional TEM images of a LEO film imaged with **g** vectors of $1\bar{1}00$ and $01\bar{1}0$, respectively. In agreement

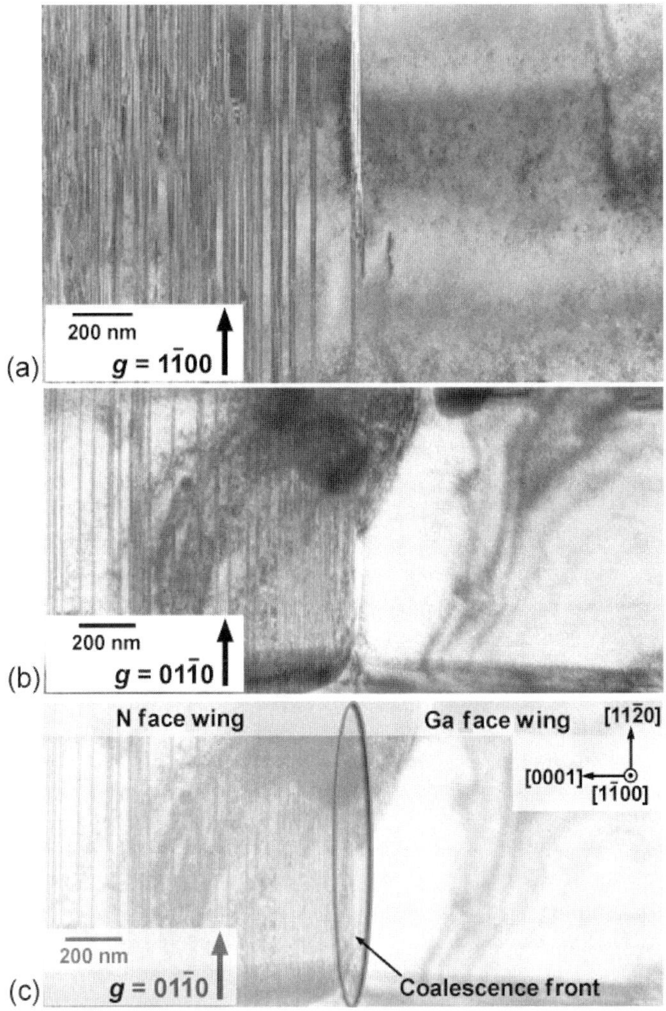

Figure 2.11 (a) Plan-view TEM of coalesced $\langle 1\bar{1}00\rangle$ stripes. (b) Corresponding cross-sectional TEM image with (c) explanation of salient features.

with observations from AFM, the window regions exhibited high threading dislocation ($\sim 9 \times 10^9$ cm^{-2}) and stacking fault ($\sim 4 \times 10^5$ cm^{-1}) densities. In contrast, the Ga-face wing region was essentially free of both dislocations and stacking faults, with densities below the images' resolutions of $\sim 5 \times 10^6$ cm^{-2} and $\sim 3 \times 10^3$ cm^{-1}, respectively. The N-face wing region was also threading dislocation-free, though basal-plane stacking faults and partial dislocations terminating the faults remained prevalent.

The results for *a*-plane GaN LEO films grown with $\langle 1\bar{1}00 \rangle$-oriented stripes have demonstrated that substantial reduction in morphological and structural defects in *a*-plane GaN may be readily achieved by LEO with HVPE. The reduction in threading dislocation density in the overgrown GaN is accompanied by a significant improvement in surface morphology compared with non-LEO planar *a*-plane GaN films.

2.4 Planar *m*-plane GaN Heteroepitaxy

The growth of planar $\{1\bar{1}00\}$ *m*-plane GaN thin-films has predated their *a*-plane analogs for several years [5]. MBE-grown films of Waltereit *et al.* exhibited adequate surface roughness for the growth of quantum wells and other heterostructures. While significant progress has been made in the fabrication of *a*-plane GaN quantum structures and devices by vapor phase growth techniques, *m*-plane GaN film growth and devices developed at a slower pace. This development lag can be largely attributed to two key issues: the lack of suitable substrates and the difficulties in growing high-quality, low-defect density *m*-plane GaN films. In this section, the morphological and structural characteristics of planar *m*-plane GaN films on (100) γ-$LiAlO_2$ and $(1\bar{1}00)$ *m*-plane 6H-SiC, as well as reduction of bulk and surface defects, are discussed.

The single greatest challenge to the growth of high-quality *m*-plane GaN films is, arguably, the lack of suitable substrates offering a combination of adequate lattice matching and chemical and thermal stability. Lithium aluminate (γ-$LiAlO_2$) is a poor substrate for use in HVPE growth of nitrides. Lithium aluminate decomposes in the presence of water or even mild acids at room temperature, thereby complicating substrate surface preparation. The substrates are attacked by hydrogen gas and hydrogen chloride at typical HVPE growth temperatures, releasing Li^+ and O^{2-} into the growth environment. $LiAlO_2$ also exhibits poor mechanical stability, frequently fracturing upon cool-down from GaN growth temperatures. Despite these drawbacks, we demonstrated thick, planar *m*-plane GaN growth on $LiAlO_2$ substrates. Growth rates ranged from 30 to 240 μm h^{-1} at substrate temperatures of 860–890 °C and V/III ratios of 10 to 60, yielding freestanding films up to 250-μm thickness.

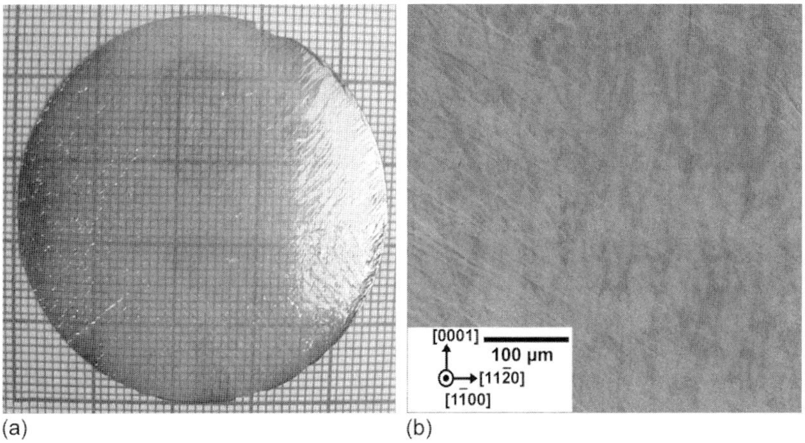

Figure 2.12 (a) Digital photograph of a 48-mm-diameter freestanding *m*-plane GaN wafer on a 1 × 1 mm grid. (b) Nomarski optical contrast micrograph of the *m*-plane GaN film shown in (a).

Figure 2.12a is a digital photograph of a 240-μm-thick freestanding *m*-plane GaN film grown on a 2-in. diameter γ-$LiAlO_2$ substrate. This wafer spontaneously separated from the $LiAlO_2$ substrate, which is consistent with previous reports [22]. Both sides of the wafer were specular and mirrorlike with no evidence of bulk or surface defects. The brownish tint of the wafer was attributed to impurity incorporation from decomposition of the $LiAlO_2$ substrate in the HVPE growth environment. Figure 2.12b is a Nomarski optical contrast micrograph of the surface of the freestanding *m*-plane GaN film pictured in Figure 2.12a. The surface was characterized by long-range noncrystallographic textures having peak-to-valley heights of less than 200 nm over greater than 500 μm lateral extents. Near-complete elimination of surface defects, such as pits and 'arrowheads', was observed in this and similar films grown on $LiAlO_2$.

Figure 2.13 shows a representative 5 × 5 μm AFM image of a planar *m*-plane GaN surface. This striated, or 'slate', morphology is dominated by ridges oriented along the GaN $\langle 11\bar{2}0 \rangle$ direction. The linear density of these ridges along the $\langle 0001 \rangle$ direction averaged $\sim 10^5$ cm^{-1}. This morphology is commonly observed for MBE-grown *m*-plane GaN films [23]. The surface exhibited an RMS roughness of 0.894 nm, which was smoother than is typically observed for MBE-grown films.

The structural quality of the planar *m*-plane GaN films was evaluated using X-ray diffraction and TEM. Figure 2.14a shows a $2\theta - \omega$ scan of a freestanding *m*-plane GaN film originally grown on (100) $LiAlO_2$. Three peaks could be observed, which were indexed to the GaN $1\bar{1}00$, $2\bar{2}00$, and $3\bar{3}00$ reflections,

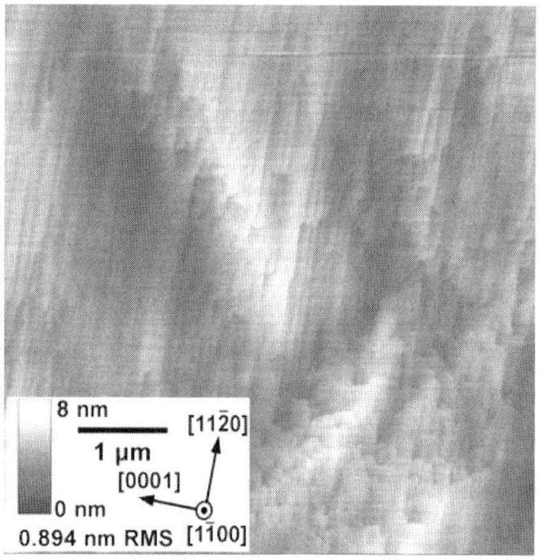

Figure 2.13 Atomic force micrograph of an *m*-plane GaN film grown by HVPE.

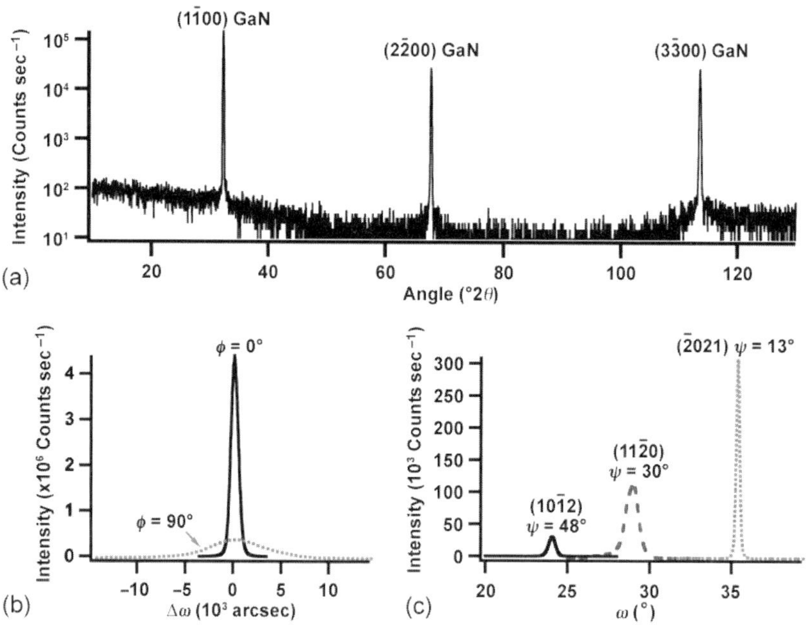

Figure 2.14 (a) $2\theta - \omega$ scan of a freestanding *m*-plane GaN film grown on (100) $\gamma - LiAlO_2$. (b) On-axis rocking curves of an *m*-plane GaN film at $\phi = 0^\circ$ and $\phi = 90^\circ$. (c) Off-axis rocking curves from the *m*-plane GaN film in (b).

demonstrating that the film was uniformly *m*-plane oriented within the sensitivity of the measurement. Figure 2.14b shows on-axis rocking curves for an *m*-plane GaN film. In contrast to the *a*-plane GaN films described above, these *m*-plane GaN films exhibited asymmetry in the peak widths of their on-axis rocking curve. This asymmetry was, however, consistent with *a*-plane GaN films grown by MOCVD [14] and with measurements of the mosaic of the MBE-grown *m*-plane GaN templates. The FWHM of the on-axis reflection of the *m*-plane GaN films when the *c*-axis was normal to the scattering plane ($\phi = 0°$) was 920 arcsec, whereas the FWHM when the GaN *c*-axis was within the scattering plane ($\phi = 90°$) was roughly 6300 arcsec. This 1:7 ratio of peak widths was comparable to that observed for freestanding *m*-plane GaN films grown on $LiAlO_2$ substrates. The asymmetry in the on-axis reflection FWHMs is a consequence of asymmetrical tilt mosaic in the crystal. The peak widths of the off-axis $\bar{2}021$, $11\bar{2}0$, and $10\bar{1}2$ reflections, all measured in a skew geometry, were 880, 3250, and 1750 arcsec, respectively.

Figure 2.15 shows representative plan view and cross-sectional TEM images of an *m*-plane GaN film. Figure 2.15a was imaged under the $\mathbf{g} = 0002$ diffraction condition showing threading dislocations having a Burgers vector component parallel to the GaN [0001] direction. The density of dislocations observed in this image was $4 \times 10^9\ cm^{-2}$, which is roughly an order of magnitude greater than what was previously reported for HVPE-grown *m*-plane GaN films [22]. The cross-sectional TEM image shown in Figure 2.15b was imaged under the $\mathbf{g} = 1\bar{1}00$ diffraction condition, thus revealing basal-plane stacking faults. The stacking fault density was $1 \times 10^5\ cm^{-1}$, which is again consistent with previous reports for both *m*-plane and *a*-plane GaN films [16, 22, 23]. This stacking fault density was comparable to the ridge density observed in the AFM images in Figure 2.13. As the intersection of the basal plane and the *m* plane is a line along the $\langle 11\bar{2}0\rangle$ direction, the close correlation in fault and ridge linear densities suggests that the surface ridges observed in the 'slate' morphology described above are caused by displacement of the *m*-plane surface caused by faulting on the basal plane.

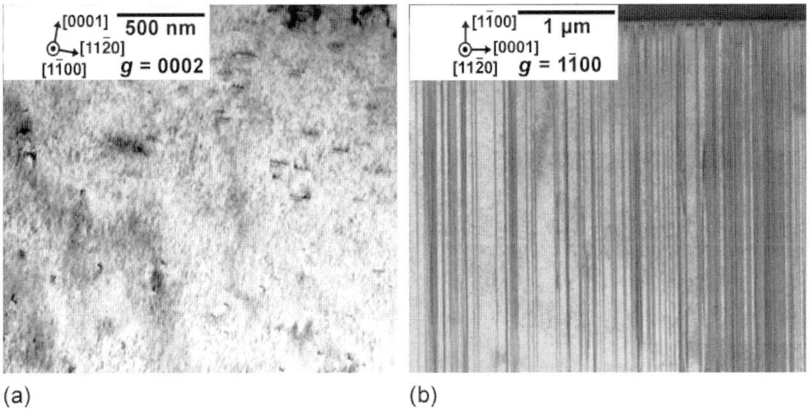

Figure 2.15 (a) Plan view and (b) cross-sectional transmission electron micrographs of an *m*-plane GaN film.

2.5 Lateral Epitaxial Overgrowth of *m*-plane GaN

While the above-described work demonstrated significant improvement in thick *m*-plane GaN growth and morphology, the high microstructural defect densities inherent to these films needed to be reduced to improve device performance. To address this concern, LEO was developed for *m*-plane GaN via HVPE. Starting instead with MBE-grown *m*-plane GaN or *m*-plane AlN templates grown on *m*-plane SiC substrates (both 4H and 6H polytypes), fully coalesced LEO films were produced, achieving comparable improvements in morphology and structural quality to those attained for *a*-plane GaN LEO.

Conventional photolithographic processing and HF wet etching were performed on SiO_2-coated *m*-plane GaN templates to create several mask patterns in a manner similar to the approach taken with the *a*-plane GaN LEO process. Ultimately, parallel stripe masks oriented along the $\langle 11\bar{2}0\rangle$ direction yielded elimination of both dislocations and stacking faults, and will therefore be the focus of the discussion below. Vertical sidewalls were again typical for this stripe orientation, terminated by $+c$ and $-c$-plane surfaces. In general, both the vertical and lateral growth rates observed were significantly faster than what have been observed for *c*- and *a*-plane GaN LEO. Typical vertical growth rates ranged from 40 to 300 $\mu m\ h^{-1}$ at substrate temperatures ranging from 860 to 1070 °C. Fully coalesced LEO films were produced by this method, yielding specular surfaces over the full area of the wafers.

The direct growth *m*-plane GaN films described above exhibited a commonly observed striated, or 'slate', morphology. This striated morphology persisted in the window regions of *m*-plane GaN LEO films, as shown in Figure 2.16a. The RMS roughness of this film's surface was 0.78 nm, which is slightly smoother than was observed for the smoothest non-LEO *m*-plane GaN films described above. Figure 2.16b shows an atomic force micrograph taken from the Ga-terminated wing of an *m*-plane LEO GaN stripe grown from $\langle 11\bar{2}0\rangle$-oriented stripes. The wing surface no longer exhibited the 'slate' morphology, rather a step-flow surface was seen. The step terraces ranged from ~100 to 150 nm across and showed no evidence of being aligned to any particular GaN crystallographic direction. The breadth and orientation of these terraces were likely dictated by the miscut of the SiC substrate. Terrace heights ranged from ~0.8 to 1.4 nm, which corresponded to roughly four to seven atomic layers in the $[1\bar{1}00]$ direction. The roughness of this surface was reduced to 0.53 nm RMS over an area of 25 μm^2. This superior surface morphology resulted from the lack of extended defects in the overgrown material; indeed TEM imaging indicated that the threading dislocation and stacking fault densities were comparable to the Ga-terminated wings in *a*-plane GaN LEO films, at $5 \times 10^6\ cm^{-2}$ and $3 \times 10^3\ cm^{-1}$, respectively.

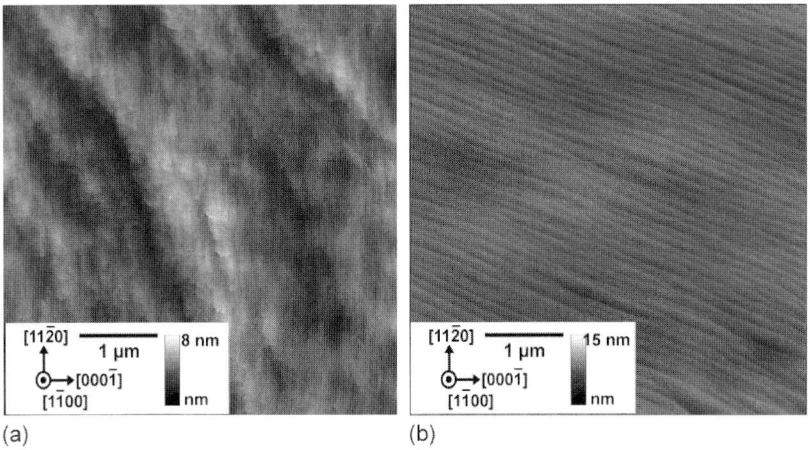

Figure 2.16 Atomic force micrographs of m-plane GaN LEO films taken from (a) the window region of a $\langle 11\bar{2}0\rangle$-oriented stripe, and (b) the Ga-terminated wing of a $\langle 11\bar{2}0\rangle$-oriented stripe.

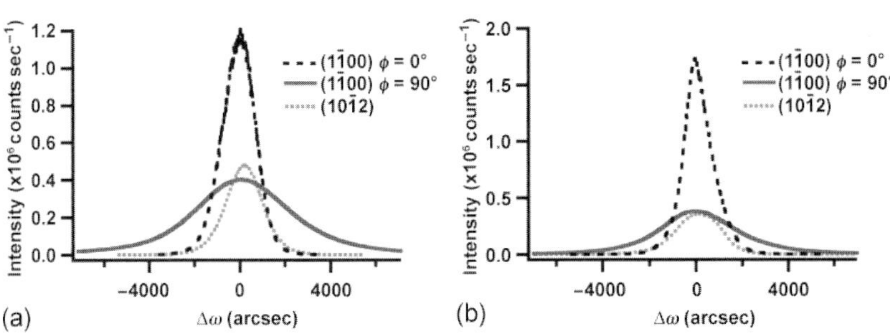

Figure 2.17 Representative on-axis (a) and off-axis (b) X-ray rocking curves of an m-plane GaN LEO film grown from $\langle 11\bar{2}0\rangle$-oriented stripes.

The structural quality of the m-plane GaN LEO films was additionally characterized by X-ray diffraction. As can be seen in Figure 2.17a, m-plane GaN LEO films grown from $\langle 11\bar{2}0\rangle$-oriented stripes typically had narrower on-axis reflections compared with non-LEO films, on the order of ~1200 arcsec at $\phi = 0$ and 3600 arcsec at $\phi = 90^\circ$. The asymmetry in the on-axis rocking curve width was independent of LEO stripe orientation; rather it was a consequence of the in-plane anisotropy of the crystalline mosaic. For comparison, the on-axis FWHM of the MBE GaN templates were typically 1100 arcsec at $\phi = 0^\circ$ and 7000 arcsec at $\phi = 90^\circ$. No splitting of the $1\bar{1}00$ reflections was observed when the scattering plane normal was parallel to the LEO stripe direction for either stripe orientation. Thus, no wing tilt was observed within the sensitivity of these X-ray measurements, which was limited by the mosaic of the underlying films and the curvature of the samples. The off-axis $10\bar{1}2$ reflections, measured at both $\phi = 0°$ and $\phi = 180°$ in a skew geometry, ranged from 1800 to 2400 arcsec, independent of ϕ or LEO stripe orientation.

2.6 Conclusion

The previous sections have reviewed the development of planar nonpolar GaN growth by HVPE. Direct heteroepitaxial growth of *a*- and *m*-plane GaN films yielded specular films possessing high threading dislocation and basal-plane stacking fault densities. LEO methods were then developed to eliminate these microstructural defects and further improve short-range surface morphology. By these methods, planar nonpolar GaN films that compare favorably with commercially available *c*-plane GaN substrates have been grown. These nonpolar GaN thick-films and substrates will facilitate a transition toward enhanced GaN-based optoelectronic devices free of the deleterious polarization effects that limit the performance of current polar nitride heterostructures.

References

1. Nishida, T. and Kobayashi, N. (2001) *Physica Status Solidi A*, **188**(1), 113.
2. Nakamura, S., Fasol, G., and Pearton, S.J. (2000) *The Blue Laser Diode*, Springer, New York.
3. Eastman, L.F. and Mishra, U.K. (2002) *IEEE Spectrum*, **39**(5), 28.
4. Deguchi, T., Sekiguchi, K., Nakamura, A. *et al.* (1999) *Japanese Journal of Applied Physics*, **38**(8B), L914.
5. Waltereit, P., Brandt, O., Trampert, A. *et al.* (2000) *Nature*, **406**, 865.
6. Maruska, H.P., Anderson, L.J., and Stevenson, D.A. (1974) *Journal of the Electrochemical Society*, **121**, 1202.
7. Sano, M. and Aoki, M. (1976) *Japanese Journal of Applied Physics*, **15**(10), 1943.
8. Shintani, A. and Minagawa, S. (1976) *Journal of Electrochemical Society*, **123**(10), 1575.
9. Madar, R., Michel, D., Jacob, G., and Boulou, M. (1977) *Journal of Crystal Growth*, **40**, 239.
10. Sasaki, T. and Zembutsu, S. (1987) *Journal of Applied Physics*, **61**(7), 2533.
11. Lei, T., Ludwig, K.F. Jr, and Moustakas, T.D. (1993) *Journal of Applied Physics*, **74**(7), 4430.
12. Melton, W.A. and Pankove, J.I. (1997) *Journal of Crystal Growth*, **178**, 168.
13. Molnar, R.J. (1999) Hydride vapor phase epitaxial growth of III-V nitrides, in Williardson, R., *et al.*, *Semiconductors and Semimetals*, Vol. 57, (eds R. Williardson, E. Weber, T. Moustakas, and J. Pankove), Elsevier Science and Technology Books, New York, pp. 1–31.
14. Craven, M.D., Lim, S.H., Wu, F. *et al.* (2002) *Applied Physics Letters*, **81**(3), 469.
15. Kung, P., Sun, C.J., Saxler, A. *et al.* (1994) *Journal of Applied Physics*, **75**(9), 4515.
16. Haskell, B.A., Wu, F., Matsuda, S. *et al.* (2003) *Applied Physics Letters*, **83**(8), 1554.
17. Paskova, T., Paskov, P.P., Darakchieva, V. *et al.* (2001) *Physica Status Solidi A*, **183**(1), 197.

18. Etzkorn, E.V. and Clarke, D.R. (2001) *Journal of Applied Physics*, **82**(2), 1025.

19. Böttcher, T., Einfeldt, S., Figge, S. *et al.* (2001) *Applied Physics Letters*, **78**(14), 1976.

20. Haskell, B.A., Wu, F., Craven, M.D. *et al.* (2003) *Applied Physics Letters*, **83**(4), 644.

21. Craven, M.D., Lim, S.H., Wu, F. *et al.* (2002) *Applied Physics Letters*, **81**(7), 1201.

22. Vanfleet, R.R., Simmons, J.A., Maruska, H.P. *et al.* (2003) *Applied Physics Letters*, **83**(6), 1139.

23. Brandt., O., Sun, Y.J., Däweritz, L., and Ploog, K.H. (2004) *Physical Review B*, **69**, 165326.

3
Nonpolar GaN Quasi-Wafers Sliced from Bulk GaN Crystals Grown by High-Pressure Solution and HVPE Methods

Izabella Grzegory, Henryk Teisseyre, Bolesław Łucznik, B. Pastuszka, Michał Boćkowski, and Sylwester Porowski

3.1
Introduction

Since recently [1], gallium nitride-based quantum structures grown in nonpolar directions are being considered as very promising physical systems for the further development of short wavelength semiconductor light emitting devices. Main advantages of the nonpolar structures are related to the elimination of strong piezoelectric fields present in the commonly used systems grown along polar (0001) axis of nitride crystal. Thus, an increased efficiency of radiative processes, stability of wavelength, and polarization of the emitted light are expected. Moreover, in the 'nonpolar' material, higher concentrations and mobilities of holes can be achieved [2].

On the other hand, the growth of high-quality structures in nonpolar directions is even more difficult than for typical (0001) orientation. 'Heteroepitaxial' approaches – that is, the use of foreign substrates such as r plane sapphire [e.g. [3, 4]], 4H-SiC [e.g. [5]], or γ-$LiAlO_2$ [e.g. [1]] – result in material containing a high number of dislocations and also stacking faults, thereby strongly influencing the optical properties of GaN [e.g. [6]]. A significant improvement can be achieved by application of epitaxial lateral overgrowth (ELOG) techniques as was shown by the University of California at Santa Barbara (UCSB) group [7].

An obvious way to obtain high-quality nonpolar GaN substrates is to slice wafers of required orientation from a high-quality bulk GaN crystal. Such an approach was recently analyzed in [8], where GaN boule grown by hydride vapor phase epitaxy (HVPE) on a free standing (0001)-oriented GaN substrate (Kyma Technologies) was sliced along the $(11\bar{2}0)$ plane, and resulting $10 \times 6\ mm^2$ nonpolar platelets were characterized by high-resolution X-ray diffraction (HR XRD) and transmission electron microscopy (TEM). A low dislocation density (about $10^5\ cm^{-2}$) and no stacking faults were found in the material.

In this work, bulk GaN crystals were also grown by HVPE but on the seeds grown previously by high-pressure solution method [9]. The seed crystals were

Nitrides with Nonpolar Surfaces. Growth, Properties, and Devices. Edited by Tanya Paskova

ISBN: 978-3-527-40768-2

in two morphological forms: thin (0001) platelets and needles elongated in c direction with well-developed nonpolar $(10\bar{1}0)$ faces.

3.2 Bulk Crystallization of GaN

3.2.1 Seed Crystals

The crystals used in this study for seeding of bulk crystallization of GaN by HVPE were grown from solutions in liquid gallium at nitrogen gas pressure of the order of 1.0 GPa [9]. This method results in crystals of two dominating morphological forms (Figure 3.1): thin hexagonal platelets, and hexagonal needles elongated into c-direction of wurtzite structure.

The needles grown usually at high supersaturations are often morphologically unstable, being partially hollow or skeletal, but they have well developed $\{10\bar{1}0\}$, which makes them good seeds for growth by HVPE. Crystals of both forms are limited in size to about 1 cm because of still insufficient control of supersaturation in gallium solutions of large volume.

The important advantage of the high-pressure solution method is that the resulting crystals of GaN are almost free of structural defects. Defect selective etching (DSE) in molten bases [10] of both (0001) polar and nonpolar faces of the crystals of both forms has shown that dislocation densities are negligible and usually do not exceed 100 cm^{-2}.

On the other hand, the crystals contain a high number of point defects owing to the high activity of oxygen always present as nonintentional impurity in the growth system. Because of this, the GaN crystals grown without intentional doping are highly n-type with free electron concentrations exceeding 10^{19} cm^{-3}. This leads to the slight expansion of the crystal lattice corresponding to a mismatch to pure GaN of the order of 10^{-4}. The free electrons can be removed

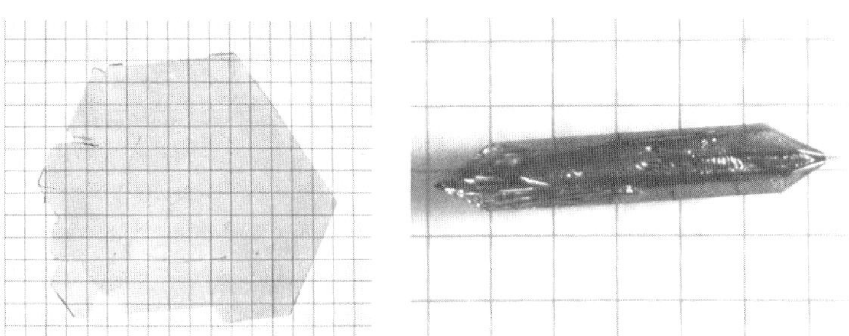

(a) (b)

Figure 3.1 Two morphological forms of GaN crystals grown under high N_2 pressure: (a) hexagonal platelet and (b) hexagonal needle.

by compensation of the oxygen donor by magnesium added to the growth solution. The lattice constants then become the same as for undoped material.

Both *n*-type GaN and highly resistive GaN : Mg platelets have been used as seeds for HVPE growth. The (0001) polar surfaces of the seed crystals were prepared for the growth by mechanical and chemical polishing. The *n*-type GaN needles were used for seeding in their 'as grown' state.

3.2.2
Bulk Crystallization of GaN by HVPE on Small Seeds

Hydride vapor phase epitaxy is currently considered as the most successful and promising method for obtaining high-quality bulk GaN substrates. This is owing to relatively high crystallization rates (about 100 $\mu m\ h^{-1}$) and high purity of the modern HVPE reactors. The method is based on the GaN synthesis reaction from gaseous NH_3 and GaCl, and deposition of the nitride on foreign or native flat substrates of desired (usually 2 inch) diameter. The growth proceeds in the direction determined by the orientation of the substrate. In case of ELOG, which occurs on patterned substrates, the growth direction changes depending on the conditions of the process. It was shown [e.g. [11]] that physical properties of GaN grown by HVPE are strongly dependent on orientation of the crystallization front. Very large nonuniformities in electron concentration (over 2 orders of magnitude) were found by photoetching also in bulk GaN, grown nominally in one direction on flat (nonpatterned) sapphire substrates [12]. This could suggest local changes in orientation of the crystallization growth during fast HVPE growth owing to the possible formation of macrosteps, growth hillocks, three-dimensional islands, or pinholes.

Small, single crystalline seeds increase their size simultaneously in different directions during the HVPE growth. So the anisotropy in physical properties is expected in both global and local scales (particularly for fast growth). This can not only lead to nonuniformity in physical properties of the resulting bulk crystal in macroscopic scale, but also generate strain and defects because of local fluctuations of lattice constants induced by, for example, large fluctuations of concentration of free electrons. As we will show in the next section, this effect is particularly visible if the seed crystal is lattice matched to the new deposited material and is almost dislocation free.

The number and type of point defects in the growing crystal depend on microscopic processes of incorporation of basic growth units and impurities on the particular crystal faces. For HVPE, these growth mechanisms can be influenced by changing temperature, flows of reacting gases, and chemical environment in the system (carrier gas, impurities).

The HVPE system used in this study was the horizontal home-built quartz reactor with the rotating-quartz susceptor. GaCl was supplied vertically, just over the surface of the susceptor. A growth temperature of about 1050 °C, temperature of GaCl synthesis of 870 °C, HCl flow in the range of

5–24 ml min^{-1}, diluted in 500 ml min^{-1} of N_2, NH_3 flow of 1200 ml min^{-1}, and 3000 ml min^{-1} of N_2 or H_2 as carrier gases were applied for runs of 4–15 hours. The growth rates observed for this geometry and the set of conditions varied from 20–200 μm h^{-1} and were dependent mainly on the flow of HCl.

3.2.3
HVPE of GaN on Platelet-shaped Seeds

For the conditions of the HVPE process listed above, the result of GaN crystallization on the thin (about 100 μm), low dislocation density pressure grown platelets depends mostly on the rate of growth and on the carrier gas. It was shown [13, 14] that in the conditions used, the crystals can grow in a stable way independent of the dimension of the seed. In Figure 3.2, some examples of GaN crystals grown on smaller and larger seeds are shown.

For the growth with a relatively high rate (100 μm or bigger), in N_2 carrier gas, the HVPE GaN-high-pressure substrate system behaves like the one with a small lattice and/or thermal mismatch even for Mg-doped substrates, which according to the X-ray estimations are lattice matched to the pure GaN. This conclusion is based on the following observations:

(1) Near dislocation-free GaN layers can be grown on the near dislocation-free high-pressure substrates only up to a 'critical' thickness of about 30–50 μm for the *n*-type substrates and of about 100 μm for the GaN : Mg substrates. In thicker layers, both dislocations and low-angle grain boundaries are revealed by DSE [13, 15]. A typical result of DSE for both polar (0001) and corresponding nonpolar ($10\bar{1}0$) faces of the HVPE-grown crystal is shown in Figure 3.3.

(2) Deposition of the HVPE layer on the high-pressure substrate results in a strong bowing of the layer-substrate system reflected by broadening of the X-ray rocking curve (XRC). The initial width of the XRC can be fully or partially recovered by removing the high-pressure

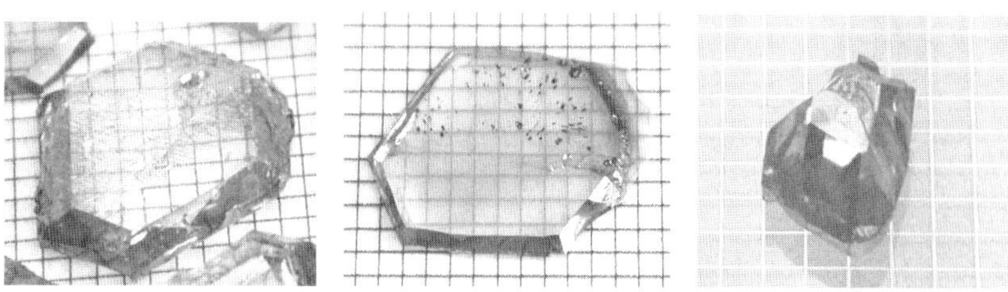

Figure 3.2 GaN crystals grown by HVPE on high-pressure GaN seeds.

substrate if the layer is of subcritical thickness. An example of such a behavior for GaN : Mg substrate is shown in Figure 3.4.

On the other hand, there are observations indicating that the reasons for strain and defect generation in the considered system are related rather to the growth mechanisms than to the mismatch of high-pressure and HVPE-grown materials. These observations are as follows:

(1) Lowering of the crystallization rate (below 40 μm h^{-1}) at otherwise the same growth conditions in N_2 atmosphere

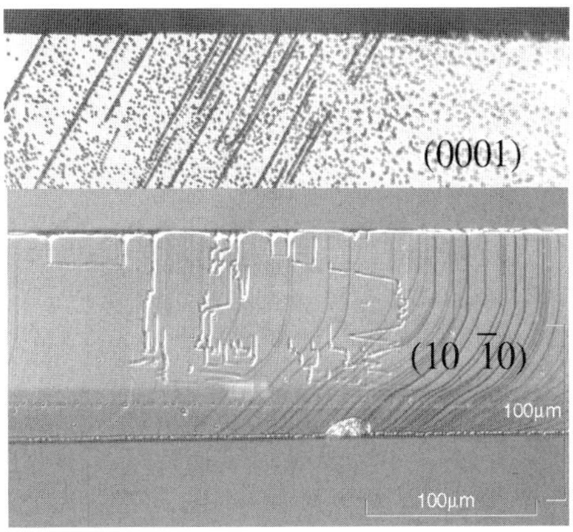

Figure 3.3 Result of DSE of (0001) face and corresponding (10$\bar{1}$0) cleavage of GaN crystal grown by HVPE with a rate of 120 μm h^{-1}, in N_2 atmosphere, on the *n*-type GaN substrate (Courtesy J. Weyher).

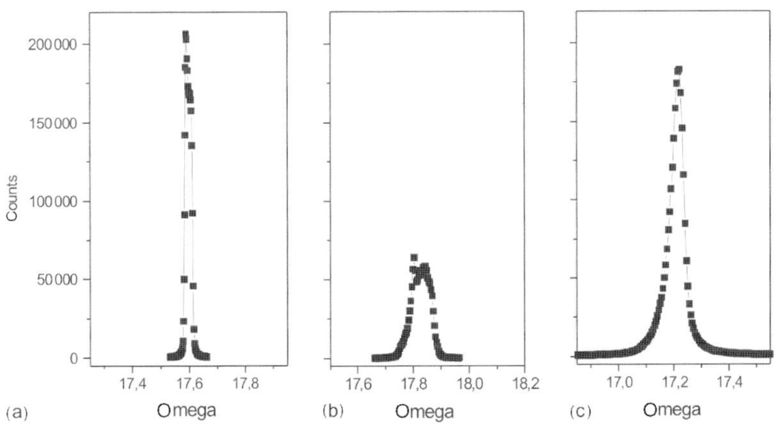

Figure 3.4 X-ray rocking curves (0002), 0.5 × 3 mm slit for 110-μm-thick GaN : Mg pressure grown crystal after technological procedures: (a) initial substrate after two side mechanical polishing and removing of subsurface damage; (b) after deposition of 100 μm HVPE GaN layer; (c) after removing of the pressure grown substrate. Both the angular and intensity scales (vertical axis) are the same for all diagrams.

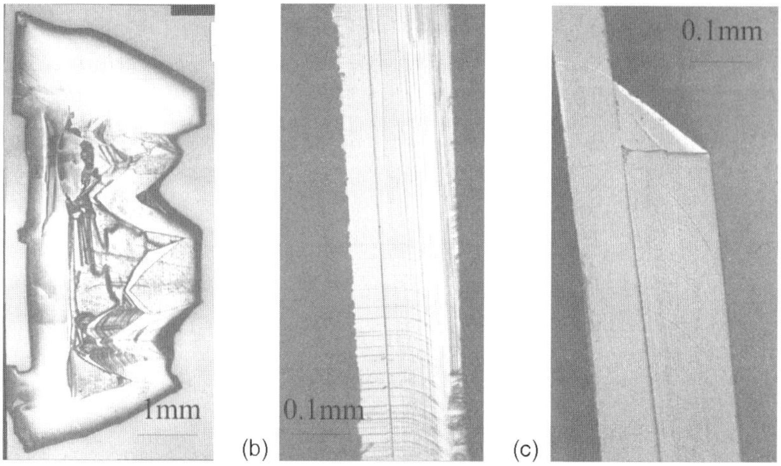

(a) (b) (c)

Figure 3.5 Results of etching in KOH under UV illumination of $(10\bar{1}0)$ cross sections of GaN crystals grown by HVPE on high-pressure *n*-type GaN substrates: (a) HVPE growth in N_2 atmosphere, rate: over 100 μm h^{-1}, high dislocation density in the resulting crystal; (b) HVPE growth in N_2 atmosphere, rate: 20 μm h^{-1}, low (10^3 cm^{-2}) dislocation density in the resulting crystal; (c) HVPE growth in H_2 atmosphere, rate: over 100 μm h^{-1}, low (10^3 cm^{-2}) dislocation density in the resulting crystal. In all three cases, the substrates were not etched owing to very high electron concentrations. In (a) and (b), the side parts resulting from lateral growth are also of high electron concentration.

changes the behavior of the XRCs: their broadening at deposition of HVPE layer of 'subcritical' thickness is no longer observed [14]. The curves remain unchanged or become even narrower than those for the initial high-pressure substrate, which leads to a significant improvement in the final quality of the HVPE material. Almost dislocation-free HVPE GaN layers with thickness exceeding 100 μm have been grown in this way on both the *n*-type and GaN : Mg substrates. Obviously, the growth with a low rate is not an optimum way to produce bulk crystals that could be sliced along the nonpolar planes.

(2) The HVPE growth in H_2 atmosphere at otherwise the same conditions on almost dislocation-free high-pressure substrates allows obtaining of almost dislocation-free HVPE material also for crystallization rates exceeding 100 μm h^{-1}. Bulk crystals with thicknesses of about 1 mm have been grown in this way on *n*-type high-pressure substrates. Very low dislocation densities ($<10^3$ cm^{-2}) were observed by DSE of both (0001) polar and nonpolar surfaces of such crystals.

(3) Photoetching under UV illumination [16] of the nonpolar surfaces (cleaved) of the considered crystals revealed a much more uniform distribution of the free electron concentration in HVPE material grown by optimized procedures – that is with a low rate and/or in H_2 atmosphere. Relevant examples are shown in Figure 3.5.

The photoetching of GaN is very useful for mapping electrical properties of large area samples. The method was calibrated with the use of complementary measurements by micro-Raman scattering, photoluminescence (PL), and profilometry [17], and a clear relation between electron concentration and depth of the photoetching has been established. In particular, it was shown that in bulk GaN crystals grown by HVPE, the inhomogeneities in electron concentration can exceed 2 orders of magnitude as they are very strongly dependent on the orientation of the crystallization front. This anisotropy in physical properties of crystals grown fast in N_2 atmosphere is schematically illustrated in Figure 3.6 where a typical nonpolar cross section of such a crystal including the high-pressure substrate is also shown. Similar studies are being

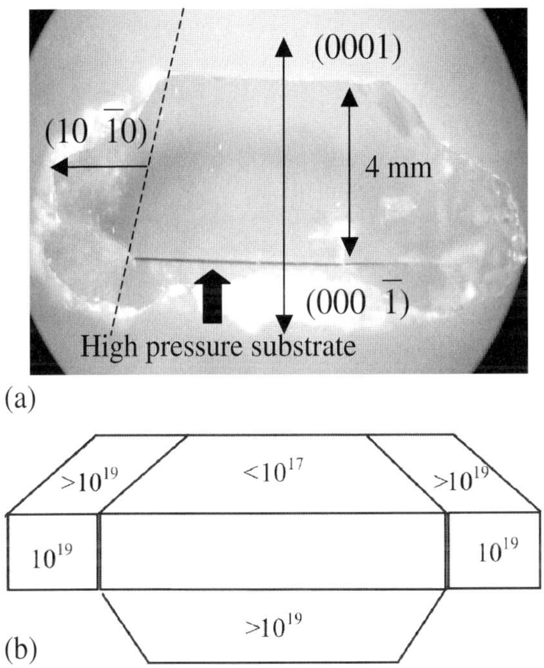

Figure 3.6 (a) Nonpolar cross section of GaN crystal grown by HVPE on a thin (<100 μm) high-pressure substrate (indicated); (b) schematic illustration of anisotropy in physical properties of GaN grown by HVPE in N_2 atmosphere; the local fluctuations observed at fast growth are not included.

performed for crystals grown in different chemical environments of the HVPE process including different composition of the carrier gas and doping with silicon during the growth.

The nonpolar GaN quasi-wafers used in this study were sliced from bulk GaN crystals grown by HVPE with a relatively fast rate in N_2 atmosphere. Therefore, according to the considerations presented above, they did contain some structural defects revealed by DSE in molten KOH–NaOH eutectics. Both dislocations of 10^4–10^5 cm^{-2} density and grain boundaries (like in Figure 3.3) were present in this material. It is, however, very probable that the structural quality of such quasi-wafers can be significantly improved by appropriate changing of growth conditions of the HVPE process as indicated above.

The platelets of nonpolar $(11\bar{2}0)$ orientation were prepared for epitaxy by both mechanical and chemical polishing. An example of the crystal and its nonpolar slices are shown in Figure 3.7a and b. Figure 3.7c shows the biggest slices currently obtained (March 2007) by this method.

3.2.4
HVPE of GaN on Needle-shaped Seeds

An alternative approach to the bulk GaN crystallization by HVPE is the growth seeded on the needle-shaped, high-pressure GaN crystals. In this case, the HVPE growth occurs mostly in the nonpolar $\langle 10\bar{1}0\rangle$ directions as the seed crystal is elongated in the direction $\langle 0001\rangle$. For conditions corresponding to the fast growth on the platelets in N_2 atmosphere, the growth on the needles results in prismatic crystals with stable morphology. Some examples are shown in Figure 3.8. The crystals can be reused for further crystallization, which allows increase in their sizes (Figure 3.8b and c). Stability of the growth front can however be perturbed by parasitic nucleation because of strain induced by macroscopic imperfections in the seed crystals. The parasitic grains are seen in Figure 3.8a and b.

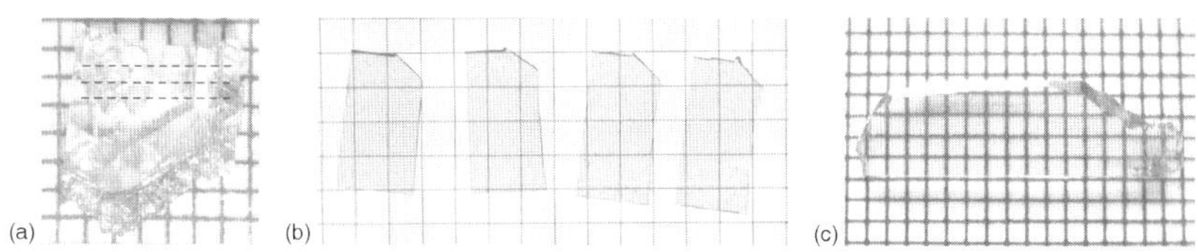

Figure 3.7 Slicing of GaN bulk crystal along the nonpolar $(11\bar{2}0)$ crystallographic planes: (a) the crystal before slicing; (b) corresponding slices; (c) one of the biggest slices obtained by this method (grid: 1 mm).

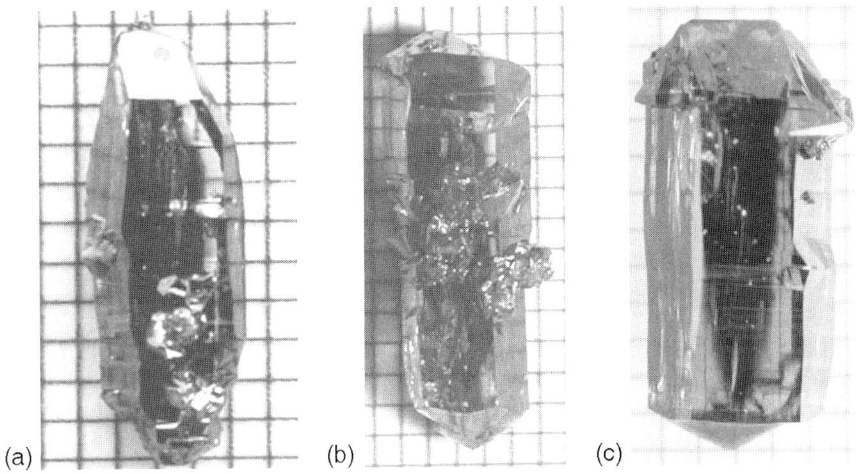

Figure 3.8 GaN bulk crystals grown by HVPE on the needle-shaped, high-pressure seeds: (a) 10-hour HVPE process; (b) twofold 10-hour HVPE process; (c) threefold 10-hour HVPE process (grid: 1 mm).

Structural defects in these newly grown crystals have been studied by DSE in molten KOH–NaOH eutectics of both polar and nonpolar cross section samples sliced from the bulk material [14]. Current results of these studies can be summarized as follows:

(1) DSE of [0001] surfaces indicates that its significant part (over 50%) is of very low defect density ($<10^4$ cm^{-2}) and the seed crystal is almost dislocation free.
(2) There is no deterioration in quality (in terms of dislocation density) between materials grown in subsequent HVPE runs.
(3) There are areas on the [0001] surfaces with nonuniformly distributed hexagonal etch pits spreading from the morphological imperfections in the seed (see Figure 3.9). In some cases, the array of dislocations converts into a misoriented parasitic grain.
(4) DSE of nonpolar [$11\bar{2}0$] surfaces, at regions close to the parasitic grains, shows lines perpendicular to the *c* axis of the crystal as probably being the basal stacking faults. These and other etch figures are being calibrated with TEM.

The crystals described in this section were of *n*-type electrical conductivity with free electron concentration of about 5×10^{18} cm^{-3}. However, as it was shown by micro-Raman scattering measurements [18], at the interfaces

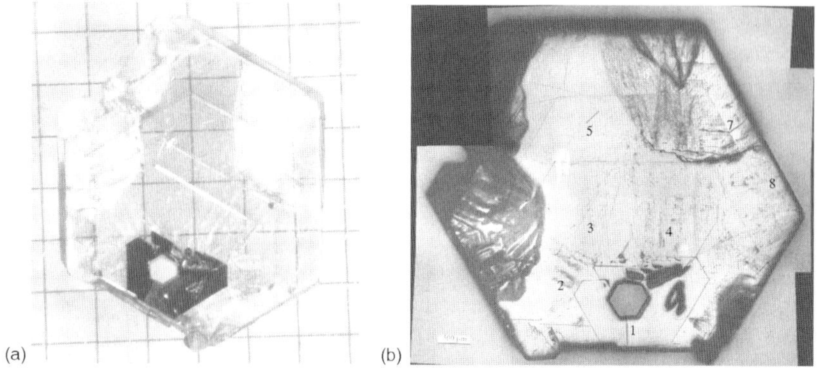

Figure 3.9 A (0001) slice of GaN crystal grown by HVPE (four runs) on high-pressure, needle-shaped seed after DSE: (a) photograph: the seed is dark, hollow crystal; (b) DIC image: the seed is almost dislocation free (1), areas (3) and (5) are of very low density of dislocations, areas (2) and (4) are arrays of etch pits linking seed imperfections with parasitic grains (grid 1 mm).

between subsequent HVPE runs the concentration of electrons was more than 1 order of magnitude higher than that in the bulk.

For obtaining low defect densities larger bulk crystals by HVPE seeded with high-pressure needles, careful selection and preparation of seeds is necessary to avoid parasitic 3D nucleation on seed imperfections. It is shown that the HVPE growth on the needle-shaped seeds is a very stable process. Therefore, it seems that there are no physical obstacles for growing much larger crystals by this method.

3.3 Nonpolar Quantum Structures

3.3.1 GaN/AlGaN Quantum Structures Grown by PA Molecular Beam Epitaxy

As already mentioned, the nonpolar GaN quasi-wafers were made by slicing the GaN crystals grown by HVPE on the pressure grown GaN substrates. Up to now, this approach has been tried for GaN crystals grown on the high-pressure platelets with relatively high rates in N_2 atmosphere. Therefore, the nonpolar wafers contained some dislocations and grain boundaries. The 300-μm-thick, $(11\bar{2}0)$-oriented slices were prepared for epitaxy by mechanical and chemical polishing. Then, 10-fold GaN/AlGaN sequences of 2-, 3-, and 4-nm-thick GaN quantum wells separated by 7-nm $Al_{0.11}Ga_{0.89}N$ barriers were deposited by plasma-assisted molecular beam epitaxy (PAMBE) [19] on both

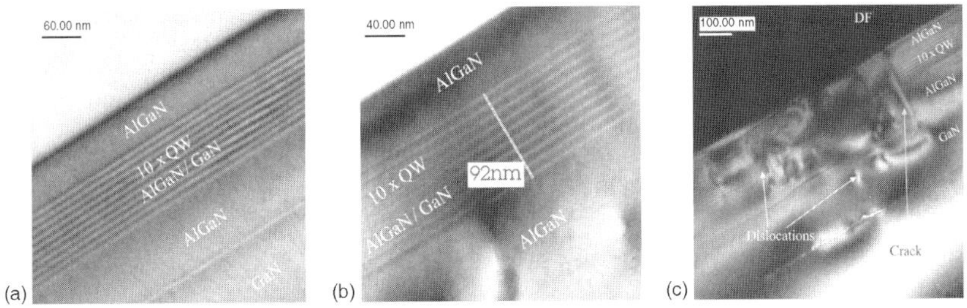

Figure 3.10 TEM images of 10-fold GaN/AlGaN MQW deposited by PAMBE on GaN bulk substrates: (a) epitaxial growth in polar ⟨0001⟩ direction; (b) defect-free area of the structure grown in nonpolar ⟨11$\bar{2}$0⟩ direction; (c) defective area of the structure grown in nonpolar ⟨11$\bar{2}$0⟩ direction (courtesy Julita Smalc).

the nonpolar and the 'usual' polar GaN substrates (for comparison of the optical properties). The structural quality of the 'nonpolar' multiple quantum well (MQW) was checked by XRD measurements and TEM [20]. TEM images of both polar and nonpolar structures deposited in the same MBE run are shown in Figure 3.10.

An advantage of the PAMBE method was that both thicknesses and compositions of the epitaxial layers comprising the MQWs were the same for both polar and nonpolar growth directions.

Because of the high structural quality of the high-pressure polar substrates, the multilayer structures of polar orientation were also almost defect free. The ones of nonpolar orientation contained areas with structural defects induced by imperfections in the substrate.

For optical pumping experiments, the GaN–AlGaN-based laser structures have also been grown by PAMBE on the nonpolar GaN quasi- wafers. Fivefold GaN–AlGaN MQWs with quantum well thicknesses of 5 and 2 nm surrounded by 12% AlGaN barriers were prepared. Then the substrates with the laser structures were cleaved along (10$\bar{1}$0) planes to get lasers of 700-, 500-, and 300-μm cavity lengths.

3.3.2
Optical Properties of Nonpolar Structures Grown on GaN Quasi-Wafers Sliced from Bulk GaN Crystals

The absence of built-in electric fields was confirmed by higher PL intensities for all the nonpolar 10-fold GaN/AlGaN sequences of 2-, 3-, and 4-nm-thick GaN quantum well samples in comparison to their polar counterparts. In addition, it was observed that the positions of the PL lines were higher in energy for samples grown along (11$\bar{2}$0) and that all these lines were located

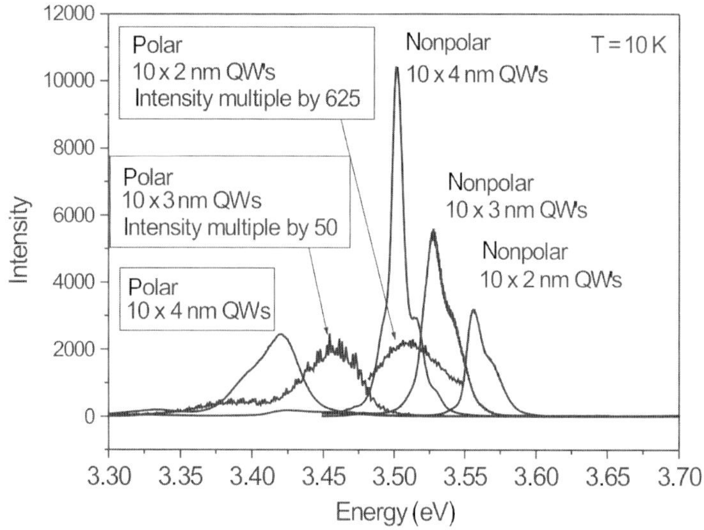

Figure 3.11 Photoluminescence spectra of GaN/AlGaN multiquantum wells of different thicknesses grown on nonpolar *a*-plane and polar *c*-plane GaN bulk substrates.

above the GaN band gap (Figure 3.11). Similar behavior was reported by other authors for structures grown on *r* plane $(11\bar{2}0)$ sapphire substrates [21].

For the polar samples, the transition energy was red-shifted because of the quantum-confined Stark effect. The position of the transition energy for the nonpolar structures strictly followed the results of the calculations for a flat band model, including corrections for exciton binding energies. It is reasonable to associate this fact with the disappearance of built-in electric fields along the $(11\bar{2}0)$ growth direction.

In the low-temperature PL spectra of the nonpolar GaN–AlGaN MQWs, well-resolved excitonic lines were visible (Figure 3.12a and b).

Detailed interpretation of the PL data required analysis of the valence band structure in GaN/AlGaN quantum wells. In case of bulk wurtzite GaN at low temperature, the *A*, *B*, and *C* excitons are all visible in, for example, photo reflection experiments. These excitons are related to three different sub-bands (Γ_{9v}, Γ^1_{7v}, and Γ^2_{7v}) from the split valence band. The situation is more complicated for GaN/AlGaN quantum wells. It was analyzed by multiband $k \cdot p$ models. We performed calculations of the valence band dispersion around $k = 0$ along crystallographic directions up to $|k| = 0.1$ by using the NextNano^3 program [22]. In the calculations, we assumed that the AlGaN barriers were strained with respect to the GaN substrate. The calculations were performed in a self-consistent way by solving the Schrödinger equations and 6×6 Hamiltonian for holes. In general, to obtain the distribution of electrostatic potential, the Poisson equation had to be solved. The program was able to perform calculations for any crystallographic orientation by using

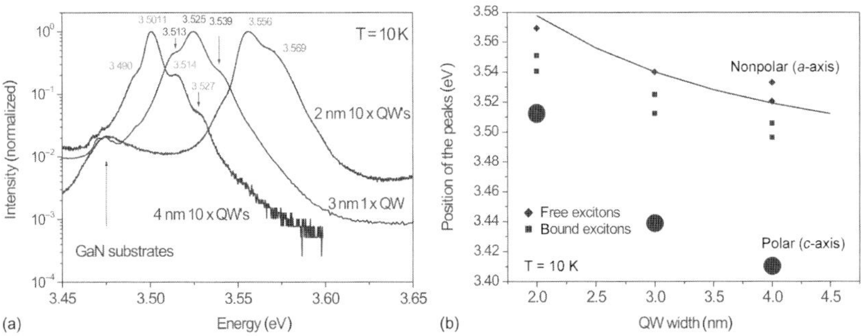

Figure 3.12 (a) Excitonic spectra of the nonpolar GaN–AlGaN MQWs with different thicknesses; (b) comparison between the energy positions of the photoluminescence bands observed in polar and nonpolar GaN/AlGaN QW samples. The lowering of energy of peak positions for polar samples is related to the quantum-confined Stark effect. For nonpolar samples, the solid line represents the calculations for rectangular quantum wells.

a rotation matrix technique [23]. In our case, the calculations were performed for the (11$\bar{2}$0) crystallographic direction corresponding to the orientation of the investigated nonpolar samples. The results are presented in Figure 3.13.

From these calculations, we found that the spacing between the upper two sub-bands did not change with the width (between 2 and 4 nm) of quantum well. At the same time, we observed a change in shape of the lowest sub-band

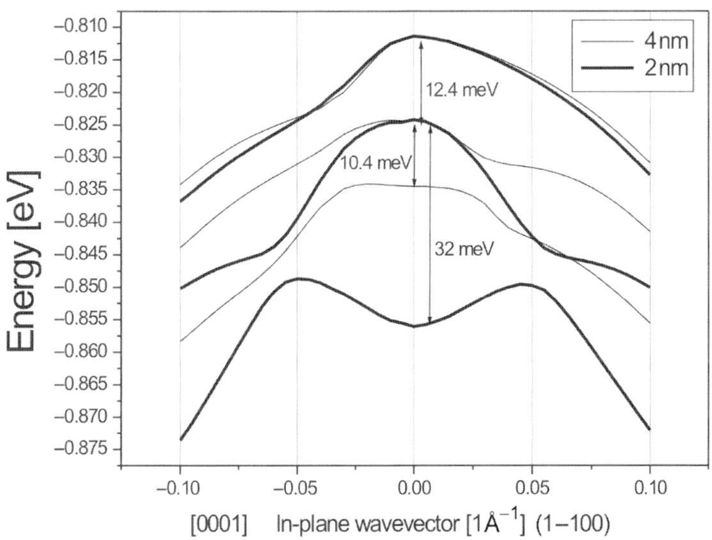

Figure 3.13 Valence band structures for the 4-and 2-nm GaN/AlGaN quantum wells grown along the (11$\bar{2}$0) axis.

and a shift of its Γ-point from 10 meV for 4 nm quantum well up to 32 meV for 2 nm QW. On the basis of these results, we were able to make the conclusion that for thinner QWs, only two excitons related to higher bands may be visible.

Furthermore, light polarization studies have been performed for both GaN homo-epitaxial layers and GaN–AlGaN quantum wells. For the nonpolar GaN homoepitaxial layers, the two bound excitons related to the *A* exciton followed group theory selection rules very closely. This means that the corresponding exciton lines were visible in one configuration (*E* perpendicular to the *c* axis), but they disappeared in the second configuration (*E* parallel to the *c* axis). The lines related to the B and C excitons were visible in both configurations with approximately equal intensities. As with the *A* exciton, the dependence of *B* and *C* exciton related lines follows the group theory selection rules and is described in detail by Bir and Pikus [24]. The results of the light polarization study in GaN homoepitaxial layers are shown in Figure 3.14.

The situation is more complicated for quantum well structures. The results of the polarization-dependent measurements for the nonpolar GaN–AlGaN quantum wells are presented in Figure 3.15. Polarization behavior of the nonpolar (11$\bar{2}$0) quantum wells is different from that for bulk material. Selection rules are changed owing to lower symmetry of system and strain in quantum wells.

In this case, we were able to see *A*, *B*, and *C* excitons directly but rather as transitions that originate from the split valence band, sometimes known as *E*1, *E*2, and *E*3 [25]. We found that the degree of light polarization in our

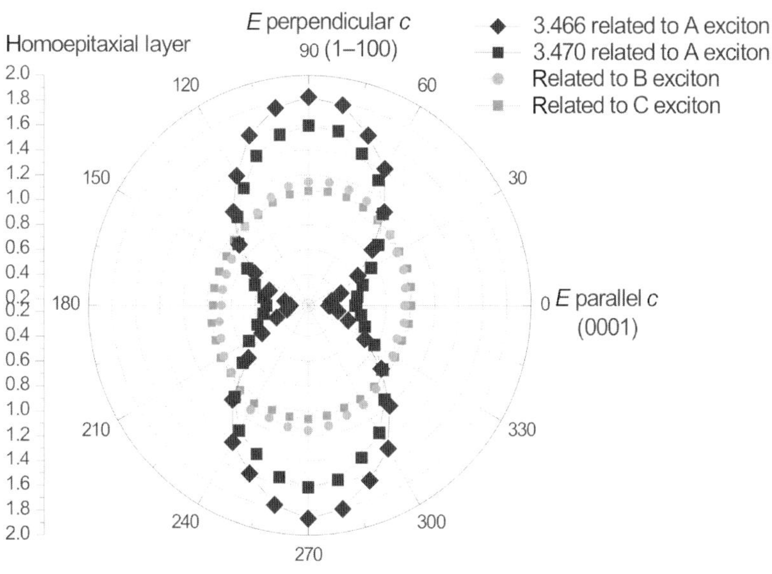

Figure 3.14 Polarization of light emitted from GaN homo-epitaxial layer grown on the nonpolar GaN substrate (cf. Color Plate XXIII).

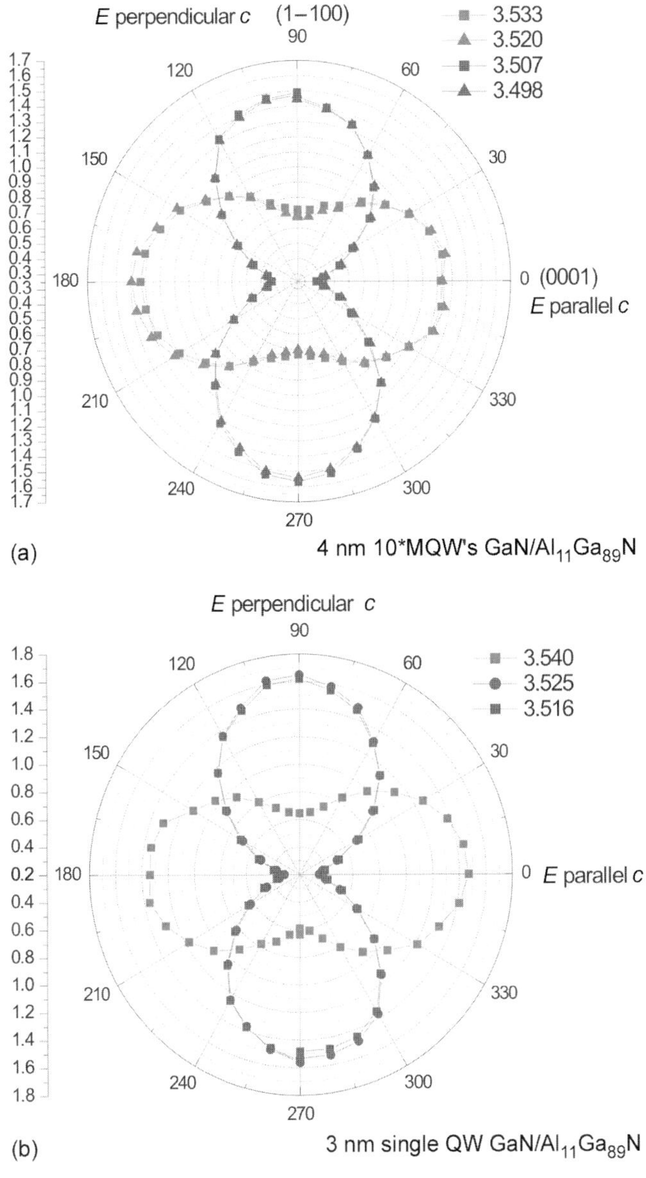

Figure 3.15 Polarization of light emitted from GaN–AlGaN MQW of different thicknesses grown on nonpolar GaN substrates (cf. Color Plate XXIV).

samples depended on the thickness of the quantum wells. For the 4-nm quantum wells, we observed four lines, two of which almost follow the selection rules of the *A* exciton in bulk originating from the *E*1 sub-band. The other two lines, probably originating from the *E*2 and *E*3 sub-bands, were highly polarized along the second crystallographic direction. It is reasonable to assume that the two lower energy lines originate from the high-energy sub-band of the valence band and that two others are related to the two other sub-bands. In contrast to the 4-nm wells, only three lines were observed for both the 3- and 2-nm wells. Two lower energy lines follow the result for the 4-nm sample, and a high-energy line is polarized in a second crystallographic direction. The disappearance of the higher energy line is probably related to the shift of the third sub-band and the negative effective mass of this sub-band at $k = 0$, which we obtained in our calculations. To confirm this interpretation, more detailed studies are necessary. The results just described can be of importance for potential design of nonpolar devices such as polarized light emitting diodes.

For the GaN–AlGaN-based laser structures, the optical pumping experiments were performed using the fourth harmonic generation of a Nd : YAG laser (266 nm). During these measurements, the laser light was focused on the surfaces of cleaved stripes by means of cylindrical lenses. The laser emission was observed in a direction perpendicular to the direction of the pumping laser pulse, and the duration was set as 10 ns with a 10-Hz repetition. A motorized optical attenuator with an Ophir power meter was used to control the laser pulse energy and finally, the light emitted from the cleaved laser stripes was collected by a SPEX 500 spectrometer.

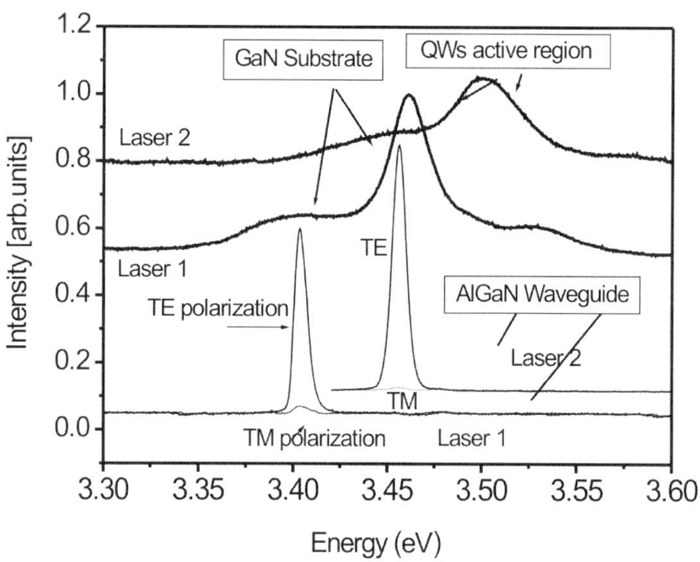

Figure 3.16 Spectra of the stimulated emission at room temperature for transverse-electric (TE) and transverse-magnetic (TM) polarization, for both types of lasers described in the text. The two upper spectra represent the spontaneous emission from both laser structures.

Figure 3.16 presents the results of the optical pumping experiments at room temperature.

Examples of the laser emission spectra just above the lasing threshold are presented for two laser structures: the first with 5-nm QWs and the second with 2-nm QWs. The emissions from both of our samples are TE polarized, which strongly confirms that stimulated emissions have been obtained. Such stimulated emissions for the thinner QWs shift toward higher energies for 53 meV. If we take into account the flat band model and perform calculations for inter-sub-band transitions, this difference should be 48 meV in cases of 5-nm and 2-nm QWs, which corresponds well with our experimental results. This also proves that the laser emissions in our samples take place in the active region of the laser structure, and not, for example, in the substrate.

Figure 3.17 summarizes the results of our optical pumping experiments. For a laser with 5-nm QWs, the determined laser threshold pump power density was equal to 260, 280, and 370 kW cm^{-2} for the 700, 500, and 300 μm cavity lengths, respectively. For the laser with 2-nm QWs, the threshold was higher: 500 kW cm^{-2}. The laser threshold power density decreased with temperature down to 60 kW cm^{-2} for the sample with 5-nm QWs and 300-μm cavity length.

There is a small number of studies concerning optical pumping experiments of GaN/AlGaN polar quantum structures. The record threshold power density reported for the polar structures was equal to 105 kW cm^{-2} at room temperature [25].

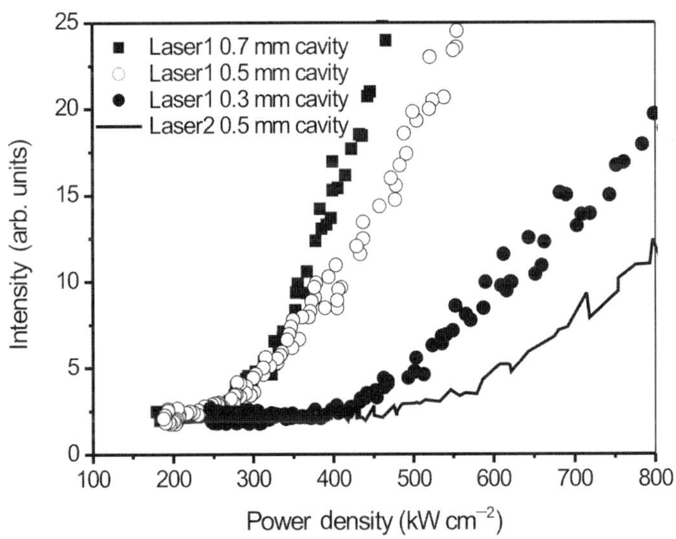

Figure 3.17 Excitation power density dependence at room temperature for the 5 and 2 nm QWs lasers. The 5-nm QWs were cleaved with 300, 500, and 700 μm laser cavities; the 2-nm QWs possessed a 500-μm laser cavity.

At this stage, it is too early to compare the optical properties of polar and nonpolar lasers. However, our results on homoepitaxial multiquantum wells and optically pumped lasers prove that it is possible to grow high-quality nitride structures along with nonpolar directions. It goes without saying that further research is necessary.

3.4 Summary

It is shown that HVPE is a very suitable method to increase the size of the high-quality pressure grown crystals. Bulk crystals of stable morphology can be grown on both platelet-shaped and needle-shaped high-pressure seeds. Therefore, it seems that there are no physical obstacles for growing much larger crystals by this method. Full reproduction of the quality of platelet-like seeds is possible at optimized growth condition; however, further studies are necessary for the growth of much larger crystals. For obtaining low defect densities, larger bulk crystals by HVPE seeded with high-pressure needles, careful selection and preparation of seeds is necessary to avoid parasitic 3D nucleation on seed imperfections. It is shown that the HVPE growth on the needle-shaped seeds is a very stable process.

The HVPE-grown bulk GaN crystals can be sliced along nonpolar surfaces and prepared for epitaxy by mechanical and chemical polishing. The GaN–AlGaN quantum structures grown by PAMBE on these substrates are of high structural quality, which is reflected by the excitonic character of the low-temperature PL spectra and by lasing under optical pumping.

Acknowledgment

The research on bulk crystallization of GaN by HVPE as well as on the epitaxial growth and studies of nonpolar GaN-based quantum structures was supported by Polish Committee for Scientific Research: GRANT Nr 3T08A 033 29 and GRANT Nr 1P03B05329.

References

1. Waltereit, P., Brandt, O., Trampert, A. *et al.* (2000) *Nature*, **406**, 865.
2. McLaurin, M., Mates, T.E., and Speck, J.S. (2005) *Applied Physics Letters*, **86**(26), 262104.
3. Ng, H.M. (2002) *Applied Physics Letters*, **80**, 4369.
4. Craven, M.D., Lim, S.H., Wu, F. *et al.* (2002) *Applied Physics Letters*, **81**, 469.
5. Zakharov, D.N. and Liliental-Weber, Z. (2005) *Physical Review B*, **71**, 235334.

6. Paskova, T., Kroeger, R., Paskov, P.P. *et al.* (2006) in *Gallium Nitride Materials and Devices,* (eds L. Cole, J.G. Grote, H. Morkoc, and A. Madhukar), Proceedings of SPIE, Vol. 6121, 612106.
7. Haskell, B.A., Wu, F., Fijii, T. *et al.* (2003) *Applied Physics Letters,* **83**, 1554.
8. Paskova, T., Kroeger, R., Figge, S. *et al.* (2006) *Applied Physics Letters,* **89**, 051914.
9. Grzegory, I., Boćkowski, M., and Porowski, S. (2005) in *Bulk Crystal Growth of Electronic, Optical and Optoelectronic Materials,* chap. 6, (P. Capper), John Wiley & Sons, Ltd., U.K., pp. 173–207.
10. Weyher, J.L., Brown, P.D., Rouviere, J.L. *et al.* (2000) *Growth,* **210**, 151.
11. Wagner, V., Parillaud, O., Buhlmann, H.J. *et al.* (2002) *Journal of Applied Physics,* **92**(3), 1307.
12. Lewandowska, R., Weyher, J.L. *et al.* (2007) *Journal of Crystal Growth,* **307**(2), 298–301.
13. Łucznik, B., Pastuszka, B., Grzegory, I. *et al.* (2005) *Journal of Crystal Growth,* **281**, 38–46.
14. Grzegory, I., Łucznik, B., Boćkowski, M., and Porowski, S. (2007) *Journal of Crystal Growth,* **300**(1), 17–25.
15. Weyher, J.L., Kamler, G., Nowak, G. *et al.* (2005) *Journal of Crystal Growth,* **281**, 135.
16. Kamler, G. to be published.
17. Weyher, J.L., Lewandowska, R., Macht, L. *et al.* (2006) *Materials Science in Semiconductor Processing,* **9**, 175–79.
18. Lewandowska, R. unpublished result.
19. Teisseyre, H., Skierbiszewski, C., Łucznik, B. *et al.* (2005) *Applied Physics Letters,* **86**, 162112.
20. Smalc, J. unpublished result.
21. Craven, M.D., Wu, F., Chakraborty, A. *et al.* (2004) *Applied Physics Letters,* **84**, 1281.
22. Birner, S., Hackenbuchner, S., Sabathil, M. *et al.* (2006) *Acta Physica Polonica A,* **110**(2), 111. For more information, visit http://www.nextnano.de/.
23. Park, S.H. and Chuang, S.L. (1999) *Physical Review B,* **59**, 4725.
24. Bir, G.L. and Pikus, G.E. (1974) *Symmetry and Strain-Induced Effects in Semiconductors,* John Wiley & Sons, Ltd, New York.
25. Bidnyk, S., Lam, J.B., Little, B.D. *et al.* (1999) *T.J. Schmidt Applied Physics Letters,* **75**, 3905.

4
Heteroepitaxial Growth of Nonpolar-face AlN on SiC Substrates by Plasma-assisted Molecular-beam Epitaxy

Jun Suda

4.1
Introduction

Group III-nitride quantum well (QW) structures grown on the *c*-plane exhibit strong internal electric fields owing to the discontinuities in piezoelectric and spontaneous polarization at the heterointerfaces. These internal fields result in reduced oscillator strength (lower internal quantum efficiency), which hinders the realization of high-efficiency green light-emitting diodes (LEDs) based on InGaN. With the goal of avoiding this problem much attention has recently been given to nonpolar or semipolar growth orientations. Extensive studies on growth of GaN with non- and semipolar orientations have been carried out using various substrates and growth methods (see other chapters in this book).

In addition to green LEDs, there is also a strong demand for development of deep ultraviolet (DUV) LEDs. DUV light is defined as light with wavelengths shorter than 300 nm. Applications of DUV light sources include sterilization, water purification, and chemical/biological analysis in medicine, environmental monitoring, and toxic substance detection.

AlN-based nitride semiconductors seem to be the most promising materials for DUV LEDs [1]. There are still many key issues to be overcome before AlN-based DUV LEDs are realized; for example growth of low defect density AlN and control of the conductivity of AlN and Al-rich alloys. The development of LEDs emitting efficiently at wavelengths less than 250 nm is considered an extremely difficult challenge. However, very recently, Taniyasu *et al.* reported room-temperature operation of an AlN *pn*-junction LED with the emission wavelength of 210 nm [2]. Although the reported external efficiency was quite low (the order of 10^{-6} %), the successful demonstration of an AlN LED clearly gives some cause for optimism regarding the future of AlN-based DUV LEDs. In the device reported by Taniyasu *et al.*, the AlN epilayer was grown on a 4H-SiC (0001) substrate by metal-organic vapor phase epitaxy (MOVPE). The

Nitrides with Nonpolar Surfaces. Growth, Properties, and Devices. Edited by Tanya Paskova

ISBN: 978-3-527-40768-2

AlN layer contained a high density (10^9 cm^{-2}) of threading dislocations (TDs). As mentioned by the above authors, it is necessary to reduce the density of TD and use a carrier confinement structure such as a double heterostructure or a QW structure to improve the internal quantum efficiency.

Various techniques such as high-temperature growth [3], alternating source supply method [4, 5] and epitaxial lateral overgrowth (ELOG) [6] have been investigated to reduce the density of TDs in AlN epilayers. These investigations have focused exclusively on the growth of AlN (0001) on the basal plane of sapphire or SiC. However, one must also pay attention to the strong internal electric fields in $Al_{1-x}Ga_xN$/GaN or AlN/$Al_{1-x}Ga_xN$ QW structures [7, 8]. Spontaneous polarization discontinuities between GaN and AlN are the main origin of the strong electric field in this system. Therefore, it is necessary to use non- or semipolar growth orientations to avoid or reduce the effects of electric fields in AlN-based DUV LEDs.

In comparison to GaN, the number of studies on non- or semipolar orientations of AlN is limited [9–16]. While there have been some reports on nonpolar AlN grown on *r*-plane ($1\bar{1}02$) sapphire for surface acoustic wave (SAW) devices [9, 10] and AlN nucleation layers for nonpolar GaN growth [11–13] only very few studies have been conducted on nonpolar AlN aiming toward AlN-based devices [14, 15]. Since 2002, we have been investigating nonpolar AlN grown on SiC substrates by plasma-assisted molecular-beam epitaxy (PAMBE) [17–24]. Our decision to use SiC substrates and the PAMBE technique was based on the following reasons: (i) It was previously shown that layer-by-layer growth of AlN from the initial stage of epitaxy is possible on SiC (0001) by PAMBE, indicating good wetting of AlN on SiC [25, 26], (ii) The lattice mismatch between AlN and SiC is much smaller than that between AlN and other substrates typically used for nonpolar nitride growth such as sapphire ($1\bar{1}02$) or $LiAlO_2$ (100), as shown in Table 4.1, and (iii) The crystal structure of SiC (4H or 6H) resembles that of AlN (wurtzite, 2H) more closely in comparison to sapphire or $LiAlO_2$ substrates.

Table 4.1 Substrates for nonpolar-face AlN.

Substrate	Growth plane	Lattice mismatch to AlN ⊥ and to *c* axis (%)
Sapphire ($10\bar{1}2$)	($11\bar{2}0$)	−13 +2.9
$LiAlO_2$ (100)	($1\bar{1}00$)	+0.7 +3.6
ZnO ($1\bar{1}00$), ($11\bar{2}0$)	($1\bar{1}00$), ($11\bar{2}0$)	+4.4 +4.5
SiC ($1\bar{1}00$), ($11\bar{2}0$)	($1\bar{1}00$), ($11\bar{2}0$)	−1.0 +0.9

In this chapter, we focus on the growth of nonpolar AlN on SiC by PAMBE. Previous studies and our latest results will be reviewed.

4.2
The Crystalline Structure of AlN and SiC

Before describing our experimental results for nonpolar AlN on SiC, it is helpful to first discuss the crystal structures of AlN and SiC and explain the defect generation processes that can be expected to occur during heteroepitaxial growth of AlN on SiC.

It is well known that the equilibrium crystal structure of AlN is wurtzite (2H). 2H-AlN is comprised of a series of Al−N bilayers with the stacking sequence (ABABAB...) along the [0001] direction as shown in Figure 4.1a. The two bilayers, A and B, are identical except for their positions within the (0001) plane.

On the other hand, SiC exhibits polytypism – that is it is found in nature to exist in many different crystal structures. More than 200 different SiC polytypes were reported [27, 28] and they are distinguished from each other by the atomic bilayer stacking sequence along the hexagonal [0001] direction (or [111] direction of a cubic structure). The most common SiC polytypes are the zincblende (3C), 4H, and 6H structures as shown in Figure 4.1b–d. 3C-SiC has a cubic structure comprising of a series of three Si−C bilayers with stacking sequence (ABCABC...) along the [111] direction, while 4H- and 6H-SiC have hexagonal structures with stacking sequences of ABCBABCB... (four bilayers) and ABCACBABCACB... (six bilayers) along the [0001] direction, respectively.

SiC substrates with the 3C, 4H, and 6H polytypes are commercially available. 3C-SiC is made by chemical vapor deposition (CVD) on Si (001) substrates [29]. 3C-SiC/Si templates and freestanding 3C-SiC wafers (obtained by thick CVD growth and Si substrate removal) are readily available in large diameters because they can be grown on large-area Si wafers. However, the crystal quality of 3C-SiC is rather poor because of the large lattice mismatch (20%)

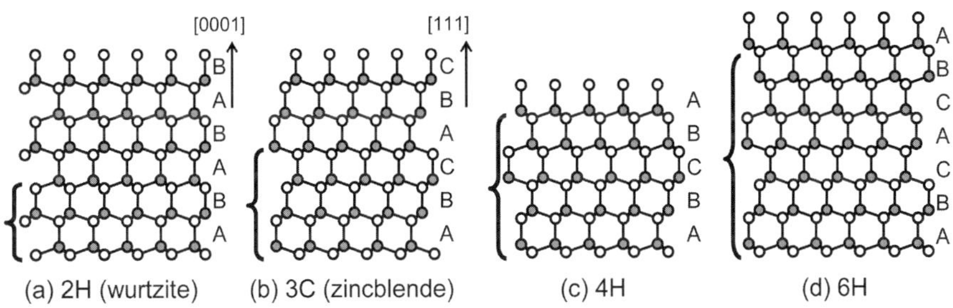

Figure 4.1 Schematic illustrations of atomic stacking sequences in 2H, 3C, 4H, and 6H polytypes.

between 3C-SiC and Si. 3C-SiC (001) has, however, been used for the growth of metastable 3C-GaN (001) [30, 31].

On the other hand, bulk crystals of 4H-SiC and 6H-SiC are made by a physical vapor transport (seeded sublimation growth) technique known as *the modified-Lely method* [32]. High-quality bulk materials (threading dislocation density (TDD) $< 10^4$ cm^{-2}) are now commercially available with relatively large diameters, up to 4" [33]. Very recently, ultra-low-dislocation density bulk SiC (TDD $< 10^2$ cm^{-2}) has been reported [34]. Moreover the conductivity of bulk SiC can be controlled, and *n*-type, *p*-type, and semiinsulating substrates are available for both 4H and 6H polytypes. Therefore, 4H- and 6H-SiC can be used as high-quality substrates for nitride devices, provided that challenges associated with heteroepitaxy are overcome.

4H- and 6H-SiC are expected to be promising substrates for 2H-AlN because of the fairly close lattice matching, especially for growth on the basal plane. The *a*-axis lattice constants of 2H-AlN and 4H- or 6H-SiC differ by about ~0.9%, as shown in Table 4.1. In addition, the polarity of AlN can be controlled by growing on a SiC {0001} surface of the appropriate polarity – for example, Al-face AlN is obtained on Si-face SiC, and N-face AlN on C-face SiC. Owing to matching of both the lattice parameter and crystal structure, it is expected that coherent (isomorphic) AlN−SiC heterointerfaces can be grown on the basal plane, as shown in Figure 4.2a.

However, for planes other than the basal plane, that is non- or semipolar growth, structural matching is not achieved as a result of differences in the [0001] stacking sequences of 2H-AlN and 4H- or 6H-SiC. For example, Figure 4.2b illustrates a 2H-AlN $(1\bar{1}00)$–6H-SiC $(1\bar{1}00)$ heterointerface. Although the (0001) interplanar distance (separation between adjacent atomic bilayers) is almost the same in AlN and SiC (~1% mismatch), three out of every six bilayers in each unit cell of 6H-SiC are not aligned with those

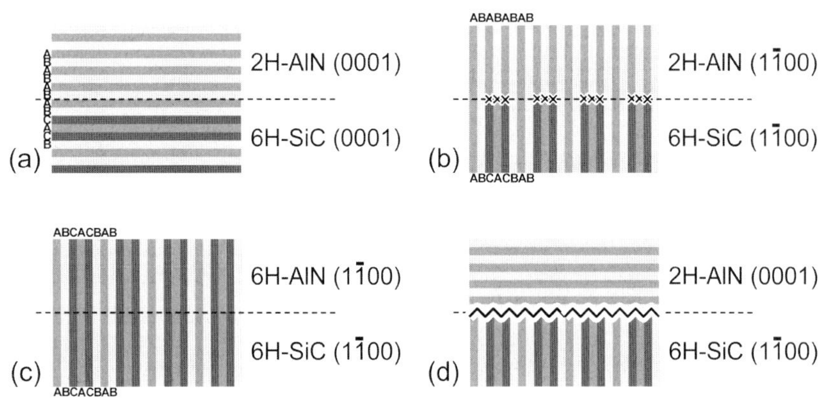

Figure 4.2 Four possible heterointerfacial configurations resulting from growth of AlN on 6H-SiC.

of 2H-AlN, and cannot form tetrahedral bonds across the heterointerface. As a result, a periodic array of partial dislocations is expected to form at the interface. In other words, because of the lack of structural matching, a high density of defects is expected at the heterointerface despite close lattice matching. The concept of structural matching (or polytype matching) is crucial for understanding the growth of nonpolar AlN on SiC.

In the case of nonpolar growth without structural matching, if the partial dislocations are localized at the interface then the upper portion of the AlN epilayer may exhibit high crystalline quality. However, if some of the partial dislocations at the interface react to form TDs, the AlN epilayer should exhibit an extremely high density of TDs because the partial dislocation density at the interface is of the order of the lattice constant. This point will be discussed further together with experimental results in Sections 4 and 5.

In the above discussion, we have assumed that AlN grows in the 2H polytype, but that does not necessarily have to be the case. In principle, a metastable AlN layer could be grown with the same crystal structure as the underlying SiC, for example a 6H-AlN $(1\bar{1}00)$–6H-SiC $(1\bar{1}00)$ heterostructure. As the basal plane (0001) does not contain any information about the atomic stacking order along the [0001] direction, growth on the basal plane should always result in thermodynamically stable 2H-AlN as shown previously in Figure 4.2a. In addition, there is a possibility to obtain metastable AlN crystal structures for nonpolar growth because the [0001] stacking order appears in the nonpolar SiC surfaces. If the AlN adatoms trace the [0001] stacking order of the substrate, the SiC crystal structure can be replicated in the AlN epilayer resulting in structures such as 6H-AlN as shown in Figure 4.2c.

When the AlN epilayer adopts the crystal structure of the underlying SiC substrate, no partial dislocations are formed at the interface, a favorable situation for high-quality heteroepitaxy of nonpolar layers. However, prior to the beginning of our work in 2002, there was no literature on growth of 6H- or 4H- group-III nitrides.[1] In 2003, we first reported the isopolytypic growth of 4H-AlN $(11\bar{2}0)$ on 4H-SiC $(11\bar{2}0)$ [20]. Details of this are discussed in Section 5.

It should be pointed out that other epitaxial relationships between AlN and nonpolar SiC are possible, such as (0001) AlN//$(11\bar{2}0)$ SiC as shown in Figure 4.2d. For example, the growth of (0001) 2H-AlN on $LiAlO_2$ (100) was reported, although $LiAlO_2$ (100) substrates favor the growth of $(1\bar{1}00)$ 2H-GaN under appropriate conditions [35]. If the desired epitaxial relationship is not achieved, we cannot obtain nonpolar AlN. Undesirable orientations such as (0001) AlN//$(11\bar{2}0)$ SiC may result when the nonpolar SiC surface energy is too low, favoring the spontaneous nucleation of AlN. Fortunately, spontaneous

1) 6-bilayer-period GaN was observed as an inclusion in 3C-GaN grown on GaAs (001) by Funato *et al.* Formation of such 6H-like GaN inclusions was attributed to the coexistence of 2H-GaN and 3C-GaN facets [36, 37].

nucleation does not occur in PAMBE growth of AlN with appropriate substrate surface treatments and growth conditions.

4.3
AlN/6H-SiC ($1\bar{1}00$)

To the best of the author's knowledge, the first work on growth of nonpolar AlN on SiC was that of Stemmer *et al.* [38]. They grew AlN on 6H-SiC ($1\bar{1}00$) substrates by PAMBE, and investigated its structure by high-resolution transmission electron microscopy (HRTEM). HRTEM clearly showed growth of single crystalline 2H-AlN with the epitaxial relationship of $(1\bar{1}00)_{AlN}//(1\bar{1}00)_{SiC}$ and $(0001)_{AlN}//(0001)_{SiC}$. This result clearly suggested that 6H-SiC ($1\bar{1}00$) substrates could be used for the growth of nonpolar $(1\bar{1}00)_{AlN}$. Because the obtained AlN layer has the 2H structure, formation of partial dislocations at the interface is expected. However, Stemmer *et al.* grew only a very thin (10 nm) AlN layer and did not mention its crystal quality. The limited information given in this report was insufficient to judge the potential of nonpolar AlN on SiC for device applications.

4.4
AlN/6H-SiC ($11\bar{2}0$)

While Stemmer *et al.* had already reported the growth of AlN on 6H-SiC ($1\bar{1}00$), we started our nonpolar AlN research with a different substrate, 6H-SiC ($11\bar{2}0$) [17–19]. The atomic arrangements on the ($1\bar{1}00$) and ($11\bar{2}0$) surfaces are very different. Thus, it is of fundamental scientific interest to see what happens when AlN is grown on ($11\bar{2}0$) SiC surfaces.

All growth experiments in our laboratory were performed in an EV-1000S MBE system (Eiko Engineering Ltd. Japan) equipped with Al and Ga Veeco (Applied Epi) SUMO effusion cells and a Veeco rf-plasma-excited nitrogen source. The substrate temperature was measured using a thermocouple located just behind the substrate. The substrates used in this study were *n*-type 6H-SiC wafers with the ($11\bar{2}0$) orientation. The substrates were cleaned in 3 : 1 H_2SO_4/H_2O_2 and 0.5% HF solutions prior to loading into the vacuum chamber. The substrate was thermally cleaned at 1000 °C for 30 minutes in a vacuum of 10^{-9} torr, and then an AlN layer was grown at 1000 °C. The growth was carried out under the following conditions: an Al flux of 4.7×10^{-7} torr, an rf power of 400 W, and a nitrogen (N_2) flow rate of 0.5 sccm, which corresponded to Al-rich growth conditions (Al droplets were observed on the grown surface). The growth rate under these conditions was 0.4 μm h^{-1}. The total thickness of the AlN layers ranged from 0.4 to 1.0 μm.

Figure 4.3a shows the reflection high-energy electron diffraction (RHEED) pattern of the 6H-SiC (11$\bar{2}$0) substrate just before AlN growth. The pattern shown indexed in Figure 4.3a is consistent with that of 6H-SiC (11$\bar{2}$0) with the [0001] azimuth. At the initial stage of the growth, a spotty AlN pattern was observed for several minutes perhaps because of three-dimensional nucleation. After approximately 5 minutes (~30 nm), the pattern changed to streaks, indicating a relatively smooth growth front. Figure 4.3b shows the RHEED pattern after AlN growth, measured with the same azimuth. The symmetry of the pattern shown in Figure 4.3b differs from that in Figure 4.3a, indicating that the AlN layer has a different crystal structure (polytype) than that of 6H-SiC. The polytype of the grown layer can be distinguished from the RHEED pattern along the [0001] azimuth – that is (1$\bar{1}$00) diffraction is not allowed for the 6H polytype, while it is allowed for the 2H polytype. The indexed pattern corresponds to 2H-AlN (11$\bar{2}$0) with the [0001] azimuth, clearly indicating that 2H-AlN and 6H-SiC (11$\bar{2}$0) have an epitaxial relationship of [11$\bar{2}$0] AlN//[11$\bar{2}$0] SiC and $[0001]_{AlN}$//[0001] SiC, which is analogous to the 2H-AlN (1$\bar{1}$00)–6H-SiC (1$\bar{1}$00) orientation relationship reported by Stemmer *et al.* [38].

Figure 4.4 shows an atomic force microscope (AFM) image of a 360-nm-thick AlN epitaxial layer grown on 6H-SiC (11$\bar{2}$0). The surface roughness of this AlN layer was low (rms roughness of 1.0 nm), however, many elongated

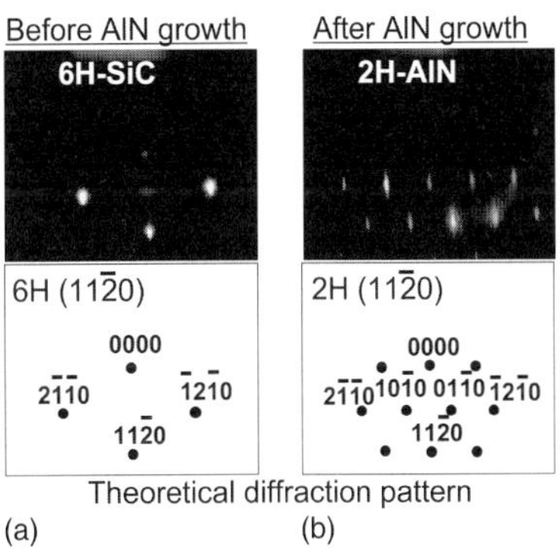

Figure 4.3 RHEED patterns (a) before and (b) after growth of AlN on 6H-SiC (11$\bar{2}$0).

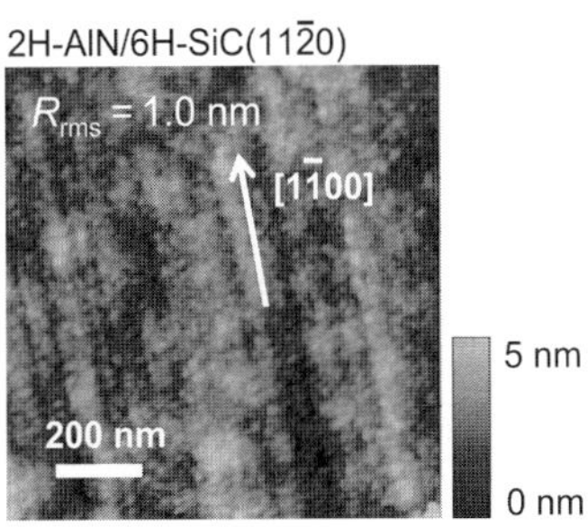

Figure 4.4 AFM image of 2H-AlN grown on 6H-SiC (11$\bar{2}$0).

surface defects were observed extending along the $[1\bar{1}00]$ direction. Since these elongated defects were not observed on the 6H-SiC substrate before growth, they suggest the formation of stacking faults (SFs) or dislocations along the $[1\bar{1}00]$ direction during AlN growth.

Figure 4.5 shows the X-ray diffraction (XRD) $2\theta/\omega$-scan curves of a 360-nm-thick AlN epitaxial layer grown on a 6H-SiC $(11\bar{2}0)$ substrate. The XRD measurements were performed with a triple crystal X-ray diffractometer. A Cu Kα1 line ($\lambda = 1.540562$ Å) was selected using a channel cut Ge (220) monochromator, and an open detector condition was used. Figure 4.5a and b shows the results for different X-ray incidence geometries perpendicular to the sample $[1\bar{1}00]$ and [0001] directions, respectively. In both cases, a diffraction peak from the 2H-AlN $(11\bar{2}0)$ plane was evidently confirmed (2θ angle $\sim 59.3°$), which agrees with the RHEED observations. No diffraction peaks corresponding to (0001)-oriented AlN could be observed even in logarithmic scale (not shown) indicating that the epilayer is single phase $(11\bar{2}0)$ within the detection limit of XRD. The AlN $(11\bar{2}0)$ diffraction peak in theX-ray incidence geometry perpendicular to the $[1\bar{1}00]$ direction (Figure 4.5a) was considerably weaker than that shown in Figure 4.5b (perpendicular to the [0001] direction). This can probably be attributed to SFs in the AlN layer, which are extended along the $[1\bar{1}00]$ direction, or dislocations running along the $[1\bar{1}00]$ direction. A high density of defects perpendicular to the X-ray incident direction is thought to weaken the diffraction of this AlN layer.

XRD ω-scan measurements were also performed as shown in Figure 4.6a and b. The full-width at half-maximum (FWHM) values were 3000 and 240 arcsec for X-ray incidence directions perpendicular to $[1\bar{1}00]$ and [0001], respectively. These values are comparable to those found for nonpolar GaN grown on sapphire $(1\bar{1}02)$ substrates [39]. Although the lattice mismatch of AlN/SiC is much smaller than that of GaN/sapphire $(1\bar{1}02)$, the FWHM values of AlN/SiC and GaN/sapphire are almost same.

The microscopic structure of the AlN/4H-SiC $(11\bar{2}0)$ interface was investigated by HRTEM. The microscopy was performed on a Hitachi

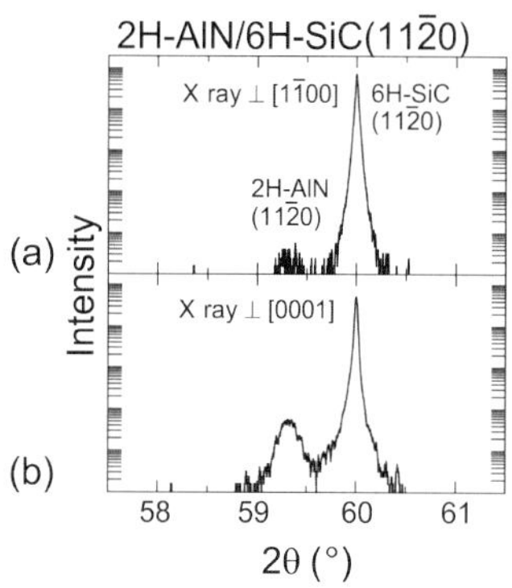

Figure 4.5 XRD $2\theta/\omega$-scan of 2H-AlN grown on 6H-SiC $(11\bar{2}0)$.

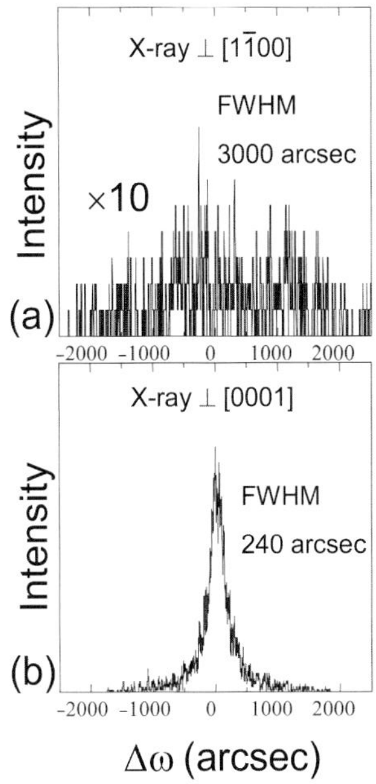

Figure 4.6 XRD ω-scan of 2H-AlN grown on 6H-SiC (11$\bar{2}$0).

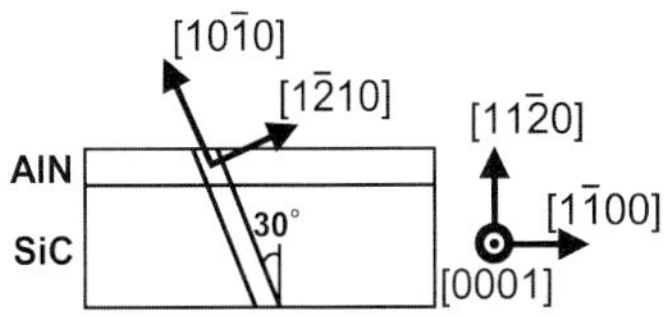

Figure 4.7 Sample geometry for cross-sectional TEM characterization.

H-9000NAR microscope operating at 300 kV. To clarify the polytype of the AlN epilayer from the stacking sequence, the incident electron beam should be directed along the ⟨11$\bar{2}$0⟩-zone axis. To achieve this geometry, a cross-sectional TEM specimen was cut from the wafer with a 30° inclination as shown in Figure 4.7. The specimen was prepared by mechanical thinning and Ar ion milling. Figure 4.8 shows the lattice image of the 2H-AlN/6H-SiC (11$\bar{2}$0). It should be noted that the AlN/SiC interface was observed as a band (projected region), as the cross section image was taken with the 30° inclination. SiC and AlN are clearly seen to have six bilayer (1.5 nm) and two bilayer (0.5 nm) periodicity, respectively. SFs can be seen in some areas of the AlN layer. The staking faults originated from the AlN−SiC interface, not from the bulk SiC. The density of SFs observed in TEM is similar to that of the elongated defects observed by AFM. The lattice image (or closed-packed plane) at the projected region of the interface is disordered. It suggests that dislocations with Burgers vectors of nonzero [0001] components exist along the AlN/SiC heterointerface.

4.5
AlN/4H-SiC (11$\bar{2}$0)

As described in the previous section, we have studied growth of AlN on 6H-SiC (11$\bar{2}$0) and obtained an epitaxial relationship analogous to that in Stemmer's work on 6H-SiC (1$\bar{1}$00). The obtained AlN crystal quality is not as good as one would expect, given the fairly close match in lattice constants. We speculated that the degradation in crystal quality originates from the crystal structure mismatch at the heterointerface (2H−6H). The severity of the structural mismatch can be reduced by changing the polytype of the SiC substrate. As shown in Figure 4.9a and b, a three bilayer mismatch for every unit cell (six

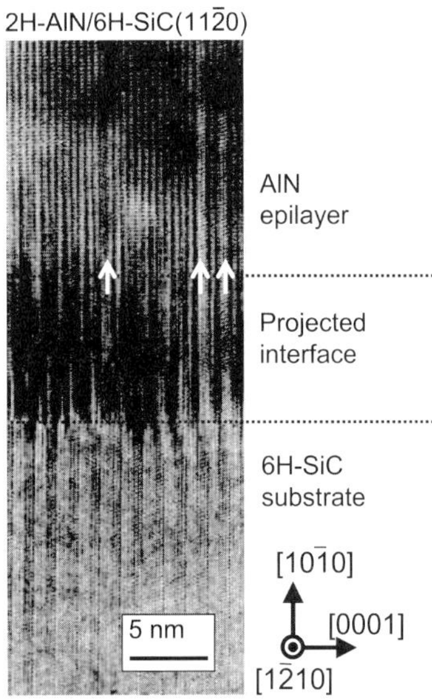

Figure 4.8 Cross-sectional HRTEM image of 2H-AlN/6H-SiC (11$\bar{2}$0) interface.

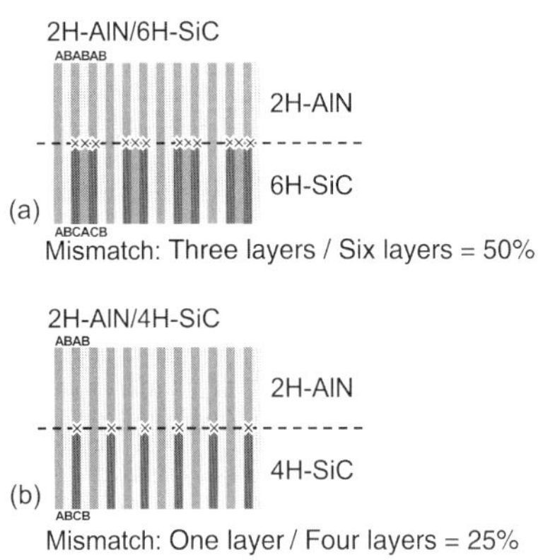

Figure 4.9 Schematic illustration of degree of atomic stacking mismatch between 2H structure and (a) 4H and (b) 6H structures.

bilayers) occurs at a 2H-AlN−6H-SiC interface, which can be described as a 50% structural mismatch. On the other hand, in a 2H-AlN−4H-SiC interface there is a one bilayer mismatch for every unit cell (four bilayers) resulting in 25% structural mismatch. Hence 4H-SiC seems to be a more promising substrate for growth of nonpolar AlN. We therefore investigated the growth of AlN on 4H-SiC (11$\bar{2}$0) next [20].

The substrates used in this section were commercially available 4H-SiC with the (11$\bar{2}$0) orientation. 380-nm-thick AlN epilayers were grown by PAMBE using the same procedure as described above for AlN/6H-SiC (11$\bar{2}$0) – that is a growth temperature of 1000 °C under Al-rich growth conditions.

In situ RHEED monitoring was conducted during AlN growth to investigate the polytype of AlN epilayer. Figure 4.10a and b shows the RHEED patterns of the 4H-SiC substrate and AlN layer along the [0001] azimuth. The two patterns are identical, indicating that growth of 4H-AlN occurred – that is the polytype of the 4H-SiC (11$\bar{2}$0) substrate was replicated in the AlN epilayer. This is a completely different result from that of 2H-AlN growth on 6H-SiC (11$\bar{2}$0) as was discussed above.

The microscopic structure of the AlN/4H-SiC (11$\bar{2}$0) interface was investigated by HRTEM. Figure 4.11 shows the lattice image of the AlN/4H-SiC heterostructure. As seen in the 4H-SiC substrate region, one set of dark and bright bands corresponds to one unit cell of the 4H structure (1.0 nm). The AlN epilayer has exactly the same dark and bright bands, indicating successful polytype replication from the 4H-SiC substrate, that is, growth of 4H-AlN. Some regions of AlN do not have the 4H structure – that is, they are faulted. Sharp dark lines were observed in the interface region (see regions marked in Figure 4.11). The length of the dark lines closely matches the product tan 30° × the TEM specimen thickness (~20 nm) and the beginnings and ends of the lines are well aligned. Therefore, it is concluded that the dark lines correspond to partial dislocations along the [1$\bar{1}$00] direction at the AlN−SiC interface. The faulted regions were located just above the line defects. In other words, the atomic arrangement of the 4H structure was not replicated from the 4H-SiC substrate across the line defects. The density of SFs is 5×10^6 cm^{-1}. Despite the high density of SFs, we obtained a very clear lattice image compared with the case of 2H-AlN/6H-SiC. If partial dislocations are localized at the interface, they result not in lattice distortions but rather only in SFs. In other words, partial dislocations located at the interface do not severely degrade the crystalline quality of AlN. Because these line defects were probably generated at the initial stage

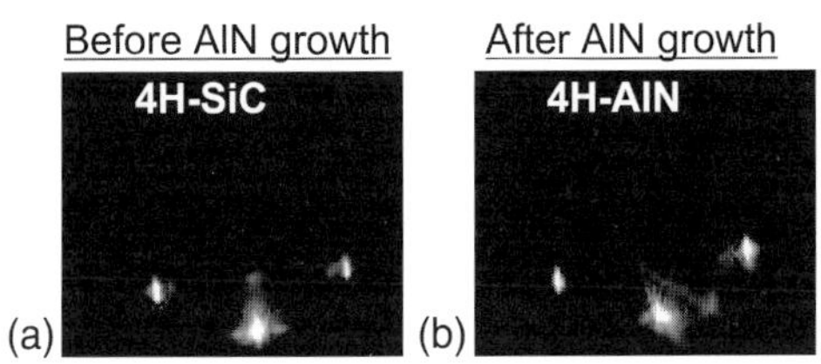

Figure 4.10 RHEED patterns (a) before and (b) after growth of AlN on 4H-SiC (11$\bar{2}$0).

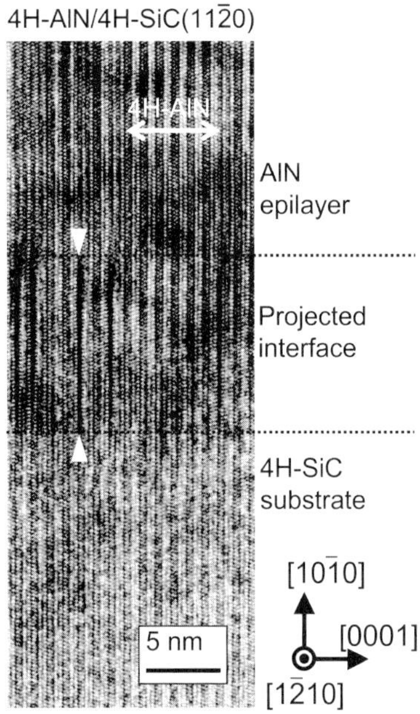

Figure 4.11 Cross-sectional HRTEM image of 4H-AlN/4H-SiC (11$\bar{2}$0) interface.

of AlN growth, it is expected that their density, as well as the density of faults in 4H-AlN, can be reduced by further optimizing the initial growth conditions.

Figure 4.12 shows an AFM image of the 4H-AlN layer. The basic features are the same as those of 2H-AlN on 6H-SiC (11$\bar{2}$0). However, 4H-AlN on 4H-SiC (11$\bar{2}$0) has a smoother surface (smaller roughness of 0.3 nm) and lower density of elongated features, suggesting that 4H-AlN/4H-SiC (11$\bar{2}$0) has lower structural defect density than 2H-AlN/6H-SiC (11$\bar{2}$0).

XRD measurements were carried out for 380-nm-thick AlN epilayers grown on 4H-SiC (11$\bar{2}$0) substrates. Figures 4.13 and 4.14 show the XRD $2\theta/\omega$-scan and ω-scan, respectively. Compared with 2H-AlN/6H-SiC (11$\bar{2}$0), the AlN peak is much stronger and sharper. More importantly, there was no large

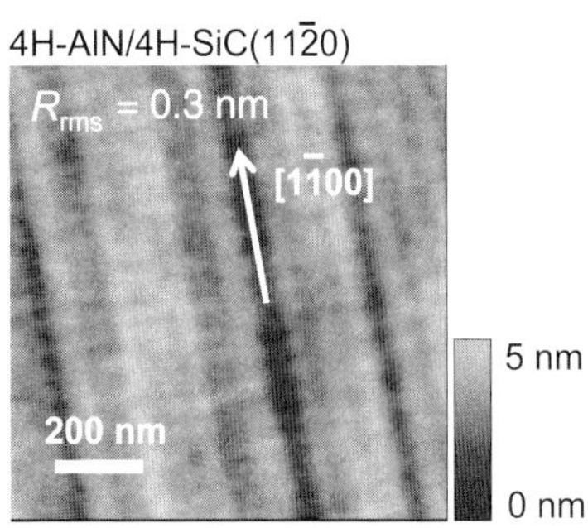

Figure 4.12 AFM image of 2H-AlN grown on 4H-SiC (11$\bar{2}$0).

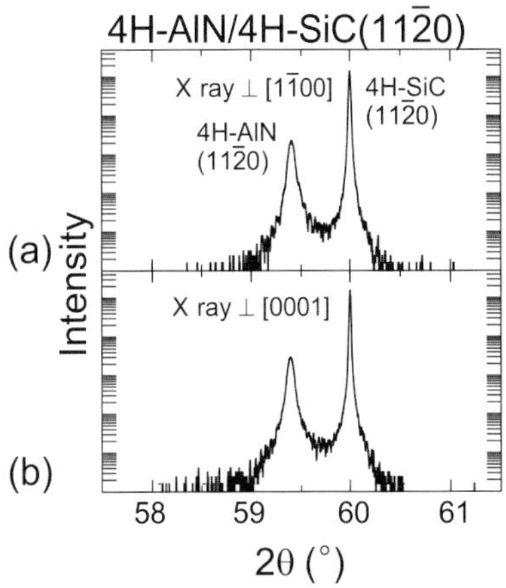

Figure 4.13 XRD $2\theta/\omega$-scan of 4H-AlN grown on 4H-SiC (11$\bar{2}$0).

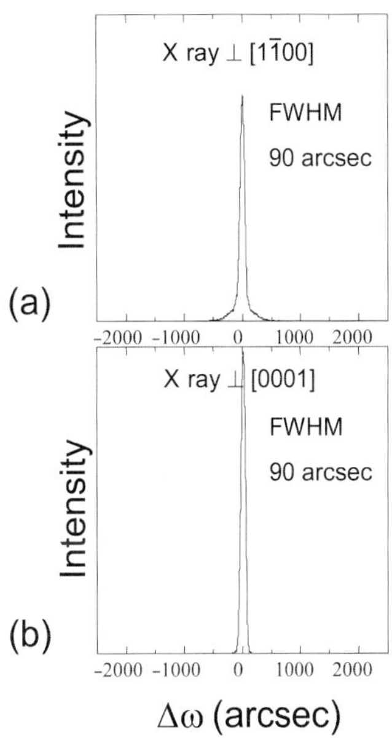

Figure 4.14 XRD ω-scan of 4H-AlN grown on 4H-SiC (11$\bar{2}$0).

difference between the two X-ray incidence geometries relative to the sample [1-100] and [0001] axes. The FWHM of ω-scan exhibited a small value of 90 arcsec for both incidence geometries (compared with 3000 and 240 arcsec for 2H-AlN/6H-SiC).

The reason polytype replication occurs successfully for the AlN/4H-SiC (11$\bar{2}$0) system but not for AlN/6H-SiC (11$\bar{2}$0) has not been elucidated yet. In our speculation, 4H-AlN (11$\bar{2}$0) can be grown on 4H-SiC (11$\bar{2}$0) owing to the similar structure of 2H (ABAB) and 4H (ABCB) – that is, only one bilayer displacement in each unit cell can transform 2H to 4H. On the other hand, to transform 2H (ABABAB) to 6H (ABCACB), three successive bilayers

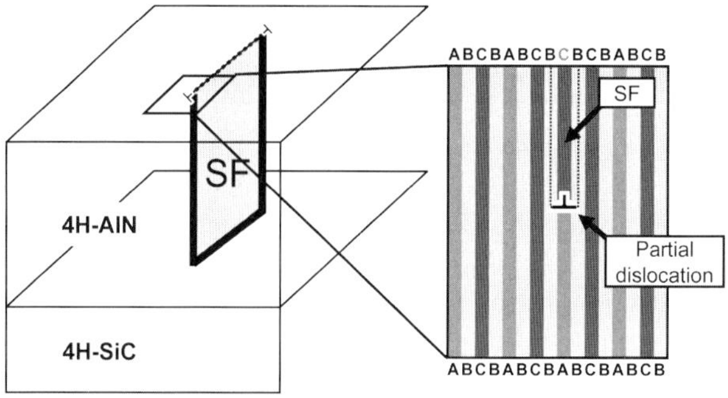

Figure 4.15 Schematic diagram illustrating termination of a stacking fault by partial dislocations.

must be displaced. This larger number of atomic displacements seems to be energetically unfavorable.

Although XRD showed excellent crystalline quality, TEM showed that 4H-AlN layer still contains high densities of defects. The main defects are SFs, as shown in cross-sectional HRTEM. Their density is 5×10^6 cm^{-1}. Plan-view TEM also revealed the existence of a high density of partial TDs with density of $\sim 10^{10}$ cm^{-2}. The partial dislocations originate from SFs – that is, each SF has two partial TDs at each end as shown in Figure 4.15. Since the Burgers vector of the dislocations is 1/3 $[1\bar{1}00]$ and the direction of dislocation line is [11-20], the partial TDs are edge type. Therefore, these partial dislocations do not contribute to broadening of symmetrical X-ray diffraction peaks, explaining why small symmetrical XRD FWHM values were obtained.

4.6
Reducing Structural Defect Densities in 4H-AlN

We have obtained 4H-AlN on 4H-SiC $(11\bar{2}0)$. Isopolytypic growth resulted in very small XRD linewidth, 90 arcsec, one of the lowest values ever reported for heteroepitaxial growth of nonpolar nitride layers. However, TEM studies showed that the 4H-AlN layers still contain high densities of SFs and partial TDs. Further reduction of the densities of these defects is desirable to develop 4H-AlN grown on 4H-SiC $(11\bar{2}0)$ for device applications.

The partial TDs originated from SFs, and the SFs originated from incomplete polytype replication of 4H-SiC to AlN. To further improve the AlN epilayer quality, it is necessary to enhance the polytype replication process (i.e. increase the fraction of AlN which is initially nucleated in the 4H structure). As clearly seen in the cross-sectional TEM images, SFs originate at the AlN/SiC interface, suggesting that control of the initial growth stage is absolutely crucial.

To enhance the polytype replication process, we have introduced two key experimental procedures [23, 24]. The first is to prepare an atomically flat,

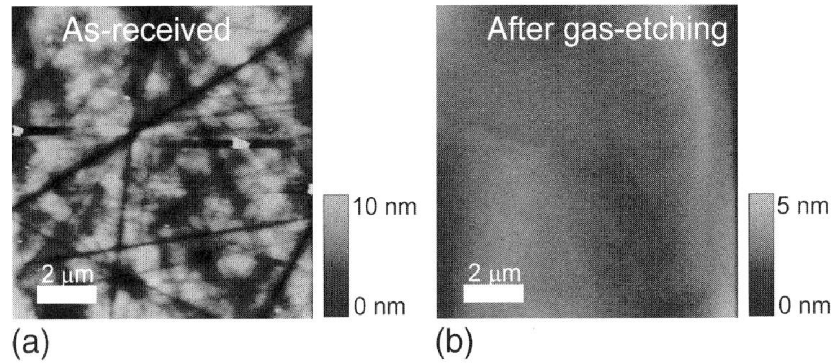

Figure 4.16 AFM images of 4H-SiC (11$\bar{2}$0) substrates: (a) as-received substrate and (b) after gas etching.

damage-free, and clean 4H-SiC substrate surface, and the second is careful optimization of the V/III ratio during AlN nucleation. As AlN naturally prefers the 2H structure, lateral overgrowth along the ⟨0001⟩ direction may result in 2H inclusions, that is SFs in 4H material. A very flat initial surface and optimized V/III ratio (slightly Al-rich conditions) ensure a very smooth AlN growth surface, which prevents lateral growth and thereby leads to high phase purity. Furthermore, substrate surface cleaning is also essential because SiC surface contamination may prevent AlN adatoms from 'sensing' the substrate stacking sequence, resulting in poor polytype replication.

The substrates used in this section were the same as in the previous section. However, two additional processes were carried out. As the polishing technology for nonpolar SiC is still under development, many polishing scratches are observed on the SiC wafer surface as shown in Figure 4.16a. To remove the scratches, the substrates were etched by a $HCl-H_2$ gas mixture at a high temperature. The etching conditions established for 6H-SiC (0001) substrates [40] were used and comprised the following: temperature of 1300 °C, H_2 including 0.3% HCl, flow rate of 1 slm, and etching time of 10 minutes. After gas etching, a very smooth SiC surface (see Figure 4.16b) was obtained, the rms roughness of which was below 0.3 nm. After this treatment, the substrates were dipped into a HF solution, rinsed by deionized water, and loaded into the MBE chamber. Prior to AlN growth, three cycles of Ga deposition and desorption processes were performed to cleanse the SiC surface *in situ* of residual oxygen. This process is known to be effective to obtain clean SiC (0001) surfaces [25, 41]. It is expected to also be effective for nonpolar SiC surfaces. AlN growth was carried out at 950 °C under various Al beam equivalent pressure (BEP) values ranging $4 \times 10^{-7} - 1 \times 10^{-6}$ torr. A N_2 flow rate of 1.0 sccm and rf-plasma power of 400 W were used. The typical growth rate was 400–500 nm h^{-1}.

After *in situ* cleaning, a sharp and intense RHEED pattern from 4H-SiC (11$\bar{2}$0) was observed as shown in Figure 4.17a. Figure 4.17b–e show the RHEED patterns observed after the growth of AlN under various Al

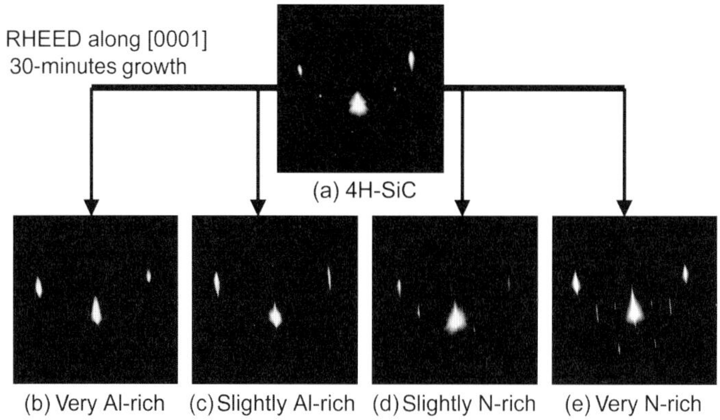

Figure 4.17 RHEED patterns of AlN layers grown on 4H-SiC (11$\bar{2}$0) with various III/V ratios.

flux values. For Al-rich conditions, the RHEED pattern was unchanged from that of the 4H-SiC substrate, indicating that the AlN layer has the 4H polytype. The RHEED pattern of the very Al-rich sample (Figure 4.17b) became faint as growth proceeded, because of the coverage of the surface by metallic Al. The pattern for the slightly Al-rich sample (Figure 4.17c) remained unchanged throughout the growth. Contrary to this, for AlN grown under N-rich conditions, (1$\bar{1}$00) diffraction gradually appeared as shown in Figure 4.17d and e. For a slightly N-rich sample (Figure 4.17d), the intensity of (1$\bar{1}$00) diffraction was weaker than that of (1$\bar{2}$10) diffraction. It suggests that the layer is a mixture of 4H and 2H polytypes – that is some portion is 4H and the rest is 2H (or a 4H structure with a very high density of SFs). The RHEED pattern of a sample highly rich in N (Figure 4.17e) appeared nearly identical to the ideal pattern expected for the 2H polytype. From these results, it can be concluded that polytype replication from the SiC substrate to AlN epilayer occurred only when AlN was grown under Al-rich conditions. With decreasing III/V ratio during growth, the fraction of 2H inclusions increased.

The crystalline quality of the AlN layer was assessed by XRD measurements. Table 4.2 shows the FWHM of XRD $2\theta/\omega$- and ω-scans of AlN samples grown under various III/V ratios. For the measurement, receiving slit (acceptance angle of 30 arcsec) was attached. Both the $2\theta/\omega$-scan and ω-scan FWHM values decreased with increasing III/V ratio.

Table 4.2 FWHM of $2\theta/\omega$- and ω-scans for AlN grown on 4H-SiC (11$\bar{2}$0) with various III/V ratios.

III/V ratio	Very Al rich	Slightly Al rich	Slightly N-rich	Very N-rich
$2\theta/\omega$-scan FWHM (arcsec)	146	175	222	214
ω-scan FWHM (arcsec)	40.5	45.2	44.6	59.5

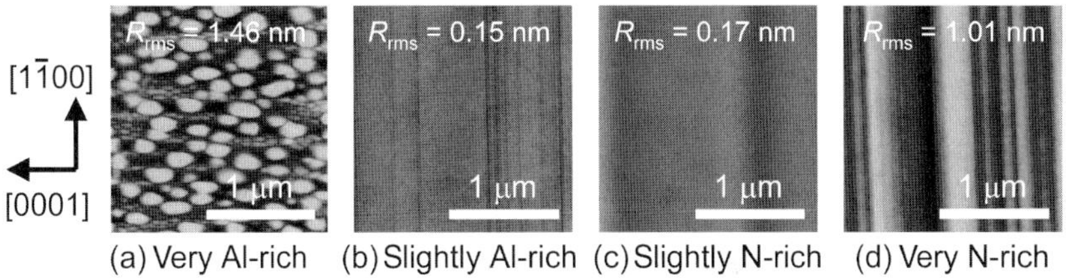

Figure 4.18 AFM images of AlN layers grown on 4H-SiC (11$\bar{2}$0) with various III/V ratios.

Figure 4.18a–d shows the surface morphology of AlN with various III/V ratios. For slightly Al-rich and slightly N-rich conditions (Figure 4.18b and c), a considerably smooth AlN layer with a rms roughness of 0.15–0.17 nm was obtained. (The rms roughness of the sample discussed in the previous section was 0.3 nm.) For slightly Al-rich conditions, several line depressions were observed in the image area (density 1×10^5 cm^{-1}). It appears that either slightly Al-rich or slightly N-rich growth conditions result in a smooth growth front. However, on the basis of RHEED observations, slightly N-rich conditions should not be used because of the formation of 2H inclusions under such conditions (incomplete polytype replication). Considering both the XRD results and surface morphology, we conclude that growth in slightly Al-rich conditions results in the 4H-AlN layers of highest quality.

For further evaluation of the crystal quality, a 470-nm-thick AlN layer grown under slightly Al-rich conditions was characterized by high-resolution XRD (HRXRD) examining the symmetrical (11$\bar{2}$0) and the asymmetrical (11$\bar{2}$4) and (03$\bar{3}$0) reflections. The results are shown in Figure 4.19a–c. For all reflections, the ω-scan FWHM is less than 100 arcsec, indicating excellent crystalline quality. For the (11$\bar{2}$0) diffraction, the minimum value observed is 40 arcsec. In the previous section nonoptimized 4H-AlN with a high density of SFs exhibited an ω-scan FWHM of 90 arcsec. The much smaller FWHM values in the present sample are indicative of lower SF

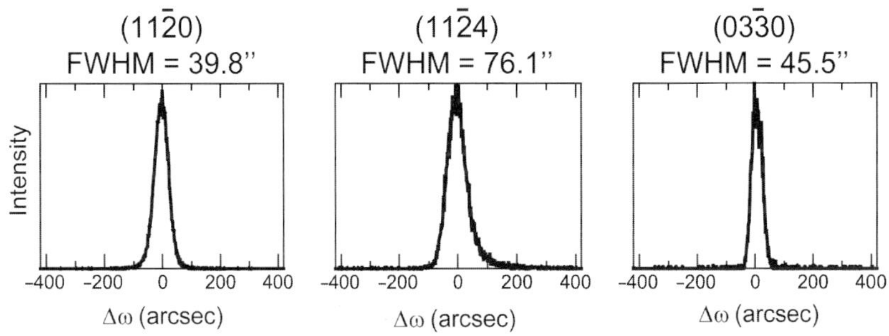

Figure 4.19 XRD ω-scan for optimized 4H-AlN layer grown on 4H-SiC (11$\bar{2}$0). (a) Symmetrical (11$\bar{2}$0) reflection, and asymmetrical (b) (11$\bar{2}$4) and (c) (03$\bar{3}$0) reflections.

and TD densities. Moreover, the small FWHM values for the asymmetrical reflections (sensitive to edge dislocations) suggest a low density of partial TDs.

A cross-sectional HRTEM image of the AlN layer is shown in Figure 4.20. To observe the stacking structure by cross-sectional HRTEM, sample cutting with a 30° inclination was used. As shown in Figure 4.20a, the periodic structure of the 4H polytype (ABCBABCB) with a lattice constant of 1.0 nm was clearly observed. It should be noted that there are no SFs within the observed area. SFs can be observed in a lower-magnification image as shown in Figure 4.20b on the basis of which the SF density was estimated to be 2×10^5 cm^{-1}. This is a factor of 25 lower than the SF density of the nonoptimized 4H-AlN in the previous section (approximately 5×10^6 cm^{-1}). Since the SF density nearly matches the density of line depressions in the AFM image, the observed line depressions may be growth pits originating from SFs.

To evaluate the TD density, plan-view TEM observations were carried out. Figure 4.21a and b shows the bright field images of a 4H-AlN epilayer with $\mathbf{g} = 1\bar{1}00$ and $\mathbf{g} = 0004$, respectively. Dark lines running along [$1\bar{1}00$] correspond to SFs in the TEM image under the $\mathbf{g} = 1\bar{1}00$ diffraction condition (Figure 4.21a). Because a partial TD exists at the end of the SF, the density

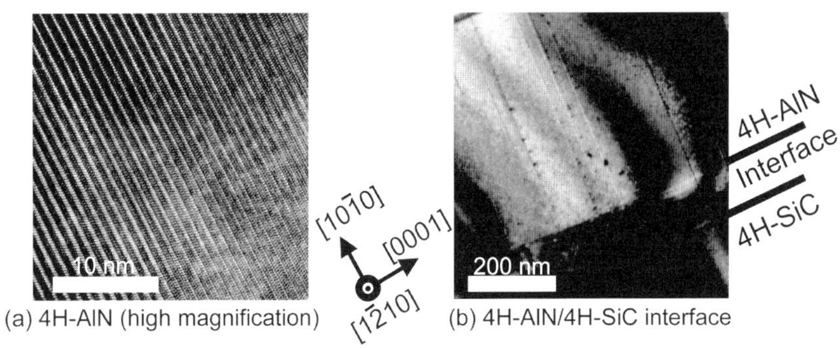

Figure 4.20 Cross-sectional TEM images of high-quality 4H-AlN grown on 4H-SiC (11$\bar{2}$0). (a) High-magnification image of 4H-AlN layer and (b) low-magnification image of 4H-AlN/4H-SiC interface.

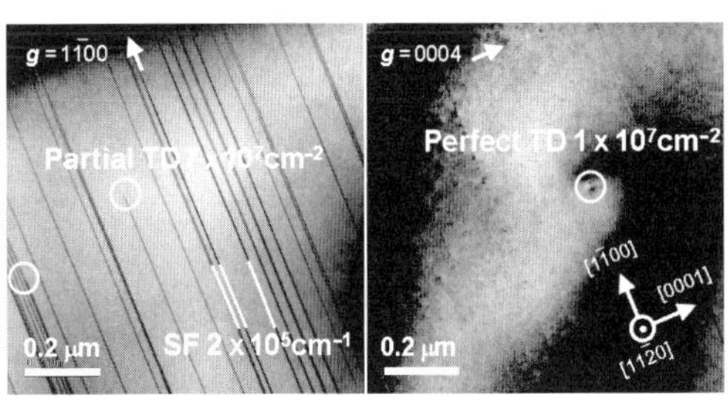

Figure 4.21 Plan-view TEM images of high-quality 4H-AlN grown on 4H-SiC (11$\bar{2}$0) with two different diffraction conditions. (a) $\mathbf{g} = 1\bar{1}00$ and (b) $\mathbf{g} = 0004$.

of partial TDs having Burgers vectors $\mathbf{b} = 1/3\ [1\bar{1}00]$ is estimated to be about $7 \times 10^7\ \text{cm}^{-2}$ by examining a larger area of $22\ \mu\text{m}^2$. A perfect TD was observed in the TEM image under the $\mathbf{g} = 0004$ diffraction condition (Figure 4.21b). Just two perfect TDs were found in an area of $22\ \mu\text{m}^2$, indicating that the density of TDs having Burgers vector of [0001] component is about $1 \times 10^7\ \text{cm}^{-2}$. The total TD density of $8 \times 10^7\ \text{cm}^{-2}$ is among the lowest ever reported for a thin heteroepitaxial nonpolar AlN layer on any substrate. The low TD density determined by TEM is consistent with the very sharp symmetrical and asymmetrical XRD ω-scan peaks.

Such a low TD density is rather surprising given the small but still significant lattice mismatch between AlN and SiC. To clarify the reason for the low TD density, detailed XRD investigations were carried out. Table 4.3 shows the lattice parameters of the 4H-SiC substrate and the AlN epilayer measured by HRXRD for both the in-plane and out-of-plane directions. (The definition of lattice parameters is shown in Figure 4.22.) The in-plane lattice constants (c and y) of AlN are equal to those of SiC (within an error of 0.1%) indicating that the AlN layer grew on the SiC substrate almost pseudomorphically. Owing to measurement uncertainty limits, we cannot distinguish whether the AlN thickness is below or just above the critical thickness. Nonetheless, pseudomorphic, or nearly pseudomorphic growth may account for the low TD density. Further investigation of the dependence of the defect densities on the AlN layer thickness is required to validate the assumption of pseudomorphic growth.

Table 4.3 Lattice parameters of high-quality 4H-AlN epilayer and 4H-SiC substrate measured by HRXRD.

Parameters	4H-SiC substrate	4H-AlN epilayer	Difference
c	10.042 Å	10.032 Å	−0.10%
$y/(\sqrt{3}/2)$	3.082 Å	3.079 Å	+0.10%
a	3.081 Å	3.105 Å	+0.78%
θ	30.00°	30.21°	0.21°

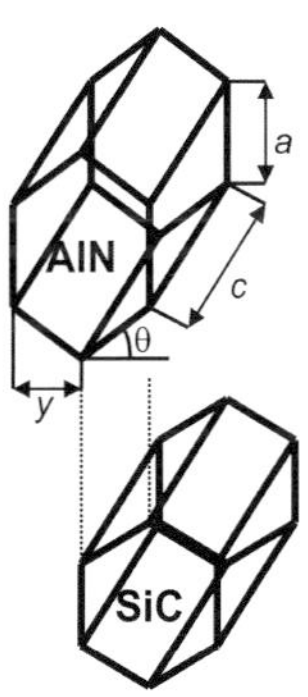

Figure 4.22 Definitions of lattice parameters, c, y, a, and θ used in Table 4.3.

4.7 AlN/4H-SiC $(1\bar{1}00)$

We found that substrate polytype replication is possible when AlN is grown on 4H-SiC $(11\bar{2}0)$. It is also worthwhile to investigate whether polytype replication can be successfully achieved for AlN on 4H-SiC $(1\bar{1}00)$.

An almost identical growth procedure was used for AlN grown on 4H-SiC $(1\bar{1}00)$ substrates as that described in the previous section for $(11\bar{2}0)$ substrates. However, it should be noted that high-temperature gas etching (to improve the SiC atomic-level smoothness before growth) was less effective for $(1\bar{1}00)$ substrates than $(11\bar{2}0)$ substrates. Although the polishing scratches were successfully removed, the gas-etched $(1\bar{1}00)$ surface shows some wavy features, which are perhaps because of the bunching of SiC atomic steps. Optimization of the gas-etching conditions specifically for the $(1\bar{1}00)$ orientation (rather than using conditions optimized for (0001) wafers) may result in $(1\bar{1}00)$ substrates with smoother surfaces at the atomic level. For AlN growth, the Al BEP was set to about 8×10^{-7} torr corresponding to the transition between Al-droplet and droplet-free growth regimes. The AlN thickness studied ranged from 40 to 720 nm.

The 4H structure of the AlN layer on 4H-SiC $(1\bar{1}00)$ is confirmed by the geometries of the RHEED patterns. Figure 4.23a and b shows RHEED patterns along the [11-20] azimuth before and after the growth of a 250-nm AlN layer. The atomic periodicity in this direction differs by a factor of two for 4H-SiC and 2H-AlN. However, the streak spacing of the AlN layer is almost identical to that of 4H-SiC, thus AlN grew in the 4H structure. RHEED patterns showing 4H symmetry are observed for AlN growth thickness up to at least 250 nm, although there is evidence for mixing of 2H and 4H phases in thicker layers. At present, the reason has not been clarified. Possible reasons are (i) insufficient control of the III/V ratio throughout growth and

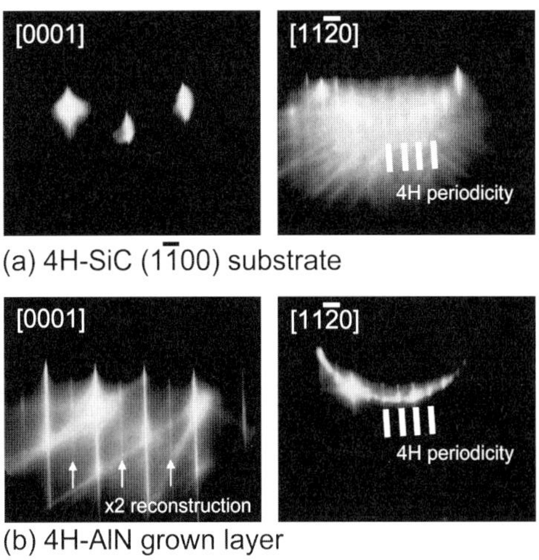

Figure 4.23 RHEED patterns (a) before and (b) after growth of AlN on 4H-SiC $(1\bar{1}00)$.

(ii) inherent instability of thicker 4H-AlN layers because of processes such as lattice relaxation.

The RHEED patterns of AlN layers grown under Al-rich conditions exhibit weak half-order reconstruction streaks along all azimuths except [11-20] after cooling to 800 °C as shown by arrows in Figure 4.23b. This corresponds to the Al/N flux ratio near the transition point, above which Al droplets are formed. The 1×2 reconstruction vanishes upon exposure to the nitrogen plasma beam, implying that it is related to Al adatoms. For higher values of the Al flux, the reconstruction streaks are more intense and can even be observed during growth at 950 °C, however, for such high Al/N flux ratios droplets are formed on the layer surfaces. If it is assumed that the observed AlN 1×2 reconstruction is analogous to that of the GaN (1$\bar{1}$00) surface, [42] then the optimal AlN growth conditions can be associated with an Al surface coverage of 1–2 monolayer (ML).

It should be noted that during the initial stages of AlN growth on 4H-SiC (1$\bar{1}$00), RHEED intensity oscillations are clearly observed as shown in Figure 4.24. The oscillation period of 2.1 seconds is in good agreement with the time (2.0 seconds) required for deposition of 1 ML ($=\sqrt{3}a/2$) of (1$\bar{1}$00) AlN at the growth rate of 8 nm min^{-1}. As intensity oscillations would not occur for a defective 2H–4H interface, this result implies successful 4H polytype replication. We confirmed the reproducibility of the RHEED oscillations for (1$\bar{1}$00) growth on gas-etched substrates. On the other hand, up to now, RHEED oscillations have not been observed for (11$\bar{2}$0) growth. The atomic-level processes involved in growth on (1$\bar{1}$00) and (11$\bar{2}$0) surfaces may be very different.

AFM images for a 250-nm-thick 4H-AlN (1$\bar{1}$00) epilayer are shown in Figure 4.25. The rms roughness is 0.9 nm and the AlN morphology is characterized by stripe features parallel to ⟨11$\bar{2}$0⟩, similar to (1$\bar{1}$00) GaN grown by MBE [42]. The stripe-feature density of the present (1$\bar{1}$00) epilayers is 10^5 cm^{-1}. An SF density of about 10^5 cm^{-1} in the AlN layer is therefore implied if the stripe features are assumed to be directly related to such faults.

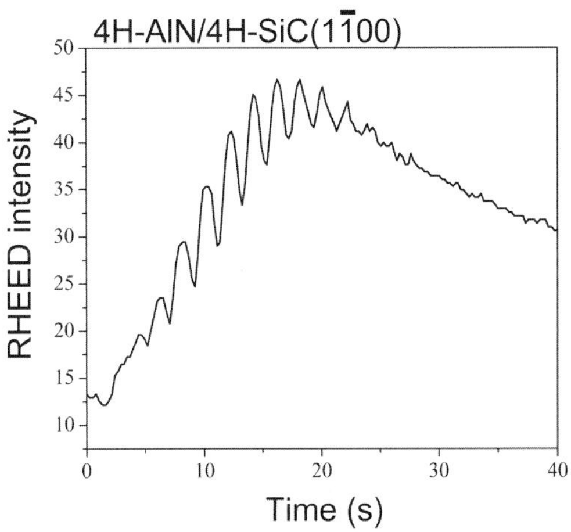

Figure 4.24 RHEED intensity oscillations during initial growth of 4H-AlN on 4H-SiC (1$\bar{1}$00).

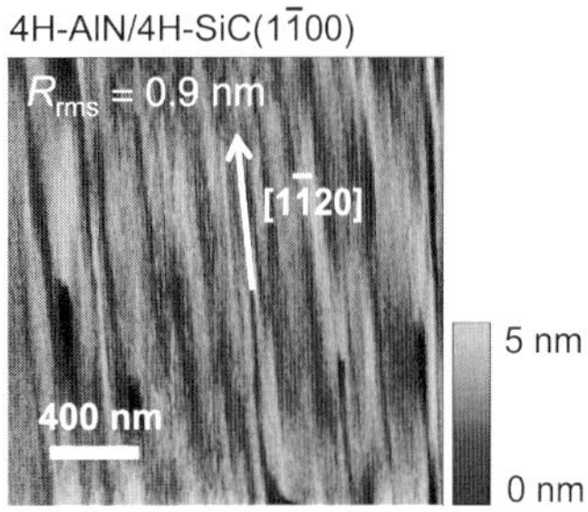

Figure 4.25 AFM image of 4H-AlN grown on 4H-SiC (1$\bar{1}$00).

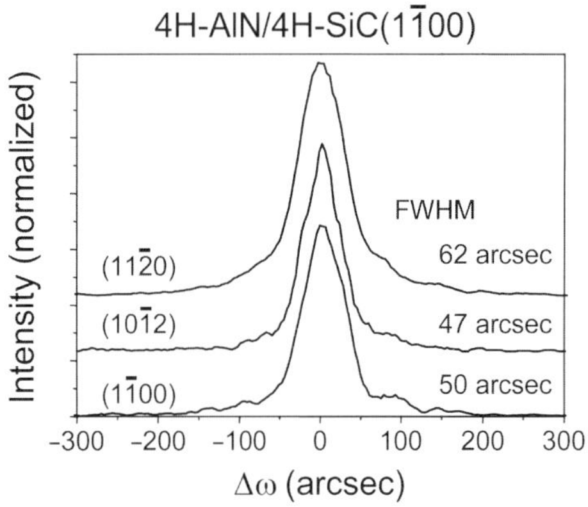

Figure 4.26 XRD ω-scans of 4H-AlN grown on 4H-SiC (1$\bar{1}$00). Symmetrical (1$\bar{1}$00) and asymmetrical (10$\bar{1}$2) and (11$\bar{2}$0) reflections are shown.

The AlN structural quality has been assessed by HRXRD. Both the symmetrical and asymmetrical reflections of the epilayers exhibit narrow ω-scan peaks, as shown in Figure 4.26. The FWHM value for the (1$\bar{1}$00) reflection is as small as 50 arcsec, a little bit larger than the best value for 4H-AlN/4H-SiC (11$\bar{2}$0). The asymmetrical reflections also show FWHM values smaller than 70 arcsec. Symmetrical FWHM values as small as those shown in Figure 4.26 have been successfully reproduced over several AlN growth runs. The symmetric ω-scan FWHM values for optimized AlN epilayers do not change when measurements are made for different orientations of the incident X-ray beam with respect to the [0001] direction.

Although TEM studies of 4H-AlN/4H-SiC (1$\bar{1}$00) are still under investigation, RHEED, XRD, and AFM results suggest that 4H-SiC (1$\bar{1}$00) substrates are also promising for high-quality nonpolar 4H-AlN growth.

4.8
Properties of 4H-AlN

The material properties of 4H-AlN are interesting from a fundamental, scientific perspective as well as for device applications. 4H-AlN can be thought of as a short-period superlattice (SPSL) of 2H-AlN and 3C-AlN. Roughly speaking, the properties of 4H-AlN are expected to be in between those of 2H-AlN and 3C-AlN. We have estimated the room-temperature energy band gap of 4H-AlN to be ~5.9 eV from reflectance spectra. This value lies in between the band gaps of 3C-AlN (5.74 eV) and 2H-AlN (6.16 eV) [43]. For DUV LED applications, the use of 4H-Al(InGa)N emitters will result in a slightly longer emission wavelength versus the same alloy composition in the 2H structure (210 nm for 2H-AlN vs. 220 nm for 4H-AlN). With an appropriate design of the QW, any desired wavelength in the deep-UV range can be achieved using either 2H- or 4H-nitride materials.

It should be noted that we can expect zone folding of the electronic band structure as well as the phonon dispersion, which is generally observed in SPSL. This might affect the effective mass, carrier transport properties, and so on. For example, we have observed additional phonon modes (folded modes) in Raman scattering of 4H-AlN. Characterization of the fundamental material properties of 4H-AlN will be described in detail in future publications.

4.9 Nonpolar AlGaN and AlGaN/AlN Heterostructures

All high-efficiency LED designs considered thus far incorporate QWs. For the DUV spectral region, high-AlN-mole-fraction, $Al_{1-x-y}Ga_xIn_yN$, QWs are required. In this section, we present some preliminary results on the growth of nonpolar AlGaN and AlGaN/AlN QW structures [22].

$Al_{1-x}Ga_xN$ layers were grown on 4H-SiC $(1\bar{1}00)$ with 50-nm-thick 4H-AlN as a buffer layer. In AlGaN layers of low Ga content, excellent XRD characteristics comparable to those of the underlying AlN buffer layer are obtained as shown in Table 4.4. However, the ω-scans become much broader (several hundred arcsec) if the Ga content is pushed above 30%. Likely reasons for this may be the increasing lattice mismatch between AlN and $Al_{1-x}Ga_xN$ and/or a possibly greater tendency for SFs to form in Ga-rich $Al_{1-x}Ga_xN$ layers. Unlike optimized AlN epilayers, Ga-rich AlGaN and GaN exhibit a large difference in ω-scan FWHM depending on the orientation of the X-ray beam with respect to the [0001] direction (as seen in 2H-AlN/6H-SiC). Data given in Table 4.4 are for the beam perpendicular to the [0001] direction. The FWHM is three to four times larger when oriented perpendicular to the [11–20] direction. Although

Table 4.4 XRD ω-scan FWHM for AlGaN layers grown on 4H-AlN/4H-SIC $(1\bar{1}00)$.

Composition	$(1\bar{1}00)$ ω-scan FWHM (arcsec)
AlN	50
$Al_{0.80}Ga_{0.20}N$	220
$Al_{0.55}Ga_{0.45}N$	1260
GaN	1000

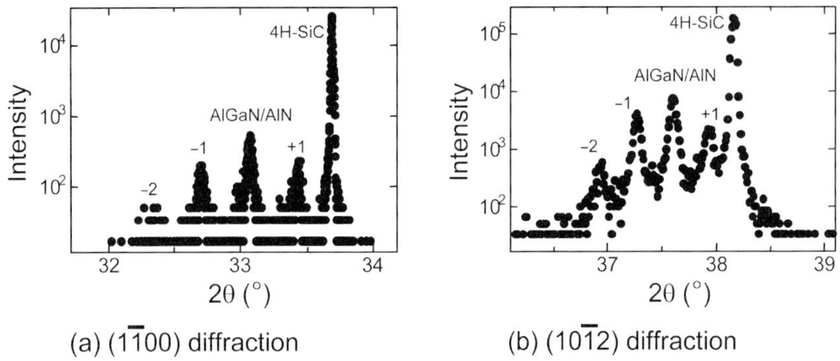

Figure 4.27 XRD $2\theta/\omega$-scan of AlGaN/AlN QW structure grown on 4H-AlN/4H-SiC (1$\bar{1}$00). (a) Symmetrical (1$\bar{1}$00) and (b) asymmetrical (10$\bar{1}$2).

the FWHM obtained for GaN (1000 arcsec) is much larger than that of AlN, it nonetheless represents material quality similar to that reported previously for nonpolar GaN epilayers obtained without lateral overgrowth methods. Details of the growth of GaN on 4H-AlN are described elsewhere [44].

On the basis of the results mentioned above, we chose $Al_{0.70}Ga_{0.30}N$ as the well composition for a preliminary test heterostructure. An $Al_{0.70}Ga_{0.30}N$/AlN multiple QW (MQW) structure of 10 periods with well and barrier thickness of 12.5 nm each was grown. Figure 4.27a shows an XRD $2\theta/\omega$-scan of the multiple MQW. Sharp and intense peaks from the MQW structure are observed, indicating high structural quality. The positions of the satellite peaks agree with the thickness expected from the growth rate. It should be noted that satellite peaks are observed even in asymmetrical diffraction as shown in Figure 4.27b, indicating a high degree of crystal perfection. The luminescence properties of nonpolar $Al_{1-x}Ga_xN$/AlN MQWs are now under investigation.

4.10 Conclusion

Growth of nonpolar AlN on 4H- and 6H-SiC with (1$\bar{1}$00) and (11$\bar{2}$0) plane surface orientations was reviewed. The first isopolytypic growth of AlN was demonstrated on 4H-SiC (11$\bar{2}$0) and (1$\bar{1}$00). The excellent structural quality of the obtained 4H-AlN layers promises a new route toward efficient nonpolar III-nitride devices: isopolytypic growth on SiC substrates.

HRTEM showed a low SF density of 5×10^5 cm^{-1} and low TDD of 8×10^7 cm^{-2} in an optimized 4H-AlN (11$\bar{2}$0) layer. However, the structural defect density obtained is still higher than that of nonpolar GaN obtained by ELOG techniques [45]. Further reduction of the densities of SFs and TDs in isopolytypic III-nitride layers on SiC remains a challenge. Evidently, structural defects originate at the initial stages of growth. Of many different AlN growth parameters, we have thus far optimized only the III/V ratio at a fixed temperature. Various methods such as two-step growth, migration-enhanced epitaxy, and so on, may prove useful for improving the layer quality. Another challenge is to grow actual device structures incorporating thicker AlN layers and AlGaN layers while maintaining low structural defect densities. Clarifying the mechanisms of defect generation in layers necessary for device structures (AlGaN, etc.) is very important.

Finally, it should be noted that high-quality 'isopolytypic' 4H-AlN−4H-SiC interfaces are also of considerable interest for future electronics applications including AlN/SiC heterojunction-field effect transistors (HFETs) [46, 47] and AlGaN/SiC heterojunction bipolar transistors (HBTs) [48, 49]. Band-gap engineering is not feasible in current SiC devices, but it can be applied if SiC is combined with AlGaN in a heterojunction. The combination of AlGaN-based band-gap engineering together with advanced SiC processing technologies (homoepitaxial growth, ion implantation, metallization, etching, and large diameter high-quality wafers) makes HFETs and HBTs very attractive for future power electronic devices.

Acknowledgments

This research would not have been possible without key contributions from Dr Norio Onojima, Dr Robert D. Armitage, and Mr. Masahiro Horita working as doctoral students or postdocs in the author's group. The author would also like to express his heartfelt thanks to Prof. Tsunenobu Kimoto and Emeritus Prof. Hiroyuki Matsunami for valuable discussions and comments. SiXON, Co. Ltd, Japan provided the 4H- and 6H-SiC $(1\bar{1}00)$ substrates used in this work.

The author's research on nonpolar AlN has been supported by several funding sources during past five years: Grant-in-Aid for the 21st Century Center of Excellence (COE) Program (#14213201), Grant-in-Aid for Young Scientists (A) (#16686002) and the Kyoto Nano-tech cluster project from the Ministry of Education, Culture, Sports, Science, and Technology of Japan (MEXT), PRESTO program from Japan Science and Technology Corporation (JST) and Grant for Industrial Technology Research (#04A48511d) from the New Energy and Industrial Technology Development Organization, Japan (NEDO).

References

1. Khan, M.A. (2006) *Physica Status Solidi A*, **203**, 1764.
2. Taniyasu, Y., Kasu, M., and Makimoto, T. (2006) *Nature*, **441**, 325.
3. Fujimoto, N., Kitano, T., Narita, G. *et al.* (2006) *Physica Status Solidi C*, **3**, 1617.
4. Zhang, J.P., Khan, M.A., Sun, W.H. *et al.* (2002) *Applied Physics Letters*, **81**, 4392.
5. Takano, T., Ohtaki, Y., Narita, Y., and Kawanishi, H. (2004) *Japanese Journal of Applied Physics*, **43**, L1258.

6. Nakano, K., Imura, M., Narita, G. *et al.* (2006) *Physica Status Solidi A*, **203**, 1632.
7. Park, S.-H. (2002) *Journal of Applied Physics*, **91**, 9904.
8. Ng, H.M., Bell, A., Ponce, F.A., and Chu, S.N.G. (2003) *Applied Physics Letters*, **83**, 653.
9. Tsubouchi, K. and Mikoshiba, N. (1985) *IEEE Transactions on Sonics and Ultrasonics*, **SU-32**, 634.
10. Masu, K., Nakamura, Y., Yamazaki, T. *et al.* (1995) *Japanese Journal of Applied Physics*, **34**, L760.
11. Ng, H.M. (2002) *Applied Physics Letters*, **80**, 4369.
12. Craven, M.D., Wu, F., Chakraborty, A. *et al.* (2004) *Applied Physics Letters*, **84**, 1281.
13. Pasokova, T., Darakchieve, V., Paskov, P.P. *et al.* (2005) *Journal of Crystal Growth*, **281**, 55.
14. Dovidenko, K., Oktyabrsky, S., Narayan, J., and Razeghi, M. (1996) *Journal of Applied Physics*, **79**, 2439.
15. Dovidenko, K., Oktyabrsky, S., and Narayan, J. (1997) *Journal of Applied Physics*, **82**, 4296.
16. Song, J.-H., Chikyow, T., Yoo, Y.Z. *et al.* (2002) *Japanese Journal of Applied Physics*, **41**, L1291.
17. Onojima, N., Suda, J., and Matsunami, H. (2002) *Japanese Journal of Applied Physics*, **41**, L1348.
18. Onojima, N., Suda, J., and Matsunami, H. (2003) *Materials Research Society Symposium Proceedings*, **743**, L3.21.
19. Onojima, N., Suda, J., Kimoto, T., and Matsunami, H. (2003) *Physica Status Solidi C*, 2502.
20. Onojima, N., Suda, J., and Matsunami, H. (2003) *Applied Physics Letters*, **83**, 5208.
21. Armitage, R., Suda, J., and Kimoto, T. (2006) *Applied Physics Letters*, **88**, 011908.
22. Armitage, R., Horita, M., Suda, J., and Kimoto, T. (2006) *Materials Research Society Symposium Proceedings*, **892**, FF28-03.
23. Horita, M., Suda, J., and Kimoto, T. (2006) *Physica Status Solidi C*, **3**, 1503.
24. Horita, M., Suda, J., and Kimoto, T. (2006) *Applied Physics Letters*, **89**, 112117.
25. Onojima, N., Suda, J., and Matsunami, H. (2003) *Japanese Journal of Applied Physics*, **42**, L445.
26. Suda, J., Onojima, N., Kimoto, T., and Matsunami, H. (2004) *Materials Research Society Symposium Proceedings*, **798**, Y3.4.
27. Verma, A.R. and Krishna, K. (1966) *Polymorphism and Polytypism in Crystals*, John Wiley & Sons, Ltd, New York.
28. Pandey, D. and Krishna, P.J. (1975) *Journal of Crystal Growth*, **31**, 66.
29. Nagasawa, H., Yagi, K., Kawahara, T., and Hatta, N. (2002) *Materials Science Forum*, **433-34**, 3.
30. Okumura, H., Ohta, K., Feuillet, G. *et al.* (1997) *Journal of Crystal Growth*, **178**, 113.
31. As, D.J., Potthast, S., Fernandez, J. *et al.* (2006) *Applied Physics Letters*, **88**, 152112.
32. Tairov, Y.M. and Tsvetkov, V.F. (1978) *Journal of Crystal Growth*, **43**, 209.
33. Hobgood, H.M., Brady, M.F., Calus, M.R. *et al.* (2003) *Materials Science Forum*, **457–60**, 3.
34. Nakamura, D., Gunjishima, I., Yamaguchi, S. *et al.* (2004) *Nature*, **430**, 1009.
35. Takayanagi, Y., Sun, Y.-J., Brandt, O., and Ploog, K.H. (2004) *Applied Physics Letters*, **84**, 4756.
36. Funato, M., Ishido, T., Fujita, S., and Fujita, S. (2000) *Applied Physics Letters*, **76**, 330.
37. Funato, M., Fujita, S., and Fujita, S. (2001) *Physical Review B*, **63**, 165319.
38. Stemmer, S., Pirouz, P., Ikuhara, Y., and Davis, R.F. (1996) *Physical Review Letters*, **77**, 1797.
39. Craven, M.D., Lim, S.H., Wu, F. *et al.* (2002) *Applied Physics Letters*, **81**, 469.
40. Nakamura, S., Kimoto, T., Matsunami,H. *et al.* (2000) *Applied Physics Letters*, **76**, 3412.
41. Kaplan, R. (1989) *Surface Science*, **215**, 111.

42. Brandt, O., Sun, Y.-J., Däweritz, L., and Ploog, K.H. (2004) *Physica E*, **23**, 339.
43. Cimalla, V., Lebedev, V., Kaiser, U. *et al.* (2005) *Physica Status Solidi C*, **2**, 2199.
44. Armitage, R., Horita, M., Suda, J., and Kimoto, T. (2007) *Journal of Applied Physics*, **101**, 033534.
45. Haskell, B.A., Wu, F., Craven, M.D. *et al.* (2003) *Applied Physics Letters*, **83**, 644.
46. Zetterling, C.-M., Wongchotigul, K., Spencer, M.G. *et al.* (1996) *Materials Research Society Symposium Proceedings*, **423**, 667.
47. Onojima, N., Kaido, J., Suda, J., and Kimoto, T. (2005) *Physica Status Solidi C*, **2**, 2643–646.
48. Pankove, J., Chang, S.-S., Lee, H.C., *et al.* (1994), in *Proceedings of International Electron Devices Meeting*, San Francisco, U.S.A, p. 389.
49. Suda, J., Nakano, Y., Shimada, S. *et al.* (2006) *Materials Science Forum*, **527-29**, 1545.

5
Metalorganic Vapor Phase Epitaxial Growth of Nonpolar Al(Ga,In)N Films on Lattice-Mismatched Substrates

Hiroshi Amano, Takeshi Kawashima, Daisuke Iida, Masataka Imura, Motoaki Iwaya, Satoshi Kamiyama, and Isamu Akasaki

5.1
Introduction

The first nonpolar GaN and their light-emitting diodes (LEDs) were grown on *r*-plane sapphire by hydride vapor phase epitaxy (HVPE) in the 1970s, which was only two or three years after the first successful growth of single crystalline GaN on *c*-plane sapphire by Pankove *et al.* [1], and Maruska and Tietjen [2]. Up to now, the mainstream of the research and development of GaN and devices such as LEDs, laser diodes, and field effect transistors were carried out using *c*-plane GaN because of its ease in growing single crystals with atomically flat surfaces compared with the growth on other planes such as the *a* or *m* planes. Detailed study of nonpolar/semipolar nitrides resumed in 1997, when the effect of piezoelectricity on the transition probability in GaInN/GaN-based strained quantum wells was estimated both theoretically and experimentally [3, 4]. Figure 5.1 shows the polar angle dependence of the piezoelectric field in a GaInN well in a violet LED, grown on a GaN template [3]. This large electric field reduces the transition probability in the GaInN well of low-injection-condition devices such as LEDs, thus increasing carrier overflow from the well to the cladding layer.

Since then, much effort has been expended on proving the advantage of nonpolar/semipolar LEDs over *c*-plane LEDs. The use of nonpolar nitrides potentially solves the problem of the large piezoelectric field, because no piezoelectric field is expected in the nonpolar heterostructures.

The characterization of the optical properties of multiquantum wells (MQWs), and the performances of LEDs using nonpolar *m* plane and semipolar bulk GaN crystals have already been reported [5–7]. Violet laser diodes formed on *m* plane GaN substrates have also been reported [8, 9]. Therefore, by using nonpolar GaN substrate, it is possible, in principle, to achieve high-efficiency and high-performance light-emitting devices, although currently, the cost of nonpolar GaN substrates is very high.

Nitrides with Nonpolar Surfaces. Growth, Properties, and Devices. Edited by Tanya Paskova

ISBN: 978-3-527-40768-2

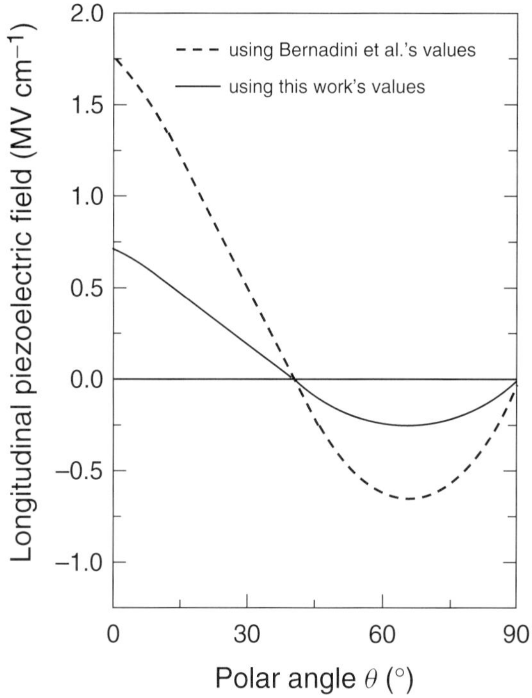

Figure 5.1 Polar angle dependence of the piezoelectric field in coherently grown $Ga_{0.93}In_{0.07}N$/GaN quantum well on GaN template. Zero degree means that the plane-normal axis is the *c* axis, while 90° means that the plane-normal axis is the *a* or *m* axis. The broken line shows the results calculated using the piezoelectric constants reported by Bernadini *et al.* [4], and the solid line shows that obtained using our experimental results [3].

As a lattice-mismatched substrate, *r*-plane sapphire, and *m*-plane and *a*-plane SiC are commonly used. Another candidate is α-$LiAlO_2$. However, in general, growing high-crystalline-quality nonpolar GaN on a lattice-mismatched substrate is known to be very difficult. For example, *a*-plane GaN grown on *r*-plane sapphire has a threading-dislocation and stacking-fault densities much higher than 10^{10} cm^{-2} and 10^5 cm^{-1}, respectively. In addition, surface defects such as pits are formed. There have been several reports on the reduction of defects in nonpolar GaN by lateral-overgrowth HVPE and metalorganic vapor phase epitaxy (MOVPE). In particular, by employing a lateral-overgrowth technique in which grooved GaN and an SiO_2 mask on terrace regions are used, called *pendeo epitaxy* [10] or *sidewall-seeded epitaxial lateral overgrowth* (SELO), it is possible to grow low-defect-density GaN on the whole area of the wafer, except the coalescence region [11, 12]. High-brightness violet and green LEDs have been fabricated by the SELO.

In this section, as examples of nonpolar GaN on lattice-mismatched substrates, the surface morphology and microstructure of *a*-plane GaN on an *r*-plane sapphire substrate and *m*-plane GaN on *m*-plane 4H-SiC are presented. Next, the SELO method of reducing threading-dislocation and stacking-fault densities is described in detail. This is followed by a description of the properties of the conductivity control of *n*-type and *p*-type nonpolar GaN, and the growth of the heterostructure/quantum well structure. Finally, the performances of the violet and green LEDs on nonpolar GaN are discussed with respect to the threading-dislocation density dependence of the output power.

5.2 Growth and Properties of *a*-plane GaN on *r*-plane Sapphire

Nonpolar GaN was grown on *r*-plane sapphire by MOVPE. First, thermal cleaning in hydrogen atmosphere was carried out for 3 minutes at 1050 °C. Subsequently, a 100-nm-thick buffer layer of AlN was grown at 1000 °C before the growth of GaN. Then, a 1.5-μm-thick GaN layer was deposited at 1000 °C.

Wide-range X-ray diffraction (XRD) $2\theta/\omega$ scans and ϕ scans have been performed to elucidate the orientation of GaN. The corresponding results are presented in Figure 5.2a and b. The X-ray diffraction results confirmed that $[11\bar{2}0]$-oriented nonpolar AlN and GaN were grown on *r*-plane sapphire substrates. According to the XRD measurement and transmission electron diffraction pattern, the (0001)-GaN is inclined at 5° to the $(10\bar{1}1)$-sapphire. This relationship is reported by several groups, irrespective of the different growth methods used [13–15]. Hence, this epitaxial relationship is specifically defined for the growth of GaN on *r*-plane sapphire.

The surface morphology of *a*-plane GaN grown on the *r*-plane sapphire substrate is strongly dependent on the growth conditions. When GaN is grown by MOVPE at atmospheric pressure, surface pits tend to be formed, as shown in Figure 5.3. To obtain GaN without pits, low-pressure growth shows better results than the growth at atmospheric pressure. The surface morphology of a-GaN on *r*-plane sapphire grown at low-pressure is shown in Figure 5.4. A mountain-range-like morphology along the *c* axis is observed. The root-mean-square (RMS) value of the surface roughness always exceeds 10 nm, which is much higher than that of GaN grown on a *c*-plane sapphire substrate with a low-temperature buffer layer.

One of the most serious problems of nonpolar GaN on lattice-mismatched substrates is the high density of defects. Cross-sectional transmission electron microscopy (TEM) images of *a*-plane GaN grown on an *r*-plane sapphire substrate are shown in Figure 5.5a and b. This is taken under the two-beam condition with $g = [1\bar{1}00]$ (Figure 5.5a) and $g = [0002]$ (Figure 5.5b), perpendicular to the *c* axis. Most of the dislocations have the Burger's vector $\mathbf{b} = \langle 0001 \rangle$. This corresponds to edge dislocation.

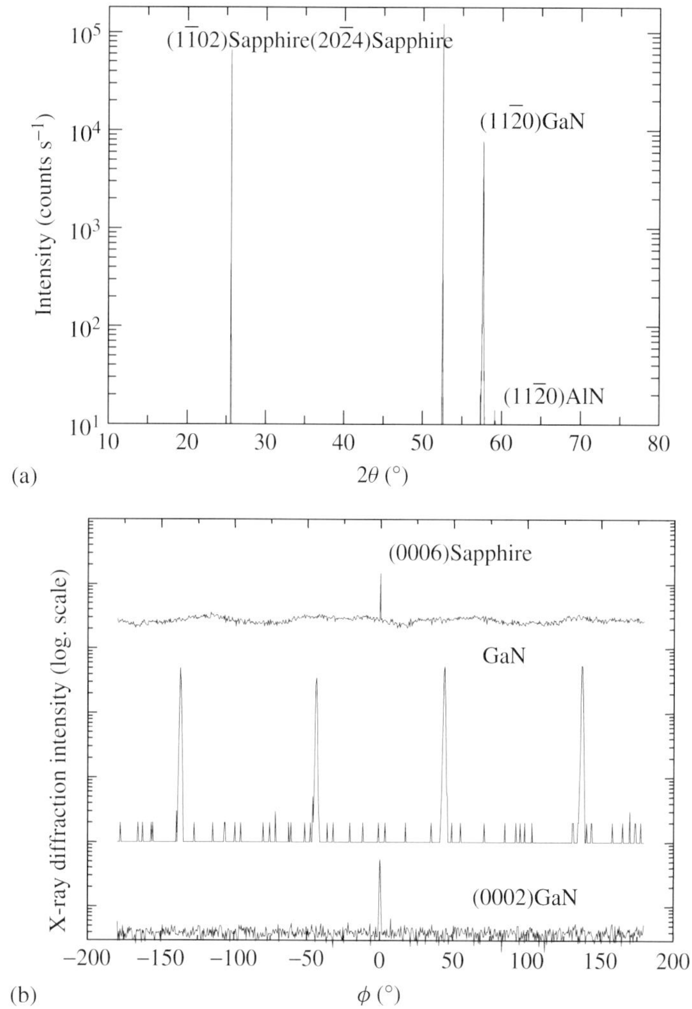

Figure 5.2 (a) Wide-range $2\theta/\omega$ scan profile of GaN on *r*-plane sapphire covered with high-temperature-grown AlN. (b)X-ray diffraction ϕ scan profile of *a*-plane GaN grown on *r*-plane sapphire.

Inserting an $Al_{0.5}Ga_{0.5}N$ interlayer between AlN and GaN is found to be effective in reducing the roughness of the surface. RMS surface roughness measured by atomic force microscopy (AFM) was reduced to 0.7 nm because of the influence of the $Al_{0.5}Ga_{0.5}N$ interlayer. A cross-sectional TEM image of GaN deposited on 1-μm-thick $Al_{0.5}Ga_{0.5}N$ interlayer is shown in Figure 5.6. Bending and coalescence of dislocations are observed. The surface of the $Al_{0.5}Ga_{0.5}N$ interlayer grown on AlN is very rough. Because of this surface

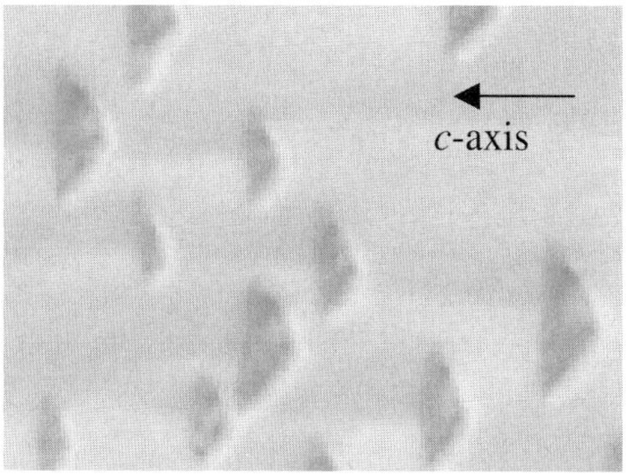

Figure 5.3 Surface pits of the *a*-plane GaN grown at atmospheric pressure on *r*-plane sapphire (cf. Color Plate XXV).

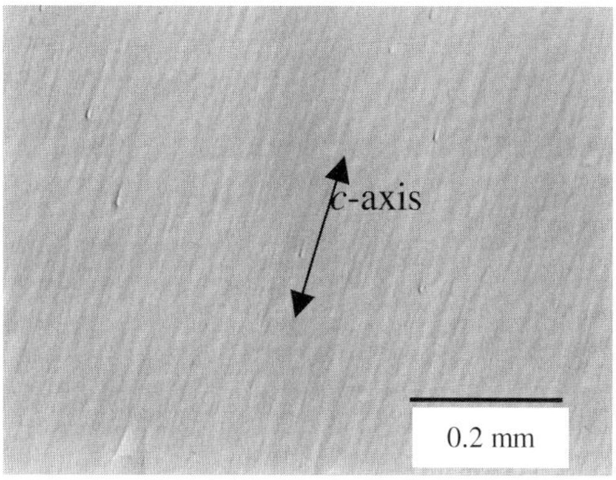

Figure 5.4 Optical microscope image of the surface of GaN grown on *r*-plane sapphire at 100 hPa.

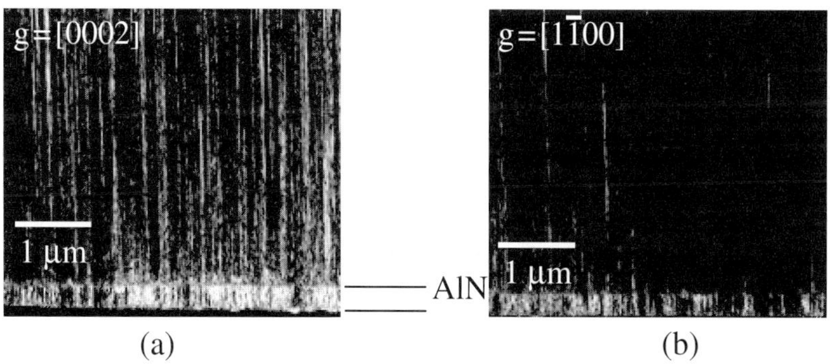

Figure 5.5 Dark-field cross-sectional TEM images of *a*-plane GaN grown on *r*-plane sapphire taken with: (a) $g = [0002]$ and (b) $g = [1\bar{1}00]$.

roughening, bending of the threading dislocations occurred. Figure 5.7 shows the plan-view TEM image of *a*-plane GaN on the $Al_{0.5}Ga_{0.5}N$ interlayer. In addition to the dislocations, stacking faults perpendicular to the *c* axis and each of the side edges terminated by the dislocations are clearly observed. The dislocation density and stacking-fault density of GaN on the AlGaN interlayer are 1.0×10^9 cm^{-2} and 5×10^4 cm^{-1}, respectively.

5.3 Growth and Properties of *m*-plane GaN on *m*-plane SiC

The process of the growth of *m*-plane GaN on *m*-plane 4H-SiC substrates by MOVPE was almost the same as that of *a*-plane GaN grown on *r*-plane sapphire substrates. GaN growth was preceded by the growth of AlN and $Al_{0.5}Ga_{0.5}N$

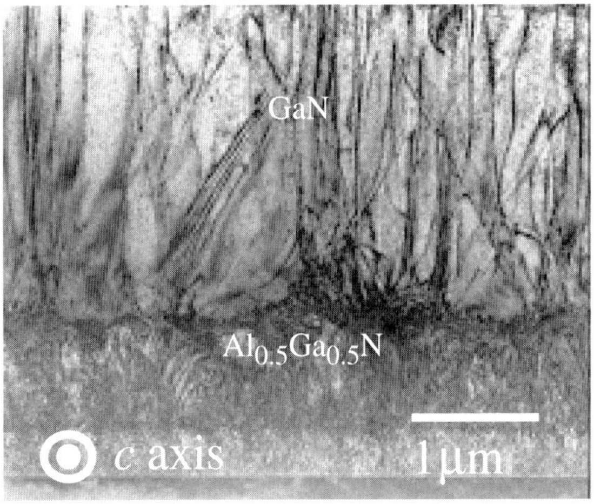

Figure 5.6 Cross-sectional TEM image of *a*-plane GaN grown on *r*-plane sapphire substrate with GaN/$Al_{0.5}Ga_{0.5}N$/AlN structures.

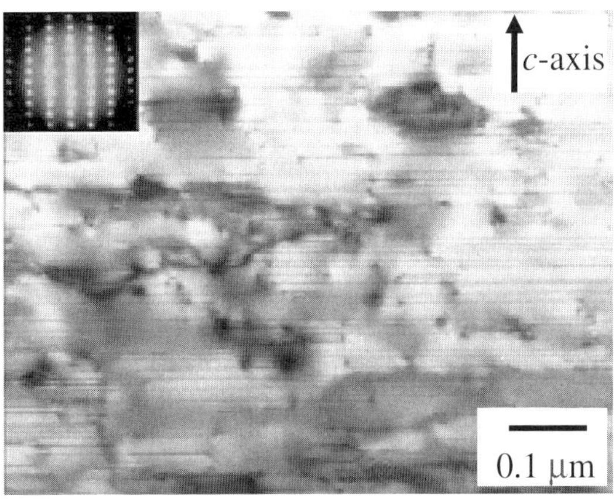

Figure 5.7 Plan-view TEM image of the *a*-plane GaN grown on *r*-plane sapphire using AlGaN interlayer. Threading dislocations and stacking faults can be clearly observed.

buffer layers having thicknesses of 200 and 300 nm, respectively. The growth temperature was 1050 °C, and the growth pressure 200 hPa.

Figure 5.8 shows a bright-field plan-view TEM image of *m*-plane GaN. High-density threading dislocations and stacking faults can be observed.

One of the particular features of *m*-plane GaN grown on *m*-plane SiC is the existence of the inversion domain boundaries. Each inversion domain is surrounded by the stacking faults as shown in Figure 5.9. Arrows in the figure shows the [0001] or $+c$ direction.

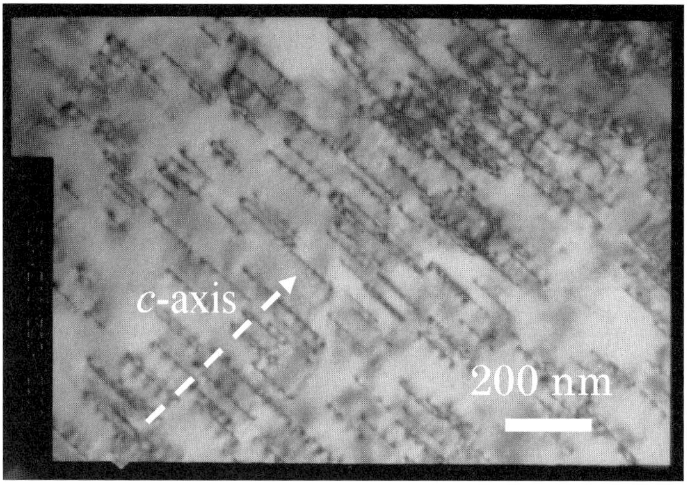

Figure 5.8 Plan view bright-field TEM image of the *m*-plane GaN grown on *m*-plane 4H-SiC with AlGaN/AlN interlayer.

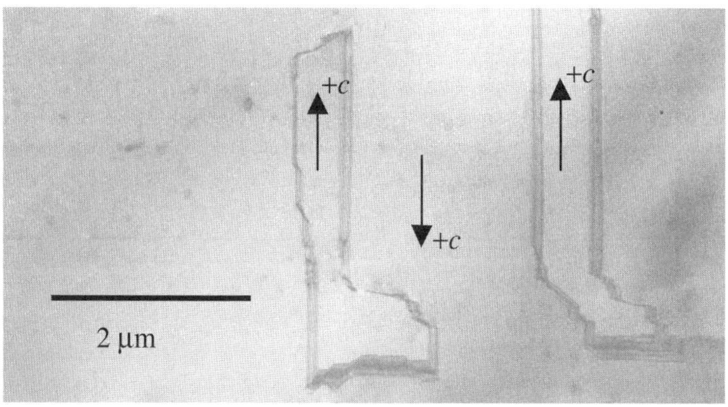

Figure 5.9 Inversion domain boundaries observed in plan view bright-field TEM image of the *m*-plane GaN grown on *m*-plane 4H-SiC with AlGaN/AlN interlayer.

5.4
Growth of GaN on Semipolar (30$\bar{3}$8) 4H-SiC Substrate

As shown in Figure 5.10, the (30$\bar{3}$8) surface of 4H-SiC is tilted 54.7° from the (0001) plane toward the ⟨10$\bar{1}$0⟩ direction, and has an atomic arrangement similar to that of 3C-SiC (001). The (30$\bar{3}$8) face has a relatively small mismatch with (11$\bar{2}$0) and (10$\bar{1}$1) AlN.

GaN was grown on (30$\bar{3}$8) 4H-SiC substrate by MOVPE at different reactor pressures of 1013, 600 and 100 hPa. First, a 100-nm-thick layer of AlN was grown at 1000 °C, then a 1.5-μm-thick GaN layer was grown. Detailed X-ray diffraction and transmission electron diffraction studies showed that the crystal orientation of GaN on 4H-SiC changed markedly with growth pressure.

For GaN grown at 1013 hPa, the GaN axis is exactly aligned with that of ⟨30$\bar{3}$8⟩ of the SiC substrate. In comparison, another relationship can be observed in GaN grown at 100 hPa. The [0001] axis of GaN grown at 100 hPa is inclined at 7.4° to the surface normal. In the case of GaN grown at 600 hPa, both orientations are found. The relationships of these two crystal orientations are schematically shown in Figure 5.11.

To maintain the crystal orientation of the substrate, high-pressure growth is preferable. However, the problem with high-pressure growth is the difficulty in growing GaN with an atomically flat surface. When GaN is grown at 100 hPa, the surface becomes flat. However, when it is grown at 1013 hPa, small pits and macroscopic steps are found. Figure 5.12 shows a SEM image of the surface of GaN at 1013 hPa. In addition, as shown in Figure 5.13, a high density of line contrasts perpendicular to the *c* axis is clearly observed in the cross-sectional TEM image. These lines are thought to be stacking faults.

5.5
Reduction of Dislocation Density and Stacking-fault Density by Sidewall Seeded Epitaxial Lateral Overgrowth

As described previously, the most serious problem of nonpolar/semipolar GaN films on lattice-mismatched substrates is that the films contain many defects.

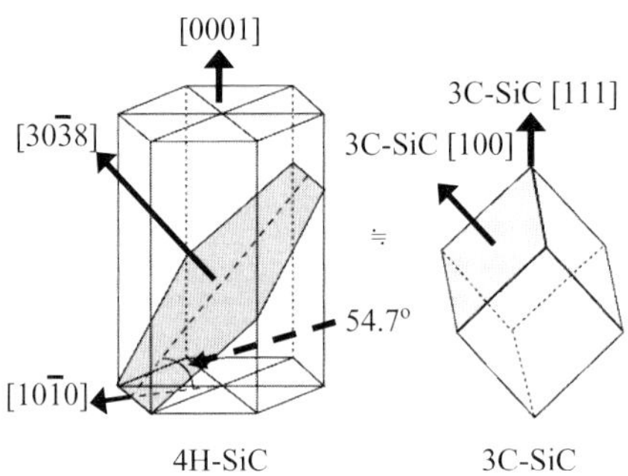

Figure 5.10 Similarity between (30$\bar{3}$8) of 4H-SiC and (100) of 3C-SiC. (cf. Color Plate XXV).

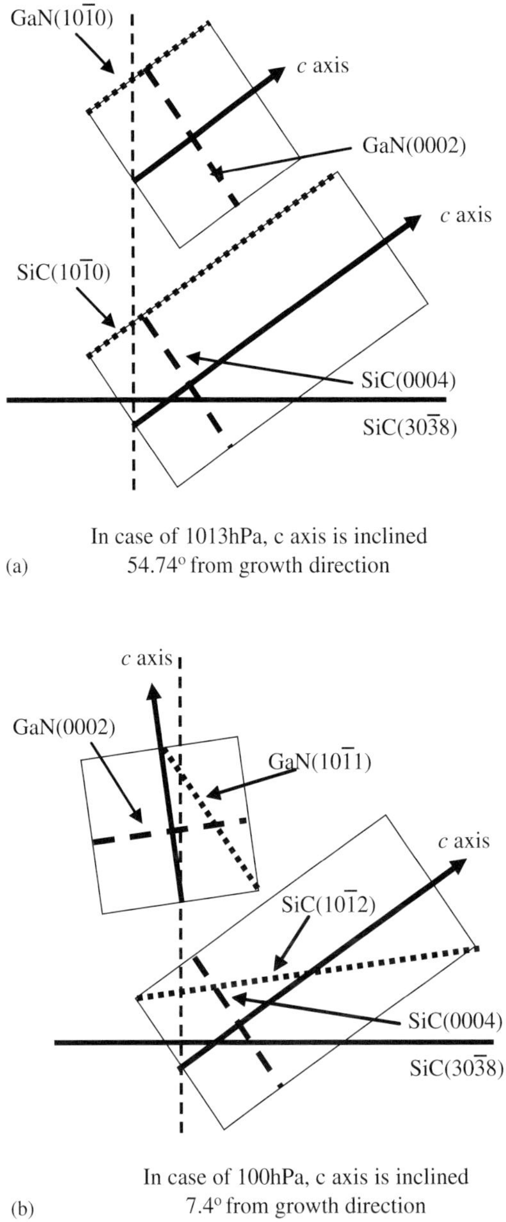

Figure 5.11 Dominant crystal orientation of GaN on (30$\bar{3}$8) 4H-SiC grown at a reactor pressure of (a) 1013 hPa and (b) 100 hPa.

In this section, the details of the growth process and optical characterization of *a*-plane and *m*-plane low-defect-density GaN are described. SELO is proposed and the resultant GaN film is characterized by TEM.

Growth process of nonpolar SELO GaN is schematically presented in Figure 5.14. First, SiO_2 is deposited over nonpolar GaN. Then, the grooved structure is formed by conventional photolithography and reactive ion etching. The preferred stripe direction is perpendicular to the *c* axis.

After etching the template, it is reloaded into the reactor. In the second step, GaN is grown from the sidewall. By controlling the flow rates of TMGa and NH_3, it is possible to grow GaN from only one sidewall because the polarities of the sidewalls are opposite to each other; one is the Ga-face while

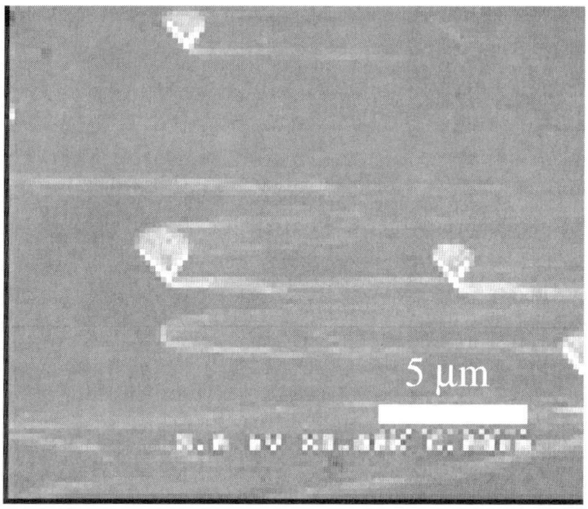

Figure 5.12 SEM image of GaN grown at 1013 hPa.

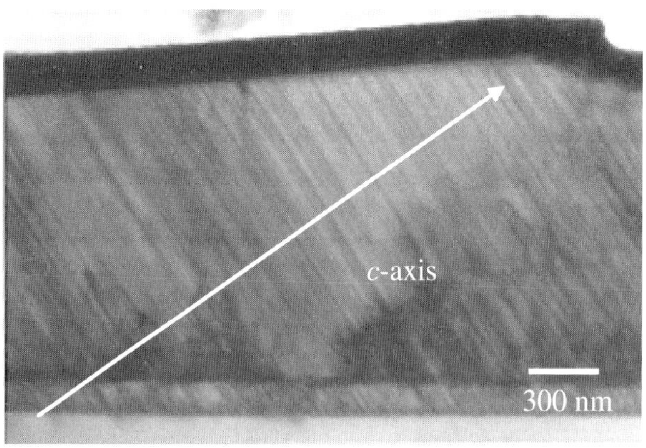

Figure 5.13 Cross-sectional TEM image of GaN grown at 1013 hPa.

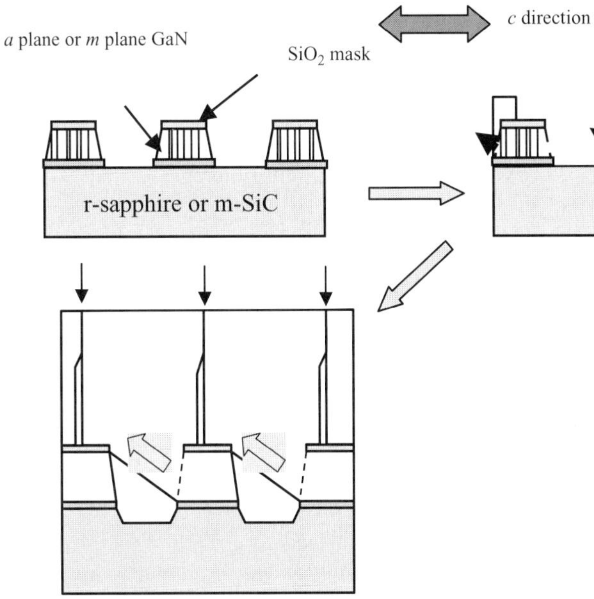

Figure 5.14 Process of sidewall-seeded epitaxial lateral overgrowth of nonpolar GaN grown on *r*-plane sapphire or *m*-plane SiC (cf. Color Plate XXVI).

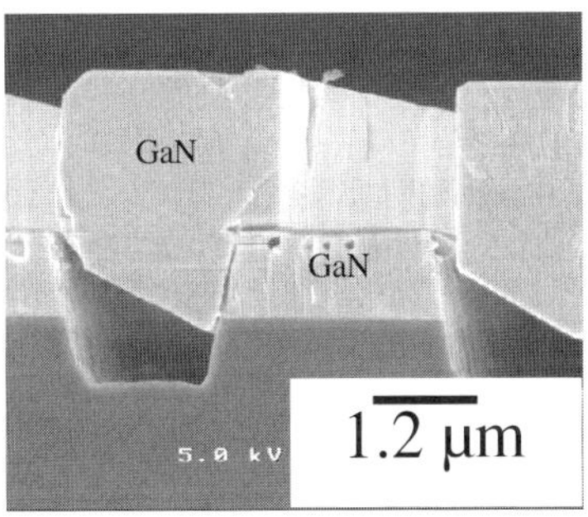

Figure 5.15 Cross-sectional SEM image of *m*-plane GaN. Cracking between GaN seed and lateral overgrown region occurred during cleavage.

the other is the N-face. Figure 5.15 shows the cross-sectional SEM image of the one-sidewall-seeded GaN growth. GaN seeds and the GaN film grown from the seeds are separated during cleavage for SEM observation.

This is followed by GaN coalescence over the groove regions and subsequent spreading over the SiO_2 region in the third step. Defects are effectively prevented from propagating to the layer surface by the SiO_2 mask.

Figure 5.16 shows a cross-sectional TEM image of SELO *m*-plane GaN (Figure 5.16a) and plan-view TEM image of SELO *a*-plane GaN (Figure 5.16b). A reduction in the number of threading dislocations and stacking faults over the whole area can be observed for both nonpolar GaN. Table 5.1 summarizes

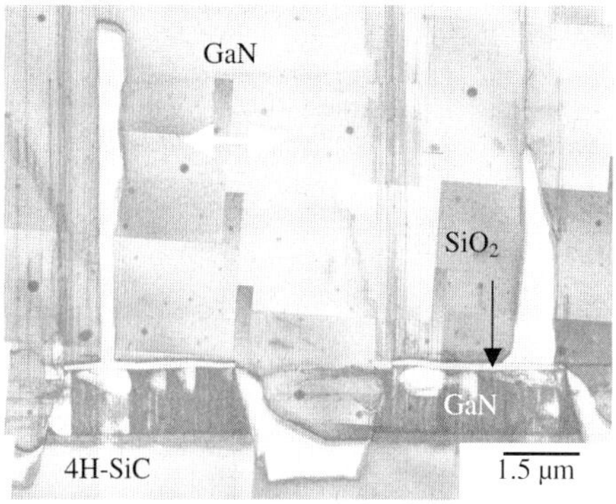

(a)

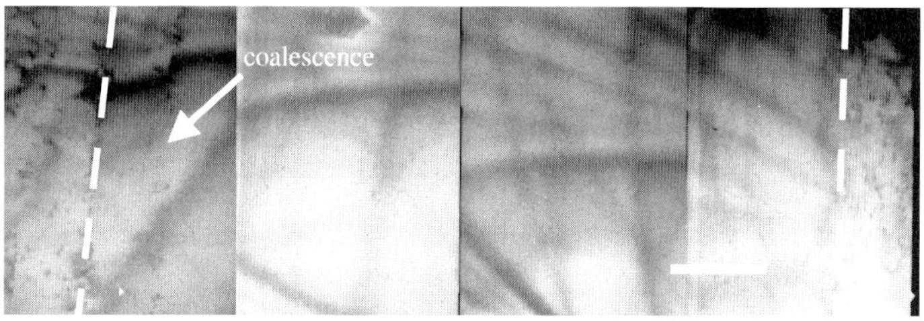

(b)

Figure 5.16 (a) Cross-sectional TEM image of *m*-plane GaN formed with SELO process. Additional threading dislocations propagating from the left wall of the GaN seed can be reduced by etching GaN vertically. (b) Plan-view TEM image of *a*-plane GaN formed with SELO process.

the RMS surface roughness, threading-dislocation density and stacking-fault density of nonpolar *a*-plane and *m*-plane GaN fabricated with and without the SELO process.

5.6
Conductivity Control of Nonpolar GaN

5.6.1
n-type GaN

The control of the electrical conductivity for both *n*-type and *p*-type GaN is the next issue for achieving high-performance LEDs. High series resistance was one of the problems that arose, when we attempted to fabricate LEDs on nonpolar GaN. The origin of the high resistance is the high series resistance of the Si-doped *n*-GaN layer and/or the contact. Figure 5.17 shows the room-temperature electron mobility of *n*-type Si-doped *a*-plane GaN grown on *r*-plane sapphire as a function of electron concentration. Electron mobility deceases rapidly with decreasing electron concentration, which is caused by the scattering at or near the threading dislocations and/or stacking faults. Low-defect-density *a*-plane GaN that is grown with the SELO process shows electron mobility similar to that of *c*-plane GaN. Therefore, it is necessary to reduce the defects density in *n*-type GaN to grow relatively high-electron-mobility GaN.

5.6.2 *p*-type GaN

Compared with electron mobility in *n*-type nonpolar GaN, the hole concentration and hole mobility of Mg-doped *p*-type GaN is not sensitive to the defect density, which is presumably due to the much shorter diffusion length of holes in GaN compared with that of electrons. Figure 5.18 shows the room-temperature hole concentrations of *a*-plane GaN grown on *r*-plane sapphire and that of *c*-plane GaN grown on *c*-plane sapphire as a function of Mg concentration. Hole concentrations become maximum at a certain Mg concentration for both GaN, although the maximum hole concentration is slightly higher in nonpolar GaN. In the case of *c*-plane GaN, inverted pyramids

Table 5.1 Summary of surface roughness, threading-dislocation density (TDD) and stacking-fault density (SFD) of template and SELO-grown nonpolar GaN on *r*-plane sapphire and *m*-plane 4H-SiC.

	a-GaN/*r*-sapphire		*m*-GaN/*m*-4H-SiC	
	Template	SELO	Template	SELO
RMS roughness (nm)	16.0	0.56	7.5	1.0
Threading-dislocation density (cm^{-2})	4×10^{10}	$<10^6$	5×10^{10}	1.2×10^7
Stacking-fault density (cm^{-1})	4×10^5	$<10^4$	6×10^5	$<3 \times 10^4$

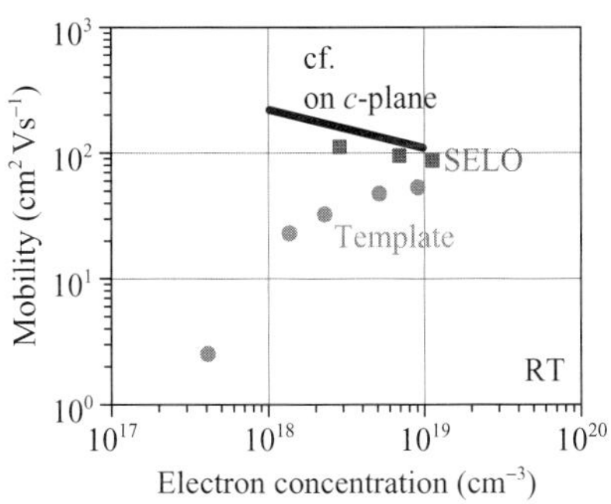

Figure 5.17 Electron mobility at room temperature of *n*-type Si-doped *a*-plane GaN grown on *r*-plane sapphire. Planar GaN grown on AlGaN/AlN intermediate layer is indicated by Template (circles). SELO indicates GaN grown with the SELO process (squares). That of *c*-plane GaN grown on sapphire using a low-temperature buffer layer is also shown for comparison (cf Color Plate XXVI).

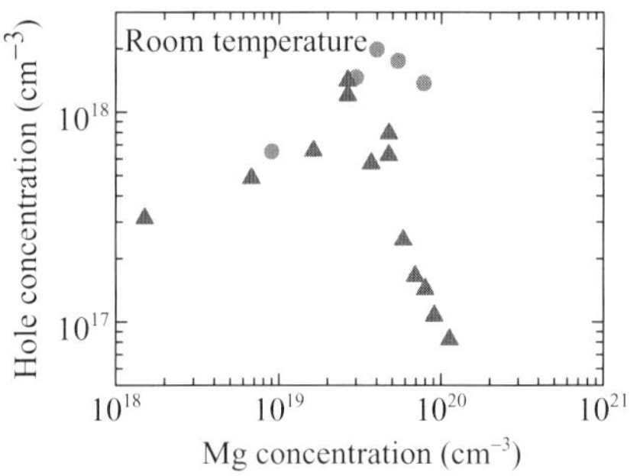

Figure 5.18 Room-temperature hole concentration of *a*-plane GaN grown on *r*-plane sapphire using AlGaN/AlN intermediate layer (red circle) and *c*-plane GaN grown on *c*-plane sapphire using low-temperature buffer layer (blue triangle) (cf. Color Plate XXVII).

are generated when Mg exceeds the optimum concentration; these pyramids are thought to affect the electrical properties of *p*-type GaN. In the case of nonpolar GaN, similar mechanism can be expected for the decrease in hole concentration beyond the optimum Mg concentration.

5.7 Heterostructures

5.7.1 GaInN/GaNMQWs

The optimization of GaInN-based quantum wells is another important issue in achieving high-performance visible LEDs. Quantum wells with a thickness of 3 to 5 nm are commonly used for the GaInN well layer in nitride-based LEDs to screen the influence of the nonradiative components induced by threading dislocations, as well as to screen the effect of piezoelectricity. Therefore, the surface roughness of films should be less than 3 nm. As described previously, RMS surface roughness can be reduced by using the SELO process. It is also important to clarify the difference in the relaxation process of lattice mismatch between the nonpolar plane and the *c*-plane heterostructures. X-ray reciprocal space mapping (RSM) measurement was performed for *a*-plane green LEDs grown with the SELO process. In the case of nonpolar heterostructures, RSM should be measured both parallel and perpendicular to the *c* axis. RSMs parallel and perpendicular to the *c* axis were measured for green LED with GaInN (3 nm)/GaN (12 nm) MQWs. The results are shown in Figure 5.19a and b. Both figures show that the horizontal axis of the GaInN/GaN diffraction spots is aligned exactly with the underlying GaN layer, which means that GaInN wells are coherently grown on the underlying layer.

Figure 5.20 shows the $2\theta/\omega$ scan profiles of the *a*-plane green LEDs grown on template GaN and on SELO-grown GaN. In the case of MQWs on SELO-grown GaN, satellite peaks up to the 5th order can be clearly observed. In comparison, in the case of MQWs on template GaN, the observed satellite

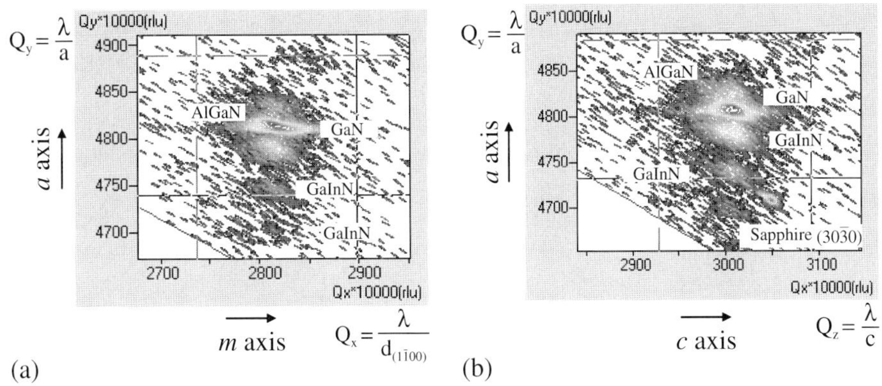

Figure 5.19 RSM of the green LED having GaInN (3 nm)/GaN (12 nm) MQWs. (a) near (20$\bar{2}$0) diffraction, and (b) near (11$\bar{2}$2) diffraction. (cf. Color Plate XXVII).

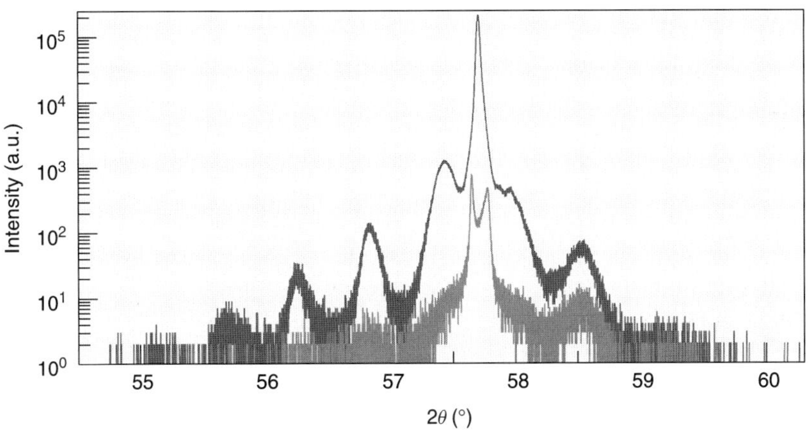

Figure 5.20 $2\theta/\omega$ scan X-ray diffraction profiles from the *a*-plane green LED having GaInN/GaN MQWs. Lower Line: MQWs on *a*-plane GaN template. Upper line: MQWs on SELO-grown GaN. (cf. Color Plate XXVIII).

peaks are 2nd order at most. Surface flatness is very important in fabricating well-controlled MQWs on nonpolar GaN.

5.7.2
AlGaN/GaN Single Heterostructure

To apply nonpolar nitrides to devices such as high-electron-mobility transistors and UV emitters/detectors, AlGaN is an essential material. However, when AlGaN is grown on GaN, cracks are generated by tensile stress when the layer

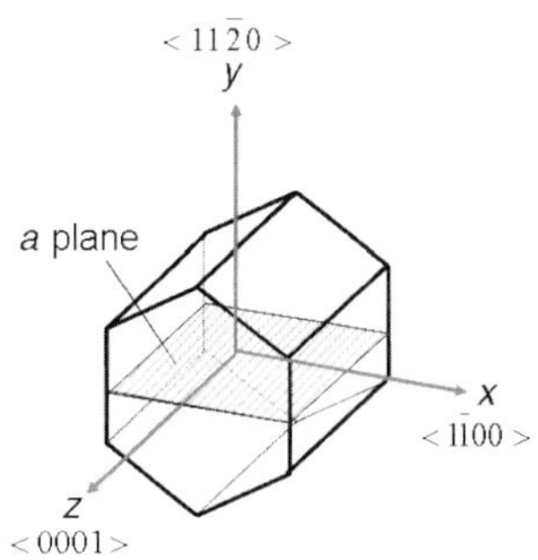

Figure 5.21 Directions of x, y and z axes corresponding to the crystal axes m, a and c. (cf. Color Plate XXVIII).

thickness exceeds a critical value. Therefore, it is essential to fully understand the stress accumulated in the AlGaN layer. The strain in the nonpolar nitride heterostructure is anisotropically biaxial in-plane and complex. In this section, results of the a-plane AlGaN/GaN heterostructure are described.

An a-plane AlGaN/GaN single heterostructure was epitaxially grown on r-plane sapphire by MOVPE using the AlGaN/GaN intermediate layer, as shown previously. The AlGaN layer grown on GaN was approximately 50-nm thick, and the underlying GaN layer was 4-μm thick. The AlN molar fraction x in $Al_xGa_{1-x}N$ alloy was varied from 0.18 to 0.31. Both AlGaN and GaN were nominally undoped. All the films were coherently grown, except for the sample having the AlN molar fraction of 0.31, which was confirmed by the two RSMs.

The directions of the x, y and z axes corresponding to the crystallographic orientations of m, a and c axes, respectively, are schematically defined in Figure 5.21.

Figure 5.22 shows the in-plane stress for a-plane heterostructures (solid squares and triangles) and c-plane heterostructure (open circles) for various AlN molar fractions in the top AlGaN layer. Stress along the c axis in the a-plane heterostructure is higher than that along the m axis. This is caused by the difference in the stiffness constants parallel and perpendicular to the c axis.

5.8
Characterization of Visible LEDs on Nonpolar GaN

LEDs with a GaInN-based MQW active layer on SELO a-plane and m-plane GaN were fabricated. LEDs with the same LED structure were also fabricated on template GaN to make a comparative analysis. Growth conditions adopted for the a-plane and m-plane LEDs were different. a-plane LEDs had 10-pairs of $Ga_{0.82}In_{0.18}N$ (3 nm)/GaN (13 nm) quantum wells, whereas m-plane LEDs had 5-pairs of $Ga_{0.90}In_{0.10}N$ (2.5 nm)/GaN (5.5 nm). The peak wavelengths of a-plane and m-plane LEDs were 500 and 398 nm, respectively. The dependence of the output power on the dislocation density is summarized in Figure 5.23.

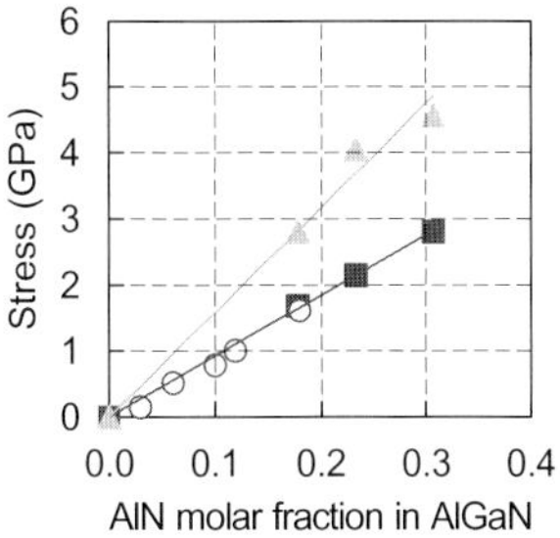

Figure 5.22 Stress in the AlGaN layer coherently grown on GaN. Solid squares: stress along *m* axis in *a* plane. Solid triangle: stress along *c* axis in *a* plane. Open circles: in-plane stress in *c* plane. (cf. Color Plate XXVIII).

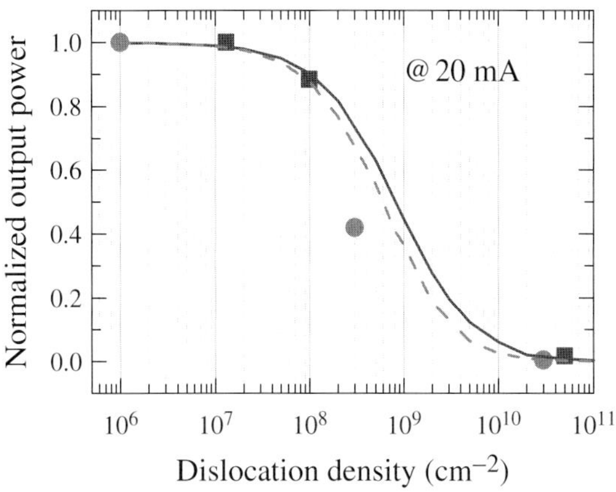

Figure 5.23 Normalized output power of *m*-plane green LEDs (circle) and *a*-plane violet LEDs (square) as a function of dislocation density. Each output power is normalized by the output power of SELO-grown LED. Solid and broken lines show the internal quantum efficiencies of violet and green LEDs as a function of dislocation density. (cf. Color Plate XXIX).

The vertical axis was normalized by the output power of SELO LEDs. The light output power of *a*-plane and *m*-plane LEDs increased with the decrease of threading dislocations from 10^{10} to 10^8 cm^{-2} and tended to saturate when the threading-dislocation density was less than 10^8 cm^{-2}. Simulation results using SiLENSe ver.3 [16], for the internal quantum efficiency of nonpolar green and violet LEDs as a function of threading-dislocation density are also shown for comparison. Simulation results coincide quite well with the experimental results. Hence, we conclude that we can achieve high-efficiency nonpolar LEDs if they are grown on a low-dislocation-density underlying layer.

5.9 Summary

MOVPE growth, electrical conductivity control, heterostructure, and LED performance of the nonpolar nitrides were overviewed. Taking into account the advantages of nonpolar planes over polar planes for use in light-emitting devices and electron devices, the most important issue is the preparation of low-defect-density bulk nitride substrates. For nonpolar nitrides to make an impact on the semiconductor industry, like blue LEDs on *c*-plane sapphire substrates, it is necessary to establish a growth method of reasonable cost for nonpolar nitride substrates.

Acknowledgments

The authors would like to thank Akira Honshio, Hideto Miyake, Hiroko Furukawa, Yoshizane Okadome, Yosuke Tsuchiya, Michinobu Tsuda, and Tetsuya Takeuchi for their great contributions to the experiments.

References

1. Pankove, J.I., Miller, E.A., Richman, D., and Berkeyheiser, J.E. (1971) *Journal of Luminescence*, **4**, 63.
2. Maruska, H.P. and Tietjen, J.J. (1969) *Applied Physics Letters*, **15**, 327.
3. Takeuchi, T., Amano, H., and Akasaki, I. (2000) *Japanese Journal of Applied Physics*, **39**, 413.
4. Bernadini, F., Fiorentini, V., and Vanderbilt, D. (1997) *Physical Review B*, **56**, R10024.
5. Chakraborty, A., Haskell, B.A., Keller, S. *et al.* (2005) *Japanese Journal of Applied Physics*, **44**, L173.
6. Funato, M., Ueda, M., Kawakami, Y. *et al.* (2006) *Japanese Journal of Applied Physics*, **45**, L659.
7. Paskova, T., Kroeger, R., Figge, S. *et al.* (2006) *Applied Physics Letters*, **89**, 051914.
8. http://techon.nikkeibp.co.jp/article/NEWS/20070130/127157/.
9. http://techon.nikkeibp.co.jp/article/NEWS/20070201/127279/.
10. Zheleva, T., Smith, S., Thomson, D. *et al.* (1999) *Journal of Electronic Materials*, **28**, L5.
11. Imer, B.M., Wu, F., DenBaars, S.P., and Speck, J.S. (2006) *Applied Physics Letters*, **88**, 061908.
12. Kawashima, T., Nagai, T., Iida, D. *et al.* (2007) *Journal of Crystal Growth*, **298**, 261.
13. Imura, M., Hoshino, A., Nakano, A. *et al.* (2005) *Japanese Journal of Applied Physics*, **44**, 7418.
14. Lei, T., Ludwig, K.F. Jr, and Moustakas, T.D. (1993) *Journal of Applied Physics* **74**, 4430.
15. Moustakas, T.D., Lei, T., and Molnar, R.J. (1993) *Physica B*, **185**, 36.
16. http://www.semitech.us/.

Further Reading

Honshio, A., Kitano, T., Imura, M. *et al.* (2004) *Materials Research Society Symposium Proceedings*, **831**, 685.

Nagai, T., Kawashima, T., Imura, M. *et al.* (2007) *Journal of Crystal Growth*, **298**, 288.

Tsuchiya, Y., Okadome, Y., Honshio, A. *et al.* (2005) *Japanese Journal of Applied Physics*, **44**, L1516.

Tsuda, M., Furukawa, H., Honshio, A. *et al.* (2006) *Physica Status Solidi B*, **243**, 1524.

6
GaN Films and Quantum Wells with Nonpolar Surfaces: Growth and Structural Properties

Oliver Brandt

6.1
Introduction

The luminous efficiencies of group III nitride light-emitting diodes (LEDs), particularly in the green or ultraviolet wavelength ranges, are significantly decreased by the presence of electrostatic fields within the active layers [1]. These fields are generated by spontaneous and piezoelectric polarization fields [2] along the polar axis of the wurtzite structure, that is, [0001]. Although this effect is useful for certain device applications, for example, heterostructure field effect transistors (HFET) [3–5], it is not desirable for light-emitting devices, as the internal electrostatic field reduces the overlap of the electron-hole wave functions (quantum-confined Stark effect) [1, 6]. The poor overlap of electron-hole wavefunctions also results in an extended radiative life time [7] and consequently low internal quantum efficiency, as there are always some competing nonradiative recombination channels at elevated temperatures.

One solution to eliminate these internal electric fields is the growth of cubic group III nitride heterostructrures [8, 9]. However, even after several efforts [10, 11], the lack of suitable substrates, and the inherent thermodynamic metastability of this phase has held up the achievement of device-quality material. Alternatively, the wurtzite structure of hexagonal GaN causes any direction orthogonal to the unique [0001] axis, such as $[1\bar{1}00]$ and $[11\bar{2}0]$, to become free of spontaneous polarization. In addition, these directions do not carry a piezoelectric polarization if shear stresses in the growth plane are absent. Therefore, due to the resulting absence of electrostatic fields, these directions constitute a potential advantage for fabricating high-efficiency GaN-based LEDs over corresponding [0001]-oriented structures on conventional substrates such as $Al_2O_3(0001)$ and SiC(0001). Waltereit *et al.* first demonstrated an improved internal quantum efficiency of GaN/(Al,Ga)N quantum wells grown along the $[1\bar{1}00]$ direction on γ-$LiAlO_2(100)$, as a result of the absence of internal electrostatic fields across the quantum wells [12]. This triggered a surge in the

Nitrides with Nonpolar Surfaces. Growth, Properties, and Devices. Edited by Tanya Paskova

ISBN: 978-3-527-40768-2

study of group III nitride heterostructures with nonpolar orientation [13–23]. Recently, ultraviolet LEDs based on GaN/(Al,Ga)N($11\bar{2}0$) quantum wells with nonpolar orientation have been realized [24].

In this chapter, we describe the investigation of the synthesis of wurtzite GaN($1\bar{1}00$) films and (In,Ga)N($1\bar{1}00$) multiple quantum wells (MQWs) on γ-$LiAlO_2$(100) by plasma-assisted molecular-beam epitaxy (MBE). First, we present a brief description of γ-$LiAlO_2$, used as substrate in this study, and its preparation prior to growth. Next, we examine how to empirically optimize the growth conditions of GaN($1\bar{1}00$) epilayers. For a better understanding, the adsorption and desorption kinetics of Ga on GaN($1\bar{1}00$) are studied using *in situ* reflection high-energy electron diffraction (RHEED). Finally, growth of (In,Ga)N/GaN($1\bar{1}00$) MQWs and a detailed investigation of their structural and optical properties are described.

6.2 Substrates

Before discussing the growth and properties of GaN($1\bar{1}00$) epilayers, a basic knowledge of γ-$LiAlO_2$, the substrate used in this study, is necessary. In this section, a brief description of its crystal structure and preparation for growth is presented. Selected properties of $LiAlO_2$ (LAO) related to the epilayers are pointed out. Finally, the merits and drawbacks of an alternative substrate [6H-SiC($1\bar{1}00$)] are discussed.

6.2.1 γ-$LiAlO_2$

6.2.1.1 Properties and Merits

γ-$LiAlO_2$ has a tetragonal structure with lattice constants $a = b = 0.51687$ nm and $c = 0.62679$ nm and belongs to the space group $P4_12_12$ [25]. The (100) plane of γ-$LiAlO_2$ exhibits a comparatively small lattice mismatch to GaN($1\bar{1}00$) (the *M*-plane), namely, $[0001]_{GaN} \| [010]_{LAO} = 0.3\%$ and $[11\bar{2}0]_{GaN} \| [001]_{LAO} = 1.7\%$, while conventional substrates such as Al_2O_3 exhibit a much larger lattice mismatch to GaN. Most interestingly, as we mentioned in the introduction, the [$1\bar{1}00$]-oriented GaN grown on γ-$LiAlO_2$(100) can prove advantageous for improving the internal quantum efficiency of light-emitting devices, as it is free of electrostatic fields along the growth direction. In addition, γ-$LiAlO_2$ can be selectively etched with respect to the GaN epilayers, allowing the fabrication of on-chip LEDs.

Figure 6.1 schematically displays the top view of γ-$LiAlO_2$(100) and SiC(0001) for comparison. The nucleation of GaN(0001) on γ-$LiAlO_2$(100) as reported previously [26] is the consequence of the near-hexagonal symmetry on the substrate surface, indicated by the dark shaded areas in Figure 6.1a, and

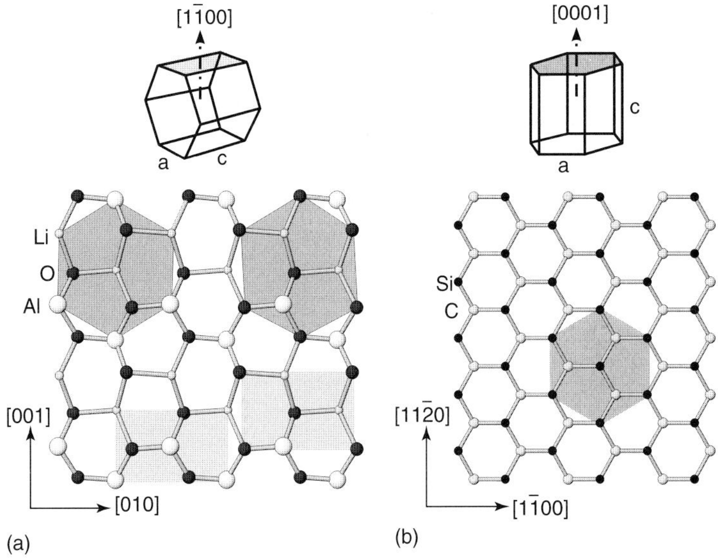

Figure 6.1 Unit cells and ball-and-stick models for the growth of GaN on (a) γ-$LiAlO_2$(100) and (b) SiC(0001). The light and dark shaded areas represent nucleation sites for GaN(1$\bar{1}$00) and GaN(0001).

is favored over GaN(1$\bar{1}$00) for certain nucleation conditions [27]. From pure energy considerations [28], GaN(1$\bar{1}$00) on γ-$LiAlO_2$(100) is more favorable than GaN(0001) on γ-$LiAlO_2$(100).

6.2.1.2 **Drawbacks**

Despite the potential merits of γ-$LiAlO_2$ as a substrate for GaN, as mentioned above, there are a number of drawbacks, which should be borne in mind. For instance, γ-$LiAlO_2$ is inhomogeneously etched by a variety of acids [29], making chemical polishing an intricate task. Even worse, γ-$LiAlO_2$ is hydrolytic [30], and care must be taken to limit the exposure to H_2O to a brief dip. Furthermore, γ-$LiAlO_2$ is thermally much less stable than, for example, Al_2O_3 or SiC, despite its high melting point of about 1700 °C [31]. Lee *et al.* [32] pointed out that the dissociation of γ-$LiAlO_2$ at the typical metal-organic chemical-vapor deposition (MOCVD) growth temperature of 1050 °C might restrict its use, and indeed incongruent evaporation of Li_2O occurs at high temperature [33]. However, we did not observe any dissociation of this material below 1300 °C in vacuum [34].

6.2.1.3 **Orientation Relationship and Microstructure**

Only films grown under certain conditions (see next section) are actually pure *M*-plane and do not contain *C*-plane inclusions. Figure 6.2a and 6.2b

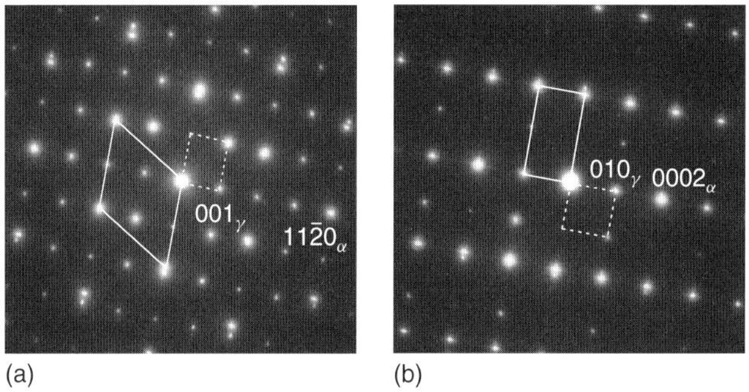

Figure 6.2 SAD pattern along $[0001]_{GaN}$ or $[010]_{LAO}$ (a) and $[11\bar{2}0]_{GaN}$ or $[001]_{LAO}$. α and γ denote the wurtzite GaN film and the γ-$LiAlO_2$ substrate, respectively. The unit meshes of the epilayer and the substrate are indicated by the solid and dashed lines, respectively.

shows selective area diffraction (SAD) patterns of such a film along the GaN[0001] and $[11\bar{2}0]$ directions, respectively. Obviously, the GaN film grows epitaxially on the substrate with the expected orientation relationship, that is, $[0001]_{GaN} \| [010]_{LAO}$ and $[11\bar{2}0]_{GaN} \| [001]_{LAO}$. The coincidence of the diffraction spots from the epilayer and substrate in both figures demonstrates the small lattice mismatch along these two azimuths. In addition, the elongated GaN spots in Figure 6.2b along the [0001] direction reveals a high density of planar defects lying in the basal plane of GaN, which are found to be stacking faults (SFs) [35]. Indeed, SFs and associated partial dislocations with a density of 10^5 cm^{-1} are the dominant structural defects in these layers, whereas perfect dislocations are found with a density of only 10^6 cm^{-2}.

6.2.2 *M*-plane 6H-SiC

6.2.2.1 Properties and Merits

6H-SiC($1\bar{1}00$) has a hexagonal structure with lattice constants $a = b =$ 0.308123 nm and $c = 1.511088$ nm, and belongs to the space group $P6_3mc$, the same as GaN and AlN. The mismatch between SiC($1\bar{1}00$) and GaN($1\bar{1}00$) (the *M*-plane) is, considering an equivalent stacking sequence, $[0001]_{GaN} \| [0001]_{SiC} = 2.9\%$ and $[11\bar{2}0]_{GaN} \| [11\bar{2}0]_{SiC} = 3.4\%$. These values are higher than those for γ-$LiAlO_2(100)$, but still considerably lower than for, for example, *A*-plane GaN on *R*-plane sapphire.

Further advantages include the fact that SiC($1\bar{1}00$) is nonpolar and that substrates are available epiready. Finally, SiC is a thermally robust and chemically inert material, and problems related to dissociation as in the case of $LiAlO_2$ are absent.

6.2.2.2 Drawbacks

Both GaN and AlN crystallize, in equilibrium, in the wurtzite structure, which exhibits an ABAB stacking sequence. 6H-SiC, however, exhibits a ABCACB stacking sequence, which is exposed to GaN nucleating on the $(1\bar{1}00)$ surface. This fact, as we will see below, constitutes a major obstacle for the realization of high-quality GaN films on SiC$(1\bar{1}00)$.

A further disadvantage is the rapid strain relaxation occurring in GaN films on SiC$(1\bar{1}00)$. The residual strain present even in thin GaN layers is negligibly small in comparison to that in GaN films on γ-LiAlO$_2$(100). As we will see in the subsequent chapter by Grahn, the comparatively large strain in the latter films is crucial for applications of these films for polarization-sensitive devices. Finally, for these applications, it is often desirable that the substrate is transparent at the energy of the GaN band gap, a property that 6H-SiC does not possess.

6.2.2.3 Orientation Relationship and Microstructure

First, we stress that we did not investigate the growth on *M*-plane SiC as systematically as that on γ-LiAlO$_2$(100). However, in all of our experiments, the GaN films were found to be pure *M*-plane regardless of nucleation conditions.

Figure 6.3 shows a high-resolution transmission electron micrograph of a GaN/6H-SiC$(1\bar{1}00)$ interface. The orientation relationship is, as expected, $[0001]_{\text{GaN}} \| [0001]_{\text{SiC}}$ and $[11\bar{2}0]_{\text{GaN}} \| [11\bar{2}0]_{\text{SiC}}$. Evidently, the stacking sequence in the GaN layer is quite disordered. Some nanometer-sized regions are actually of 6H stacking, but overall we see that the layer contains a very high density of SFs [far higher than that for GaN films on γ-LiAlO$_2$(100)]. This observation is found to be quite independent of the nucleation conditions. Even AlN nucleation layers, which are observed to undergo a polytype replication and to consequently exhibit a very high quality, do not change this massive

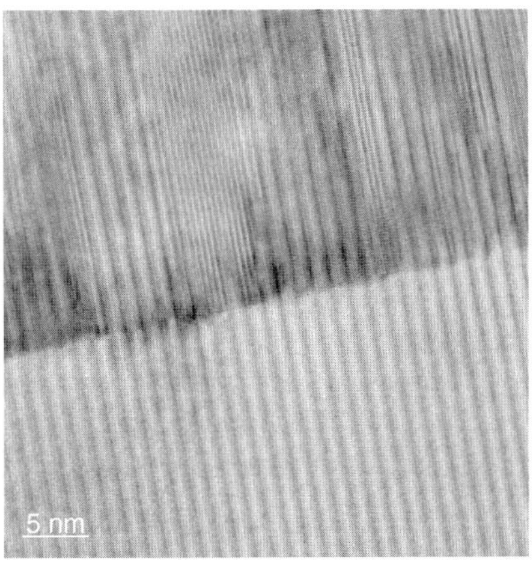

Figure 6.3 High-resolution transmission electron micrograph of a GaN/6H-SiC$(1\bar{1}00)$ interface. Note the very high density of stacking faults.

generation of SFs at the AlN/GaN interface. Apparently, GaN is not as tolerant of the 6H polytype as AlN. The solution to this problem is likely to be found in the synthesis of a 2H-AlN nucleation layer with a comparatively low density of SFs, and we have found indications that this might indeed be possible. Until then, however, the structural and morphological properties of our GaN films on SiC $(1\bar{1}00)$ are far inferior to those of their counterparts on γ-$LiAlO_2(100)$. Consequently, the rest of this chapter is devoted to films on the latter substrate, but the results are certainly generally valid.

6.3
Optimization of the Heteroepitaxy of GaN Films on γ-$LiAlO_2(100)$

There are only a few reports on group III nitride heterostructures with nonpolar orientation [13–23], and a systematic investigation of their growth is still lacking. This section presents a study of the impact of nucleation conditions on the growth of *M*-plane GaN$(1\bar{1}00)$ films with high phase purity. The optimum growth conditions for obtaining reproducibly high-quality *M*-plane GaN are then established empirically.

6.3.1
Substrate Preparation and Impact of Polarity

The γ-$LiAlO_2$ substrates used in this study were grown by a modified Czochralski method at the Institute for Crystal Growth in Berlin, Germany, and were chemo-mechanically polished. The peak-to-valley (P/V) roughness and root-mean-square (RMS) roughness of the polished substrates have values of 1.0 nm and 0.08 nm, respectively, across a $3 \times 3\ \mu m^2$ area. These values compare favorably with those obtained for state-of-the-art polished SiC or Al_2O_3 substrates.

Substrate preparation deserves utmost attention for obtaining reproducibly high-quality epilayers. Usually, the substrate is subject to chemical preparation before it is loaded into the growth reactor. For example, for SiC(0001) substrates, a dramatic improvement of the surface quality is attained upon annealing the wafers to about 1600 °C in flowing H_2 [36]. Etched SiC wafers exhibit a well-ordered stepped surface.

Since LAO is attacked by various acids and even water, well-established cleaning processes used for other substrates are not applicable. In fact, we have not yet found any effective method, such as thermal or chemical etching, to improve the surface morphology of LAO. The as-received substrates are thus only degreased sequentially in trichloroethylene (10 min), acetone (5 min), and methanol (5 min) in an ultrasonic bath. Finally, the substrate is dipped briefly (30–60 s) in deionized water. The LAO substrate is then fixed with In onto a Si wafer clipped to a Mo holder. Prior to growth, the substrate is outgassed in the load-lock chamber for 1 h at 200–300 °C. During mounting and loading, the substrate surface is again exposed to air and thus recontaminated. Although an outgassing procedure can remove part of these contaminants, especially water, it would be desirable to have a method for a further cleaning of the substrate inside the MBE system. For SiC(0001), Kaplan proposed that Ga deposition and subsequent desorption is effective in removing O-related contaminants [37]; this was later confirmed by Strite *et al.* [38] The procedure consists of the deposition of Ga onto the surface where it reacts with contaminants to form a volatile compound, for example, Ga_2O. An annealing step at 750 °C desorbs these products and excess Ga remaining on the surface. However, for LAO, this procedure was found to merely induce an excessive roughening of the substrate, and the structural properties of GaN films grown on such a substrate were actually degraded.

Considering the polar nature of LAO(100), we first investigate the influence of its polarity on the growth of *M*-plane GaN. While we cannot yet provide a microscopic identification of the two polarities, it is at this stage, perhaps, more important to be able to phenomenologically distinguish between them by simple chemical etching. Figure 6.4 shows the surface morphology of either face after a 30 min exposure to HCl(32%) at 60 °C, recorded by a Zeiss interference contrast microscope. The most obvious features in both micrographs are rhomboidal etch pits, which we believe to occur at locations where dislocations in LAO intersect the surface. The elongated side of the etch pits is along the [010] direction. Indeed, the distinct linear arrangement of these etch pits was attributed to subgrain boundaries formed by dislocations [39]. More important, however, is the fact that in between these etch pits, the surface of face A remains relatively smooth, whereas face B becomes clearly corrugated. Note, however, that a longer treatment or higher temperatures tend to obscure this difference due to the resulting excessive roughening of both faces.

The influence of the polarity of the substrate on the epilayer is checked by growth on two opposite sides of LAO(100). Figure 6.5 shows triple-crystal ω–2θ X-ray scans for two films grown on each face under the same conditions.

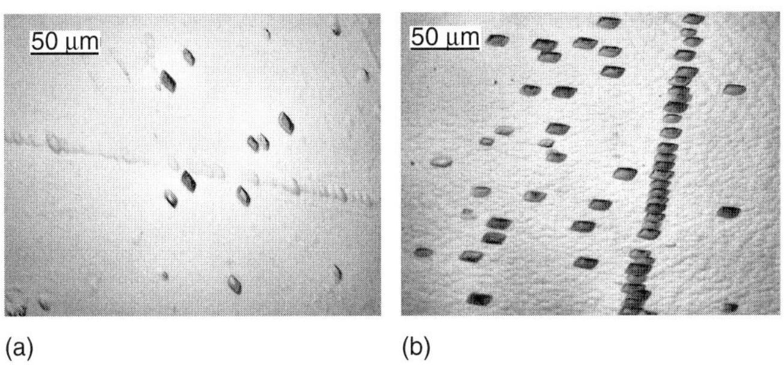

Figure 6.4 Interference contrast micrographs of the surface of (a) side A and (b) side B of LAO(100) exposed to HCl(32%) at 60 °C for 30 min.

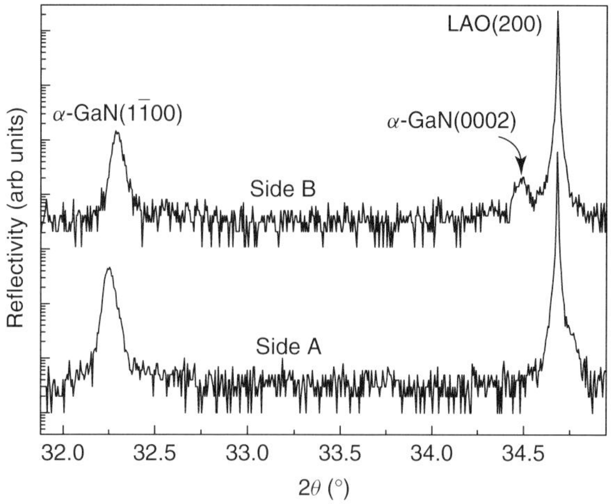

Figure 6.5 ω–2θ triple-crystal X-ray scans of two samples grown on two opposite sides of LAO(100), but under the same conditions.

In both profiles, the peaks located at $2\theta = 34.682°$ and $32.254°$ are due to the LAO(200) and GaN(1$\bar{1}$00) reflections, respectively. The film grown on face B exhibits an additional weak peak to the left of the LAO(200) diffraction peak, namely, at 34.5°. While this angular position coincides with that of either wurtzite GaN(0002) or cubic GaN(111), the RHEED pattern of this sample shows additional reflections stemming from GaN(0001). However, no such peak is detected on the film grown on face A, indicating a high phase purity of this film. Pure *M*-plane GaN epilayers can be reproducibly synthesized on face A if the nucleation conditions are appropriately chosen (see Section 6.3). On the basis of this finding, we have concentrated on layers grown on face A.

6.3.2
Impact of Nucleation Conditions on Phase Purity

Nucleation is generally believed to be the most critical step for obtaining high-quality *C*-plane GaN layers [40–42]. This is thought to be also applicable for the growth of *M*-plane films.

6.3.2.1 Growth

Growth was carried out in a custom-designed, solid-source three-chamber MBE system equipped with a radio-frequency plasma source. The N source is operating at a plasma power of 250–500 W with an N_2 flow of 0.3–0.4 sccm, yielding a GaN growth rate of 0.25–0.4 μm/h. The base pressure in the growth chamber is kept below 10^{-10} Torr by an ion-getter pump and a Ti-sublimation pump. During growth, the chamber is cryo-pumped and has a pressure around 1.5×10^{-5} Torr for a typical N_2 flow of 0.3 sccm. The growth front is monitored *in situ* by RHEED using a 20 kV electron gun. The diffraction pattern is recorded using a CCD camera.

6.3.2.2 RHEED

First of all, it should be stressed that RHEED allows the observation of the phase purity of the layers in real time. During nucleation, a spotty pattern appears, which in itself allows for a judgment of the success of the nucleation procedure, since the symmetries of *M*- and *C*-plane GaN are distinctly different. Further growth at Ga-stable conditions yields a streaky RHEED pattern, reflecting the progressive smoothening of the surface. Phase mixture manifests itself in the occurrence of additional reflections (typically spots), which are readily distinguishable from the 'clean' *M*-plane pattern. In addition, RHEED patterns can also reveal the in-plane orientation relationship between epilayer and substrate.

6.3.2.3 XRD

Triple-crystal high-resolution X-ray diffraction (HRXRD) provides an *ex situ* nondestructive tool to check the relationship of corresponding out-of-plane orientations between the epilayer and the substrate and simultaneously, the phase purity as well. In the text that follows, we describe the phase purity of the samples and their surface orientation as examined by XRD.

6.3.2.4 Impact of Nucleation Conditions on Phase Purity

We have described how specific nucleation conditions are required for the growth of *M*-plane GaN [27, 34]. The results of our investigations are summarized in Table 6.1, which also includes particulars of the respective nucleation condition and phase purity obtained.

The nucleation regime can be divided into three groups, namely, the insertion of an AlN buffer, nitridation prior to growth, and direct deposition. The growth of AlN buffer was initiated at 580 °C under different stoichiometry. Then, the growth of GaN was performed under slightly Ga-rich conditions at high temperatures. Nitridation was carried out by exposing the substrate to the N plasma at 350 W and 660 °C for 40 min. Subsequent nucleation was executed under either N-rich or Ga-rich conditions. The initial stage of direct deposition of GaN (nucleation) was performed at various temperatures and under different stoichiometry. In all cases, the nucleation temperature T_N can be divided into four regimes: high (740 °C, actually the growth temperature T_G), intermediate (660 °C), low (580 °C) and very low (510 °C). The thickness of the AlN buffer or the GaN nucleation layers was about

Table 6.1 Nucleation conditions for the samples investigated. The table provides information on whether nitridation was employed or not (yes √/no –), on the stoichiometry (N-rich/Al-rich or Ga-rich), and on the temperature T_N during nucleation. The phase purity of the layers (M: *M* plane, C: *C* plane) and the RMS roughness obtained by atomic force microscopy (AFM) over 5 × 5 μm^2 are also indicated. All films listed have a thickness of 500–600 nm.

Sample	AlN buffer	Nitridation	Stoichiometry	T_N (°C)	Phases	RMS (nm)
A	√	–	N	580	C	n/a
B	√	–	Al	580	C	n/a
C	–	√	N	660	M/C	80
D	–	√	Ga	660	M/C	15
E	–	–	N	660	C/M	38
F	–	–	Ga	740	M	23
G	–	–	Ga	660	M	7.5
H	–	–	Ga	580	M	0.8
I	–	–	Ga	510	M	3.3

100 nm. For all samples, the growth temperature T_G of 740 °C was kept constant during the subsequent growth, and the stoichiometry was set to slightly Ga-rich. During the initial deposition, regardless of the nucleation conditions, a spotty RHEED pattern first appears, while further growth at Ga-stable conditions yields a streaky RHEED pattern [27], reflecting the progressive smoothening of the surface. RHEED allows us to judge the orientation of the growth front, since the symmetries of *M*- and *C*-plane GaN are distinctly different. A phase mixture manifests itself in the occurrence of additional reflections (typically spots), which are readily distinguishable from the 'clean' *M*-plane pattern. Several conclusions can be drawn from this table.

(1) It is often reported that the use of an AlN nucleation layer has a positive impact on the properties of subsequently grown *C*-plane GaN [43, 44]. However, as can be seen from Table 6.1, the samples grown on an AlN buffer layer had a predominant *C*-plane orientation regardless of the nucleation conditions. To investigate the origin of this effect, 200-nm thick AlN layers deposited at the same temperature as the buffer used in sample A and B were grown and measured by XRD. It was found that the AlN layers are [0001]-oriented, while no peak pertinent to the *M*-plane orientation was detected [34]. AlN nucleation layers (at least for our nucleation conditions) are thus clearly detrimental for the growth of *M*-plane GaN layers.

(2) Considering that nitridation of Al_2O_3 substrates prior to nucleation is well established to improve the crystal

quality of the subsequent GaN film by the formation of an AlN template [45–48], we have studied the nitridation of LAO as a potential means to improve epitaxial growth also in this case. Apparently, nitridation of LAO prior to growth invariably induces the phase mixture of *C* and *M* plane, as seen from Table 6.1.

(3) Though N-rich growth has been established to be disadvantageous for the PAMBE growth of *C*-plane GaN epilayers on either Al_2O_3 [49], SiC [50] or GaN [51], particularly concerning the surface morphology, it is not clear whether N-rich nucleation is unfavorable for *M*-plane films. As demonstrated, the immediate N-rich nucleation conditions indeed leads to predominant *C*-plane GaN [34].

(4) Clearly, the nucleation conditions are crucial for the out-of-plane orientation of GaN epilayers on LAO(100). Our study shows that only immediate Ga-rich nucleation results in the growth of pure *M*-plane GaN [27, 34].

6.3.3
Influence of Nucleation Temperature on Surface Quality

The surface morphologies and microstructure of *M*-plane GaN films with different nucleation temperatures are studied by AFM (cf. Ref. 34) and transmission electron microscopy (TEM), respectively [35].

6.3.3.1 Roughness versus T_N

The P/V and RMS roughness of the pure *M*-plane GaN are shown versus the nucleation temperature in Figure 6.6. Obviously, the impact of T_N on the surface morphology is pronounced. Films grown at low nucleation temperatures (580 °C) are significantly smoother than those grown at high nucleation temperatures. A nucleation temperature of 580 °C is apparently the optimum one. A further reduction of the nucleation temperature to 510 °C does not result in a further reduction of the surface roughness, on the contrary, the films are significantly rougher than those nucleated at 580 °C. Chen *et al.* [17] recently reported an RMS roughness of 0.7 nm over $2 \times 2\ \mu m^2$ for a 4-μm-thick homoepitaxial *M*-plane GaN layer grown by MOCVD. This value is similar to the value of 0.6 nm for the best samples in Figure 6.6 over the same area, demonstrating a virtually identical degree of roughness. This degree of roughness is close to that obtained on *C*-plane films and is well suited for the growth of heterostructures, such as quantum wells [52].

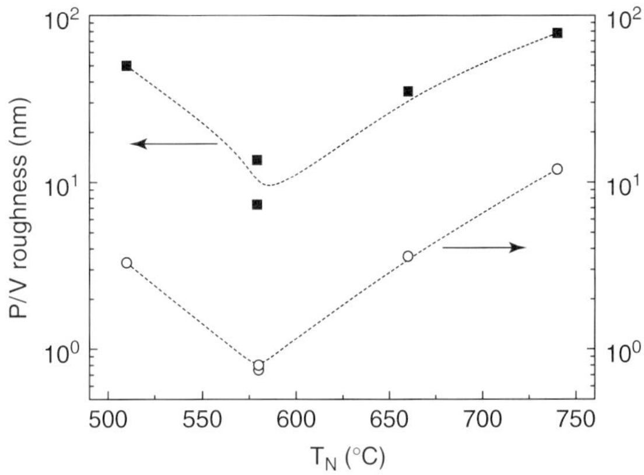

Figure 6.6 P/V (solid circles) and RMS (open circles) roughness over an area of $5 \times 5\ \mu m^2$ as a function of the nucleation temperature. The dashed lines are a guide to the eye. The thicknesses of all samples are 500–600 nm.

6.3.3.2 TEM

To examine the microstructure of the layers, we employ cross-sectional TEM performed in a JEOL3010 microscope operating at 300 kV. Figure 6.7 shows a two-beam cross-sectional bright-field micrograph of one typical *M*-plane GaN epilayer with $\mathbf{g} = 1\bar{1}00$ near the $[11\bar{2}0]$ zone axis. The microstructure of *M*-plane GaN is quite different from that of *C*-plane GaN. The density of perfect threading dislocations is found to be below $1 \times 10^6\ cm^{-2}$. The most dominant defects are basal plane SFs, most of which intersect the entire thickness of the layer. The partial dislocations terminating these basal plane SFs have been determined to be of Shockley type with Burgers vectors $\mathbf{b} = \frac{1}{3}[10\bar{1}0]$ or $\frac{1}{3}[01\bar{1}0]$ [53]. A recent investigation suggests that coalescence of two nucleation regions with different stacking sequences is the reason for the formation of SFs [54], as three $\langle 1\bar{1}00\rangle$ faces have ABAB stacking and the other three have ACAC stacking. Nevertheless, it is clear that a low nucleation temperature results in a significantly improved microstructure and surface morphology [35].

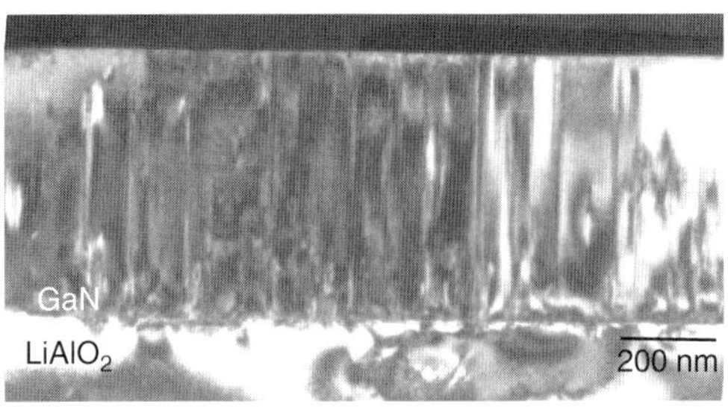

Figure 6.7 Cross-sectional $\mathbf{g} = 1\bar{1}00$ two-beam bright-field TEM micrograph near the $[11\bar{2}0]$ zone axis of a GaN film grown on an LAO substrate.

6.4
Ga Adsorption and Desorption Kinetics

In the case of GaN(0001), several studies have established the optimum growth conditions in MBE and related them to the existence of a Ga bilayer on the growth front [55–59]. The lateral mobility of the Ga atoms within the bilayer is predicted to be high, consistent with the high morphological quality of GaN(0001) grown under conditions giving rise to bilayer coverage. Empirically, we have found that Ga-rich conditions results in GaN($1\bar{1}00$) layers with a comparatively smooth surface morphology, whereas N-rich conditions lead to a roughening of the growth front. Recently, Lee *et al.* [18] observed a reconstruction of approximate symmetry '(4 × 5)' for Ga-rich grown GaN($1\bar{1}00$) on ZnO by scanning tunneling microscopy. The surface was found to be metallic in nature, and the authors proposed that it consists of more than 2 monolayers (MLs) of Ga based on the fact that Ga-adlayer structures with coverages of 2–3 MLs on GaN($1\bar{1}00$) are energetically favorable for Ga-rich conditions. In this section, we present a study of the adsorption and desorption kinetics of Ga on GaN($1\bar{1}00$) using RHEED.

6.4.1
Surface Reconstruction

In the course of our previous work [27], we noticed that all smooth layers exhibited a complex RHEED pattern when the surface was quenched to about 150 °C with a cooling rate of 40 °C/min and the N flux shut off immediately at the termination of growth. The RHEED pattern observed exhibits a fourfold periodicity along all major azimuths, consistent with a (4 × 4) reconstruction with respect to the unit mesh of the GaN($1\bar{1}00$) surface. Note that this reconstruction is not identical to the pseudo-(4 × 5) reconstruction reported by Lee *et al.* [18] which gave rise to a fourfold pattern along the [0001] azimuth only. In the following, we establish a relation between the adsorbed Ga coverage and reconstructions of the surface.

6.4.1.1 Relation between Ga Coverage and Surface Reconstructions

The growth front was monitored *in situ* by RHEED at 20 kV energy and an angle of incidence of 1.5°. The diffraction patterns along all major azimuths of GaN($1\bar{1}00$) , that is, [$11\bar{2}0$] (0°), [$22\bar{4}3$] (≈39°), [$11\bar{2}3$] (≈58°), and [0001] (90°), were recorded using a CCD camera. The intensity of the specular spot and other regions of interest were recorded using a digital RHEED analysis system [60], which was set to sample data at a rate of 25 Hz. Growth was initiated by nucleation at 580 °C under Ga-rich conditions. The growth temperature was then raised to 740 °C and kept constant during the subsequent adsorption/desorption studies and growth. All of these studies were performed using a 500-nm thick GaN(1$\bar{1}$00) layer grown under standard conditions, that is, Ga-stable growth ($\phi_{Ga} = 0.475$ ML s^{-1}, $\phi_N = 0.275$ ML s^{-1}) and a substrate temperature of 740 °C.

During growth of *M*-plane GaN under standard conditions, RHEED shows the characteristic (1 × 1) pattern [27]. The pattern is entirely streaky, but dim, very similar to the RHEED pattern observed during growth of *C*-plane GaN. Upon interrupting the growth for about 30 s at constant temperature with the N flux turned on, the intensity of the pattern increases strongly, while the (1 × 1) symmetry does not change. The RHEED pattern of the recovered surface does not exhibit any appreciable change when the N flux is switched off, nor does it change upon subsequently cooling the sample down to 100 °C. Upon deposition of up to 4 MLs of Ga on this surface at 100 °C, the only noticeable change is that the RHEED pattern becomes dim, similar to that observed during growth. This finding is in marked contrast to the observation of a (4 × 4) reconstruction for the surface quenched to low temperatures immediately after growth. Apparently, the thermal energy at 100 °C is insufficient for the Ga adatoms to condense into this reconstruction.

To study the formation of the (4 × 4) reconstruction even further for this sample, the excess Ga deposited at 100 °C is first flashed off by heating the sample to 740 °C. The sample is then cooled down to 580 °C, at which the desorption of Ga is negligible. Next, we deposit 1 ML of Ga onto this surface. Upon cooling to 100 °C, clear half-order streaks are observed along all azimuths except for [11$\bar{2}$0], consistent with a (1 × 2) reconstruction of GaN(1$\bar{1}$00) . Upon heating, the half-order streaks abruptly disappear at a temperature of 250 °C, but the pattern stays dim in comparison to that of the recovered surface. An additional ML of Ga is then deposited at 580 °C, which gives rise to a more intense (1 × 2) reconstruction after cooling to 100 °C as shown in Figure 6.8. However, we often observe the formation of 1/4-order streaks [i. e. next to the

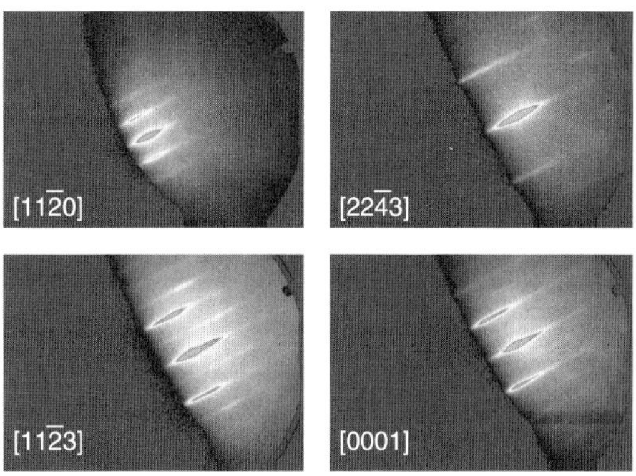

Figure 6.8 RHEED patterns of the GaN(1$\bar{1}$00) film recorded at 100 °C after a two-step deposition of 2 MLs Ga at 580 °C. Note the intense half-order streaks along all azimuths (cf. Color Plate XXIX).

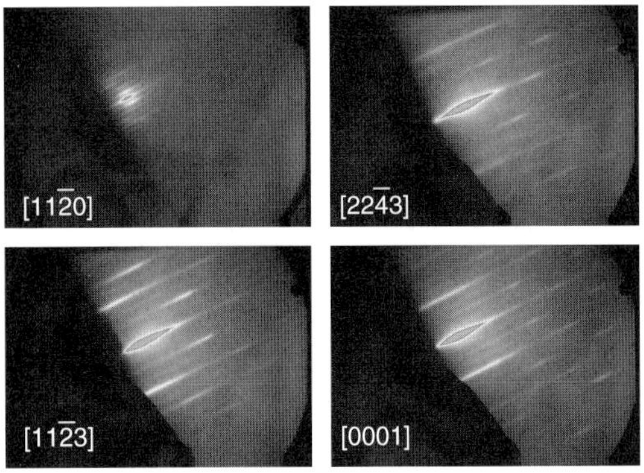

Figure 6.9 RHEED patterns of the GaN(1$\bar{1}$00) film recorded at 100 °C after a three-step deposition of 3 MLs Ga at 580 °C. Note the intense 1/4-order streaks along all azimuths (cf. Color Plate XXX).

(00) streak] along the [11$\bar{2}$0] azimuth at this coverage, which would indicate a (4 × 2) reconstruction. This observation might indicate a chainlike ordering of (1 × 2)-reconstructed domains.

The sample is then heated to 580 °C again. Finally, the procedure is repeated for the deposition of an additional ML of Ga, resulting in a total of 3 MLs of Ga on the surface. Figure 6.9 shows the resulting RHEED pattern at 100 °C, which exhibits an intense (4 × 4) reconstruction, identical to that observed when quenching the sample after growth at standard growth conditions. We have checked that the deposition of 2 and 3 MLs of Ga in one step at 580 °C results in RHEED patterns which are indistinguishable from those displayed in Figures 6.8 and 6.9, respectively. Deposition of noninteger amounts, such as 2.5 MLs, often results in a superposition of both reconstructions, in that a fourfold periodicity is observed along [11$\bar{2}$0], but a twofold one along all other azimuths. Deposition of more than 3 MLs reduces the overall intensity of the pattern, but does not change its symmetry. In all cases, the 1/4-order streaks get diffuse and eventually disappear at temperatures above 250 °C.

It is clear from these experiments that two stable Ga coverages exist, which condense at low temperatures into a (1 × 2)/(4 × 2) reconstruction at bilayer coverage and a (4 × 4) reconstruction at trilayer coverage. On surfaces with noninteger amounts of adsorbed Ga, both reconstructions may coexist (presumably in the form of domains), although we have no experimental evidence for such a coexistence. Furthermore, the recovery behavior of the surface at elevated temperatures provides the opportunity to study the adsorption/desorption kinetics of Ga in real time.

6.4.2
Ga Adsorption/Desorption Kinetics

First, we investigate Ga adsorption and desorption kinetics in vacuum. The sample was held at a constant temperature of 740 °C. Ga adsorption/desorption isotherms were recorded by monitoring the intensity of the specularly

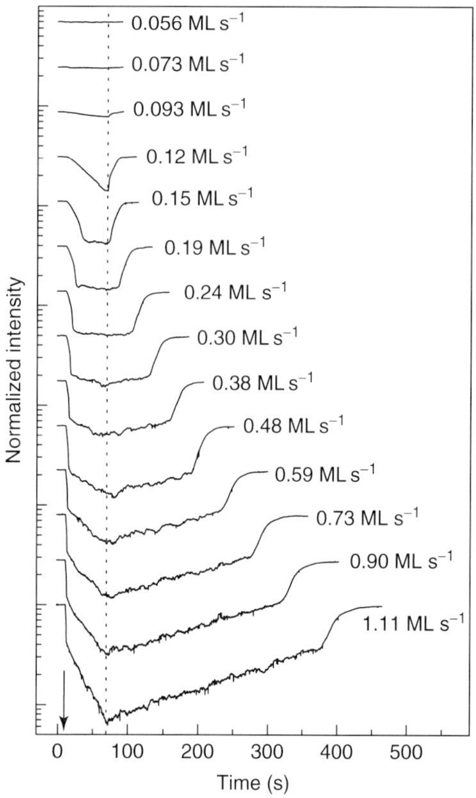

Figure 6.10 Variation of the specular RHEED intensity upon adsorption of Ga for 60 s and subsequent desorption for Ga fluxes as indicated. The arrow at $t = 10$ s indicates the opening, the dashed line at $t = 70$ s the closing of the Ga shutter.

reflected beam along the [$22\bar{4}3$] azimuth, starting with 60 s of adsorption and allowing the surface to recover completely prior to the next experiment.

Figure 6.10 shows the variation of the specular RHEED intensity I_s for Ga fluxes between 0.056 and 1.11 ML s^{-1}. For the two lowest fluxes, no appreciable change in the RHEED intensity is observed, reflecting that Ga desorption at this temperature inhibits the adsorption of significant amounts of Ga on the surface. At higher fluxes (0.12 ML s^{-1}), the RHEED intensity exhibits a clear drop, once the Ga deposition is initiated, and an immediate recovery, once the Ga supply is stopped. For $\phi_{Ga} = 0.15$ and 0.19 ML s^{-1}, the RHEED intensity saturates after the initial drop, and recovers without (0.15 ML s^{-1}) or with a very short (0.19 ML s^{-1}) delay. Apparently, this situation corresponds to a balance between impinging and desorbing Ga flux, thus establishing an equilibrium Ga coverage at the surface independent of the deposition time. For a Ga flux higher than this value, the recovery is progressively delayed. Furthermore, the RHEED intensity continues to decrease after the initial, rapid drop, albeit at a much slower rate.

We interpret these observations according to the following four-stage scenario as depicted in Figure 6.11. Initially (stage I), Ga adsorbs in a

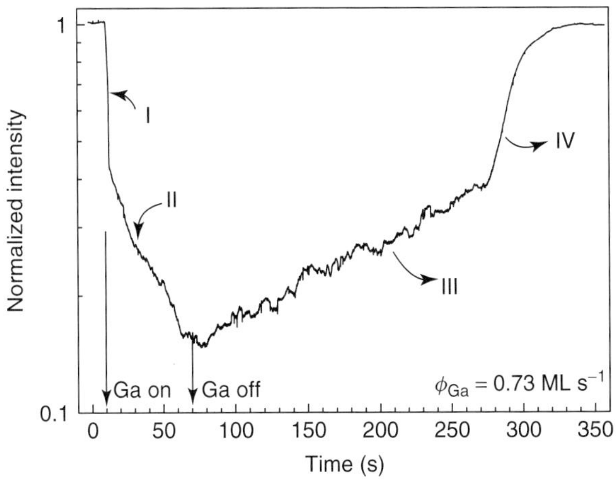

Figure 6.11 The four stages of Ga adsorption and desorption for the example of $\phi_{Ga} = 0.73$ ML s^{-1}.

layer-by-layer fashion, resulting in the initial, rapid drop of the RHEED intensity until full coverage is established. If the impinging Ga flux exceeds the desorbing flux, Ga accumulation takes place, eventually leading to the formation of droplets, which results in a further but slower decrease of the RHEED intensity (stage II). Once the Ga supply is stopped, the Ga adlayer starts to desorb. Excess Ga, however, acts as a reservoir, thus stabilizing the Ga adlayer and delaying the recovery (stage III). Once the excess Ga is depleted, the Ga adlayer desorbs, accounting for the rapid final recovery (stage IV). An approximate analysis of these transients indicate an adsorbed amount of Ga at the end of stage I slightly below 2 MLs for $\phi_{Ga} \geq 0.15$ [61].

Next, Ga adsorption and desorption in the presence of an active N flux is investigated. The sample was held at a constant temperature of 740 °C and exposed to an active N flux of 0.275 ML s^{-1}. Ga adsorption/desorption isotherms were recorded exactly as described above. Ga fluxes range from 0.19 to 1.66 ML s^{-1}. The behavior observed is analogous to that shown in Figure 6.10, and we thus interpret it in the same way. Quantitatively, the adsorbed amount of Ga at the end of stage I was found to be 3 MLs, that is, higher than for adsorption in vacuum [61].

To confirm the four-stage scenario mentioned earlier, we have developed a quantitative model. The model is required to account for adsorption and desorption of Ga, for the formation of excess Ga interacting with the Ga adlayer and for GaN growth in the case of the simultaneous presence of Ga and N. Our model, in units normalized to the maximum Ga-adlayer coverage, is described by the following equations:

$$\frac{d\theta}{dt} = j_{Ga}(1-\theta) + d_N(1-\theta)n - \gamma_\theta\theta - j_N\theta \qquad (1)$$

$$\frac{dn}{dt} = j_{Ga}\theta - d_N(1-\theta)n - \gamma_n n \tag{2}$$

where θ denotes the Ga-adlayer coverage, n the amount of excess Ga, j_{Ga} and j_N the Ga and N delivery rate, d_N the diffusion rate of excess Ga adatoms impinging onto θ, and γ_θ (γ_n) the rates of Ga-adlayer (excess Ga) desorption.

The first term of each equation accounts for adsorption of Ga, building up the Ga adlayer θ and subsequently excess Ga n. In much the same way, the second term accounts for diffusion of excess Ga impinging onto the Ga adlayer. The third term describes the desorption of θ and n. The last term in Equation 1, finally, stands for the incorporation of Ga and N adatoms into the crystal, that is, actual growth. Note that the model described by Equations 1 and 2 is a linearized and highly simplified version of microscopic theories of the surface kinetics such as the one developed in Ref. 62. This simplification is necessary to reduce the number of free parameters and thus to guarantee stable fits, but also means that there are certainly a variety of more refined models which could equally well describe the data. Finally, we have to relate the surface coverage calculated by means of Equations 1 and 2 to the experimentally observed RHEED intensity. Since the Ga adlayer at the growth temperature does not condense into a reconstruction, but rather represents a liquid, it serves merely to dampen the RHEED intensity. We thus follow the well-established relation between surface coverage and intensity I in Auger attenuation experiments [63], which reads

$$I(\theta, n) = e^{-\theta - bn} \tag{3}$$

Both the adlayer coverage θ and the excess coverage n are assumed to damp the intensity exponentially. The ripening of excess Ga to clusters and droplets leads to a different rate with respect to the Ga-adlayer coverage as expressed by the prefactor b.

Figure 6.12a shows three experimental Ga adsorption/desorption isotherms in vacuum together with simulations based on Equations 1–3. Except for the Ga flux, which was taken to be equal to the experimentally determined flux as indicated in the figure, we used the same parameters for these simulations, namely, a diffusion coefficient of 2×10^{-15} cm^2 s^{-1}, a Ga-adlayer desorption flux of 0.137 ML s^{-1}, an excess Ga desorption flux of 0.003 ML s^{-1}, and a prefactor b of 0.06 ± 0.003. Evidently, the agreement between the experimental transients and the simulations is very good. Most importantly, this agreement can only be obtained with the above set of parameters, since the slope of the transient in stage III is mainly determined by γ_n, while γ_θ determines the slope of the transient in stage IV. Furthermore, the onset of stage IV depends essentially on the sum of γ_n and γ_θ, resulting in unique fits to the experimental data. This result indicates that our general understanding of the adsorption/desorption processes is essentially correct. The coexistence of θ (1.64 ML) and n (0.42 ML) deduced from the simulations at the end of stage I is a direct result of the finite diffusion rate of the excess Ga. For

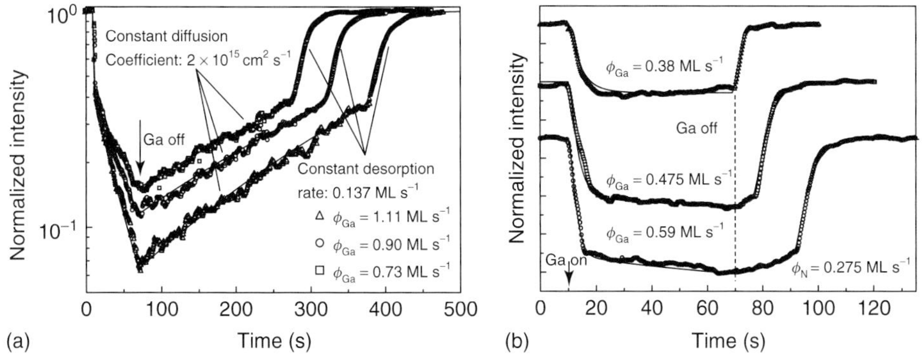

Figure 6.12 Experimental (symbols) and simulated (lines) RHEED transients for three different Ga fluxes as indicated in the figure: (a) in vacuum, (b) under an active N flux of 0.275 ML s^{-1} (the transients are vertically offset for clarity).

$\theta < 1$, the Ga adlayer is not completed, and will thus not form a continuous film, but two-dimensional islands. This inevitable fact has several important consequences. First of all, our model treats θ as a continuous variable and neglects its inherently discrete nature at low coverage. Second, the desorption kinetics of two-dimensional islands is known to be dominated by desorption from the islands' edges, which changes the first order (monomolecular) desorption kinetics assumed here to half-order (exponent 1/2). Third, the desorption rate γ_θ will be enhanced in this case, essentially resulting in a coverage-dependent desorption rate, which is no longer a constant, but a function of θ. In fact, for Ga fluxes lower than 0.19 ML s^{-1}, we had to increase γ_θ significantly to account for the early and comparatively rapid recovery observed. In view of the above mentioned facts, it is clear that this simple measure is purely phenomenological and only approximates the actual processes.

We next turn to a comparison between experimental Ga adsorption/desorption isotherms under active N flux and simulations based on Equations 1–3. As above, we take the experimentally determined Ga flux for the simulations, and, of course, the additional N flux. For a Ga flux larger than 0.475 ML s^{-1}, the parameters required to obtain a good agreement between simulated and experimental transient are identical to those given above, except for γ_θ, which turns out to be slightly larger (0.155 ML s^{-1}). For lower Ga fluxes, the Ga adlayer is not completed, and γ_θ increases drastically as discussed above for the case of desorption in vacuum. Figure 6.12b shows the three RHEED transients at the transition between N-stable and Ga-stable conditions. The desorption fluxes required to fit them are 0.155, 0.21, and 0.28 ML s^{-1} from low to high Ga flux. The good agreement between experimental and simulated transients shows that the adsorption/desorption kinetics is not fundamentally different in the presence of an N flux compared to that under vacuum. The N

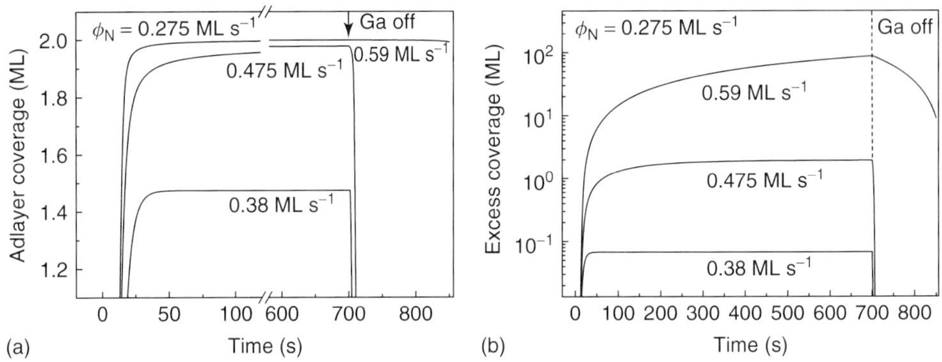

Figure 6.13 Simulated (a) Ga adlayer and (b) excess Ga coverage during growth. The parameters used are those deduced from the simulations shown in Figure 6.12, but the Ga supply is now terminated at $t = 700$ s.

flux seems merely to consume Ga and thus to shift the adsorption/desorption balance toward higher Ga fluxes. For the sake of completeness, we determine again the Ga adlayer and excess Ga coverage predicted by the simulations at the end of stage I. This coverage is found to be independent of Ga flux for $\phi_{Ga} \geq 0.475$ ML s^{-1} and amounts to $\theta = 1.94$ MLs and $n = 0.88$ MLs, respectively. Note that these values are, in fact, significantly larger than those obtained for adsorption in vacuum. At this point, we shall examine the predictions of our model for prolonged growth using the parameters used for the simulations in Figure 6.12b.

Figure 6.13a shows the Ga bilayer coverage versus growth time for the three different Ga fluxes at the transition between N-stable and Ga-stable conditions. As expected, full bilayer coverage is not reached for the lowest Ga flux. In contrast, the Ga-adlayer coverage slowly converges toward a full bilayer for the intermediate Ga flux and is obtained rapidly for the highest Ga flux. However, as shown in Figure 6.13b, the price one pays for this rapid bilayer completion is a diverging Ga excess coverage, which increases without bound for prolonged growth. In fact, layers grown under these conditions exhibit a high density of Ga droplets.

In contrast, the Ga excess coverage saturates at 2 MLs for the intermediate Ga flux, reflecting that a dynamic equilibrium is established under these conditions. Note that the desorption flux for this case is 0.21 ML s^{-1}, which in fact is almost exactly equal to the difference of the Ga and N fluxes. The total Ga coverage during growth is thus 4 MLs, which is consistent with the observation of the Ga-trilayer-related (4×4)-reconstruction observed for this growth condition after quenching the sample to 100 °C (cf. Figure 6.9). For the lowest Ga flux, the excess coverage is (as expected) negligible.

Finally, Figure 6.14 shows an AFM micrograph of a GaN$(1\bar{1}00)$ layer grown under our standard conditions (see above), which has been judged previously

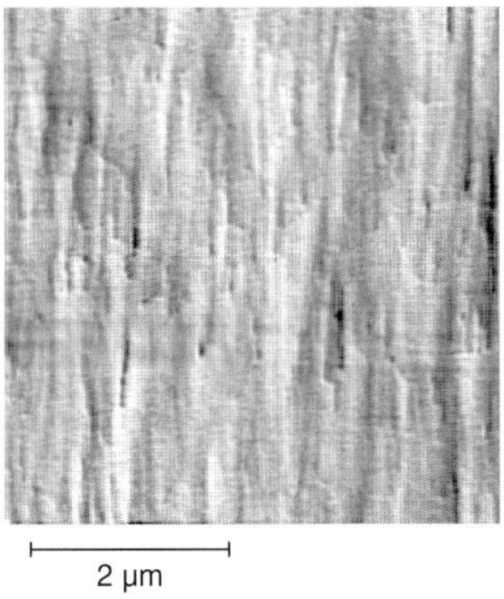

Figure 6.14 AFM micrograph of a 500-nm thick GaN(1$\bar{1}$00) layer grown under standard conditions. The vertical stripes are oriented along [11$\bar{2}$0]. The P/V and RMS roughness of this layer are 6.6 and 0.6 nm, respectively, over an area of $5 \times 5\ \mu m^2$.

to be most suitable on empirical grounds [27]. In particular, these conditions result in droplet-free and microscopically smooth surfaces. The surface of the layer exhibits the characteristic highly anisotropic, stripelike morphology of GaN(1$\bar{1}$00). The orientation of these stripes, investigated by Raman scattering, is found to be along ⟨11$\bar{2}$0⟩, that is, perpendicular to the *C*-axis. The origin of this surface might be a consequence of the growth kinetics. The migration length of adatoms most likely differs along [11$\bar{2}$0] and [0001], which exhibit distinct step structures.

6.5 *M*-plane (In,Ga)N/GaN MQWs

Wurtzite group III nitride MQWs with nonpolar orientation are expected to be free of electrostatic fields along the growth direction. To experimentally confirm this presumption, one usually relies on the results of spectroscopic studies. The correct interpretation of these results, however, requires an accurate knowledge of the structural parameters of the samples under investigation. This statement is particularly true for the case of (In,Ga)N/GaN quantum wells, in which the surface segregation of In may severely distort the intended composition profile as indeed found for the case of *C*-plane structures [64–66]. This section details the growth of *M*-plane (In,Ga)N/GaN MQWs on LAO(100). This is followed by a description of the investigation of their structural and optical properties.

6.5.1 In Incorporation and Surface Segregation

To study the impact of the orientation on the In incorporation during the growth of MQWs, a set of 20-period (In,Ga)N/GaN MQWs was grown on γ-$LiAlO_2$(100) substrates.

6.5.1.1 Growth

For all the samples, growth was initiated by a 100-nm-thick, low-temperature (580 °C) GaN nucleation layer, after which the temperature was increased to 740 °C for the growth of a 500-nm-thick GaN buffer layer, which has an RMS roughness of only 0.6 nm as shown in Section 6.3. For the subsequent growth of the (In,Ga)N/GaN MQWs, the substrate temperature was decreased to 600 °C. In, Ga, and N fluxes were identical for all MQWs, which differ only in the growth times for the wells, namely, 25 s for samples A and B, and 40 s for samples C and D. The RHEED patterns remained streaky during the entire growth of the MQWs. In fact, the RMS roughness of the samples as determined by AFM lies between 1.2 and 1.7 nm. The samples are further characterized by HRXRD and secondary-ion-mass-spectroscopy (SIMS).

6.5.1.2 HRXRD

To determine the structural parameters of these *M*-plane MQWs by HRXRD, we follow the same procedure as employed in Ref. 67. Symmetric X-ray ω–2θ scans were taken with a Philips X'Pert PRO triple-axis diffractometer with a Cu$K_{\alpha 1}$ source, a Ge(220) hybrid monochromator, and a Ge(220) three-bounce analyzer crystal. We first analyzed the angular positions of the satellite peaks kinematically based on the nominal growth times to obtain the structural parameters of the sample, implicitly assuming that segregation does not occur. Next, we employed simulations based on the dynamical diffraction theory [67] and varied the kinematically obtained parameters, until the intensity of the satellites matched the experiment in addition to their position. The simulations were performed for perfect interfaces and are convoluted with the instrumental resolution (25″) only. As an example, Figure 6.15 shows ω–2θ scans for sample MQW-A across the (a) GaN($1\bar{1}00$) and (b) GaN($2\bar{2}00$) reflection.

Several important conclusions can be drawn from these scans: (i) No additional signal is detected at the position of either GaN(0002) or in between GaN(0002) and (In,Ga)N(0002), demonstrating that the entire structure has the expected orientational relationship, that is, $[1\bar{1}00]_{\text{(In,Ga)N}} \| [100]_{\text{LAO}}$. Raman measurements (not shown here) demonstrate that the structure also fulfills the expected in-plane orientation relationship, namely, $[0001]_{\text{(In,Ga)N}} \| [010]_{\text{LAO}}$ and $[11\bar{2}0]_{\text{(In,Ga)N}} \| [001]_{\text{LAO}}$ [68]. (ii) The angular separation between the LAO(200) and GaN($1\bar{1}00$) reflections or LAO(400) and GaN($2\bar{2}00$) indicates a lattice dilation of 0.6% along the growth direction, that is, [$1\bar{1}00$], for the GaN buffer. This fairly high strain is at least partly a result of the coherent growth of GaN($1\bar{1}00$) on LAO(100) along the [0001] direction [28], where the lattice mismatch amounts to only 0.3%. The mismatch in thermal expansion, of course, is also compressive and enhances this effect. (iii) Superlattice satellites up to the 2nd order are observed for all samples, demonstrating the high periodicity of the structures and allowing us to determine their structural properties.

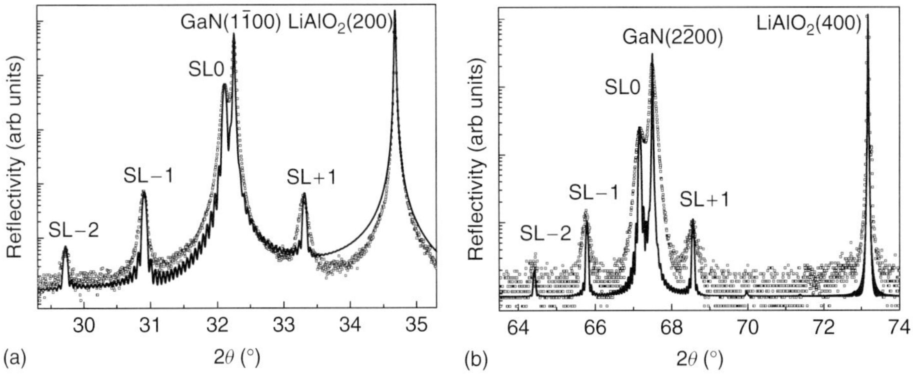

Figure 6.15 Experimental (□) and simulated (—) triple-axis ω–2θ scans across the (a) GaN(1$\bar{1}$00) and (b) GaN(2$\bar{2}$00) reflection of a 20-period *M*-plane (In,Ga)N/GaN MQW grown on $LiAlO_2$(100) (sample MQW-A). The superlattice satellites are denoted by SL $\pm$ *n*.

In fact, the agreement between the experimental and simulated profiles is quite satisfactory, whereas simulations with the kinematically obtained parameters (not shown here) exhibit satellite intensity ratios clearly inconsistent with those experimentally observed. The results of this analysis are summarized in Table 6.2 and indicate a very significant In surface segregation for all samples under investigation. Interestingly, the In content obtained from this analysis is 7%, while *C*-plane samples grown under identical conditions exhibit an In content of about 15%.

The broadening of the satellites in the experiment with respect to the simulation, which is particularly evident in Figure 6.15b, may shed some

Table 6.2 Parameters of the *M*-plane (In,Ga)N/GaN MQWs under investigation as determined by the kinematical analyses based on growth time (kin), the dynamical simulations taking the real depth profile of In into account (dyn), and SIMS. The In content and the well width are denoted by x and d_w (nm), respectively. The lower and upper margins for the well thickness in the SIMS column correspond to the value obtained directly from the In content and that obtained from the FWHM of the In profile, respectively.

Sample	kin		dyn		SIMS	
	x	d_w	x	d_w	x	d_w
MQW-A	0.14	1.35	0.068	2.7	0.070	2.7–3.1
MQW-B	0.16	1.47	0.070	3.3	0.071	3.3–4.3
MQW-C	0.19	2.43	0.066	7.0	0.072	6.4–7.2
MQW-D	0.20	2.48	0.068	7.4	0.071	7.1–7.6

doubt on the validity of our approach. Indeed, if this broadening were caused by interface roughness, the intensity ratio between the satellites would have changed and could not have been described correctly by the present model. However, the widths of the 0th and the -1th order satellites are virtually identical for all samples, suggesting that the broadening arises from heterogeneous strain rather than from interface roughness. A Williamson–Hall plot [69] of the width of the SL0 reflection results in an estimated heterogeneous strain along the growth direction of 1.2×10^{-3}, larger than that obtained typically for bare GaN layers. This result is not unexpected, as compositional fluctuations in the ternary (In,Ga)N alloy, which are believed to be largely driven by In surface segregation [70], will inevitably result in a certain degree of inhomogeneous strain.

6.5.1.3 **SIMS**

In order to independently confirm the actual In content and structural parameters, we performed high-resolution SIMS measurements. SIMS was carried out with a CAMECA IMS 4f-E6 system, employing Cs^+ primary ions with an impact energy of 5.5 keV and $InCs^+$ ions as species for detection. The depth resolution under these conditions is about 1 nm/decade and 2 nm/decade at the leading and trailing edge, respectively.

As an example, the In depth profile of sample MQW-A is shown in Figure 6.16. The leading and trailing edges of this depth profile are 1.4 and 2.2 nm/decade, respectively, that is, close to the instrumental resolution. The progressive damping and broadening of the signal with depth (cf. inset of Figure 6.16) is caused by roughening due to the sputtering process. The actual well width can be deduced from this profile in two ways. First, we may take

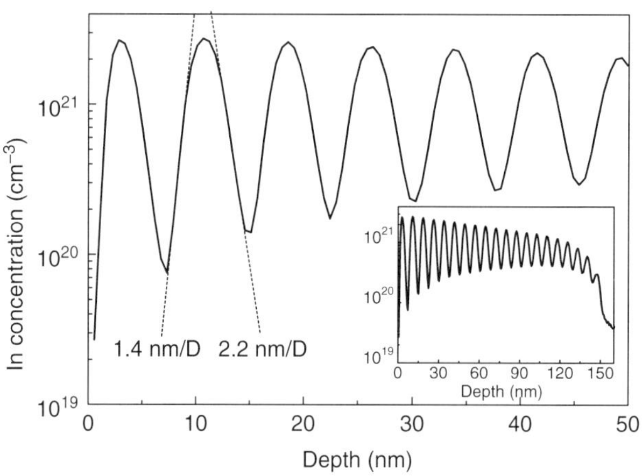

Figure 6.16 SIMS depth profile of the top 50 nm of sample MQW-A. The depth resolution is indicated by the dashed lines. The entire depth profile is shown in the inset.

the In content as measured by SIMS for granted (note that $InCs^+$ detection is indeed known to be quite accurate in this regard) and combine this value with the average In content as measured by HRXRD. Second, we may take the full-width at half-maximum (FWHM) of the In peaks to reflect the actual well width, if the instrumental broadening is Gaussian and is significantly smaller than the well width itself (otherwise the FWHM will overestimate the well width). The In contents and well thicknesses thus determined for the samples under investigation are listed in Table 6.2. The good agreement with those deduced from the HRDXD simulations is evident. This result signifies that In surface segregation is present on the *M* plane as well as on the *C* plane of GaN. The lower In content found for the former even suggests that it is stronger in the present case. In other words, the In incorporation efficiency on the *M* plane, phenomenologically lumping together processes such as segregation and desorption, seems to be lower when compared to the *C* plane.

An interesting point is the symmetrical 'top-hat' shape of each QW in Figure 6.16, which we observed for all samples under investigation. This finding is analogous to the case of *C*-plane (In,Ga)N/GaN MQWs [65] and indicates a 0th-order In segregation process. This process implies an In incorporation limit (here 7%) at a given growth temperature and has the consequence that the incorporation of In does not depend on the actual amount of In floating on the growth front.

6.5.2 Recombination Mechanism

The spontaneous emission of *C*-plane (In,Ga)N quantum wells is determined by both the electron-hole wavefunctions separation due to the built-in internal electrostatic field (quantum-confined Stark effect) and exciton localization caused by potential fluctuations [71–74]. The realization of *M*-plane (In,Ga)N/GaN MQWs allows us to investigate the impact of exciton localization on radiative recombination without the influence of internal electrostatic fields. To study the recombination mechanism of *M*-plane (In,Ga)N/GaN MQWs, continuous-wave photoluminescence (cw-PL) spectroscopy and time-resolved (TR) PL were carried out.

6.5.2.1 cw-PL

cw-PL spectra at various temperatures were recorded using a 325 nm He-Cd laser with an excitation intensity of 10 W/cm^2. Figure 6.17a shows the evolution of the cw-PL spectra from 5 to 300 K for an *M*-plane (In,Ga)N/GaN MQW (MQW-A). The dominant emission centered close to 3 eV stems from the quantum wells. Interestingly, we also observe emission near 3.36 eV, originating from excitons bound to SFs in GaN [35] and at 3.497 eV from the (D^0,X) transition in GaN. The energy position of the latter indicates a large compressive strain [75, 76], which is consistent with the results of the XRD measurements described above. However, the observation of luminescence from GaN alone is interesting, since we never detected any signal from GaN in the case of *C*-plane (In,Ga)N/GaN MQWs with a total thickness, as the present sample, exceeding the penetration depth of the laser. The lack of GaN luminescence in this case is plausible – the internal electrostatic fields in the barriers of *C*-plane (In,Ga)N/GaN MQWs rapidly separate electrons and holes and direct them to the wells, thereby enhancing the capture efficiency. The observation of rather strong luminescence from the GaN barriers in the present sample may thus be taken as a manifestation of the absence of internal electrostatic fields.

Figure 6.17b shows Arrhenius plots of the integrated cw-PL intensity of MQW-A and MQW-D. The total PL intensity of these two *M*-plane MQWs is reduced by a factor of 17 and 44, respectively, when the temperature is increased from 5 to 300 K. In fact, all samples under investigation exhibit a strong and broad PL band between 400 and 420 nm at room temperature. For all of our *M*-plane MQWs, the luminescence quenches by a factor of 15–50 when the temperature increases from 5 to 300 K. The PL of these MQWs is still detectable up to 500 K. This finding is in contrast to *C*-plane MQWs of well widths similar to, for example, MQW-D (7 nm), where the emission intensity decreases drastically with well width and is hardly detectable at room temperature.

6.5.2.2 **Anomalous Temperature-dependent PL Width**

In both *C*-plane (In,Ga)N/GaN MQWs [77–79] and cubic (In,Ga)N/GaN MQWs [80, 81], exciton localization caused by In compositional fluctuation has

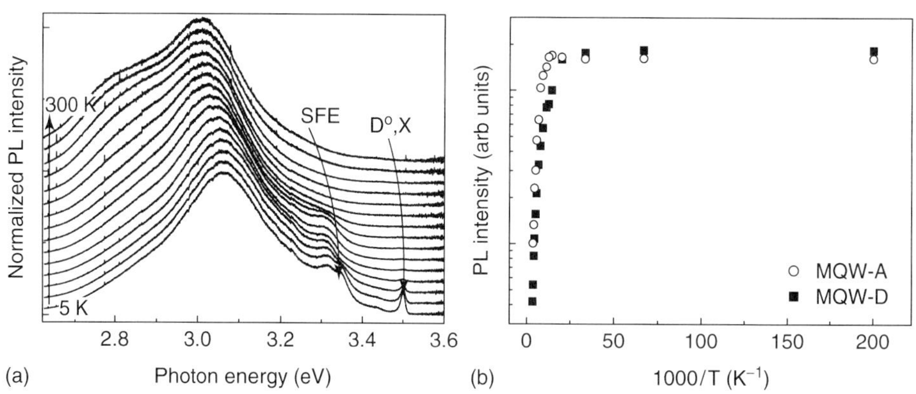

Figure 6.17 (a) PL spectra of MQW-A in the temperature range from 5 to 300 K as indicated. All spectra are normalized and vertically offset for clarity. (b) Arrhenius representation of the integrated PL intensity of MQW-A and MQW-D.

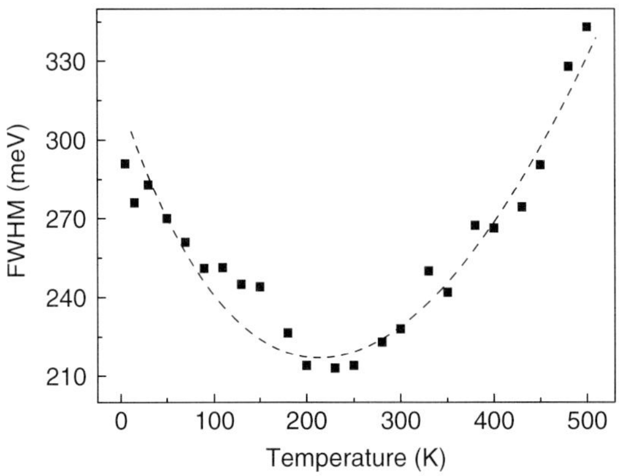

Figure 6.18 FWHM of the PL spectrum (squares) as a function of temperature for the *M* plane MQW-A. The dashed line is a guide to the eye.

a crucial influence on the radiative recombination process. A clear indication of the presence of localized states in the *M*-plane (In,Ga)N/GaN MQWs is obtained by the anomalous temperature dependence of the FWHM of the PL spectrum, as shown in Figure 6.18, for MQW-A. The narrowing of the spectrum upon increasing temperature has been observed in several materials systems in which strong localization occurs, such as InAs/GaAs quantum dots [82], (In,Ga)As/GaAs QWs [83], and (In,Ga)N QWs [71, 84]. At low temperatures, excitons are randomly trapped by the localizing centers, and the PL linewidth basically corresponds to the width of the distribution of local potential minima. With increasing temperature, excitons are thermally activated and become mobile within the band of localized states, allowing them to relax further into deeper potential minima. This thermal activation will thus result in the distinct narrowing of the linewidth. At even higher temperature, the thermal distribution of excitons will force them to also populate higher-energy states. As a result, the PL band broadens, and its width once again approaches the band width.

6.5.2.3 **PL Transition Energy as a Function of Well Thickness**

With the accurate structural parameters obtained as described in Section 6.5.1, a quantitative comparison of the transition energy as a function of the well thickness can be performed. Figure 6.19 shows a comparison between the experimental PL peak positions at room temperature, and the calculated transition energies versus the well widths. The calculated values of the transition energy are obtained by a self-consistent Schrödinger–Poisson calculation assuming a constant In content of 7%, a band gap for $In_{0.07}Ga_{0.93}N$ of 3.14 eV [85], a band offset ratio of 70 : 30, [86] effective electron and hole masses of 0.2 and 1 [86], a nonlinear spontaneous piezopolarization following

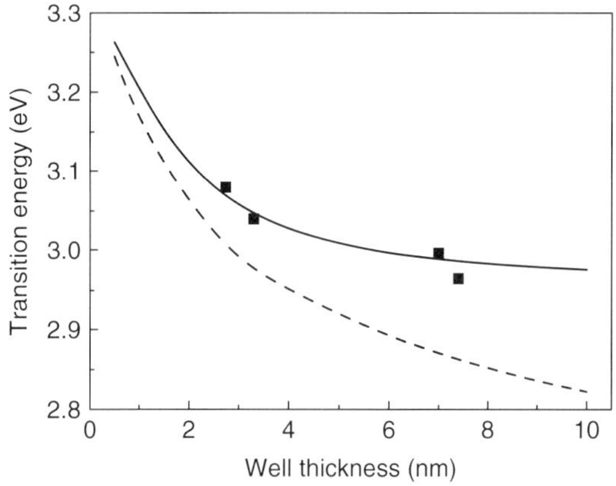

Figure 6.19 Transition energy as a function of the well thickness for samples MQW-A to MQW-D, assuming a constant *x* of 7%. The symbols are experimental data, while the solid and dashed lines are calculated on the basis of polarization charges of 0 and 9.7×10^{12} C cm^{-2}, respectively. Note that both curves are rigidly shifted down to lower energies by 110 meV as explained in the text.

Ref. 87, and assuming a uniform background carrier density of 1×10^{18} cm^{-3}, which is a typical value for bare GaN layers grown on $LiAlO_2$. The dashed curve takes the full polarization field into account, while the solid curve is obtained assuming no polarization. The calculation neither takes into account excitonic effects, which may amount to 40–60 meV in the present structures [88], nor exciton localization. Consequently, the calculated values are systematically higher in energy than the experimental ones, and are rigidly shifted down by 110 meV for clarity in Figure 6.19. The experimental transition energies are seen to follow closely the prediction for this quasi-flat-band case. Note that all of the above parameters, except for the polarization constants, merely rigidly shift these curves up or down, but do not change their shape. These results thus provide evidence for the fact that the structures under investigation are free from internal electrostatic fields encountered in corresponding *C*-plane structures.

6.5.2.4 **Recombination Dynamics**

To understand the dynamics of radiative recombination of the *M*-plane MQWs, time-resolved PL measurements at 7 K and at various excitation fluences were performed. We used a frequency-doubled Ti : sapphire laser with a photon energy of 3.4 eV, a pulse width of about 200 fs, and a repetition rate of 4.75 MHz. For detection, a Hamamatsu C5680 streak camera was used. Figure 6.20a shows TRPL transients for a 20-period *M*-plane (In,Ga)N/GaN MQW after direct excitation of the wells for different excitation fluences. The initial decay is rapid, faster than that observed for *C*-plane MQWs with similar

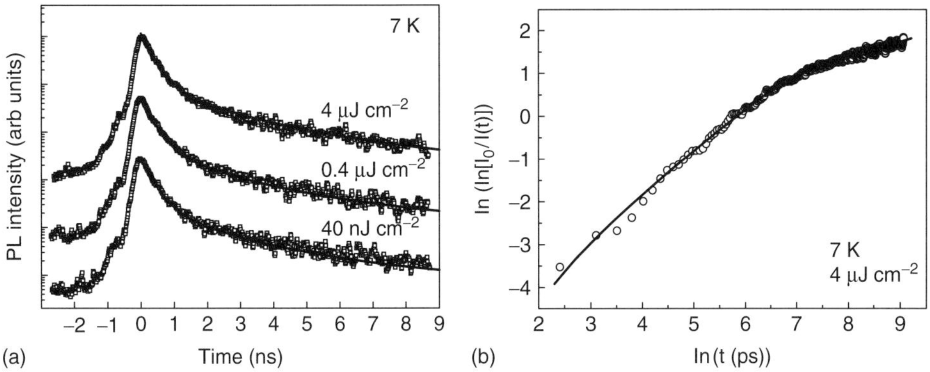

Figure 6.20 (a) Spectrally integrated TRPL intensity transients for three different excitation fluences for a 20-period *M* plane (In,Ga)N/GaN MQW. Symbols denote the experimental data, lines indicate fits with identical parameters except for the amplitude. The apparent step in the data prior to the pulse is an experimental artifact. (b) Doublelog-log representation of the PL transient at the highest fluence. Circles indicate the experimental data, the line denotes the fit.

structural parameters. The transients, however, could not be fit by either a biexponential or a stretched exponential [80–93]. Note that the transients exhibited an identical behavior for different excitation fluences. This finding is in complete contrast to *C*-plane MQWs, where screening of the electric field occurred at these carrier densities (10^{18} cm^{-3}), and as such provided a further sign for the actual absence of internal electrostatic fields.

Figure 6.20b shows the transient obtained with the highest excitation fluence plotted as $\ln(\ln[I_0/I(t)])$ versus $\ln(t)$, in which a stretched exponential would be a straight line. Obviously, the data cannot be represented by a stretched exponential. The key for understanding the recombination dynamics for this sample is displayed in Figure 6.21. The PL band exhibits rapid spectral diffusion during the first 500 ps after excitation. For longer times, the PL band continues to redshift, but is significantly slower.

We interpret this finding as follows. After excitation at 3.4 eV, the photo-generated excitons thermalize and initially occupy extended states in the QW. These free excitons then rapidly relax toward lower-lying localized states, which is the reason for the rapid spectral diffusion as well as for the rapid initial decay of the PL transient. Once they reach the localized states, the excitons are subject to a much slower redistribution within the band of the localized states, described by a stretched exponential [89]. Mathematically, we thus deal with the following situation:

$$\frac{\mathrm{d}n_f(t)}{\mathrm{d}t} = G(t) - \frac{n_f(t)}{\tau_f} - \frac{n_f(t)}{\tau_r} \tag{4}$$

$$\frac{\mathrm{d}n_b(t)}{\mathrm{d}t} = \frac{n_f(t)}{\tau_r} - t^{\beta-1}\frac{n_b(t)}{\tau_b} \tag{5}$$

where $G(t)$ denotes the excitation pulse, n_f and n_b the densities of free and bound excitons, respectively, τ_f and τ_b the corresponding radiative lifetimes, τ_r the relaxation time, and β the dimensionality parameter.

This rate-equation system has no analytical solution, but we have found that the sum of an exponential [$\propto \exp(-t/\tau_r')$, representing the relaxation from free excitons toward localized states with $1/\tau_r' = 1/\tau_f + 1/\tau_r$] and a stretched exponential [$\propto \exp(-(t/\tau_b')^{\beta'})$, accounting for the slow communication between the localized states] is a sound approximation, although the values for the individual lifetimes and the dimensionality parameter β are then modified. The fits shown in Figure 6.20a and 6.20b were achieved with identical parameters (except for the amplitude), namely, $\tau_r' = 0.2$ ns, $\tau_b' = 1$ ns, $\beta' = 0.68$. The former value is a superposition of both relaxation and radiative recombination of free excitons and thus cannot be decomposed. The latter parameters, however, roughly correspond to $\tau_b = 3$ ns and $\beta = 0.5$ as judged from our numerical simulations of Equations 5 and 6. These values are similar to those found for cubic (In,Ga)N [9, 80]. Finally, it is interesting to note that the energy shift displayed in Figure 6.21b is logarithmic in nature and can be described by $\Delta E = -kT_{ac} \ln(\nu t)$, with the acoustic phonon temperature T_{ac} and the attempt-to-escape rate $\nu \approx 10^{12}$ s^{-1}. This logarithmic relaxation has been previously observed both for the case of multiple-trapping relaxation of carriers within a band of localized states [94] and for relaxation of excitons toward localized states in quantum wells [95]. The temperatures required to fit our data are found to be significantly higher than the actual lattice temperature, namely, 50 K for the short-time and 20 K for the long-time region. This behavior is identical to that observed by Göbel and Graudszus [94] and has been attributed to the fact that the relaxation process itself proceeds by the generation of phonons, thus increasing the phonon population above the thermal

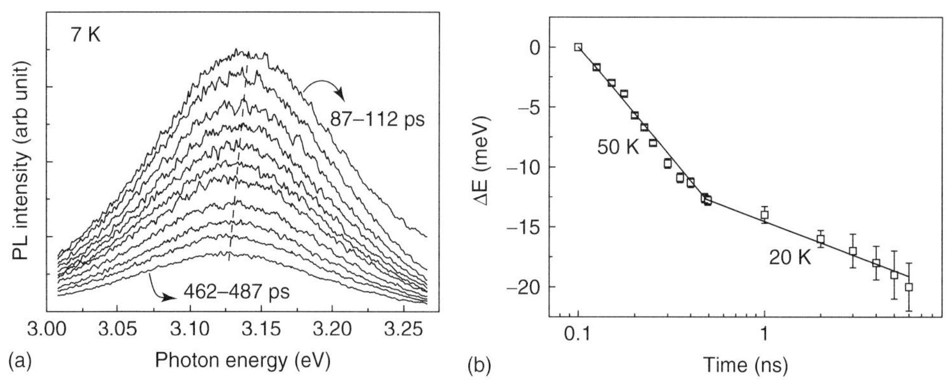

Figure 6.21 (a) Spectral diffusion of the PL band for the first 500 ps and (b) temporal variation of the photon energy at the PL spectral peak. Note the different slopes for short and long times. The lines are fits according to the equation $\Delta E = -kT_{ac} \ln(\nu t)$.

equilibrium value. In this context, the difference of the phonon temperatures extracted from our experiments is easy to understand – while the relaxation of free excitons toward localized states proceeds exclusively by the generation of phonons, excitons, once occupying localized states, have at least partly to rely on existing phonons to assist them to migrate within the band of localized states.

6.6 Conclusion and Outlook

In this chapter, we have described how high-quality *M*-plane GaN films can be reproducibly synthesized on γ-$LiAlO_2(100)$. The phase purity and surface quality of these *M*-plane GaN epilayers are very sensitive to the nucleation conditions. Studies of the Ga adsorption and desorption kinetics provided a guide for establishing the optimum growth conditions of *M*-plane GaN films, which is crucial for a further investigation of this promising material. Successful growth of (In,Ga)N/GaN$(1\bar{1}00)$ MQWs has been demonstrated on the basis of high-quality *M*-plane GaN buffer layers. Significant In surface segregation was determined by both HRXRD and SIMS. This knowledge is of importance for realizing higher In incorporation in the future. We have demonstrated that these heterostructures are, as expected, free of internal electrostatic fields along the growth direction. As a consequence, a high PL efficiency of these structures, even at elevated temperatures (500 K) and for wide wells, was observed. This finding indicates the potential of (In,Ga)N/GaN $(1\bar{1}00)$ for the active region of light-emitting devices. Future work on this promising material should concentrate on the following issues: first, the surface quality of γ-$LiAlO_2$ is still inferior when compared with commercially available SiC(0001) and $Al_2O_3(0001)$ wafers. For further improving the surface quality of *M*-plane GaN, it is desirable to find a way to improve the surface morphology of γ-$LiAlO_2$. Second, the In incorporation efficiency is quite low in our *M*-plane (In,Ga)N/GaN MQWs. Exploring appropriate growth conditions for higher In contents is another interesting topic. Finally, we note that *M*-plane bulk AlN substrates are now commercially available.[1] Regarding the chemical and thermal stability, AlN is certainly superior to γ-$LiAlO_2$ as a substrate. With an identical structure, *M*-plane AlN has the same stacking sequence as *M*-plane GaN. Thus, GaN films grown on *M*-plane AlN substrates might be free from the planar defects observed in those grown on structurally incommensurable substrates, such as γ-$LiAlO_2(100)$. In addition, AlN is an O-free material, unlike $LiAlO_2$. One thus does not need to worry about O contamination originating from the substrate during growth, especially for MOCVD. Therefore, the epitaxial growth of GaN films and relevant structures on *M*-plane AlN substrates will be technologically and commercially interesting in the future.

1) AlN substates with a diameter up to 2 inch are available from Crystal IS, Inc. 70 Cohoes Avenue, Green Island, NY 12183, USA.

Acknowledgments

The author is indebted to his coworkers for their vital contributions to this work, particularly, P. Waltereit, Y. J. Sun, A. Trampert, T. Y. Liu, S. Cronenberg, S. Dhar, and H. T. Grahn. Special thanks are due to H.-P. Schönherr for his expert technical assistance.

References

1. Deguchi, T., Sekiguchi, K., Nakamura, A. *et al.* (1999) *Japanese Journal of Applied Physics, Part 2*, **38**, L914.
2. Bernardini, F., Fiorentini, V., and Vanderbilt, D. (1997) *Physical Review B*, **56**, 10024(R).
3. Bykhovski, A., Gelmont, B., and Shur, M. (1993) *Journal of Applied Physics*, **74**, 6734.
4. Kaiser, S., Jakob, M., Zweck, J. *et al.* (2000) *Journal of Vacuum Science and Technology B*, **18**, 733.
5. Arulkumaran, S., Egawa, T., Ishikawa, H., and Jimbo, T. (2003) *Journal of Vacuum Science and Technology. B*, **21**, 888.
6. Grandjean, N., Damilano, B., Dalmasso, S. *et al.* (1999) *Journal of Applied Physics*, **86**, 3714.
7. Langer, R., Simon, J., Ortiz, V. *et al.* (1999) *Applied Physics Letters*, **74**, 3827.
8. Chichibu, S.F., Sugiyama, M., Kuroda, T. *et al.* (2001) *Applied Physics Letters*, **79**, 3600.
9. Chichibu, S.F., Sugiyama, M., Onuma, T. *et al.* (2001) *Applied Physics Letters*, **79**, 4319.
10. Brandt, O. (1998) *Group III Nitride Semiconductor Compounds: Physical and Applications*, (B. Gil), Oxford University Press, New York.
11. Feng, Z.H., Yang, H., Zheng, X.H. *et al.* (2003) *Applied Physics Letters*, **82**, 206.
12. Waltereit, P., Brandt, O., Trampert, A. *et al.* (2000) *Nature*, **406**, 865.
13. Ng, H.M. (2002) *Applied Physics Letters*, **80**, 4369.
14. Craven, M.D., Lim, S.H., Wu, F. *et al.* (2002) *Applied Physics Letters*, **81**, 469.
15. Craven, M.D., Lim, S.H., Wu, F. *et al.* (2002) *Applied Physics Letters*, **81**, 1201.
16. Kuokstis, E., Chen, C.Q., Gaevski, M.E. *et al.* (2002) *Applied Physics Letters*, **81**, 4130.
17. Chen, C.Q., Gaevski, M.E., Sun, W.H. *et al.* (2002) *Applied Physics Letters*, **81**, 3194.
18. Lee, C.D., Feenstra, R.M., Northrup, J.E. *et al.* (2003) *Applied Physics Letters*, **74**, 1793.
19. Craven, M.D., Waltereit, P., Wu, F. *et al.* (2003) *Japanese Journal of Applied Physics, Part 2*, **42**, L235.
20. Ng, H.M., Bell, A., Ponce, F.A., and Chu, S.N.G. (2003) *Applied Physics Letters*, **83**, 653.
21. Haskell, B.A., Wu, F., Craven, M.D. *et al.* (2003) *Applied Physics Letters*, **83**, 644.
22. Lu, H., Schaff, W.J., Eastman, L.F. *et al.* (2003) *Applied Physics Letters*, **83**, 1136.
23. Sun, W.H., Yang, J.W., Chen, C.Q. *et al.* (2003) *Applied Physics Letters*, **83**, 2599.
24. Chen, C.Q., Adivarahan, V., Yang, J.W. *et al.* (2003) *Japanese Journal of Applied Physics, Part 2*, **42**, L1039.
25. Marezio, M. (1965) *Acta Crystallographica*, **19**, 396.
26. Hellman, E.S., Liliental-Weber, Z., and Buchanan, D.N.E. (1997) *MRS Internet Journal of Nitride Semiconductors Research*, **2**, 30.
27. Sun, Y.J., Brandt, O., and Ploog, K.H. (2003) *Journal of Vacuum Science and Technology B*, **21**, 1350.
28. Trampert, A., Liu, T., Waltereit, P. *et al.* (2002) *Institute of Physics Conference Series*, **169**(169), 277.

29. Hsu, C.H., Ip, K.P., Johnson, J.W. *et al.* (2001) *Electrochemical and Solid-State Letters*, **4**, C35.
30. Beckerman, S.J., Ford, R.B., and Nemeth, M.T. (1996) *Powder Diffraction*, **11**, 312.
31. Cockayne, B. and Lent, B. (1981) *Journal of Crystal Growth*, **54**, 546.
32. Lee, J.W., Pearton, S.J., Abernathy, C.R. *et al.* (1997) *Materials Research Society Symposium Proceedings*, **449**, 1041.
33. Novoselov, A. and Pajaczkowska, A. (1998) *Crystal Research and Technology*, **6**, 949.
34. Sun, Y.J. (2004) *Growth and Characterization of M-plane GaN and (In,Ga)N/GaN multiple quantum wells*, Ph.D. Thesis, Humboldt-Universität zu Berlin, Berlin.
35. Sun, Y.J., Brandt, O., Jahn, U. *et al.* (2002) *Journal of Applied Physics*, **92**, 5714.
36. Ramahandran, V., Brady, M.F., Smith, A.R. *et al.* (1998) *Journal of Electronic Materials*, **27**, 308.
37. Kaplan, R. (1989) *Surface Science*, **215**, 111.
38. Strite, S., Lin, M.E., and Morkoç, H. (1993) *Thin Solid Films*, **31**, 197.
39. Xu, K., Deng, P.Z., Xu, J. *et al.* (1998) *Journal of Chinese Ceramic Society*, **26**, 281.
40. Yi, M.S., Lee, H.H., Kim, D.J. *et al.* (1999) *Applied Physics Letters*, **75**, 2187.
41. Yun, F., Reshchikov, M.A., Visconti, P. *et al.* (2001) *Materials Research Society Symposium Proceedings*, **639**, G3.17.
42. Davidsson, S.K., Andersson, T.G., and Zirath, H. (2002) *Applied Physics Letters*, **81**, 664.
43. Tanaka, S., Iwai, S., and Aoyagi, Y. (1997) *Japanese Journal of Applied Physics, Part 1*, **170**, 329.
44. Nikishin, S.A., Faleev, N.N., Antipov, V.G. *et al.* (1999) *Applied Physics Letters*, **75**, 2073.
45. Ebel, R., Fehrer, M., Figge, S. *et al.* (1999) *Journal of Crystal Growth*, **201–202**, 433.
46. Namkoong, G., Doolittle, W.A., Brown, A.S. *et al.* (2002) *Journal of Vacuum Science and Technology B*, **20**, 1221.
47. Heinlein, C., Grepstad, J., Berge, T., and Riechert, H. (1997) *Applied Physics Letters*, **71**, 341.
48. Pécz, B., Di Forte-Poisson, M.A., Huet, F. *et al.* (1999) *Journal of Applied Physics*, **86**, 6059.
49. Sampath, A.V., Bhattacharyya, A., Sandeep, I. *et al.* (2001) *Materials Research Society Symposium Proceedings*, **639**, G6.56.
50. Brandt, O., Muralidharan, R., Waltereit, P. *et al.* (1999) *Applied Physics Letters*, **75**, 4019.
51. Heying, B., Averbek, R., Chen, L.F. *et al.* (2000) *Journal of Applied Physics*, **88**, 1855.
52. Sun, Y.J., Brandt, O., Jenichen, B., and Ploog, K.H. (2003) *Applied Physics Letters*, **83**, 5178.
53. Liu, T.Y., Trampert, A., Sun, Y.J. *et al.* (2004) *Philosophical Magazine Letters*, **84**, 435.
54. Vanfleet, R.R., Simmons, J.A., Maruska, H.P. *et al.* (2003) *Applied Physics Letters*, **83**, 1139.
55. Smith, A.R., Feenstra, R.M., Greve, D.W. *et al.* (1998) *Journal of Applied Physics*, **16**, 2242.
56. Northrup, J.E., Neugebauer, J., Feenstra, R.M., and Smith, A.R. (2000) *Physical Review B*, **61**, 9932.
57. Mula, G., Adelmann, C., Moehl, S. *et al.* (2001) *Physical Review B*, **64**, 195406.
58. Adelmann, C., Brault, J., Jalabert, D. *et al.* (2002) *Journal of Applied Physics*, **91**, 9638.
59. Adelmann, C., Brault, J., Mula, G. *et al.* (2003) *Physical Review B*, **67**, 165419.
60. Schwarz, V.T.S. *GmbH, SAFIRE Hindenburgstrasse 12*, D-76332, Bad Herrenalb, Germany.
61. Brandt, O., Sun, Y.J., Däweritz, L., and Ploog, K.H. (2004) *Physical Review B*, **69**, 165326.
62. Kreuzer, H.J. (1988) in *Chemistry and Physics of Solid Surfaces VII*, (eds R. Vanselow and R.F. Howe), Vol. 7, Springer, Berlin-Heidelberg, p. 259.
63. Zinke-Allmang, M. (1992) *Surface Science Reports*, **16**, 377.

64. Brandt, O., Waltereit, P., Dhar, S. *et al.* (2002) *Journal of Vacuum Science and Technology B*, **20**, 1626.
65. Waltereit, P., Brandt, O., Ploog, K.H. *et al.* (2002) *Physical Review B*, **66**, 165322.
66. Monroy, E., Gogneau, N., Jalabert, D. *et al.* (2003) *Applied Physics Letters*, **82**, 2242.
67. Brandt, O., Waltereit, P., and Ploog, K.H. (2002) *Journal of Physics D*, **35**, 577.
68. Waltereit, P., Brandt, O., Ramsteiner, M. *et al.* (2000) *Journal of Crystal Growth*, **218**, 143.
69. Williamson, G.K. and Hall, W.H. (1953) *Acta Metallurgica*, **1**, 22.
70. Brandt, O., Sun, Y.J., Schönherr, H.-P. *et al.* (2003) *Applied Physics Letters*, **83**, 90.
71. Naranjo, F.B., Sánchez-García, M.A., Calle, F. *et al.* (2002) *Applied Physics Letters*, **80**, 231.
72. Jahn, U., Dhar, S., Brandt, O. *et al.* (2003) *Journal of Applied Physics*, **93**, 1048.
73. Dhar, S., Jahn, U., Brandt, O. *et al.* (2002) *Applied Physics Letters*, **81**, 673.
74. Chichibu, S., Sota, T., Wada, K., and Nakamura, S. (1998) *Journal of Vacuum Science and Technology B*, **16**, 2204.
75. Skromme, B.J., Jayapalan, J., Vaudo, R.P., and Phanse, V.M. (1999) *Applied Physics Letters*, **74**, 2358.
76. Kornitzer, K., Ebner, T., Thonke, K. *et al.* (1999) *Physical Review B*, **60**, 1471.
77. Narukawa, Y., Kawakami, Y., Fujita, S. *et al.* (1997) *Physical Review B*, **55**, R1938.
78. Chichibu, S.F., Marchand, H., Minsky, M.S. *et al.* (1999) *Applied Physics Letters*, **74**, 1460.
79. O'Donnell, K., Martin, R., and Middleton, P. (1999) *Physical Review Letters*, **82**, 237.
80. Holst, J.-C., Hoffmann, A., Rudloff, D. *et al.* (2000) *Applied Physics Letters*, **76**, 2832.
81. Chichibu, S.F., Onuma, T., Sota, T. *et al.* (2003) *Applied Physics Letters*, **93**, 2051.
82. Patanè, A., Garassi Alessi, M., Intonti, F. *et al.* (1997) *Physica Status Solidi A*, **164**, 493.
83. Garassi Alessi, M., Fragano, F., Patanè, A. *et al.* (2000) *Physical Review B*, **61**, 10985.
84. Lefebvre, P., Taliercio, T., Morel, A. *et al.* (2001) *Applied Physics Letters*, **78**, 1538.
85. Wu, J., Walukiewicz, W., Yu, K.M. *et al.* (2002) *Applied Physics Letters*, **80**, 4741.
86. Mayrock, O., Wünsche, H.-J., and Henneberger, F. (2000) *Physical Review B*, **62**, 16870.
87. Fiorentini, V., Bernardini, F., and Ambacher, O. (2002) *Applied Physics Letters*, **80**, 1204.
88. Mayrock, O. (2001) *Localization, Disorder, and Polarization Fields in Wide-Gap Semiconductor Quantum Wells*, Ph.D. Thesis, Humboldt-Universität zu Berlin, Berlin.
89. Chen, X., Henderson, B., and O'Donnell, K.P. (1992) *Applied Physics Letters*, **60**, 2672.
90. Li, J.Z., Lin, J.Y., Jiang, H.X. *et al.* (1996) *Applied Physics Letters*, **69**, 1474.
91. Scher, H., Shlesinger, M.F., and Bendler, J.T. (1991) *Physics Today*, **44**(1), 26.
92. Vertikov, A., Ozden, I., and Nurmikko, A.V. (1999) *Journal of Applied Physics*, **86**, 4697.
93. Pophristic, M. Long, F.H. Tran, C. *et al.* (1998) *Applied Physics Letters*, **73**, 3550.
94. Göbel, E.O. and Graudszus, W. (1982) *Physical Review Letters*, **48**, 1277.
95. Collet, J.H., Kalt, H., Dang, L.S. *et al.* (1991) *Physical Review B*, **43**, 6843.

PART II
Properties

Nitrides with Nonpolar Surfaces. Growth, Properties, and Devices. Edited by Tanya Paskova

ISBN: 978-3-527-40768-2

7
GaN Films and (In,Ga)N/GaN Multiple Quantum Wells with Nonpolar Surfaces: Optical Polarization Properties

Holger Grahn

7.1
Introduction

The interest in nonpolar orientations of the wurtzite crystal structure originates from the absence of internal electric fields in group-III-nitride-based heterostructures, which are expected to improve the efficiency of light emitters. However, GaN films and quantum wells with nonpolar orientations can also be used for novel device concepts based on polarization sensitivity. GaN films grown along the polar direction (*C* plane), which are usually unstrained or isotropically strained, do not exhibit any in-plane polarization anisotropy. In contrast, unstrained *M*- and *A*-plane GaN films, where the *c* axis lies in the film plane (cf. Figure 7.1a), exhibit a significant in-plane polarization anisotropy with a degree of linear polarization of one for the A exciton, while the polarization anisotropy is much smaller for the B and C excitons (cf. Figure 7.1b).

The in-plane polarization anisotropy can be enhanced by anisotropic strain. This can be achieved by choosing a nonpolar orientation with an appropriate substrate. In the extreme case of *M*-plane GaN on $LiAlO_2$, the degree of linear polarization can be increased to its maximum value of one for all three transitions between the three uppermost valence bands (VBs) and the conduction band (CB), corresponding to complete linear polarization for all three transitions. This optical anisotropy can be observed in transmission (absorption) and reflection, as well as photoluminescence (PL) and photoreflectance (PR) spectroscopy. It can therefore be used for polarization filtering, polarization-sensitive photodetectors (PSPDs), and polarized light emitters. For anisotropically strained *C*-plane GaN films on $(11\bar{2}0)$ sapphire, the in-plane polarization properties have been previously reported in Refs. 1–3.

The electrical and optical properties of semiconductors are mainly governed by the electronic band structure (EBS) in the vicinity of the absolute VB

Nitrides with Nonpolar Surfaces. Growth, Properties, and Devices. Edited by Tanya Paskova

ISBN: 978-3-527-40768-2

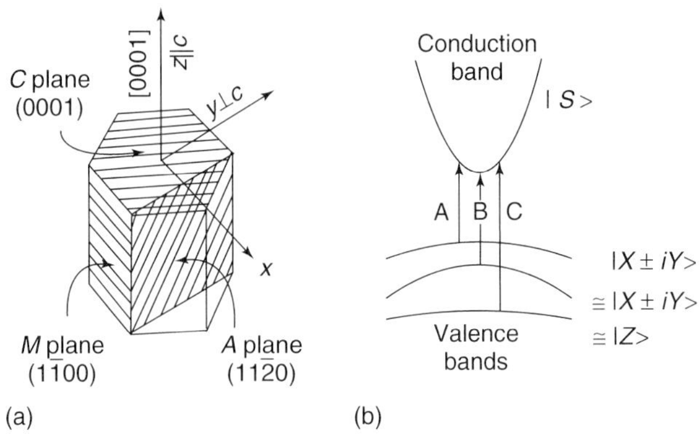

Figure 7.1 (a) The wurtzite unit cell of GaN showing the *C*, *M*, and *A* planes and the choice of coordinates. (b) Schematic EBS of unstrained GaN, showing the CB and VB wave-function symmetries.

maximum and CB minimum. In heteroepitaxial layers, the EBS of the film material can be strongly modified because of strain induced by the mismatch of the lattice constants as well as the thermal expansion coefficients between the film and substrate. To correctly predict the optical transition energies and polarization properties of the strained film, it is important to determine the dependence of the EBS modification on in-plane strain. For the wurtzite crystal structure, the strain-induced EBS modification in the vicinity of the Γ point can be theoretically estimated by adopting the $\mathbf{k}\cdot\mathbf{p}$ perturbation approach outlined by Bir and Pikus [4] and applied to GaN films on nonpolar surfaces in Ref. 5. For the transition energies at the Γ point, the number of required parameters in the $\mathbf{k}\cdot\mathbf{p}$ perturbation approach is reduced to the crystal-field (Δ_{cr}) and spin-orbit splitting (Δ_{so}) energies and six distinct deformation potentials α (CB) as well as D_1 to D_5 (VB).

For materials exhibiting the wurtzite crystal structure, Δ_{cr} and Δ_{so} are usually deduced from the differences of the A-, B-, and C-exciton transition energies (for strained samples, we use the notation T_1, T_2, and T_3 in order of increasing energy), while the deformation potentials are determined from strain-dependent measurements. For GaN, a large number of values for Δ_{cr}, Δ_{so}, and D_5, are reported in the literature, which are derived from both first-principle calculations and experimental data [5–10]. The literature values for the other deformation potentials α as well as D_1 to D_4 vary much less.

After describing the samples, we first derive values for Δ_{cr} and Δ_{so} using an unstrained *C*-plane GaN film grown on a GaN template. Then, we discuss the in-plane polarization anisotropy for an unstrained *A*-plane GaN film grown on *A*-plane GaN. Using the determined values for Δ_{cr} and Δ_{so}, we present the theoretically expected dependence of the transition energies and oscillator strengths for polar and nonpolar orientations as a function of in-plane strain derived from the **k·p** model. The experimentally obtained optical polarization anisotropy recorded at low and room temperature by PR spectroscopy is then discussed for four different GaN films with a nonpolar orientation and compared with the calculated results. The next two subsections are devoted to the polarized PL spectra of nonpolar oriented GaN films and (In,Ga)N/GaN multiple quantum wells (MQWs). Finally, we briefly discuss several examples of device applications: polarization filtering, PSPDs, very narrow-band photodetectors, and polarized light-emitting devices.

7.2 Experimental Details

The investigated GaN films of different orientations are listed in Table 7.1 according to their number, orientation, substrate, growth method, film thickness, and measured out-of-plane dilation, which was determined by high-resolution triple-axis X-ray diffraction (HRXRD) measurements at 295 K. As shown in Figure 7.1a, the coordinates x, y, and z are parallel to the $[11\bar{2}0]$, $[1\bar{1}00]$, and [0001] directions, respectively. Sample I was grown by plasma-assisted molecular-beam epitaxy (PAMBE) on a commercially available 200-μm-thick freestanding GaN template. The *A*-plane GaN substrate for sample II was obtained by first growing 0.1-mm GaN by the high-pressure synthesis method. In the next step, the plates were overgrown using hydride vapor phase epitaxy (HVPE) to achieve a thickness of 2–3 mm. Finally, these plates were

Table 7.1 Sample number, orientation, substrate, growth method, film thickness, and measured out-of-plane dilation. $LiAlO_2$ refers to γ-$LiAlO_2(100)$.

#	Orientation	Substrate	Growth method	Thickness (μm)	Out-of-plane dilation (%)
I	*C*	GaN template	PAMBE	1.0	$\lvert\varepsilon_{zz}\rvert < 0.01$
II	*A*	*A*-plane GaN	PAMBE	0.27	$\lvert\varepsilon_{xx}\rvert < 0.01$
III	*M*	$LiAlO_2$	PAMBE	1.22	$\varepsilon_{yy} = 0.29$
IV	*M*	$LiAlO_2$	PAMBE	0.7	$\varepsilon_{yy} = 0.42$
V	*A*	*A*-plane 6H-SiC	MOVPE	1.0	$\varepsilon_{xx} = 0.04$
VI	*A*	*R*-plane sapphire	MOVPE	1.0	$\varepsilon_{xx} = 0.09$
VII	*M*	$LiAlO_2$	PAMBE	0.4	$\varepsilon_{yy} = 0.43$
VIII	*M*	$LiAlO_2$	PAMBE	0.7	$\varepsilon_{yy} = 0.35$
IX	*M*	$LiAlO_2$	PAMBE	0.4	$\varepsilon_{yy} = 0.39$

PAMBE: plasma-assisted molecular-beam epitaxy, HVPE: hydride vapor phase epitaxy, MOVPE: metal-organic vapor phase epitaxy.

precisely cut along an exact crystallographic direction perpendicular to the c axis and mechanochemically polished, obtaining substrates in the form of stripes with a length and width of a few millimeters. The growth on γ-$LiAlO_2$(100) substrates by PAMBE has been described in Refs. 11 and 12 and Chapter 6. Samples V and VI were grown by metal-organic vapor phase epitaxy (MOVPE) on a 0.2-μm-thick AlN nucleation layer followed by the undoped GaN film. A detailed description of the growth conditions and the postgrowth characterization of these films has been reported in Refs. 13 and 14. Samples III and IV were grown in one machine, samples VII and VIII in another one at different times, and sample IX in a third machine; as a result, sample VIII (IX) and sample IV (VII) were grown at different times and in different machines.

The $In_{0.1}Ga_{0.9}N$/GaN MQW (sample X) was also grown by PAMBE on γ-$LiAlO_2$(100). The MQW consists of 20 periods of 1.9-nm $In_{0.1}Ga_{0.9}N$ wells and 6.6-nm GaN barriers. A detailed description of the growth conditions can be found in Ref. 15 and Chapter 6. The orientation of the MQW was verified using HRXRD. For the (In,Ga)N/GaN MQW, a biaxial compressive strain with an out-of-plane dilation of $\varepsilon_{yy} = 1.26\%$ is deduced for the $In_{0.1}Ga_{0.9}N$ layers [15].

The reflectance and PR measurements of samples I–II and samples IV–VI, respectively, were carried out at near normal incidence between 6 and 10 K and at 295 K. For the PR measurements, the pump beam at 300 nm (4.13 eV) was obtained from a combination of a 150 W Xe lamp with a 1-m monochromator, while the beam for both the reflectance and PR probe was derived by dispersing the light from a 75 W Xe lamp with a 0.64-m monochromator and polarizing it with a Glan–Taylor prism. The reflected light was detected by a silicon photodiode using standard lock-in techniques.

The continuous-wave (cw) PL spectra of the M-plane GaN films (samples VII and VIII) were excited at 244 nm by a frequency-doubled Ar^+ laser, dispersed by a 1-m monochromator, and detected by a UV-enhanced GaAs photomultiplier. The polarization properties of the PL spectra were studied by polarizing the emitted PL signal before it entered the monochromator. All spectra were recorded under normal incidence so that the c axis of the GaN films was oriented perpendicular to the propagation direction of the incoming light. Excitation power densities of up to 50 W cm^{-2} were used.

The polarization anisotropy for the (In,Ga)N/GaN MQW sample (sample X) was measured in emission with a μ-PL setup in backscattering geometry, utilizing the 325-nm line of a He–Cd laser for excitation and a cooled charge-coupled-device array for detection. The linear polarization of the PL signal at 10 K was analyzed by rotating a polarizer in front of a single monochromator between a polarization angle $\phi_i = 0^\circ$ and 360°, where $\phi_i = 0^\circ$ refers to $\mathbf{z}\|\mathbf{E}\|\mathbf{c}$. The experimental data were carefully corrected for the polarization dependence of the setup, as confirmed by the fact that the luminescence spectra of C-plane structures remained unpolarized as expected.

7.3 The Effect of Strain on the Exciton Transition Energies and Oscillator Strengths

In this section, we first describe the exciton transition energies for the three uppermost VBs and the corresponding oscillator strengths for zero strain in GaN films with polar and nonpolar planes. Following this, the effect of strain is considered. We apply the **k·p** perturbation approach for the wurtzite crystal structure as described in Refs. 4 and 5 with a single set of parameters. For the calculations, the following parameters have been used: the low-temperature energy gap $E_g = 3.504$ eV of GaN, an energy gap at room temperature reduced by 70 meV [16], an exciton binding energy of $E_{EB} = 26$ meV [9], the crystal-field $\Delta_{cr} = 9.2$ meV and spin-orbit splitting energies $\Delta_{so} = 18.9$ meV [17], the deformation-potential parameters $\alpha = -44.5$ eV for the CB and $D_1 = -41.4$ eV, $D_2 = -33.3$ eV, $D_3 = 8.2$ eV, and $D_4 = -4.1$ eV [5, 9] as well as $D_5 = -3.6$ eV [17] for the VBs. We used the elastic stiffness constants $C_{11} = 390$, $C_{12} = 145$, $C_{13} = 106$, and $C_{33} = 398$ [18]. The momentum matrix elements parallel and perpendicular to **z** are equal according to Ref. 19. There are several approaches to mathematically solve the equations, a numerical diagonalization of the Hamilton operator [5] or an analytical solution [20].

7.3.1 Unstrained GaN with Polar and Nonpolar Orientations

To obtain accurate values of the crystal-field and spin-orbit splitting energies (Δ_{cr} and Δ_{so}), the reflectance spectrum for sample I was recorded at 10 K as shown in Figure 7.2. It exhibits very sharp resonances owing to the free A, B, and C excitons. The reflectance spectrum is fitted with a Lorentzian line shape [21]

$$R(E) = R_0 + R_X \mathfrak{Re}\left[\frac{E_X - E + i\Gamma_X}{\Gamma_X^2 + (E - E_X)^2} e^{i\Theta_X}\right] \tag{1}$$

where X denotes the exciton A, B, or C, $\mathfrak{Re}$ the real part of the function, R_X the reflectance, E_X the corresponding transition energy, and Γ_X a broadening parameter. Θ_X is a phase factor, which accounts for the mixture of the real and imaginary components of the dielectric function as well as the influence of nonuniform electric fields, while R_0 corresponds to the background. The transition energies obtained for sample I are $E_A = 3.478$, $E_B = 3.483$, and $E_C = 3.501$ eV. These values agree quite well with corresponding exciton transition energies measured for unstrained films grown on GaN substrates [22–24]. These values are also reproduced from the calculations using the parameters Δ_{cr} and Δ_{so} given above within an accuracy of 1 meV ($E_A^{cal} = 3.478$, $E_B^{cal} = 3.483$, and $E_C^{cal} = 3.501$ eV).

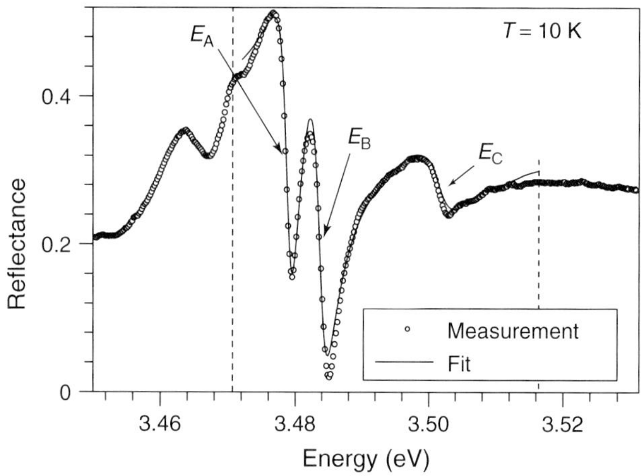

Figure 7.2 Reflectance spectrum (circles) of sample I measured at 10 K. The solid line indicates a fit to Equation 1 for the energy range marked by the vertical dashed lines, resulting in $E_A = 3.478$, $E_B = 3.483$, and $E_C = 3.501$ eV.

To determine Δ_{cr} and Δ_{so} from the exciton transition energies for the unstrained *C*-plane GaN film, we apply Hopfield's quasicubic model developed for the wurtzite crystal structure [25]. The relation between the differences $\delta_{X,Y} = E_X - E_Y$ in the exciton transition energies and the splitting energies is given using $\delta_+ = \delta_{C,A}$ and $\delta_- = \delta_{B,A}$ by

$$\delta_\pm = \frac{\Delta_{cr} + \Delta_{so}}{2} \pm \sqrt{\left(\frac{\Delta_{cr} + \Delta_{so}}{2}\right)^2 - \frac{2}{3}\Delta_{cr}\Delta_{so}} \quad (2)$$

Solving the above equations for $\Delta_- = \Delta_{cr}$ and $\Delta_+ = \Delta_{so}$ leads to

$$\Delta_\pm = \frac{1}{2}\left(\delta_{C,A} + \delta_{B,A} \pm \sqrt{2\delta_{C,B}^2 - \delta_{B,A}^2 - \delta_{C,A}^2}\right) \quad (3)$$

Using the values for E_X given above, we obtain $\Delta_{cr} = 9.2$ and $\Delta_{so} = 18.9$ meV. While the spin-orbit splitting energy is in good agreement with values in the literature, there is a wide variation in the values obtained for the crystal-field splitting energy, since this energy is more strongly affected by strain [10]. However, our value for Δ_{cr} is in very good agreement with the ones given by Stępniewski *et al.* [22] and Kornitzer *et al.* [24], which were also obtained for homoepitaxial GaN.

For unstrained GaN, the crystal symmetry results in equal oscillator strengths for the x and y directions, that is, $f_x = f_y$. For the A exciton, the calculated oscillator strengths are $f_x = 0.50$ and $f_z = 0$, for the B exciton $f_x = 0.29$ and $f_z = 0.42$, and for the C exciton $f_x = 0.21$ and $f_z = 0.58$. Consequently, there is no in-plane polarization anisotropy for unstrained *C*-plane GaN films. This property holds even for isotropically strained *C*-plane GaN films. If we define the calculated degree of linear in-plane polarization for k and l denoting x, y, or z as

$$\rho_{kl} = \left| \frac{f_k - f_l}{f_k + f_l} \right| \tag{4}$$

we obtain for all three excitonic transitions $\rho_{xy} = 0$.

For unstrained GaN films with nonpolar planes, the transition energies are the same. However, the in-plane polarization properties are completely different. First of all and most importantly, the A exciton becomes completely polarized, that is, $\rho_{yz}(\rho_{xz}) = 1$, for both *A*- and *M*-plane GaN films. This result has been experimentally verified using sample II. This is shown in Figure 7.3, where its low-temperature reflectance spectrum is shown for two in-plane polarization directions. For linear polarization parallel to the c axis ($\mathbf{E}\|\mathbf{c}$), the A exciton is completely absent, while the transition becomes very strong for linear polarization perpendicular to the c axis ($\mathbf{E}\perp\mathbf{c}$). The exciton transition energies derived from a fit to two ($E_B^{fit} = 3.484$ and $E_C^{fit} = 3.502$ eV) and three Lorentzian lines ($E_A^{fit} = 3.479$, $E_B^{fit} = 3.484$, and $E_C^{fit} = 3.502$ eV) for $\mathbf{E}\|\mathbf{c}$ and $\mathbf{E}\perp\mathbf{c}$, respectively, are again in excellent agreement with the values obtained for the unstrained *C*-plane film, and the calculated values.

Second, the B and C excitons are partially polarized in *M*-plane (*A*-plane) GaN films with values of $\rho_{xz} = 0.18$ ($\rho_{yz} = 0.18$) and $\rho_{xz} = 0.47$ ($\rho_{yz} = 0.47$), respectively. If the A and B excitons are not spectrally resolved, the degree of linear in-plane polarization becomes $\rho_{xz} = 0.31$ ($\rho_{yz} = 0.31$) for both transitions together. Therefore, polarized light emission could be obtained

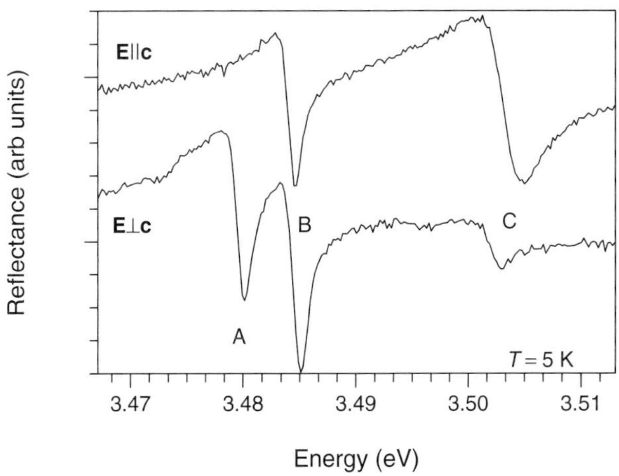

Figure 7.3 Reflectance spectrum of sample II measured for light polarization $\mathbf{E}\|\mathbf{c}$ (upper curve) and $\mathbf{E}\perp\mathbf{c}$ (lower curve) at 5 K. The transitions energies determined from a fit are $E_A = 3.479$, $E_B = 3.484$, and $E_C = 3.502$ eV. The curves have been shifted vertically for better readability.

using unstrained GaN films, if they are grown on nonpolar planes and the linewidths of the A and B excitons are much smaller than their energy separation.

7.3.2
Strain Dependence for Polar Orientations of GaN

In the case of the polar orientation, the GaN films are usually grown on *C*-plane sapphire or 6H-SiC so that the strain due to the lattice mismatch between the film and the substrate is isotropic in the film plane. The calculated transition energies T_1, T_2, and T_3, as well as the difference between T_1 and T_2 are shown in Figure 7.4a–d, respectively, as a function of the in-plane strain ε_{xx} and ε_{yy} for *C*-plane GaN. Note that due to the symmetry of the in-plane crystal structure, the transition energies are symmetric with respect to the interchange of ε_{xx} and ε_{yy}. The solid lines in Figure 7.4a–c denote the condition $\varepsilon_{xx} = \varepsilon_{yy}$. For zero strain, the separation between the T_1 and T_2 transitions amounts only to about 5 meV, while the energy difference between the transitions T_1 and T_3 is 23 meV. The transition energies T_1 and T_2 become equal for $\varepsilon_{xx} = \varepsilon_{yy} = 0.08\%$, that is, for an isotropic tensile strain of 0.08%. In contrast, for large isotropic positive strain values (tensile) as well as as large anisotropic positive (tensile), negative (compressive), and mixed strain values,

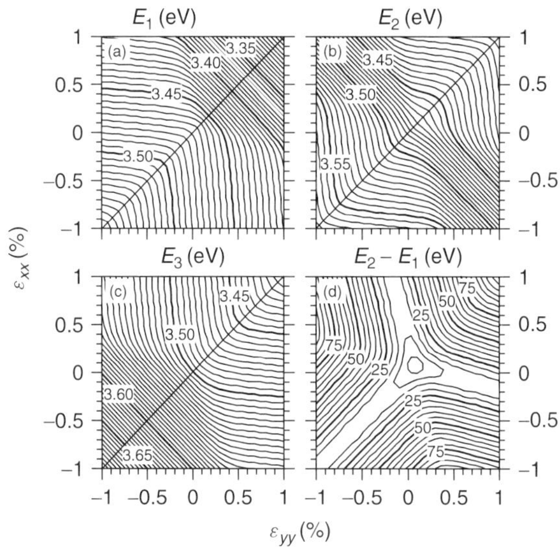

Figure 7.4 Calculated values of the energies (a) E_1, (b) E_2, and (c) E_3 for the transitions T_1, T_2, and T_3, respectively, as a function of the in-plane strain ε_{xx} and ε_{yy} for a *C*-plane GaN film at 5 K. The thin diagonal lines indicate isotropic in-plane strain, that is, $\varepsilon_{xx} = \varepsilon_{yy}$. (d) Energy difference $E_2 - E_1$ as a function of the in-plane strain ε_{xx} and ε_{yy}.

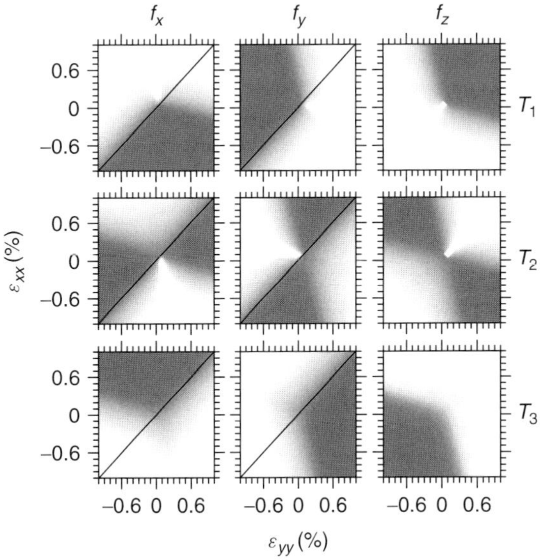

Figure 7.5 Relative oscillator strength components f_x, f_y, and f_z for the T_1, T_2, and T_3 transitions as a function of the in-plane strain ε_{xx} and ε_{yy} for a C-plane GaN film. The oscillator strength is shown on a logarithmic scale, where white corresponds to $f_k = 0$ and black to $f_k = 1$. The thin diagonal lines in the panels for f_x and f_y indicate isotropic in-plane strain, that is, $\varepsilon_{xx} = \varepsilon_{yy}$.

one could achieve a separation of the transitions T_1 and T_2 of more than 50 meV.

The calculated oscillator strengths f_k, where k denotes x, y, or z, for the three transitions T_1, T_2, and T_3 are shown in Figure 7.5 as a function of the in-plane strain ε_{xx} and ε_{yy} for C-plane GaN. For isotropic strain, the oscillator strengths f_x and f_y are always equal. Therefore, there are no strain values for which all three transitions become completely linearly polarized. However, for anisotropic strain, there are strain values for a particular transition, for which the oscillator strength becomes one for one direction, while it is zero for the other two directions. For the same strain values, the oscillator strength of another transition becomes one for a different direction and zero for the two orthogonal ones. Therefore, for anisotropic strain, all three transitions become completely linearly polarized for certain values of the in-plane strain.

7.3.3
Strain Dependence for Nonpolar Orientations of GaN

For nonpolar orientations, the GaN films are usually grown on $LiAlO_2$, R-plane sapphire, and A- or M-plane SiC so that the strain due to the lattice mismatch

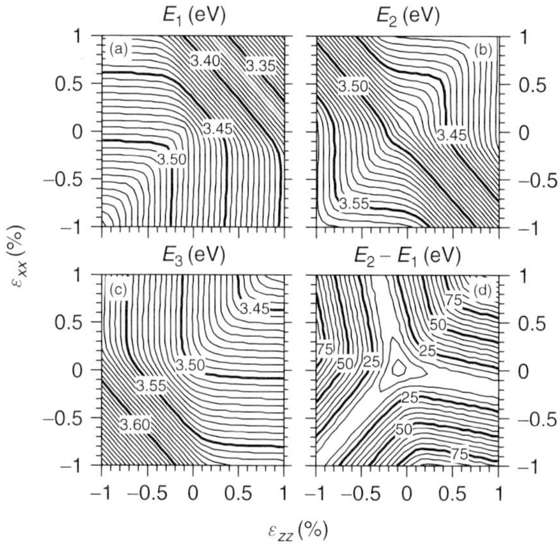

Figure 7.6 Calculated values of the energies (a) E_1, (b) E_2, and (c) E_3 for the transitions T_1, T_2, and T_3, respectively, as a function of the in-plane strain ε_{xx} and ε_{zz} for an *M*-plane or *A*-plane GaN film at 5 K. (d) Energy difference $E_2 - E_1$ as a function of the in-plane strain ε_{xx} and ε_{zz}.

between the film and the substrate is always anisotropic in the film plane. The calculated transition energies T_1, T_2, and T_3 as well as the difference between T_1 and T_2 are shown in Figure 7.6a–d, respectively, as a function of the in-plane strain ε_{xx} and ε_{zz} for *M*-plane GaN. Due to the symmetry of the wurtzite crystal structure, the Hamiltonian is symmetric with regard to interchanging the x and y directions. Therefore, the results calculated for *M*-plane GaN are also valid for *A*-plane GaN. Note that due to the lack of symmetry of the in-plane crystal structure, the symmetry of the transition energies with respect to the interchange of the in-plane strain components ε_{xx} (ε_{yy}) and ε_{zz} is lost. For zero strain, the results are identical to the ones in Figure 7.4. Note that Figure 7.6d differs also from Figure 7.4d in the sense that the transition energies T_1 and T_2 become equal for $\varepsilon_{xx} = 0.02\%$ and $\varepsilon_{yy} = -0.09\%$, which is anisotropic. As for *C*-plane GaN, one can achieve a separation of the transitions T_1 and T_2 of more than 50 meV for large anisotropic positive (tensile), negative (compressive), and mixed strain values.

The calculated oscillator strengths f_k for the three transitions T_1, T_2, and T_3 are shown in Figure 7.7 as a function of the in-plane strain ε_{xx} and ε_{zz} for *M*-plane GaN. For example, one can achieve almost complete linear polarization of all three transitions, that is, $f_x \approx 1$ for T_1, $f_z \approx 1$ for T_2, and $f_y \approx 1$ for T_3, for in-plane strain values $\varepsilon_{xx} = -0.91\%$ and $\varepsilon_{zz} = -0.30\%$, which corresponds to very anisotropic compressive strain. We discuss specific calculated values of the oscillator strengths in greater detail in the next section.

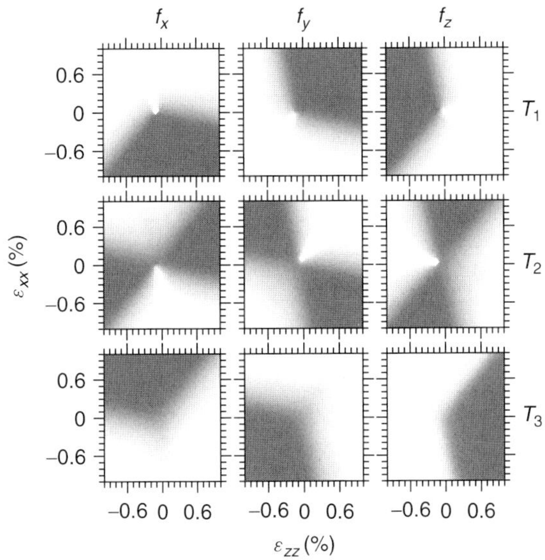

Figure 7.7 Relative oscillator strength components f_x, f_y, and f_z for the T_1, T_2, and T_3 transitions as a function of the in-plane strain ε_{xx} and ε_{zz} for an *M*-plane or *A*-plane GaN film. The oscillator strength is shown on a logarithmic scale, where white corresponds to $f_k = 0$ and black to $f_k = 1$.

7.4 Optical Polarization Anisotropy in GaN Films and (In,Ga)N/GaN Multiple Quantum Wells

7.4.1 Strained GaN Films with Different Nonpolar Orientations

Figure 7.8a–c displays the PR spectra for samples IV, V, and VI, respectively, measured at 10 K for $\mathbf{E}\|\mathbf{c}$ and $\mathbf{E}\perp\mathbf{c}$ in the film plane [17]. In addition to the z axis, the x axis (y axis) also lies in the film plane for *M*-plane (*A*-plane) GaN films. The PR spectra have a complex line shape, which is dominated by excitonic features. However, on the low-energy and high-energy sides of the PR spectra, these features are superimposed by interference fringes and Franz–Keldysh oscillations, respectively. To extract the exciton transition energies from the PR spectra, we use the first derivative of a Gaussian oscillator and exclude the regions of interference fringes and Franz–Keldysh oscillations for the fit. For sample IV, the PR spectra in Figure 7.8a reveal a single transition at 3.504 eV for $\mathbf{E}\perp\mathbf{c}$, while another transition appears at 3.558 eV for $\mathbf{E}\|\mathbf{c}$. For sample V, the situation is altogether different. Here the measurements shown in Figure 7.8b suggest a single transition at 3.463 eV for $\mathbf{E}\perp\mathbf{c}$ and two transitions at 3.493 and 3.514 eV for $\mathbf{E}\|\mathbf{c}$. For sample VI, the measurements shown in Figure 7.8c reveal that there is a single transition at 3.493 eV for $\mathbf{E}\perp\mathbf{c}$ and

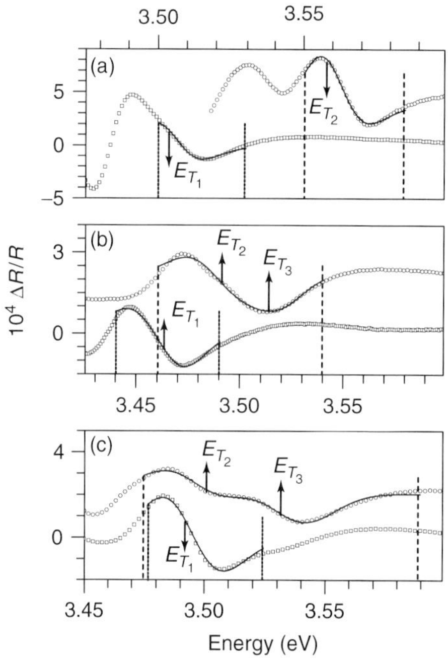

Figure 7.8 PR spectra of (a) sample IV, (b) sample V, and (c) sample VI measured for light polarization **E**$\|$**c** (upper curves) and **E**$\perp$**c** (lower curves) at 10 K. The solid lines indicate the fits described in the text for the energy ranges marked by the vertical dashed lines. The resulting energies for the T_1, T_2, and T_3 transitions are marked by E_{T_1}, E_{T_2}, and E_{T_3}, respectively, and are listed in Table 7.2. For clarity, the PR spectra for **E**$\|$**c** in (a) are shifted vertically by 4×10^{-4}, while the shift in (b) and (c) is 2×10^{-4}.

two transitions at 3.499 and 3.531 eV for **E**$\|$**c**. All experimental values for the transition energies at low temperatures and the out-of-plane dilation at room temperature of samples IV–VI are listed in Table 7.2. In addition, the values for sample III taken from Ref. 5 are also included.

A systematic approach to accurately determine the deformation potential D_5 is the reproduction of the low-temperature transition energies and polarization properties of the resonant features observed in the PR spectra for all four films with a nonpolar orientation listed in Table 7.2 using the $\mathbf{k}\cdot\mathbf{p}$ perturbation approach presented in Ref. 5 with a single set of parameters. We vary the deformation potential D_5 over a range of values and find that $D_5 = -3.6$ eV results in the best agreement of the measured and calculated transition energies taking into account the oscillator strengths and the out-of-plane dilation. The crystal-field and spin-orbit splitting energies (Δ_{cr} and Δ_{so}) are taken from Section 7.3.1. The low-temperature value of $E_g = 3.504$ eV has been derived from the reflectance spectra of the unstrained *C*-plane GaN film in Figure 7.2. The resulting transitions energies and relative oscillator strengths f_k are compiled in Table 7.2.

For sample III, the transition energies calculated for the strain values of $\varepsilon_{xx} = -0.73\%$ and $\varepsilon_{zz} = -0.22\%$ agree quite well with the measured ones. The calculated oscillator strength of $f_x = 0.98$ ($f_z = 0.94$) suggests that the T_1 (T_2) transition is almost completely x polarized (z polarized). Because of the smaller thickness of sample IV in comparison to sample III, ε_{yy} in sample IV

Table 7.2 Measured (E_{exp}) and calculated energies (E_{cal}) for 10 K of the transitions T_i ($i = 1$, 2, and 3) for samples III, IV, V, and VI for **E**⊥**c** (⊥) and **E**‖**c** (‖), measured out-of-plane dilation ε_{opd}, in- and out-of-plane strain values $\varepsilon_{\beta\beta}$ obtained from the calculated transition energies, and calculated oscillator strengths f_k for the corresponding linear polarization. For the calculations, we used the parameters $\Delta_{cr} = 9.2$ meV, $\Delta_{so} = 18.9$ meV, $D_5 = -3.6$ eV, and $E_g = 3.504$ eV. The experimental values for sample III were taken from Ref. 5.

#	Experiment		k·p calculations					
	E_{exp} (eV)	ε_{opd} (%)	$\varepsilon_{\beta\beta}$ (%)	T_i	E_{cal} (eV)	f_x	f_y	f_z
III	3.498 (⊥)		$\varepsilon_{xx} = -0.73$	T_1	3.498	0.98	0.00	0.02
(*M*)	3.546 (‖)		$\varepsilon_{zz} = -0.22$	T_2	3.544	0.01	0.05	0.94
		$\varepsilon_{yy} = 0.29$	$\varepsilon_{yy} = 0.33$	T_3	3.578	0.01	0.95	0.04
IV	3.504 (⊥)		$\varepsilon_{xx} = -0.91$	T_1	3.504	0.98	0.01	0.01
(*M*)	3.558 (‖)		$\varepsilon_{zz} = -0.30$	T_2	3.557	0.01	0.02	0.97
		$\varepsilon_{yy} = 0.42$	$\varepsilon_{yy} = 0.42$	T_3	3.602	0.01	0.97	0.02
V	3.463 (⊥)		$\varepsilon_{yy} = -0.29$	T_1	3.463	0.04	0.95	0.01
(*A*)	3.493 (‖)		$\varepsilon_{zz} = 0.21$	T_2	3.489	0.89	0.03	0.08
	3.514 (‖)	$\varepsilon_{xx} = 0.04$	$\varepsilon_{xx} = 0.05$	T_3	3.517	0.07	0.02	0.91
VI	3.493 (⊥)		$\varepsilon_{yy} = -0.17$	T_1	3.492	0.03	0.85	0.12
(*A*)	3.499 (‖)		$\varepsilon_{zz} = -0.14$	T_2	3.503	0.30	0.04	0.66
	3.531 (‖)	$\varepsilon_{xx} = 0.09$	$\varepsilon_{xx} = 0.10$	T_3	3.521	0.67	0.11	0.22

is significantly larger than in sample III (cf. Table 7.2). The calculated and measured transition energies are in excellent agreement for $\varepsilon_{xx} = -0.91\%$ and $\varepsilon_{zz} = -0.30\%$ (comparable to the lattice mismatch between the film and substrate along the c axis), revealing that sample IV has a larger biaxial, anisotropic in-plane strain than sample III. As expected, the transition energies of sample IV are shifted to higher values due to the larger compressive strain. As for sample III, the calculated oscillator strength $f_x = 0.98$ ($f_z = 0.97$) of sample IV suggests that the T_1 (T_2) transition is completely x polarized (z polarized). The transition between the CB and the third highest VB, which gives rise to the T_3 transition, is y polarized for samples III and IV. Under normal incidence, this transition is not accessible for M-plane films.

The polarization properties of samples V and VI are quite different from those of samples III and IV, because the in-plane strain in the former samples is considerably smaller than in the latter ones. In samples V and VI, two transitions appear for **E**‖**c** in contrast to a single transition in samples III and IV, while one transition is again measured for **E**⊥**c**. The calculated energies agree with the measured ones within the experimental accuracy of about 5 meV. According to the calculated oscillator strengths, the single transition for **E**⊥**c** is almost completely y polarized (note the different in-plane axis because of the A-plane orientation), while the two transitions observed for **E**‖**c** are due to a significant oscillator strength of the transitions T_2 and T_3

for z polarization. Note, however, that for sample V f_z for T_2 is still rather small, which needs further investigation. The resulting in-plane strain values for sample V (VI) are $\varepsilon_{yy} = -0.29\%$ (-0.17%) and $\varepsilon_{zz} = 0.21\%$ (-0.14%).

The PR spectrum of the C-plane GaN sample at 295 K is quite broad, and the A and B exciton cannot be resolved [16]. A fit using a single oscillator (superposition of A and B exciton) yields a transition energy of $E = 3.414$ eV. Using the energy separation of the A and B exciton at $T = 10$ K of about 5 meV, we estimate a transition energy of 3.411 eV for the A exciton and 3.416 eV for the B exciton at 295 K. Therefore, the experimentally determined difference in the energy gap between $T = 10$ and 295 K is about 68 meV. Figure 7.9a–c displays the PR spectra for samples IV, V, and VI, respectively, measured at 295 K for **E**$\|$**c** and **E**$\perp$**c**. For sample IV, the PR spectra in Figure 7.9a reveal a single transition at 3.428 eV for **E**$\perp$**c**, while another transition appears at 3.478 eV for **E**$\|$**c**. For sample V, the measurements shown in Figure 7.9b suggest a single transition at 3.421 eV for **E**$\perp$**c** and two transitions at 3.441 and 3.446 eV for **E**$\|$**c**. For sample VI, the measurements shown in Figure 7.9c reveal that there is a single transition at 3.432 eV for **E**$\perp$**c** and two transitions at 3.439 and 3.462 eV for **E**$\|$**c**. All experimental values for the room-temperature transition energies and out-of-plane dilation of samples IV–VI are listed in Table 7.3. In addition, the values for sample III taken from Ref. 5 are also included. Note that for samples III and IV the difference in the energy gap between room and low temperatures is about 70 meV, which is almost

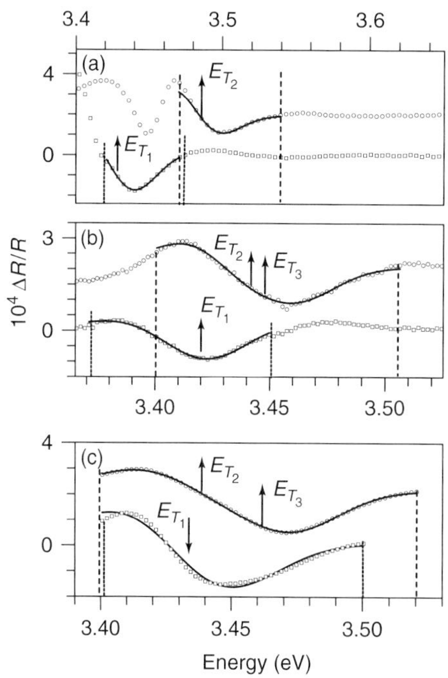

Figure 7.9 PR spectra of (a) sample IV, (b) sample V, and (c) sample VI measured for light polarization **E**$\|$**c** (upper curves) and **E**$\perp$**c** (lower curves) at 295 K. The solid lines indicate the fits described in the text for the energy ranges marked by the vertical dashed lines. The resulting energies for the T_1, T_2, and T_3 transitions are marked by E_{T_1}, E_{T_2}, and E_{T_3}, respectively, and are listed in Table 7.3. For clarity, the PR spectra for **E**$\|$**c** are shifted vertically by 2×10^{-4}.

Table 7.3 Measured (E_{exp}) and calculated energies (E_{cal}) for 295 K of the transitions T_i ($i = 1, 2$, and 3) for samples III, IV, V, and VI for **E**⊥**c** (⊥) and **E**∥**c** (∥), measured out-of-plane dilation ε_{opd}, and in- and out-of-plane strain values $\varepsilon_{\beta\beta}$ obtained from the calculated transition energies, and calculated oscillator strengths f_k for the corresponding linear polarization. For the calculations, we used the parameters $\Delta_{cr} = 9.2$ meV, $\Delta_{so} = 18.9$ meV, $D_5 = -3.6$ eV, and $E_g = 3.434$ eV. The experimental values for sample III were taken from Ref. 5.

#	Experiment		k·p calculations					
	E_{exp} (eV)	ε_{opd} (%)	$\varepsilon_{\beta\beta}$ (%)	T_i	E_{cal} (eV)	f_x	f_y	f_z
III	3.428 (⊥)		$\varepsilon_{xx} = -0.64$	T_1	3.426	0.98	0.00	0.02
(*M*)	3.468 (∥)		$\varepsilon_{zz} = -0.19$	T_2	3.468	0.01	0.07	0.92
		$\varepsilon_{yy} = 0.29$	$\varepsilon_{yy} = 0.29$	T_3	3.496	0.01	0.93	0.06
IV	3.428 (⊥)		$\varepsilon_{xx} = -0.89$	T_1	3.426	0.99	0.00	0.01
(*M*)	3.478 (∥)		$\varepsilon_{zz} = -0.20$	T_2	3.485	0.01	0.04	0.95
		$\varepsilon_{yy} = 0.42$	$\varepsilon_{yy} = 0.39$	T_3	3.520	0.00	0.96	0.04
V	3.421 (⊥)		$\varepsilon_{yy} = -0.21$	T_1	3.421	0.03	0.89	0.08
(*A*)	3.441 (∥)		$\varepsilon_{zz} = -0.13$	T_2	3.436	0.28	0.03	0.69
	3.446 (∥)	$\varepsilon_{xx} = 0.04$	$\varepsilon_{xx} = 0.11$	T_3	3.453	0.69	0.08	0.23
VI	3.432 (⊥)		$\varepsilon_{yy} = -0.23$	T_1	3.428	0.00	0.81	0.19
(*A*)	3.439 (∥)		$\varepsilon_{zz} = -0.22$	T_2	3.440	0.16	0.13	0.71
	3.462 (∥)	$\varepsilon_{xx} = 0.09$	$\varepsilon_{xx} = 0.15$	T_3	3.463	0.84	0.06	0.10

identical to the experimental value obtained on the unstrained GaN film of sample I.

A similar analysis as performed for low temperatures leads to the results at room temperature listed in Table 7.3. Note that the strain values are different from the ones at low temperature owing to the different thermal expansion coefficients of the film and substrate material.

7.4.2
Polarized Photoluminescence Spectroscopy of *M*-plane GaN Films

Figure 7.10a and b shows polarized PL spectra of samples VII and VIII for detection **E**⊥**c** and **E**∥**c**, respectively, at 9 K. For sample VII (sample VIII), the PL peaks at 3.505 eV (3.491 eV) and 3.561 eV (3.549 eV) are attributed to the transitions T_1 and T_2, respectively. These energies are blue shifted with respect to the corresponding transitions of unstrained GaN due to the presence of in-plane biaxial compressive strain. From the measured transition energies and out-of-plane dilation, we can determine the in-plane strain components as discussed in the previous subsection for PR spectroscopy.

The measured degree of linear in-plane polarization as defined by

$$\rho = \frac{I_{\perp}^{PL} - I_{\parallel}^{PL}}{I_{\perp}^{PL} + I_{\parallel}^{PL}} \quad (5)$$

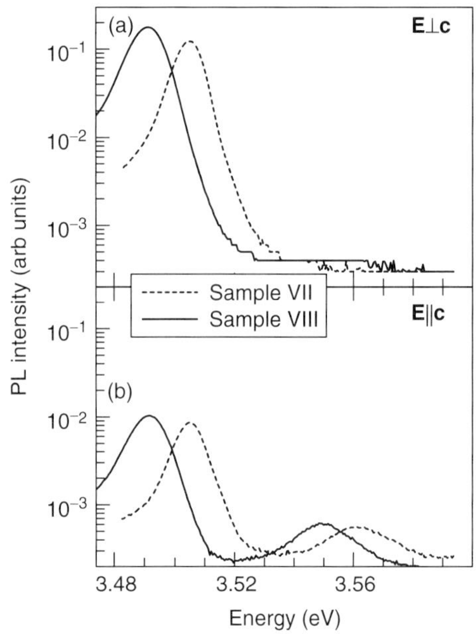

Figure 7.10 Polarized PL spectra of the transitions T_1 and T_2 in samples VII (dashed line) and VIII (solid line) detected for (a) $\mathbf{E}\perp\mathbf{c}$ and (b) $\mathbf{E}\|\mathbf{c}$ at 9 K. Note that the scale for the PL intensity is identical in (a) and (b) in order to directly compare the intensities.

where $I_{\perp}^{\mathrm{PL}}$ ($I_{\|}^{\mathrm{PL}}$) denotes the PL intensity for $\mathbf{E}\perp\mathbf{c}$ ($\mathbf{E}\|\mathbf{c}$) obtained from the data in Figure 7.10 is 0.87 (0.89) for T_1 in sample VII (sample VIII). These values agree quite well with the result of the theoretical calculations. However, for T_2, the degree of polarization appears to be -1.0 for both samples, since this transition is only visible for $\mathbf{E}\|\mathbf{c}$. According to Figure 7.10, the intensity of the transition T_2 for $\mathbf{E}\perp\mathbf{c}$ was at least 500 times smaller than the one of T_1.

The transition energies determined at 9 K together with the out-of-plane dilation measured at room temperature lead to the in-plane strain components for sample VII (sample VIII) of $\varepsilon_{xx} = -0.93\%$ ($\varepsilon_{xx} = -0.85\%$) and $\varepsilon_{zz} = -0.30\%$ ($\varepsilon_{zz} = -0.12\%$). Corresponding results for sample VII (sample VIII) at room temperature are $\varepsilon_{xx} = -0.89\%$ ($\varepsilon_{xx} = -0.77\%$) and $\varepsilon_{zz} = -0.36\%$ ($\varepsilon_{zz} = -0.24\%$). The recalculated transition energies agree within an error of less than 2 meV with the measured ones. The strain values at low and room temperature are quite similar. However, ε_{xx} decreases with increasing temperature, while ε_{zz} increases, resulting in a smaller VB splitting at room temperature. This effect can be more clearly seen for sample VIII. Comparing the strain values for samples VII and VIII, the change in ε_{zz} between low and room temperature appears to be larger for sample VIII than for sample VII, which in turn exhibits a larger VB splitting. The strain, which is already anisotropic, also seems to relax in an anisotropic fashion due to an anisotropic mismatch of the thermal expansion coefficients.

Taking into account the rather large splitting between the two uppermost VBs, most photoexcited holes will relax from the second uppermost to the uppermost VB before recombination. Therefore, at low temperatures, the intensity of T_2 will be determined by the competition between the radiative recombination time τ_{rad} and the relaxation time τ_{rel}. In particular, the much larger separation of the two uppermost VBs of more than 50 meV in comparison to a few meV volts in unstrained C-plane GaN is responsible for the large difference in the PL intensity for the T_1 and T_2 transitions. As the temperature increases, the average hole population in the second uppermost VB also increases. Hence, the PL intensity of T_2 is expected to behave differently from the one of T_1. Figure 7.11 shows the temperature dependence of the integrated PL intensity for both samples and polarization directions. In Figure 7.11b, the relative intensity of T_2 with respect to T_1 increases with increasing temperature, since a certain fraction of holes does not contribute to the relaxation process and the nonradiative recombination must compete with the hole-relaxation process. The intensity for the T_1 transition is about two hundred times smaller at room temperature than at low temperatures, whereas the change is only five times for the T_2 transition.

The PL intensity of the T_2 transition is related to the radiative recombination time τ_{rad} and the relaxation time τ_{rel} in the following way

$$I^{PL} = \frac{I_0^{PL}}{1 + \tau_{rad}/\tau_{rel}} \tag{6}$$

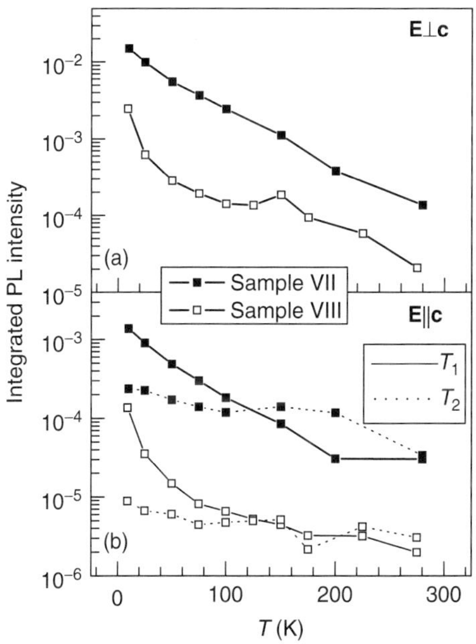

Figure 7.11 Integrated PL intensities versus temperature T in samples VII and VIII for (a) **E**⊥**c** of the transition T_1 and for (b) **E**||**c** of the transitions T_1 and T_2.

where I_0^{PL} denotes the intensity for maximum efficiency. The effect of temperature on the PL intensity of T_2 can be included in Equation 6 in a straightforward manner separating the contribution of the holes involved in the relaxation process from the ones populating the band. The hole relaxation to the highest VB states can be considered as a nonradiative channel. As the temperature increases, the hole relaxation competes with other nonradiative mechanisms. The rate equation for the hole density p_2 in the second uppermost VB in this simple model can be written in steady state as

$$\frac{\mathrm{d}p_2}{\mathrm{d}t} = G - \left(\frac{1}{\tau_{\mathrm{rad}}} + \frac{1}{\tau_{\mathrm{nrad}}} + \frac{1}{\tau_{\mathrm{rel}}}\right) p_{21} - \left(\frac{1}{\tau_{\mathrm{rad}}} + \frac{1}{\tau_{\mathrm{nrad}}}\right) p_{22} = 0 \qquad (7)$$

where G denotes the optical generation rate and τ_{nrad} the nonradiative recombination time. The hole population p_2 can be separated into holes being affected by the relaxation (p_{21}) and holes occupying the band in steady state (p_{22}). Both p_{21} and p_{22} depend on temperature. Increasing the temperature results in higher values of p_{22} and, therefore, in lower values of p_{21}. Since the spontaneous emission intensity related to the second uppermost VB is proportional to p_2/τ_{rad}, Equation 7 can be expressed in terms of the PL intensity I as

$$I^{\mathrm{PL}} = \frac{I_0^{\mathrm{PL}}}{1 + \tau_{\mathrm{rad}} \left(\frac{p_2}{p_{21}\tau_{\mathrm{nrad}}} + \frac{1}{\tau_{\mathrm{rel}}}\right) \frac{p_{21}}{p_2}} \qquad (8)$$

The temperature dependence in Equation 8 is mainly accounted for by the activation of nonradiative channels (that is τ_{nrad} decreases as a function of temperature) and the hole population distribution in steady state. We assume that the relaxation times are nearly temperature independent in the measured temperature range. Since the separation between the two uppermost VBs can be as large as 50 meV, the effect of temperature on the PL intensity through the hole populations will not be very noticeable (the hole population of the second uppermost VB will change from a negligible population at 9 K to about 15% at 295 K). Therefore, the temperature dependence will be dominated by τ_{nrad}. Nevertheless, the relaxation process will be the most limiting factor at low temperatures. The hole relaxation at low temperatures on an ultrafast timescale is discussed in Section 7.5.1.

7.4.3
Polarized Photoluminescence Spectroscopy of *M*-plane (In,Ga)N/GaN Multiple Quantum Wells

The transmittance of (In,Ga)N/GaN MQWs also shows a dependence on the polarization angle ϕ_i. However, due to the larger energy gap of the GaN layers,

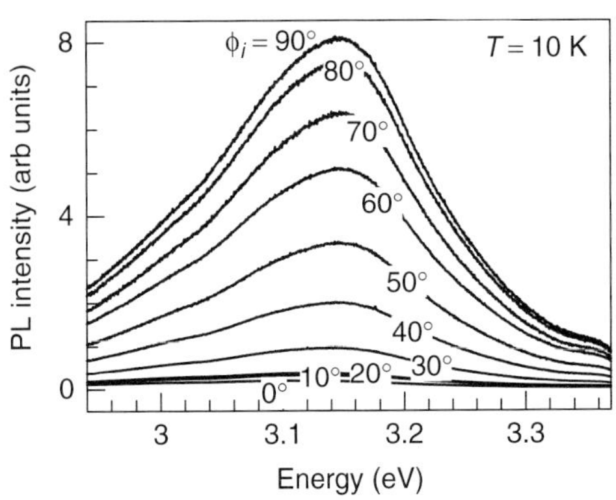

Figure 7.12 μ-PL spectra for sample X (*M*-plane (In,Ga)N/GaN MQW) at 10 K for polarization angles between $\phi_i = 0^\circ$ and 90° in steps of 10°.

the expected features of the (In,Ga)N layers are in the transparent spectral range of the GaN layers and may ride on interference fringes from the GaN layers. This complicates the extraction of the optical constants for the (In,Ga)N layers. We therefore investigated the polarization anisotropy in emission, that is, of the PL intensity. Figure 7.12 shows the low-temperature PL spectra of the (In,Ga)N/GaN MQW for polarization angles between 0° and 90° in steps of 10°. The dominant PL band centered at 3.14 eV originates from the MQW, while the higher energy shoulder near 3.36 eV is due to excitons bound to stacking faults (SFs) [26]. The PL spectrum is strongly polarized. The angular dependence of the PL intensity is described by

$$I^{\mathrm{PL}}(\phi_i) = I_{\parallel}^{\mathrm{PL}} \cos^2(\phi_i) + I_{\perp}^{\mathrm{PL}} \sin^2(\phi_i) \tag{9}$$

However, as can be seen from Figure 7.12, only one transition contributes to the PL intensity, namely, the one lowest in energy, which corresponds to T_1 being polarized perpendicular to the c axis. Therefore, in emission $I_{\parallel}^{\mathrm{PL}}$ is zero, because the states of the second uppermost VB are not occupied. Figure 7.13 shows the angular dependence of the PL intensity as a function of ϕ_i confirming Equation 9 with $I_{\parallel}^{\mathrm{PL}} = 0$. As shown in Figure 7.14, the peak energy of the PL band varies with polarization angle, developing a second peak at 3.1 eV with decreasing ϕ_i. The peaks at 3.14 and 3.10 eV are attributed to more extended-like states and more localized-like states, respectively [27]. A detailed analysis of the energy dependence of the degree of polarization ρ results in a maximum value of 0.96, which is observed at higher energies as shown in Figure 7.14. This degree of polarization is even larger than the value of 0.90 for *M*-plane GaN/$Al_{0.1}Ga_{0.9}N$ MQWs as previously reported [28]. The large in-plane polarization anisotropy in *M*-plane (In,Ga)N/GaN MQWs has probably the same origin as the one for *M*-plane GaN films. As a result of the much larger anisotropic strain in the *M*-plane (In,Ga)N wells in comparison to corresponding GaN films, the VB structure is strongly changed with respect to the one in unstrained or strained *C*-plane (In,Ga)N/GaN MQWs so that the character of the VBs with respect to linear polarization becomes comparable to the one for *M*-plane GaN films.

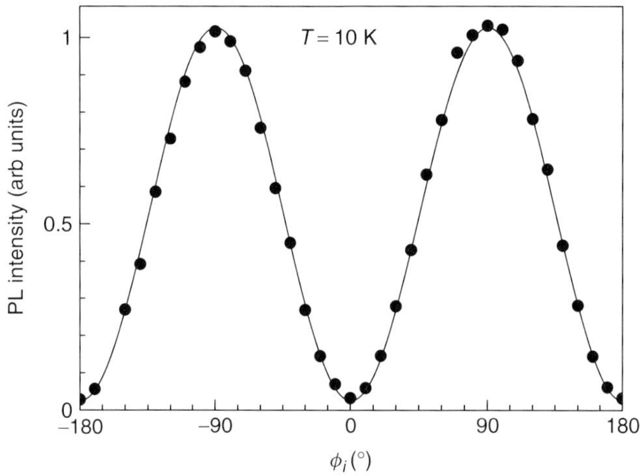

Figure 7.13 Normalized PL intensity versus ϕ_i for sample X (*M*-plane (In,Ga)N/GaN MQW) recorded at 10 K. The dots indicate the experimental data, the solid line the calculation using Equation 9 with $I_{\parallel}^{PL} = 0$.

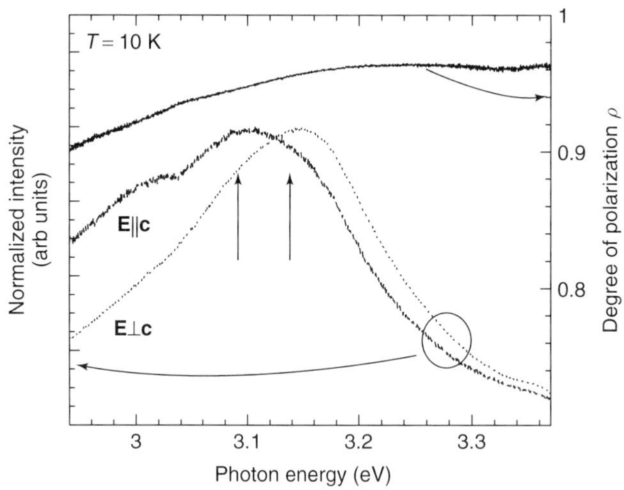

Figure 7.14 Normalized μ-PL spectra for **E**$\parallel$**c** and **E**$\perp$**c** as well as $\rho(E)$ for sample X (*M*-plane (In,Ga)N/GaN MQW) at 10 K. The arrows indicate the positions for the PL maxima in the spectra for **E**$\parallel$**c** and **E**$\perp$**c**.

7.5 Applications of Optical Polarization Anisotropy

7.5.1 Static and Dynamic Polarization Filtering

Optical polarization anisotropy results in a polarization filtering [29], which can be measured by the effective rotation of the polarization angle from ϕ_i to ϕ_f (see lower right inset of Figure 7.15a). At the maximum of the polarization filtering (around 360 nm), the birefringence of the film can be neglected [30]. The squares in Figure 7.15a indicate the measured polarization rotation $\Phi = \phi_i - \phi_f$ for 360 nm at 300 K with $\phi_f = \arctan(\sqrt{I_\perp/I_\parallel})$ measured in sample IV, where $I_\perp$ and $I_\parallel$ denote the corresponding transmitted intensities. ϕ_f can be also expressed as $\arctan[\tan(\phi_i)\sqrt{T_\perp/T_\parallel}]$, where $T_\perp$ and $T_\parallel$ denote the corresponding transmittances. The dashed line in Figure 7.15a indicates the calculated value of Φ using this relation with a value of $T_\perp/T_\parallel = 1/37$, which was independently measured by detecting the transmittances without an analyzer for $\phi_i = 0$ and 90°. Both, the squares and the dashed line, agree quite well over the whole range of the polarization angle of the incoming light beam. The maximum static rotation of 40° is observed for $\phi_i = 60^\circ$. Note that the maximum static rotation is solely determined by the value of the ratio $T_\perp/T_\parallel = a^2$, since $\Phi_{\max} = \arctan[(1-a)/(2\sqrt{a})]$, which becomes about 46° for $a^2 = 1/37$.

Figure 7.15b shows the dynamic rotation $\Phi(t)$, which is determined from the transmitted probe intensities $I_\perp(t)$ and $I_\parallel(t)$ using the relation for ϕ_f given

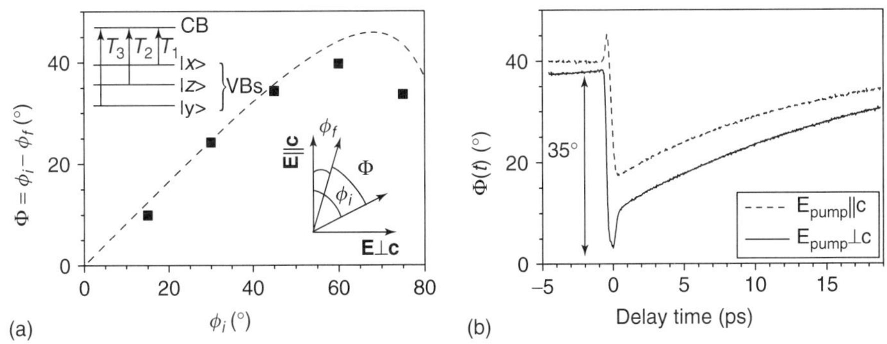

Figure 7.15 (a) Measured (squares) and calculated (dashed line) static rotation angle Φ for $T_\perp/T_\parallel = 1/37$ versus ϕ_i in sample IV for 360 nm at 300 K. The upper left inset connects the transitions T_1, T_2, and T_3 from the three uppermost VBs into the CB to their respective polarizations x, z, and y. The lower right inset shows the definition of the angles and polarization directions. (b) Dynamic rotation $\Phi(t)$ versus time delay in sample IV for two pump polarizations **E**‖**c** (dashed line) and **E**⊥**c** (solid line) recorded for $\phi_i = 60^\circ$ at 300 K.

above, as a function of time delay for two pump polarizations at $\phi_i = 60^\circ$ and 300 K in a pump-and-probe time-dependent experiment. The initial response decays very rapidly due to the hole redistribution between the two uppermost VBs and a coherent artifact as described in Ref. 31. After the hole redistribution is completed within less than 1 ps, $|\Phi_\perp(t) - \Phi_\perp^0|$ is only about 20% larger than $|\Phi_\parallel(t) - \Phi_\parallel^0|$, where Φ_j^0 with $j = \perp$ or $\parallel$ denotes the static rotation angle. While the average value of $\Phi_\perp(t)$ and $\Phi_\parallel(t)$ is determined by the redistribution time constants of the holes between the two uppermost VBs, the difference between $\Phi_\perp(t)$ and $\Phi_\parallel(t)$, which is almost time independent, originates from the different density of states of the two uppermost VB states. The actual value of $\Phi(t)$ and in particular its maximum value near $t = 0$ strongly depends on the pump intensity. From the excitation density dependence of the rotation angle recorded at 1 ps, the excited carrier density for $\mathbf{E}_{\text{pump}} \parallel \mathbf{c}$ can be estimated to be about 40% of the density for $\mathbf{E}_{\text{pump}} \perp \mathbf{c}$. The dynamical aspects of the hole distribution are described in detail in Refs. 31 and 32.

7.5.2
Polarization-sensitive Photodetectors

We have recently realized the first PSPD based on *M*-plane GaN films using sample VIII [33]. The measured photoresponse changes linearly as a function of excitation power in the measured range from 2×10^{-5} to $1\ \text{W cm}^{-2}$. Its dependence on the in-plane polarization angles ϕ_i is shown in Figure 7.16 for Schottky barrier photodiodes in the photovoltaic mode. Figure 7.16 clearly

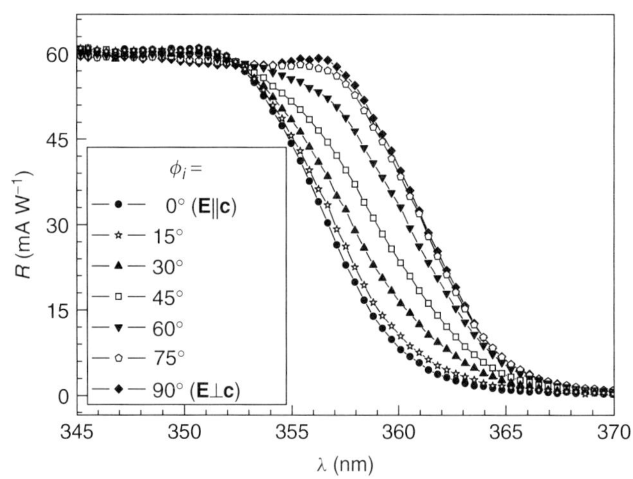

Figure 7.16 Photocurrent responsivity versus wavelength λ for different in-plane polarization angles ϕ_i under normal incidence at room temperature for a photodetector realized with sample VII.

demonstrates that the photodetection edge is blue shifted as the polarization of the incoming light changes from **E**⊥**c** to **E**∥**c**. The peak value of the responsivity R for **E**⊥**c** is as high as 60 mA W^{-1}, being slightly higher for **E**∥**c**. Calculations support this result, since the joint density of states is also slightly higher (about 5%) for the T_1 state than for the T_2 state [33]. We found modulation features in the photocurrent spectra, which can be associated with band mixing effects at higher energies (e.g. between the T_1 and T_2 states). In the case of **E**⊥**c**, the photocurrent has a relative minimum at 353 nm, whereas for **E**∥**c** a similar modulation is observed at 347 nm.

To determine the polarization sensitivity near the band gap, we measured the spectral dependence of the $R_\perp/R_\parallel$ contrast as shown in Figure 7.17 for a photodetector realized with sample VII, where $R_\perp$ and $R_\parallel$ are the components of the responsivity for **E**⊥**c** and **E**∥**c**, respectively. A maximum contrast $R_\perp/R_\parallel = 7.25$ is reached at 363 nm. Note that the maximum contrast occurs below the maximum of $R_\perp$, but in a wavelength range where $R_\parallel$ is already rather small. This maximum value of the contrast is limited by the absorption below the GaN band gap (the theoretical oscillator strength ratio for **E**⊥**c** and **E**∥**c** is larger in this region). The origin of the absorption below the energy gap can be defect related as frequently observed in GaN [34]. Note that, while the dislocation density in M-plane GaN grown on $LiAlO_2$ is much lower than in conventional C-plane GaN films, the density of SFs is much higher for M-plane GaN [26, 35]. However, SFs do not affect the optical response of the photodetector in a significant way. The polarization-sensitive bandwidth

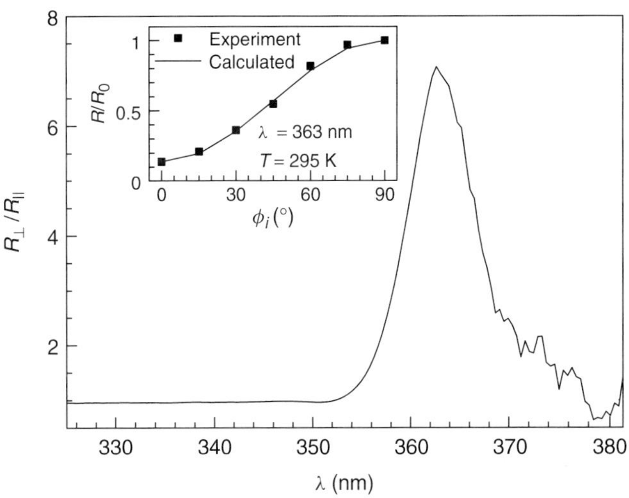

Figure 7.17 Contrast of the responsivities versus wavelength λ for polarization perpendicular ($R_\perp$) and parallel ($R_\parallel$) to the c axis at room temperature for a photodetector realized with sample VII. Inset: Responsivity R as a function of the in-plane polarization angle ϕ_i (squares) and fit (lines) using Equation 10.

is defined at the half-maximum responsivity, resulting in a room-temperature value of (7.7 ± 0.4) nm (72.6 meV) at 363 nm for Schottky photodetectors. This bandwidth cannot be directly compared with the energy splitting between the T_1 and T_2 states, since other extrinsic factors such as the absorption below the band gap and the layer thickness must be taken into account for the PSPD photoresponse. However, since the photoresponse is proportional to the incident power, we can express the angular dependence of the in-plane polarization of the responsivity as follows

$$R = R_{\perp} \sin^2(\phi_i) + R_{\parallel} \cos^2(\phi_i) \tag{10}$$

The inset of Figure 7.17 shows that the experimental result can be well described by Equation 10 at the peak wavelength of the contrast. Consequently, we can characterize the full photoresponse for an arbitrary polarization angle ϕ_i by its $R_{\perp}$ and $R_{\parallel}$ components.

7.5.3 Very Narrow-band Photodetectors

We have very recently realized a very narrow-band photodetector by combining a conventional PSPD as described in the previous subsection with a polarization filter with in-plane crystal orientation rotated by 90° with respect to the photodetector [36]. Figure 7.18a shows the transmission spectrum of sample IX for normally incident light with $\mathbf{E}\perp\mathbf{c}$ and $\mathbf{E}\parallel\mathbf{c}$. The shift in the transmission edge is due to the difference in the optical band gap for the two polarizations. This in turn leads to a polarization-dependent spectral shift in the responsivity (R_{det}) of the Schottky barrier photodetector fabricated from sample IX as shown in Figure 7.18b, which is analogous to the one in Figure 7.16. The wavelength λ_{short} (λ_{long}) corresponds to the effective excitonic band gap for $\mathbf{E}\parallel\mathbf{c}$ ($\mathbf{E}\perp\mathbf{c}$). Note that the responsivity remains high at shorter wavelengths.

To achieve a narrow-band spectral responsivity, we use another piece of the as-grown sample IX as a filter, placing it front of the photodetector such that $\mathbf{c}_{\text{filt}}$ is perpendicular to $\mathbf{c}_{\text{det}}$ (cf. left inset in Figure 7.19). This combination of detector and filter operates in the following way. For $\mathbf{E}\perp\mathbf{c}_{\text{det}}$, light with $\lambda \leq \lambda_{\text{short}}$ is absorbed by the filter and consequently prevented from reaching the photodetector, while light with $\lambda_{\text{short}} \leq \lambda \leq \lambda_{\text{long}}$ is transmitted by the filter and can therefore reach the detector. In contrast, all light with $\lambda \leq \lambda_{\text{long}}$ will be absorbed by the filter for $\mathbf{E}\parallel\mathbf{c}_{\text{det}}$ so that no light of this polarization will be detected. The presence of the small signal for $\mathbf{E}\parallel\mathbf{c}_{\text{det}}$ is essentially due to absorption at energies below the band gap [37], which reduces the anisotropy. Thus, the narrow-band detection configuration (NBDC) detects light of polarization $\mathbf{E}\perp\mathbf{c}_{\text{det}}$ in the wavelength band $\Delta\lambda = \lambda_{\text{long}} - \lambda_{\text{short}}$ as shown in Figure 7.19. In the present case, the responsivity peaks at 360 nm with a bandwidth of $\Delta\lambda = 6$ nm. The measured dependence of the NBDC

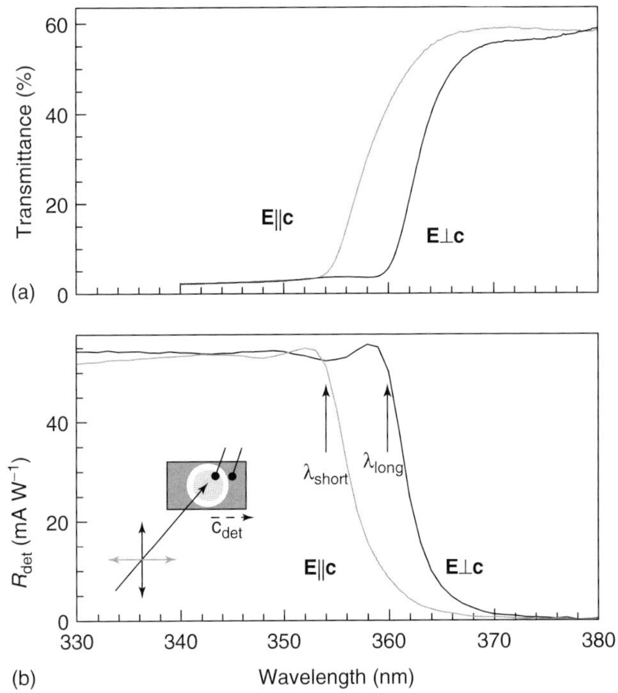

Figure 7.18 (a) Transmission spectrum of sample IX for **E**||**c** and **E**⊥**c** at 297 K. (b) Spectral responsivity R_{det} of a Schottky diode photodetector processed from sample IX for **E**||**c** and **E**⊥**c**. The inset depicts a schematic of the device and the measurement geometry.

responsivity (R_{NBDC}) at 360 nm on the in-plane polarization angle ϕ_i, which is defined relative to $\mathbf{c}_{det}$, is well described by $R_{NBDC} = R^{max} \sin^2(\phi_i)$ (cf. right inset in Figure 7.19). For this type of NBDC, one could also use *A*-plane GaN films grown on *R*-plane sapphire or *A*-plane SiC [17, 38]. However, only the substrates $LiAlO_2$ and sapphire are transparent down to 200 nm, thus facilitating operation in the UV spectral range. For efficient operation of the NBDC, high absorption is necessary for one polarization direction (e. g. **E**⊥**c**), while no absorption should occur for the orthogonal polarization (e.g. **E**||**c**) in the wavelength range λ_{long} to λ_{short}. Next, for $\lambda < \lambda_{short}$, we need a high absorption for the orthogonal polarization (**E**||**c**) so that the filter efficiently blocks such light from reaching the detector. The above considerations imply that, if λ_{long} (λ_{short}) is associated with T_1 (T_2), the oscillator strengths must be such that the quantity $[f_{1x} - f_{1z}] \times f_{2z}$ or $[f_{1z} - f_{1x}] \times f_{2x}$ is positive and maximal. In the former (latter) case, the polarization $\mathbf{E}\perp\mathbf{c}_{det}$ ($\mathbf{E}\|\mathbf{c}_{det}$) will be detected. Due to absorption of light below the band gap [37], the central operating wavelength λ_c of the NBDC is likely to be closer to λ_{long}.

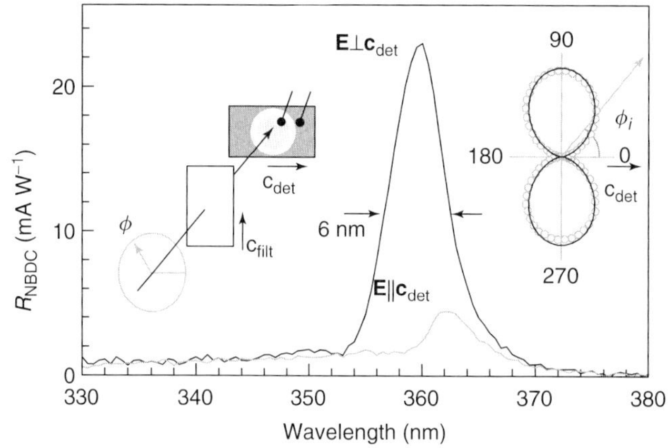

Figure 7.19 Spectral responsivity R_{NBDC} of the narrow-band detection configuration (NBDC) based on sample IX for $\mathbf{E}\|\mathbf{c}_{\text{det}}$ and $\mathbf{E}\perp\mathbf{c}_{\text{det}}$ at 297 K. The left inset displays the measurement configuration. The right inset depicts a polar plot of the measured relative R_{NBDC} at 360 nm (circles) as a function of the in-plane polarization angle ϕ_i and the line represents a fit to $R_{\text{NBDC}} = R^{\text{max}} \sin^2(\phi_i)$.

Comparing the HRXRD and optical spectroscopy data with the results of the EBS calculations, we can assign the transition edge in sample IX for $\mathbf{E}\perp\mathbf{c}$ ($\mathbf{E}\|\mathbf{c}$) to T_1 (T_2) and estimate the in-plane strain to be $\varepsilon_{xx} \approx -0.9\%$ and $\varepsilon_{zz} \approx -0.22\%$. The corresponding calculated values of $\Delta\lambda = 6$ nm and $\lambda_c = 361.5$ nm agree fairly well with the measured values for the NBDC. The utilization of thicker films in the NBDC, for example, a 1-μm-thick *M*-plane GaN film on γ-$LiAlO_2$, can further improve the rejection at $\lambda < \lambda_{\text{short}}$ by a factor larger than 250. Due to the symmetry of the wurtzite crystal structure, these results are also valid for anisotropic strain in *A*-plane GaN films with x and y interchanged everywhere [38]. For a shorter (longer) λ_c, one can use suitable $Al_xGa_{1-x}N$ ($In_xGa_{1-x}N$) alloy compositions, for which similar calculations are necessary to determine the strain required for optimal performance.

7.5.4
Polarized Light Emitters

In the same way as for photodetectors, it would be desirable to have polarized light-emitting diodes (LEDs) with a built-in polarization filter so that the emitted light may be completely linearly polarized. This application is particularly useful for flat-panel displays of laptops or notebooks. Gardner *et al.* [39] fabricated (In,Ga)N/GaN MQW LEDs on $(10\bar{1}0)$ *M*-plane GaN films by low-pressure MOVPE on $(10\bar{1}0)$ *M*-plane 4H-SiC substrates. The electroluminescence spectra showed a strong polarization anisotropy, with

73 times more light emitted with polarization perpendicular to the c axis compared to that parallel to the c axis. The polarization ratio exceeds 0.8 and is persistent to a drive current density of 800 A cm^{-2} and a temperature of 300 °C. An Arrhenius fit of the polarization ratio indicates that there is a 49 meV difference in the energy gap between the two polarization states. This suggests that a high polarization ratio can be maintained even at temperatures higher than 150 °C and drive current densities required for high-power LED applications.

Chakraborty *et al.* [40] reported on the DC and pulsed performance of nonpolar *M*-plane (In,Ga)N/GaN MQW LED lamps. The LED structures were grown by MOVPE on freestanding *M*-plane GaN substrates with a reduced defect density. The LEDs with an area of 300 × 300 mm^2 were packaged into conventional epoxy-encapsulated lamps. The emission wavelength of the LEDs was 452 nm. A cw output power as high as 0.6 mW was achieved at a drive current of 20 mA, corresponding to an external quantum efficiency of 1.09%. Pulsed power measurement on the lamps yielded an output power of 23.5 mW at a drive current of 1 A for a duty cycle of 0.25%. There was no information given on the polarization state of the emitted light.

Koyama *et al.* [41] investigated the internal quantum efficiency for the blue emission peak of nonpolar *M*-plane $(1\bar{1}00)$ $In_xGa_{1-x}N/GaN$ MQW LEDs fabricated by MOVPE on freestanding *M*-plane GaN substrates. Although the measured value for internal quantum efficiency of 34% at 300 K is yet lower than the one of conventional *C*-plane blue LEDs (>70%), the results encourage one to realize high-performance green, amber, and red LEDs by reducing the concentration of nonradiative defects, according to the absence of the quantum-confined Stark effect due to the polarization fields parallel to the quantum well normal. The electric field component of the blue surface emission was polarized perpendicular to the c-axis with the in-plane polarization ratio of 0.58 at 300 K.

7.6 Summary

The transition energies for two *M*- and two *A*-plane GaN films, which are all anisotropically strained, have been determined from a line-shape analysis of the polarized PR spectra measured at low and room temperature. In all cases, the anisotropic strain together with the reduced crystal symmetry of the *M*- and *A*-planes films results in a strong optical anisotropy. Due to the large strain in at least one of the in-plane directions for the *M*-plane GaN films, the polarization anisotropy is much more pronounced in the *M*-plane than in the *A*-plane films. The polarized PL spectra of *M*-plane GaN films and *M*-plane (In,Ga)N/GaN MQWs clearly exhibit a strong polarization anisotropy with degrees of polarization of about 0.90 for the films and 0.96 for the MQW. However, as a result of the large separation of the two uppermost

VBs, the transition related to the second uppermost VB usually exhibits a PL intensity, which is orders of magnitude smaller than the transition related to the uppermost VB.

The optical polarization anisotropy in GaN films with nonpolar orientations can be used for static as well as dynamic polarization filtering. Furthermore, a PSPD has been realized, which has been extended to a narrow-band photodetector by combining it with a polarization filter having the in-plane crystal structure rotated by 90°. The concept of polarization anisotropy has also been applied to achieve polarized emission in (In,Ga)N/GaN LEDs grown along nonpolar orientations.

Acknowledgments

The author would like to thank U. Behn, O. Brandt, T. Flissikowski, S. Ghosh, Y. Kawakami, P. Misra, E. Muñoz, K. Omae, K. H. Ploog, C. Rivera, J. S. Speck, Y. J. Sun, and H. Teisseyre for their numerous invaluable contributions to this work. The author would also like to thank T. Hashizume and the Research Center for Integrated Quantum Electronics at Hokkaido University in Sapporo, Japan, for their hospitality during the final stages of the preparation of the manuscript.

References

1. Alemu, A., Gil, B., Julier, M., and Nakamura, S. (1998) *Physical Review B*, **57**, 3761.
2. Julier, M., Campo, J., Gil, B. *et al.* (1998) *Physical Review B*, **57**, R6791.
3. Paskov, P.P., Paskova, T., Holtz, P.O., and Monemar, B. (2002) *Physica Status Solidi A*, **190**, 75.
4. Bir, G.L. and Pikus, G.E. (1974) *Symmetry and Strain Induced Effects in Semiconductors*, John Wiley & Sons, Ltd, New York.
5. Ghosh, S., Waltereit, P., Brandt, O. *et al.* (2002) *Physical Review B*, **65**, 075202.
6. Suzuki, M., Uenoyama, T., and Yanase, A. (1995) *Physical Review B*, **52**, 8132.
7. Gil, B., Briot, O., and Aulombard, R.-L. (1995) *Physical Review B*, **52**, R17028.
8. Shan, W., Hauenstein, R.J., Fischer, A.J. *et al.* (1996) *Physical Review B*, **54**, 13460.
9. Shikanai, A., Azuhata, T., Sota, T. *et al.* (1997) *Journal of Applied Physics*, **81**, 417.
10. Vurgaftman, I. and Meyer, J.R. (2003) *Journal of Applied Physics*, **94**, 3675.
11. Waltereit, P., Brandt, O., Ramsteiner, M. *et al.* (2000) *Journal of Crystal Growth*, **218**, 143.
12. Sun, Y.J., Brandt, O., and Ploog, K.H. (2003) *Journal of Vacuum Science and Technology B*, **21**, 1350.
13. Craven, M.D., Lim, S.H., Wu, F. *et al.* (2002) *Applied Physics Letters*, **81**, 469.
14. Craven, M.D., Wu, F., Chakraborty, A. *et al.* (2004) *Applied Physics Letters*, **84**, 1281.
15. Sun, Y.J., Brandt, O., Cronenberg, S. *et al.* (2003) *Physical Review B*, **67**, 041306(R).
16. Behn, U., Misra, P., Grahn, H.T. *et al.* (2007) *Physica Status Solidi A*, **204**, 299.

17. Misra, P., Behn, U., Brandt, O. *et al.* (2006) *Applied Physics Letters*, **88**, 161920.
18. Polian, A., Grimsditch, M., and Grzegory, I. (1996) *Journal of Applied Physics*, **79**, 3343.
19. Suzuki, M. and Uenoyama, T. (1996) *Japanese Journal of Applied Physics Part 1*, **35**, 543.
20. Rivera, C., Misra, P., Pau, J.L. *et al.* (2007) *Journal of Applied Physics*, **101**, 053527.
21. Korona, K.P., Wysmołek, A., Pakuła, K. *et al.* (1996) *Applied Physics Letters*, **69**, 788.
22. Stępniewski, R., Korona, K.P., Wysmołek, A. *et al.* (1997) *Physical Review B*, **56**, 15151.
23. Mayer, M., Pelzmann, A., Kamp, M. *et al.* (1997) *Japanese Journal of Applied Physics Part 2*, **36**, L1634.
24. Kornitzer, K., Ebner, T., Thonke, K. *et al.* (1999) *Physical Review B*, **60**, 1471.
25. Hopfield, J.J. (1960) *Journal of Physics and Chemistry of Solids*, **15**, 97.
26. Sun, Y.J., Brandt, O., Jahn, U. *et al.* (2002) *Journal of Applied Physics*, **92**, 5714.
27. Sun, Y.J., Brandt, O., Ramsteiner, M. *et al.* (2003) *Applied Physics Letters*, **82**, 3850.
28. Rau, B., Waltereit, P., Brandt, O. *et al.* (2000) *Applied Physics Letters*, **77**, 3343.
29. Omae, K., Flissikowski, T., Misra, P. *et al.* (2005) *Applied Physics Letters*, **86**, 191909.
30. Misra, P., Sun, Y., Brandt, O., and Grahn, H.T. (2004) *Journal of Applied Physics*, **96**, 7029.
31. Flissikowski, T., Omae, K., Misra, P. *et al.* (2006) *Physical Review B*, **74**, 085323.
32. Flissikowski, T., Misra, P., Brandt, O., and Grahn, H.T. (2007) *Physica Status Solidi C*, **4**, 175.
33. Rivera, C., Pau, J.L., Mu~noz, E. *et al.* (2006) *Applied Physics Letters*, **88**, 213507.
34. Reshchikov, M.A. and Morkoç, H. (2005) *Journal of Applied Physics*, **97**, 061301.
35. Liu, T.Y., Trampert, A., Sun, Y.J. *et al.* (2004) *Philosophical Magazine Letters*, **84**, 435.
36. Ghosh, S., Rivera, C., Pau, J.L. *et al.* (2007) *Applied Physics Letters*, **90**, 091110.
37. Lebedev, V., Cimalla, I., Kaiser, U., and Ambacher, O. (2004) *Physica Status Solidi C*, **1**, 233.
38. Ghosh, S., Misra, P., Grahn, H.T. *et al.* (2005) *Journal of Applied Physics*, **98**, 026105.
39. Gardner, N.F., Kim, J.C., Wierer, J.J. *et al.* (2005) *Applied Physics Letters*, **86**, 111101.
40. Chakraborty, A., Haskell, B.A., Masui, H. *et al.* (2006) *Japanese Journal of Applied Physics Part 1*, **45**, 739.
41. Koyama, T., Onuma, T., Masui, H. *et al.* (2006) *Applied Physics Letters*, **89**, 091906.

8
Luminescence of GaN Layers Grown in Nonpolar Directions

Plamen P. Paskov and Bo Monemar

8.1
Introduction

Wurtzite group III nitrides (GaN, AlN, and InN) and their alloys are natural pyro- and piezoelectric materials and exhibit a high macroscopic polarization in the polar [0001] direction (*c* axis). The total polarization is the sum of the spontaneous and piezoelectric polarizations and depends on the chemical composition and the strain in the crystal. Thus, in heterostructures grown along the [0001] (or [000$\bar{1}$]) direction, the polarization discontinuity at the interfaces produce a large internal electric field, which has a significant effect on the properties of the entire structure [1]. Although advantageous for two-dimensional electron gas formation in high electron mobility transistors (HEMTs), the polarization-induced field is detrimental for quantum well (QW)-based light-emitting devices because it gives rise to the quantum-confined Stark effect (QCSE). The QCSE causes a red shift of the optical transitions, a strong current dependence of the emission energy, as well as reduces the emission efficiency as a result of the spatial separation of the electrons and holes across the wells.

A way to overcome the effects of the built-in electric field is growth in nonpolar directions (perpendicular to the *c* axis), for example along the [1$\bar{1}$00] or [11$\bar{2}$0] directions. Since the first demonstration of nonpolar GaN–AlGaN multiple-quantum-well (MQW) structures free of polarization-induced field [2], there has been an emerging interest in the nonpolar group III nitrides. All known growth techniques have been used for development on these new types of material: hydride vapor phase epitaxy (HVPE), either alone or combined with high-pressure high-temperature solution growth or low-pressure solution growth, has been used for substrate development, while thin epitaxial layers and QW structures have been grown by both molecular beam epitaxy (MBE) and metalorganic chemical vapor deposition (MOCVD). Heteroepitaxial growth along the [1$\bar{1}$00] direction (*m*-plane layers) was realized on γ-$LiAlO_2$(100) [2] or on *m*-plane SiC [3], while for growth along the [11$\bar{2}$0] direction (*a*-plane layers) *r*-plane sapphire [4] or *a*-plane SiC [5] were used. Another way of reducing

Nitrides with Nonpolar Surfaces. Growth, Properties, and Devices. Edited by Tanya Paskova

ISBN: 978-3-527-40768-2

or possibly eliminating the polarization effect is to perform growth along directions inclined to the *c* axis (so-called semipolar directions). Semipolar $(10\bar{1}\bar{1})$, $(10\bar{1}\bar{3})$, and $(11\bar{2}2)$ GaN layers have been grown on $MgAl_2O_4$ (spinel) and *m*-plane sapphire [6, 7].

Epitaxial growth of GaN along the nonpolar and semipolar directions has appeared to be quite challenging. The difficulty in producing high-quality layers lies in the planar anisotropic nature of the growing surface [8]. The anisotropy of both the in-plane strain and the adatom diffusion results in stripelike surface features as well as in a high density of extended structural defects. To better understand the growth processes, surface morphology and crystal microstructure of nonpolar and semipolar layers grown on different substrates and by different techniques have been widely studied. Several approaches for growth optimization and structural defect reduction have been used and GaN templates and GaN–AlGaN and InGaN–GaN MQW structures with reliable quality have been produced. A successful demonstration of nonpolar and semipolar light-emitting devices has also been reported. As the emission efficiency in MQWs is entirely determined by the density of the nonradiative defects, improving the crystalline quality of the structures was found to be pivotal for achieving high-performance devices. From this point of view, a thorough investigation of the relation between emission and structural properties of nonpolar and semipolar GaN and related compounds is extremely important.

In this chapter, we summarize the current understanding of the emission properties of GaN layers grown along nonpolar and semipolar directions. We particularly concentrate on *a*-plane GaN and present a detailed study of luminescence in layers grown by MOCVD and HVPE on *r*-plane sapphire. A variety of photoluminescence (PL) techniques, namely polarization-dependent PL, time-resolved photoluminescence (TRPL), and spatially resolved micro-PL, are applied to clearly distinguish between the recombination mechanisms and to properly assign the different emissions. Cathodoluminescence (CL) measurements are also performed aiming to establish the spatial distribution of the emissions and to correlate them with particular microstructure features. In addition, we review the emission properties of *m*-plane GaN layers grown on γ-$LiAlO_2$ (100) and *m*-plane SiC, as well as of semipolar GaN grown on *m*-plane sapphire. For comparison, the PL spectra of bulk GaN substrates with nonpolar *a*- and *m*-plane surfaces sliced from boules grown along the [0001] direction are also discussed. All results on the luminescence in nonpolar and semipolar GaN layers are compared with the emission properties of conventional *c*-plane layers, which are also briefly summarized.

8.2
Luminescence in GaN Layers Grown along the [0001] Direction (*c*-Plane Layers)

The luminescence properties of *c*-plane-oriented GaN have been reviewed in several recent papers [9–11], and are briefly summarized here as a reference

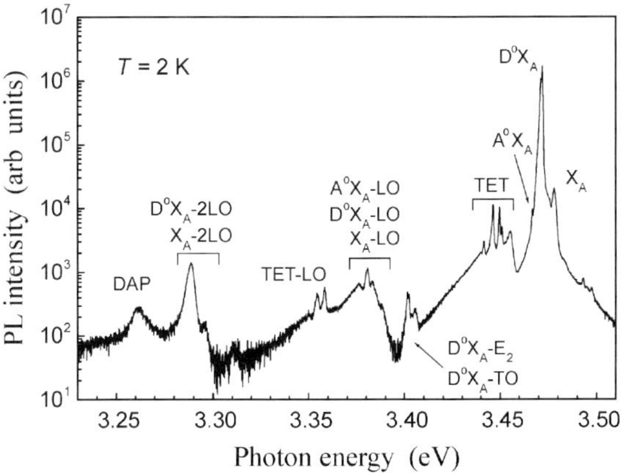

Figure 8.1 Low-temperature PL spectrum of a 1-mm-thick *c*-plane GaN layer grown by HVPE.

to the discussion of data for nonpolar and semipolar GaN. Most available data are from PL measurements with above band-gap laser excitation, and we shall restrict our description to such results. In the near band-gap region, low-temperature PL spectra are dominated by free excitons (FEs) and bound excitons, as illustrated by a typical spectrum from a thick, nominally undoped HVPE layer in Figure 8.1. The lowest FE peak at about 3.478 eV is related to the A valence band (X_A), while the FEs related to the higher valence bands (X_B and X_C) split off by the spin–orbit interaction and the crystal field appear as excited states. In the best samples, the typical linewidth of the FE lines is about 1 meV. The spin-exchange interaction splits the X_A exciton into Γ_5 and Γ_6 states and both X_B and X_C excitons into Γ_1, Γ_2, and Γ_5 states. The three Γ_5 states are optically allowed for $\mathbf{E}\perp\mathbf{c}$ polarization while the two Γ_1 states are allowed in $\mathbf{E}\|\mathbf{c}$ polarization. The Γ_1 and Γ_6 states are forbidden for both polarizations. Due to the strong coupling with photons all optically active states form mixed exciton–polariton states. The substructure related to spin-exchange interaction and polariton effects can be shown by careful measurements at different polarizations in high-quality samples [12, 13]. At low temperature, the emission from the lower polariton branch of the X_A exciton dominates in the $\mathbf{E}\perp\mathbf{c}$ polarized spectra, while the emission from the lower polariton branch of the X_A exciton is much stronger for $\mathbf{E}\|\mathbf{c}$ polarization (see Figure 8.2). The finiteness of the exciton wave vector, small uniaxial strain, or slight deviation from $\mathbf{k}\perp\mathbf{c}$ geometry can result in mixing of exciton states of different symmetry and emission lines of the Γ_6 state and the mixed longitudinal-transverse state of the X_A exciton can be seen in the $\mathbf{E}\|\mathbf{c}$ polarized spectra [12, 13]. At elevated temperatures, emission from X_B and X_C excitons gains in intensity for $\mathbf{E}\perp\mathbf{c}$ and $\mathbf{E}\|\mathbf{c}$ polarization, respectively, because of thermal population of these states. The FE emission is usually weaker at low temperatures where the donor-bound exciton (DBE) peaks are stronger

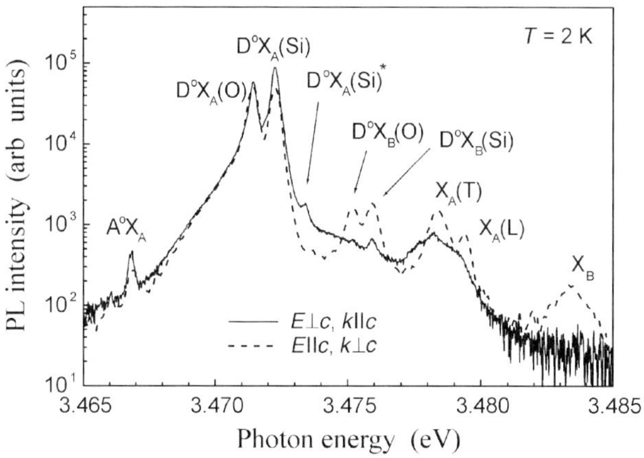

Figure 8.2 Low-temperature PL spectrum of the same 1-mm-thick *c*-plane GaN layer obtained in different polarization geometries.

but is dominant at high temperatures (such as room temperature). From the temperature dependence of the FE PL, the value of the band-gap energy of unstrained and undoped GaN has been estimated to be 3.503 eV at 2 K [14] and 3.448 eV at 300 K [15]. Phonon replicas related to emission of 1-longitudinal optical (LO) and 2-LO phonons are also strong, and exhibit characteristic temperature dependence [16].

The bound exciton spectra show narrow lines (<0.5 meV for residual doping $<10^{16}$ cm^{-3}) at an energy characteristic for each shallow dopant. In Figure 8.2, two DBE lines at 3.4714 eV (D^oX_A(O)) and 3.4723 eV (D^oX_A(Si)) due to the binding of X_A excitons to the shallow O and Si donors, respectively, are shown [17]. In the $\mathbf{E}\|\mathbf{c}$ polarized spectra, the Si and O bound exciton line related to X_B are also observed. The DBE lines are present up to 100 K. Weaker excited states ($D^oX_A(Si)^*$) exist at slightly higher energies because of hole excitations (rotator states [18]). Orbital excited states ($n = 2$, 3 etc.) are observed close to the band gap at about 3.50 eV for both FEs and DBEs [19]. At lower energies, the so-called two-electron transitions (TETs) (Figure 8.1), related to DBE recombination leaving the neutral donors in excited final states, are prominent in the 3.44–3.46 eV region, thereby allowing an accurate determination of the donor binding energies [17, 20]. Phonon replicas involving different optical phonons are also observed in the range 3.28–3.46 eV (Figure 8.1). The main neutral acceptor-bound exciton (A^oX_A) is observed as a narrow line at about 3.466 eV at 2 K, related to an unidentified complex acceptor [21]. In Zn-doped GaN, a deeper acceptor-bound exciton PL line is observed at about 3.456 eV [22], a line at a similar energy 3.455 eV is seen in highly Mg-doped GaN [23], possibly related to the Mg acceptor. Acceptor-bound exciton PL lines are typically thermally quenched at low temperatures (<50 K).

The recombination dynamics observed in TRPL experiments show a very fast transient of the order 100 ps for the FE with above band-gap excitation, believed to be dominated by nonradiative surface recombination [24]. With below band-gap bulk excitation, the corresponding decay time for the FE has been observed as much longer, about 2 ns at 4 K, presumably close to the bulk radiative lifetime of the FE in GaN [25]. At room temperature, the corresponding value seems to be >10 ns [25]. The room-temperature decay time is regarded as a measure of the quality of the material, the longer the decay time the weaker is the nonradiative defect contribution to the recombination. The radiative decay times of the DBEs are also found to be above 1 ns, for example about 1.1 ns for the D^oX_A(Si) and 1.8 ns for the D^oX_A(O), as derived from the TETs [20].

At photon energies further below the band-gap, impurity- or defect-related PL transitions dominate [11]. These may also be described in a bound exciton picture, like the stacking-fault-related emissions, where one carrier is strongly localized, while the other is usually weakly bound [10]. Such defects are not prominent in high-quality GaN material though, and essentially absent in freestanding thick HVPE bulk GaN layers. More common are transitions of the donor–acceptor pair (DAP) type. Some of these show up as well-defined no-phonon peaks, with distinct LO-phonon replicas, as is the case with the often observed so-called ultraviolet luminescence (UVL) at 3.27 eV at 2 K in GaN [11]. The acceptor associated with this emission is not identified. Substitutional group II acceptors appear to create broad DAP PL bands owing to a much stronger phonon coupling [9]. Intrinsic defects like vacancies are not stable at the growth temperature, but can be frozen in as complexes with impurities or other defects. One example of such a defect-related PL is the so-called yellow luminescence (YL) peak at about 2.3 eV in GaN [26]. It has been assigned as involving a V_{Ga}–O pair, a double acceptor in GaN. The broad PL would then be because of a DAP recombination of holes at this deep acceptor with electrons at shallow donors. There are many broad PL bands in the range 2–3 eV, believed to be related to intrinsic defects in GaN, as described by Reshchikov and Morkoc [11]. These deep level defects are generally associated with rather slow radiative recombination, and are therefore not the main nonradiative centers in the material.

8.3 Luminescence in GaN Layers Grown along the [11$\bar{2}$0] Direction (*a*-Plane Layers)

8.3.1 *a*-Plane GaN Layers Grown on *r*-Plane Sapphire

In earlier days of the GaN research, the *r*-plane sapphire has often been used as substrate for epitaxial growth mainly because this form of the sapphire was cheaper and most readily available at that time. There was also an expectation that lower lattice mismatch along the GaN [0001] direction would result in films of better quality. Layers on *r*-plane sapphire substrates have been grown by HVPE [27], MBE [28, 29], and MOCVD [30, 31], but in neither case the quality of GaN grown on (0001) sapphire was reached. Despite some controversy in the reported results, it was generally found that the surface morphology is quite rough and the crystallinity as shown by the broadening of the X-ray rocking curves is worse than that in the layer grown on (0001) sapphire. The detailed examinations of the crystal orientation and the epitaxial relationships [27, 30, 32] showed that GaN on *r*-plane sapphire grew along the nonpolar $[11\bar{2}0]$ direction. At that time, however, the researchers were not aware of the polarity issue in nitrides and the growth on the *r*-plane sapphire was abandoned. There were no detailed optical studies of these early *a*-plane GaN layers. Only Sasaki and Zembutsu [30] have reported room-temperature PL spectra of GaN grown by MOCVD, which showed a lower intensity of the YL at 2.2 eV and blue luminescence (BL) at 2.9 eV in layers on *r*-plane sapphire compared with those grown on *c*-plane sapphire. Later, a low-temperature PL spectrum of GaN layers grown by MBE on *r*-plane sapphire has been reported by Tripathy *et al.* [33]. The spectrum is very similar to that in *c*-plane GaN and shows narrow donor-bound and free-exciton emission peaks with energies blue shifted compared with the strain-free positions. However, these results should be taken with a great caution because the X-ray diffraction clearly showed (0002) peak with intensity higher than that of the $(11\bar{2}0)$ peak [34] indicating that the layers are with a mixed orientation with predominant *c*-plane domains.

Since the year 2000, with realizing the advantages of the nitride heterostructures grown along nonpolar directions, the interest in growth of GaN on *r*-plane sapphire has been renewed. Applying a two-step growth process with low-temperature GaN or high-temperature AlN nucleation layers and optimizing growth conditions, which appeared to be different from those used for the conventional *c*-plane GaN, layers with improved surface and structural quality compared with previously reported GaN on *r*-plane sapphire have been grown by MOCVD [4, 35, 36], HVPE [37, 38], and MBE [39, 40].

Despite the progress in the growth of *a*-plane GaN, the layers still suffer from threading dislocations (TDs) with a density of $\sim 10^{10}$ cm^{-2} typical for heteroepitaxial GaN and basal-plane stacking faults (BSFs) with a density in the range 10^5–10^6 cm^{-1} [4, 37]. This high density of structural defects has a significant impact on optical properties of the material. The low-temperature luminescence spectra of *a*-plane GaN are dominated by several emission bands in the spectral region 3.29–3.42 eV, which are not common for high-quality *c*-plane GaN [41–47]. These emissions appear at different energy positions and with different intensities depending on the growth technique, type of the nucleation layer and III/V ratio and have been interpreted as impurity related [41, 42] or structural defect related [43, 44]. On the other hand, in all reported spectra, the characteristic bound exciton emission is rather weak or completely absent.

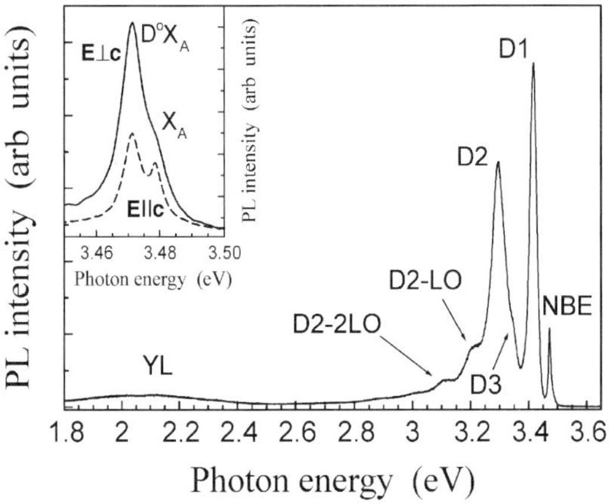

Figure 8.3 Low-temperature (2 K) PL spectrum of a 1-μm-thick *a*-plane GaN layer grown by MOCVD on *r*-plane sapphire with low-temperature GaN nucleation layer. The inset shows the high-resolution spectra in the exciton region for **E**⊥**c** and **E**||**c** polarizations.

A typical low-temperature PL spectrum of a nonpolar *a*-plane layer grown by MOCVD is shown in Figure 8.3. The dominant emissions, labeled NBE (near-band edge), D1, and D2, are peaked at 3.474, 3.42, and 3.29 eV, respectively. The weak low-energy sidebands separated by ~91 meV from the D2 peak are presumably LO phonon replicas. At high excitation intensity, the shoulder at 3.35 eV (D3) develops as a separate emission band not related with D1 and D2. The NBE emission is quite broad, but high-resolution polarized spectra (see the inset) show that it comprises two emission lines unambiguously identified as donor-bound exciton (D^oX_A) and free-exciton (X_A) recombination [44]. In the lower energy part of the spectrum, a weak broad YL peaked at 2.1 eV is also observed. Clearly, the YL, which is typical for MOCVD-grown *c*-plane GaN, is preserved in *a*-plane layers as well.

For this particular sample, the D^oX_A and X_A peaks appear at slightly higher energies than those in the strain-free bulk and homoepitaxial GaN [14, 17]. Considering the effect of the strain on the optical transitions in GaN [48], the blue shift of 2–3 meV implies that the layer is under small compressive in-plane strain. For GaN layers on *r*-plane sapphire, however, the strain-induced band-gap shift cannot be predicted so straightforward as in the case of GaN grown on *c*-plane sapphire because it is a function of two independent components of the strain tensor. (For layers on *c*-plane sapphire GaN, there is only one independent strain component [48]). The reason for this complication is that the in-plane strain in *a*-plane GaN is highly anisotropic owing to the anisotropy in the lattice and thermal mismatches with the substrate. The epitaxial relationship between the GaN and *r*-plane sapphire are $[0001]_{GaN} \| [\bar{1}101]_{sapphire}$ and $[\bar{1}100]_{GaN} \| [11\bar{2}0]_{sapphire}$ [4, 30]. Taking the room-temperature values for the lattice constants and the thermal expansion coefficients of GaN and sapphire [49], the lattice and thermal mismatches are estimated ≈ 1 and $\approx -72\%$ along the GaN [0001] direction and ≈ 16 and $\approx -25\%$ along the GaN $[\bar{1}100]$ direction, respectively. As a result of the in-plane strain anisotropy, the strain in the basal plane of the GaN hexagonal unit cell is also anisotropic. Thus, the crystal point group symmetry is reduced from the hexagonal (C_{6v}) to the orthorhombic one (C_{2v}) and the common optical selection rules for unstrained or strained GaN grown along *c* axis are no longer valid. For instance, calculations of the oscillator strengths of the band-to-band transitions show that for *a*-plane GaN on *r*-plane sapphire, the transitions from the three valence bands in order of increasing energy are polarized predominantly along the $[\bar{1}100]$, [0001], and $[11\bar{2}0]$ directions, respectively [50]. Concerning the exciton states, the strain anisotropy splits the original Γ_5 states to Γ_2 and Γ_4 states polarized along the two orthogonal basal-plane directions and splits the Γ_6 state of the A exciton to Γ_1 and Γ_3 states [51]. The Γ_1 state is dipole- allowed and polarized along *c* axis. This explains why the X_A exciton emission is observed for $\mathbf{E}\|\mathbf{c}$ polarization (the inset of Figure 8.3).

To elucidate the origin of the unusual emissions at 3.42, 3.35, and 3.29 eV, their behavior with excitation intensity and temperature, as well as their polarization properties have been studied in detail [44, 45, 52]. At low temperature, the D1 (3.42 eV) and D2 (3.29 eV) emissions blue shift with increasing excitation. The shift of D1 is only 3–4 meV over 3 orders of magnitude of the excitation intensity, while the D2 shifts by more than 10 meV. The spectrally integrated intensity of D1 increases linearly with the excitation intensity, while the D2 emission saturates at high excitation. The D3 (3.35 eV) emission can be clearly resolved only at the highest excitation levels. With increasing temperature, D1, D2, and D3 emissions quench leaving the FE emission to dominate at room temperature (Figure 8.4). The quenching behavior, however, is different for different emissions. The D1 emission decreases very fast in the 2–30 K temperature interval, but can still be seen as a shoulder on the FE emission up to 260 K. In contrast, D2 is quite stable up to 100 K, then rapidly quenches and is hardly observable above 200 K. The D3 emission completely disappears above 50 K. A prominent characteristic of the D1 emission is the S-shaped temperature dependence of the peak energy position [44]. Between 2 and 80 K, the D1 peak first red shifts and then blue shifts by about 10 meV. Above 120 K, the peak follows the evolution of the FE peak X_A. Such behavior is typical for a recombination of an ensemble of localized excitons or carriers with varying localization energies. On the other hand, the D2 peak energy is almost temperature independent up to 80 K and then red shifts slightly faster than the X_A.

Polarization-dependent PL measurements (Figure 8.5) show that all the emissions are $\mathbf{E}\perp\mathbf{c}$ polarized, presumably reflecting the symmetry of the topmost valence band in GaN (A band) [45]. The polarization degree defined as $\rho = (I_\perp - I_\|)/(I_\perp + I_\|)$, where $I_\perp (I_\|)$ is the PL intensity for $\mathbf{E}\perp\mathbf{c}$ ($\mathbf{E}\|\mathbf{c}$), is, however, different for different emissions. In these spectra, the fine structure

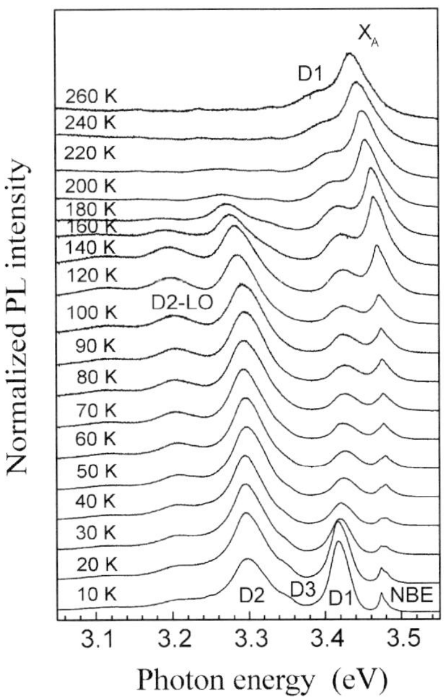

Figure 8.4 PL spectra of *a*-plane MOCVD GaN layer measured at different temperatures.

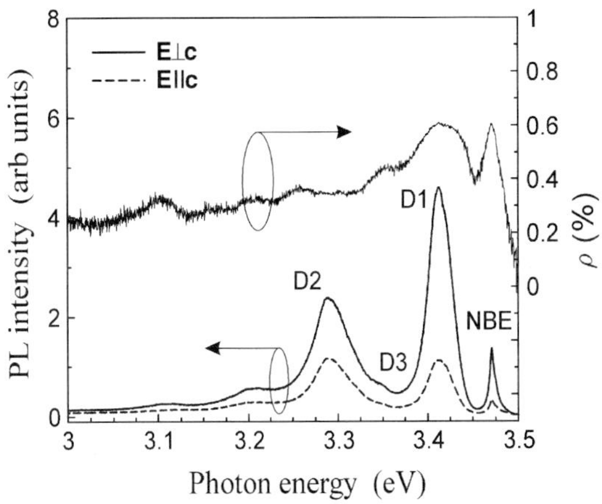

Figure 8.5 Low-temperature (2 K) polarized PL spectra and polarization degree ρ for an MOCVD-grown *a*-plane GaN layer.

of the NBE emission is not resolved but the mean polarization degree amounts to 60%. As already discussed, the anisotropic strain induces a mixing of the excitonic states from the different valence bands and a polarization degree less than 100% is not surprising. The D1 emission is also strongly $\mathbf{E} \perp \mathbf{c}$ polarized, while the polarization degree of D2 is much lower. It is quite likely that D2 emission arises from strongly localized excitons/carriers, which are expected to have more 'symmetrized' wave functions. With increasing temperature, the polarization degree of the NBE emission decreases as a result of the thermal population of the B valence band. The decrease of ρ for the D1 emission occurs with much slower rate indicating that the D1 polarization is not affected by the symmetry of the valence band holes alone. The polarization degree of the D3 emission remains almost constant apparently because of the deeper localization.

All PL spectra of *a*-plane MOCVD layers discussed so far have been taken with a broad area excitation (spot size of about 100 μm). Although the behavior of the D1, D2, and D3 emission bands with temperature, excitation intensity, and polarization is generally similar in a large number of studied samples, their peak energies in the low-temperature spectra are slightly different. Moreover, for the same layer, the micro-PL mapping shows an energy shift of the D1 peak as much as 10 meV between different excitation spots. At the same time, the exciton emission line shifts by only 1–2 meV excluding the possibility of a large local strain variation. In some spots, the D1 emission band comprises at least two separate emissions. The D2 emission band also shows a multiple-line structure, even better pronounced in some case. Figure 8.6 depicts the CL spectra taken for two different excitation areas with the same acceleration voltage and current of the electron beam, thus the two spectra also correspond to two different excitation intensities [44]. In the case of a larger area (lower excitation level) two lines at 3.29 and 3.32 eV are resolved. For the smaller spot, a single line at 3.306 eV, which is perhaps the 3.29-eV line but upshifted because of the higher excitation intensity, is observed. In the later spectrum, the D3 emission is also clearly seen at 3.35 eV. From these spectra, it is also evident that only the 3.306-eV emission line exhibits a pronounced phonon coupling. Obviously, the D2 emission band observed in the broad area spectra consists of at least two emissions with different origin.

In *c*-plane GaN, the low-temperature emission lines at 3.41–3.43 eV have been commonly related with BSFs [53, 54]. Having in mind the high density of BSF in *a*-plane GaN layers, it is plausible to associate the D1 emission also with BSFs. Indeed, a direct spatial correlation between the emission at

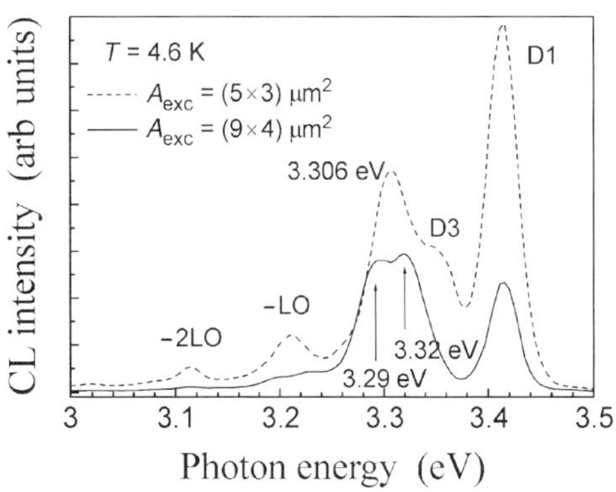

Figure 8.6 CL spectra of an *a*-plane GaN layer taken over two different excitation areas.

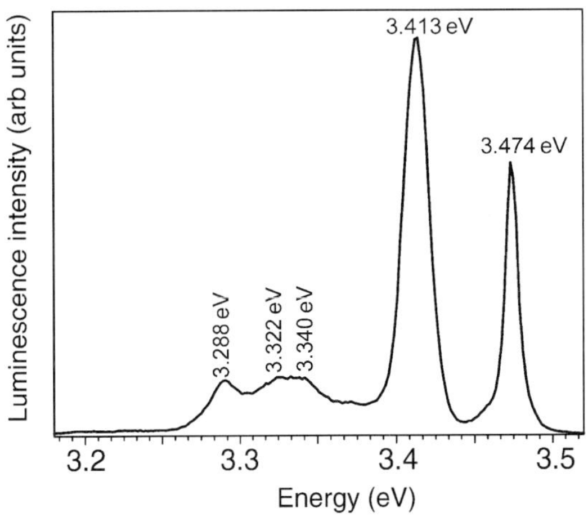

Figure 8.7 Low-temperature (5 K) CL spectrum of an *a*-plane GaN layer taken over an area of $10 \times 10\ \mu m^2$. (Reprinted with permission from [43]. Copyright 2005, American Institute of Physics.)

3.41 eV in *a*-plane GaN and the BSFs has been established by combined CL and transmission electron microscopy (TEM) measurements of a layer grown by MOCVD using a selective area lateral epitaxy (SALE) technique [43]. The low-temperature CL spectrum of this layer (Figure 8.7) shows strong NBE and D1 emissions at 3.474 and 3.413 eV, respectively. Weak emissions are also observed at 3.340, 3.322, and 3.288 eV. Figure 8.8 shows the plan-view TEM image and monochromatic CL images taken on the exact same location of a thinned foil from the sample. The dark lines running across the TEM image are identified as BSFs of type I_1 [43]. In the CL image taken at 3.41 eV, these lines (marked as 1, 2, and 3) appear as bright regions, which is a clear evidence that BSFs are responsible for this emission. One can also see an anti-correlation between the 3.47 and the 3.41 eV CL images implying that the BSFs suppress the NBE emission.

The BSFs in GaN do not introduce localized states in the band gap, but can be considered as a thin zinc-blend (ZB) layer embedded in the wurtzite (W) matrix. The theory predicts that this polytypic QW has a type II band alignment with a conduction band offset of about 270 meV [55]. The D1 emission can be described as a recombination of electrons confined in the ZB region with holes residing in the W barrier. At low temperatures, the holes are expected to be localized in a triangular potential notch formed in the barrier because of the discontinuity of the spontaneous polarization across the ZB–W interface [55]. Thus, the emission occurs between separately localized but closely spaced electrons and holes. At higher temperatures, the holes become delocalized and the emission is attributed to a recombination between electrons still localized in ZB QW and holes are attracted to them by Coulomb interaction. This

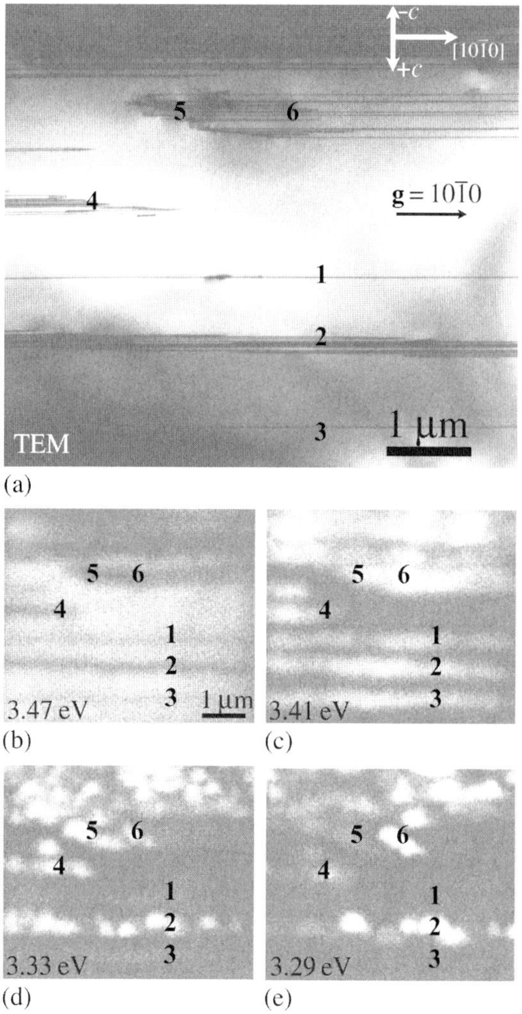

Figure 8.8 Plan-view TEM image (a) and monochromatic CL images (b–e) taken at the same location on a TEM specimen of an *a*-plane GaN layer. (Reprinted with permission from [43]. Copyright 2005, American Institute of Physics.)

interpretation is consistent with the observed S-shaped temperature evolution of the D1 peak energy [44]. The relatively large linewidth of the emission is most likely due to the wide energy distribution of localized holes. As for the variation of the energy position and the lineshape of the emission, it is worth noting that in principle, two transitions having very close energies but an opposite shift with the excitation intensity could be expected from a single BSF [56]. It is also possible that the D1 emission arises from a bundle of

BSFs rather than from a single one. Such a complex defect structure can be considered as polytypic multiple QWs, which could have different total width as well as different barrier thickness and consequently, the emission will take place at slightly different energies. The suggestion that the D1 emission arises from the QW-like structures formed along the *c* axis is in agreement with its strong $\mathbf{E}\perp\mathbf{c}$ polarization. In PL measurements with backscattering geometry from the layer surface, the $\mathbf{E}\perp\mathbf{c}$ polarized component corresponds to the transverse electric (TE) mode in such QWs, which is known to be much stronger than the transverse magnetic (TM) mode that is the $\mathbf{E}\|\mathbf{c}$ component in this case.

The detailed analysis of the TEM and CL images in Figure 8.8 shows that the emission lines in the region of 3.29–3.35-eV are also connected with stacking faults [43]. In the CL image taken at 3.33 eV (including both the 3.322 and the 3.340-eV lines), the bright spots correspond to the regions where prismatic stacking faults (PSFs) are seen in the TEM image. Therefore, the emission can be associated either with the PSFs or with the stair-rod dislocations at the intersection between BSFs and PSFs. On the other hand, the 3.29 eV emission correlated with the partial dislocations terminating BSFs. An impurity decoration of the core of stair-rod and partial dislocations is also suggested as a possible origin of these emissions. In fact, the overall behavior of the 3.29-eV emission band resembles a DAP emission. It exhibits distinct blue shift and saturation with increasing excitation intensity and relatively strong LO-phonon coupling [44]. In addition, the 3.29 eV emission is found to correlate with the triangle-shaped pits occurring on the layer surface. These pits are believed to arise from the large difference in the growth rate of the N and Ga faces of the GaN and seems to be common for *a*-plane MOCVD layers. The surface pits are part of three-dimensional hexagonal pyramids lying in the growth plane with the tips pointing at the [000$\bar{1}$] direction [57]. The vertical facets opposite to the tips coincide with the *c* plane and form a N-face wall. The monochromatic CL image taken on a region with a high density of pits reveals an enhancement of the emission intensity on the side of pit bases, where N-face growth occurs (Figure 8.9). The N-terminated surface in GaN is highly reactive and easier incorporation of impurity is expected as compared with the more stable Ga-terminated face [58].

Figure 8.10 shows a low-temperature PL spectrum of a thick *a*-plane GaN layer grown by HVPE directly on *r*-plane sapphire [45]. Like in the MOCVD layers, the spectrum is dominated by the D1 emission at ~3.42 eV. The NBE emission is also resolved but now it is rather weak. The main difference, however, is the absence of D2 emission band at 3.29 eV. The emission at 3.35 eV (D3), if present, is very weak and presumably overlapped with the long low-energy tail of the D1 emission. As shown in the inset of Figure 8.10, both the D1 and NBE emissions exhibit a red shift with increasing layer thickness. The NBE peak shifts from 3.482 to 3.466 eV in the thickness interval studied. The later value is somewhat surprising because it is lower than the D^oX_A position in thick HVPE *c*-plane GaN (3.471–3.472 eV) [17]. In analogy with

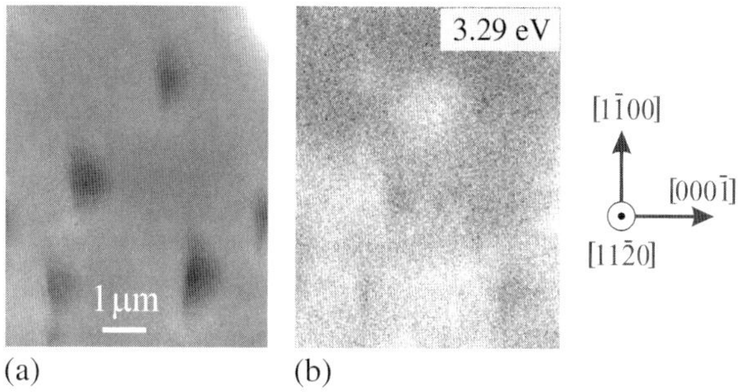

Figure 8.9 Plane-view scanning electron microscopy (SEM) image (a) and low-temperature monochromatic CL image at 3.29 eV (b) for an MOCVD-grown *a*-plane GaN layer exhibiting a high density of surface pits.

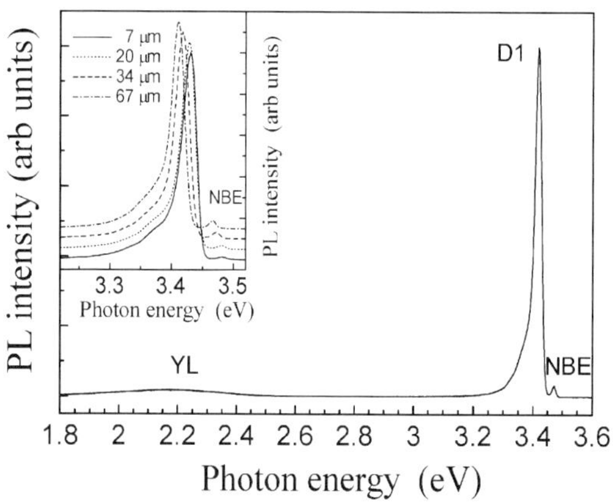

Figure 8.10 Low-temperature (2 K) PL spectrum of a 34-µm-thick *a*-plane GaN grown by HVPE. The inset shows the emission spectra in HVPE layers with different thickness.

the *c*-plane GaN layers, the shift could be explained by the relaxation of the compressive in-plane strain. However, as was already mentioned, the strain in *a*-plane layers on *r*-plane sapphire is strongly anisotropic resulting in a more complex strain-induced shift of the exciton energies. Moreover, the X-ray diffraction measurements on this set of layers show that the strain in the [0001] direction relaxes more effectively with the thickness than the strain in the [1$\bar{1}$00] direction [59]. The bending of the layers could also contribute to the offset of the NBE peak from the strain-free position of D^oX_A.

The emission spectra in *a*-plane HVPE layers grown with AlN buffer [60], MOCVD GaN buffer [45], or by a two-step growth process [47] also show a single defect-related emission band at 3.41–3.43 eV. The emission is certainly associated with the BSFs because its monochromatic CL images show bright strings oriented along the [0001] direction [45]. The TEM studies show that the defect microstructure in HVPE layers is similar to that in the MOCVD layers and the density of the BSFs of type I_1 is of the same order [37, 61]. At the same time, however, the behavior of the D1 emission in HVPE layers is quite different compared to that in MOCVD layers [62]. The emission does not shift with increasing excitation intensity over more than 4 orders of magnitude, and exhibits a slow temperature quenching and a weakly pronounced S-shaped temperature dependence of the peak position. In addition, the emission distribution is quite homogeneous on a micrometer scale, and micro-PL mapping does not reveal any change in the energy position or lineshape. The recombination dynamics of the D1 emission is also found to be different in the two types of layers. Figure 8.11 show the PL transients for different emission energies within the emission band measured at 2 K. At the high-energy side of the emission band in the MOCVD layer, the decay curves are strongly nonexponential with a fast component of about 100 ps. With decreasing emission energy, this fast component becomes slower and the decay curves become closer to a single exponent (Figure 8.11a). The effective decay time measured as a delay for which the maximum PL intensity drops 10 times increases monotonically from 0.2 ns at 3.43 eV to 2.1 ns at 3.40 eV. In the HVPE layer, the transients are also nonexponential but with a longer fast component and with a weaker dependence on the emission energy (Figure 8.11b). The effective decay time now varies from 0.8 to 1.2 ns in the same energy interval. The nonexponential character of the PL decay is a clear indication for a recombination of localized excitons or separately localized electrons and holes. The spread of localization energies, however, is much smaller in the HVPE layer as shown by the weaker energy dependence of the decay time. As the electrons involved in the D1 emission are strongly localized in the thin ZB QW-like region, the difference is likely to arise from the hole localization.

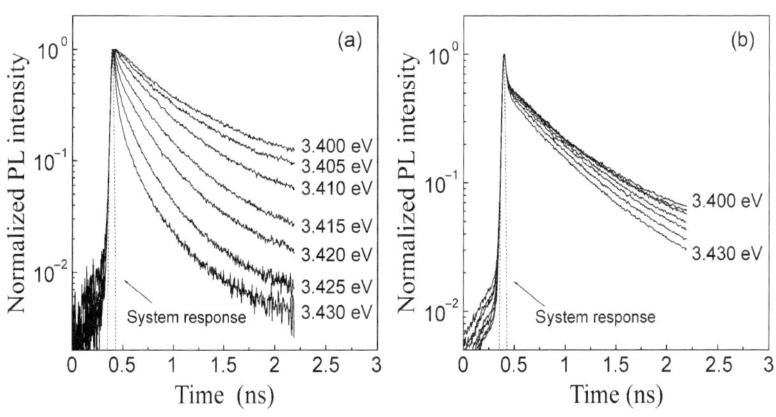

Figure 8.11 Low-temperature (2 K) PL transients at different emission energies for the D1 emission band in an *a*-plane GaN layer grown by MOCVD (a) and HVPE (b).

The different localization of carriers involved can explain the difference in the thermal quenching of the D1 emission in the two types of layers. In MOCVD layers, the emission quenches very fast at low temperature (thermal activation energy of ~7–10 meV), presumably because of the thermal delocalization of the hole from a potential in the BSF vicinity and recapturing by the defects responsible for 3.29-eV emission. For the high-temperature interval (>80 K), an activation energy of ~30 meV is extracted [52]. On the contrary, in the HVPE layers, the emission is quite stable (thermal activation energy of ~50–60 meV) and persists up to room temperature. In some layers, the D1 emission is even stronger than the FE emission at 300 K. In fact, the intensity ratio between the D1 and FE emissions is found to be polarization dependent. Figure 8.12 shows room-temperature spectra in a 67-μm-thick HVPE layer measured at different polarization orientations. The D1 emission dominates the **E**⊥**c** spectrum but almost vanishes for **E**||**c** polarization. The high polarization degree (≈90%) cannot be simply explained assuming that only the holes from the A valence band are involved and confirms the suggestion that the D1 emission arises from MQW-like structures formed along the *c* axis by a bundle of BSFs. The polarization dependence of the FE emission is governed by the common selection rules in GaN. For **E**⊥**c** polarization, the emission is dominated by the X_A exciton recombination, while only the X_B excitons are seen for **E**||**c** polarization. The X_B excitons (Γ_5 states) also contribute to the **E**⊥**c** polarized spectrum, but with a lower strength. The X_A and X_B excitons are very closely spaced (~5 meV in a strain-free GaN) and cannot be resolved owing to the large PL broadening at room temperature. The shift of the FE emission between the two polarizations is, however, clearly seen in Figure 8.12.

All PL and CL results discussed so far clearly show that the optical properties of *a*-plane GaN layers grown on *r*-plane sapphire are considerably affected by the presence of extended structural defects. The stacking faults (SFs) appear to be optically active defects and give rise to emission bands, which always

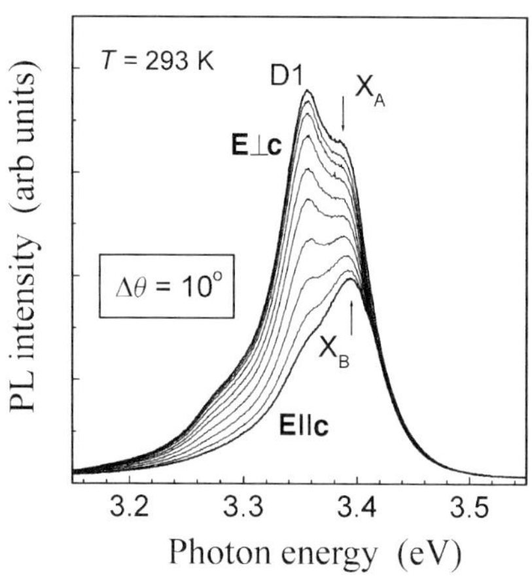

Figure 8.12 Polarized room-temperature PL spectra of a 67-μm-thick *a*-plane GaN layer grown by HVPE. The set of spectra is taken by varying the angle θ between the polarization of detected PL signal and the [0001] direction (*c* axis) from 0° to 90° with a step of 10°.

dominate the luminescence spectra. On the other hand, TDs mainly act as nonradiative recombination centers and thus reduce the efficiency of the exciton or band-to-band emission. Therefore, reducing or preferably complete elimination of extending defects is essential to improve the optical quality of the *a*-plane GaN. Epitaxial lateral overgrowth (ELO) has been proven to be an effective approach for the reduction of the TD density in GaN layers grown on *c*-plane sapphire. The ELO technique relies on selective epitaxy on patterned GaN templates and is implemented in both MOCVD and HVPE (for a detailed review of the ELO technique see [63]). The ELO templates are typically fabricated by depositing a thin SiO_2 mask over a few micrometer thick GaN layer followed by forming a pattern with parallel stripes separated by open window areas by conventional photolithography and etching. During the regrowth process, initially a vertical growth over the window regions takes place and once the film reaches the top of the stripes, lateral growth over the mask starts. Continuing growth under proper conditions leads to a fully coalesced layer. The reduction of the average TD density results from the fact that the mask blocks the TD propagation and the laterally overgrown areas are practically defect free.

The ELO technique has also been used for improvement of the crystalline quality of *a*-plane GaN layers grown by MOCVD [64–67] and HVPE [68]. Together with a reduced TD density, the fully coalescent ELO *a*-plane layers possess a low density of BSFs as well as smooth surface morphology free of pits and ridges. It was, however, found that the crystallographic orientation of the mask stripes is of a crucial importance for successful ELO. Only the stripes along the [1$\bar{1}$00] direction (perpendicular to the *c* axis) allowed a significant defect reduction [66, 68]. In this case, the lateral growth occurs in the polar [0001] (Ga-face wing regions) and [000$\bar{1}$] (N-face wing regions) directions and TDs propagate only in the window regions. The density of SFs in the overgrowth region is also reduced because the BSFs predominantly occur on the N face [57], which grows slower than the Ga face by a factor of ~6–10 depending on the growth conditions [64, 68]. For example, in the Ga-face wing regions of coalescent HVPE ELO layers, the density of TDs and BSFs is found to be below $5 \times 10^6\ cm^{-2}$ and $5 \times 10^3\ cm^{-1}$, respectively, which is much lower than the values in the window regions ($\sim 9 \times 10^{10}\ cm^{-2}$ and $\sim 5 \times 10^5\ cm^{-1}$) [68]. The N-face wing regions are also almost TD free but the BSFs are still prevalent.

As a result of the lower average density of the BSFs, the intensity of the 3.42- and 3.29-eV emissions is reduced and low-temperature PL spectra of ELO layers with [1$\bar{1}$00]-oriented stripes are dominated by the NBE emission [67, 69]. At room temperature, only the FE emission is observed in the NBE spectral region in both MOCVD and HVPE ELO layers [65, 68]. The enhanced efficiency of the exciton emission in the overgrown regions is evidenced by the spatially resolved CL [68] and near-field scanning optical microscopy (NSOM) [69] imaging, where the wings appear as bright stripes, while the window regions exhibiting a high density of TDs remain dark. Room-temperature TRPL data also showed improved emission efficiency in ELO *a*-plane layers [69, 70]. As typical for GaN, the PL decay exhibits a biexponential character and two decay time components are extracted from the measured transients. Both components, fast and slow, are limited by the nonradiative recombination. A fast decay component of $\sim$80 ps is determined in MOCVD [69] and HVPE [70] ELO layers, which has to be compared with $<$40 ps in the corresponding planar layers. Nearly the same ratio is found for the slow components. The longer decay components in ELO layers are a clear indication for a reduced density of nonradiative centers, namely point defect complexes and TDs. In highly excited MOCVD ELO layers, a single PL decay time of $\sim$430 ps is measured for the electron–hole plasma (EHP) recombination, which is almost 50 times longer than the decay time in layers grown directly on *r*-plane sapphire [71]. The increased EHP lifetime in these layers allows optically pumped stimulated emission to be demonstrated [65]. Interestingly, the stimulated emission threshold ($\sim$1 MW cm^{-2}) is approximately the same as that observed in high-quality *c*-plane GaN layers despite the pump beam size being much larger compared with the wing/window widths and covers 'low-defective' and 'high-defective' regions.

Generally, in order to enlarge the low defect density area in ELO *a*-plane GaN layers, a pattern with a high wing/window width ratio should be used. In such a case, however, achieving a fully coalesced layer will be very challenging because of the large coalescence thickness and difficulties to control the uniformity. Several modifications of ELO technique have been developed to reduce the defect density over a larger area and to improve the uniformity of *a*-plane layers grown by MOCVD: selective area lateral epitaxy (SALE) [72], trenched epitaxial lateral overgrowth (TELO) [73], and sidewall epitaxial lateral overgrowth (SELO) [74, 75]. An ELO pattern with stripes oriented along the [$1\bar{1}00$] direction is used in all techniques. The SALE is very similar to two-step ELO, while TELO and SELO are single-step processes resembling pendeoepitaxy (PE) [76]. In the first SALE step, pillars with a high height/width ratio are grown over the windows by an enhanced vertical growth rate. Then, the top surface of the pillars is covered by SiO_2 and in the second step the growth proceeds from the sidewalls. In the TELO and SELO, after transferring the ELO pattern the GaN in the window regions is completely etched by reactive ion etching. Thus, a template with grooves and terraces along the [$1\bar{1}00$] direction is formed similar to the pillar structure obtained after the first SALE step. In TELO, the SiO_2 mask is removed aiming at a fast coalescence. Owing to the suppressed vertical growth, the defect densities in layers grown by these modified ELO techniques are effectively reduced in the window regions as well. The defective regions are restricted to the regions where only the N-face growth occurs.

Figure 8.13 shows spatially resolved low-temperature spectra of a fully coalesced *a*-plane layer grown by SELO with 3 µm/15 µm (groove/terrace) pattern. The initial planar *a*-plane GaN layer was grown with a high-temperature AlN nucleation layer followed by an AlGaN interlayer [75]. The

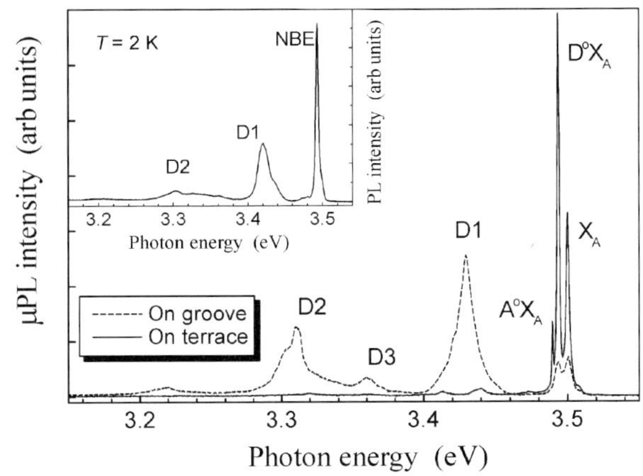

Figure 8.13 Spatially resolved micro-PL spectra at 2 K of a fully coalescent *a*-plane GaN layer grown by sidewall epitaxial lateral overgrowth (SELO). The inset shows the conventional PL spectrum taken with a large excitation spot (~100 μm).

micro-PL spectrum taken at the center of the terraces shows a clear NBE emission with three well-resolved peaks; these are the acceptor-bound exciton (A^oX_A), donor-bound exciton (D^oX_A), and free-exciton (X_A) emission lines. No defect-related emissions are observed. The exciton lines are blue shifted by ~20 meV compared with the strain-free position indicating a large compressive in-plane strain in the layer. The full width at half maximum (FWHM) of the D^oX_A line is ~1.5 meV, a value typical for heteroepitaxial *c*-plane layers grown by MOCVD. Although the D1 and D2 emissions are dominant, the spectrum from the region above the grooves also exhibits narrow D^oX_A and X_A lines. Besides, the intensity of the exciton lines is comparable to that of the D1 and D2 emissions. Owing to the strain-induced shift, all defect-related emissions appear at energies higher than those in planar *a*-plane MOCVD layers. The PL spectrum measured with a large excitation spot (thc inset in Figure 8.13) is also shown. The spectrum, in fact, represents an overlap of the emission from the terrace and groove regions and is apparently dominated by the D^oX_A line. These PL results clearly demonstrate that by using the SELO techniques, a significant reduction in the structural defect density, particularly in the BSF density, can be achieved over a large area of *a*-plane GaN layers.

8.3.2
a-Plane GaN Layers Grown on *a*-Plane SiC

From a crystallographic point of view, (11$\bar{2}$0) *a*-plane SiC is a better substrate for epitaxial growth of *a*-plane GaN than the *r*-plane sapphire. First, the layer and substrate have identical crystal orientation and the epitaxial relationships are $[0001]_{GaN} \| [0001]_{SiC}$ and $[1\bar{1}00]_{GaN} \| [1\bar{1}00]_{SiC}$. Second, the lattice mismatch is smaller and almost isotropic: ~2.9% along the [0001] direction and ~3.5% along the $[1\bar{1}00]$ direction. *a*-Plane GaN has been grown on *a*-plane 6H-SiC [5] and 4H-SiC [77] by MOCVD using a high-temperature AlN buffer. Despite the closer lattice match, no improvement of the structural quality compared to layers on *r*-plane sapphire has been achieved. The densities of the TDs and SFs were found to be the same or even higher, $\sim 3 \times 10^{10}$ cm^{-2} and $\sim 7 \times 10^{5}$ cm^{-1}, respectively [5]. Using ELO techniques such as SALE [78] and PE [79] have resulted in a smoother surface morphology and reduced defect density in the wing regions. In general, however, the quality of layers on *a*-plane SiC still remains far from that achieved on *r*-plane sapphire.

Only room-temperature PL spectra have been reported for *a*-plane GaN grown on SiC [78]. In planar layers on 4H-SiC, the emission occurs at 3.39 eV, while common FE emission at 3.422 eV is observed from the ridges grown by the SALE. The 3.39-eV band is presumably the D1 emission related to the BSFs of type I_1, which as in the case *a*-plane layers on *r*-plane sapphire are the prevalent structural defect [5, 77].

8.4
Luminescence in GaN Grown along the $[1\bar{1}00]$ Direction (*m*-Plane Layers)

8.4.1
m-Plane GaN Layers Grown on γ-$LiAlO_2$

Lithium aluminate, γ-$LiAlO_2$ lithium aluminate (LAO), has been suggested as an alternative to sapphire and SiC for GaN heteroepitaxy because of its very small mismatch to GaN and without concern of the surface orientation on the polarization issue in nitrides [80]. It has been found that the (100) plane of LAO has a superior lattice and structural match to the $(1\bar{1}00)$ *m* plane of GaN with orientation relationships $[0001]_{GaN} \| [010]_{LAO}$ with ~1.7% misfit and $[11\bar{2}0]_{GaN} \| [001]_{LAO}$ with only ~0.3% misfit. The first attempts to grow GaN on LAO(100) have resulted in (0001) oriented films [80, 81], but later on *m*-plane layers have successfully been deposited by MOCVD [82] and MBE [2]. Then, realizing the benefits of the nonpolar surface orientation, the first nitride MQW structures free of internal electric field have also been grown on LAO(100) [2].

A detailed study of emission properties of *m*-plane GaN grown on LAO(100) and its relation to the structural properties of the layers have been reported by Sun *et al.* [83]. Figure 8.14 shows low-temperature spectra of two 500-nm-thick layers grown by plasma-assisted MBE. Sample A was grown at a constant temperature (750 °C), while for sample B, a two-step growth process with a low-temperature nucleation layer (~100 nm) was used. In both samples, a strong emission at 3.48 eV attributed to a donor-bound exciton (D^oX) is observed. The D^oX emission is blue shifted by ≈10 meV with respect to the position in bulk GaN implying a large compressive strain. The in-plane strain in these layers

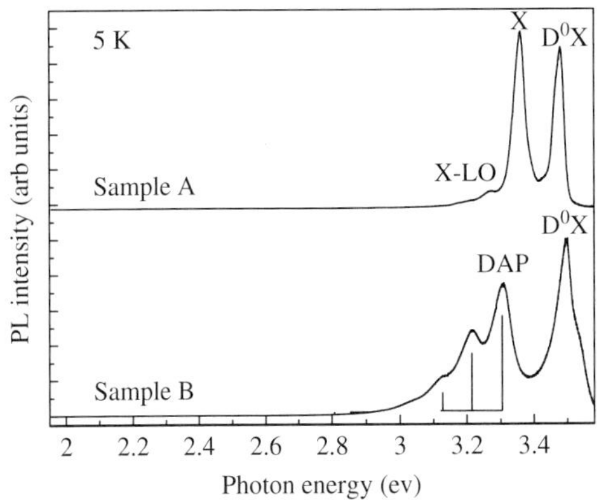

Figure 8.14 PL spectra of *m*-plane GaN layers grown by MBE on γ-$LiAlO_2$. Sample B is grown by a two-step process with a low-temperature nucleation layer, while sample A is grown in a single growth step. (Reprinted with permission from [83]. Copyright 2002, American Institute of Physics.)

is mainly thermally induced. Note that the mismatch between the thermal expansion coefficients of GaN and LAO is quite large $\approx -79\%$ along the $[0001]_{GaN}$ direction and $\approx -21\%$ along the $[11\bar{2}0]_{GaN}$ direction. In sample A, an additional strong emission at 3.356 eV (labeled X) is observed. The excitation-intensity-dependent measurements show that the energy position and the width of the X emission are independent of the excitation intensity [83]. At the same time, the spectrally integrated intensity of the emission increases linearly with an increase of the excitation intensity over 5 orders of magnitude. This behavior is very similar to that observed for the D1 emission in *a*-plane GaN on *r*-plane sapphire. Furthermore, the PL decay time measured at the peak of the X emission ($\sim$1.1 ns) is comparable to the decay of the D1 emission. The X emission does not appear in the spectrum of sample B. Instead, an emission at 3.30 eV followed by several LO-phonon replicas is observed. The emission of 3.30 eV was attributed to a DAP emission despite the fact that its energy position (even accounting for the strain shift) is rather high compared to the common one in GaN ($\sim$3.27 eV). The arguments for this assignment are the saturation and the large blue shift of the emission with increasing the excitation intensity as well as its very long decay time ($\sim$100 ns) [83]. The YL was not detected in any of the two layers.

The plan-view CL image of the X emission revealed a stripelike contrast with stripes oriented along the GaN [11$\bar{2}$0] direction of GaN [83]. A similar pattern consisting of smooth ridges and rough trenches was observed in the scanning electron microscopy (SEM) image of the sample A. Despite the spatial correlation between the bright spots in the CL image and the trench regions, the cross-sectional CL has showed that the emission actually occurs within the entire depth of the layer, and not preferentially from the surface. The X emission was associated with BSFs, which as in the case of *a*-plane GaN were found to be the most dominant structural defects. The TEM study has revealed that the BSFs in these layers are terminated by Shockley partial dislocations and then have to be identified as BSFs of type I_2. In GaN, the energy of type I_2 BSFs is twice the energy of the BSFs of type I_1 [55] but obviously their formation in *m*-plane layers grown on LAO(100) is favored. It is believed that the type I_2 BSFs being concentrated within the trench regions are growth-induced defects and are caused by the morphology of the substrate rather than misfit stress [84]. Actually the surface morphology of sample A was found to replicate the surface of the substrate [83]. In contrast, the sample B that is not showing the X emission has exhibited a significantly smoother and less anisotropic surface morphology and a significantly smaller BSF density compared to sample A. Therefore, the low-temperature nucleation layer is crucial for reducing the structural defect density and improving the surface morphology in *m*-plane layers. An improved structural quality has been reported for *m*-plane layers grown by ion-beam-assisted MBE directly on LAO(100), that is without buffer layer [85]. Low-temperature PL spectra of these layers also show the X emission, but it is much weaker than the dominant D^oX emission.

The recombination mechanism involved in the type I_2 BSF-related X emission is essentially the same as that of the D1 emission. The only difference is the thickness of the QW-like structure formed by the BSFs. The type I_2 BSFs consist of two ZB bilayers and then the confinement energy of the electrons will be smaller than that in the case of type I_1 BSFs (containing one ZB bilayer). Consequently, the recombination of the electrons and holes localized at the type I_2 BSFs is expected to occur at a lower energy compared to the D1 emission. It is worth noting that the peak energy of the X emission matches the energy of the D3 emission (3.35 eV) observed in *a*-plane layers. Thus, we can speculate that the D3 emission is also related to type I_2 BSFs, which are also present in *a*-plane GaN but with a lower density compared with the BSFs of type I_1 [61].

Recently, polarization-dependent PL measurements in *m*-plane GaN grown by plasma-assisted MBE have been reported [86]. The improved structural quality of the layers has allowed exciton transitions arising from the two uppermost valence bands to be resolved. The large polarization anisotropy in the PL spectra was attributed to the in-plane anisotropy of the thermally induced strain. The experimental results were found to be in a good agreement with the predictions of the band structure calculations for strained *m*-plane GaN [87].

HVPE has also been utilized for growth of *m*-plane GaN on LAO(100). Thick layers (up to 300 μm) as well as freestanding wafers produced by wet chemical etching have been reported [88, 89]. Surprisingly, the PL spectra of HVPE-grown *m*-plane GaN are quite different from those of MBE layers. Low-temperature PL spectra of a freestanding wafers have shown a dominant emission band at 3.42 eV [90]. A broad NBE emission at 3.47 eV and an emission band at 3.36 eV have also been observed. The relative intensities of the three emissions were found to vary with sample surface treatment (polishing, chemical cleaning, or etching) as well as across the wafer. Both the 3.42- and 3.36-eV emissions have shown similar excitation intensity and temperature dependences, which in turn are similar to those of the D1 emission in *a*-plane layers. Apparently, the 3.42- and 3.36-eV emissions have to be related to the BSFs of type I_1 and I_2, respectively. The dominance of the 3.42-eV emission in the spectra merely suggests a large density of type I_1 BSFs. Indeed, TEM studies have identified the most dominant BSFs in *m*-plane HVPE layer as being of type I_1 [91].

Emission spectra of homoepitaxial layers grown by MOCVD on a freestanding *m*-plane HVPE substrate have also been reported [92]. At room temperature, the FE emission at 3.43 eV and the YL at 2.1 eV with almost equal intensities have been observed. A strong reduction of the YL was found with increase of the V/III molar ratio. In the low-temperature spectrum, two peaks have been resolved and attributed to the donor-bound and FE emission. However, the energy positions of the peaks are lower than those in homoepitaxial *c*-plane layers, and no polarization dependence is observed, so this assignment is quite uncertain.

With the aim of reducing the densities of TDs and BSFs, which in *m*-plane HVPE GaN were found to be of the order $\sim 4 \times 10^{10}$ cm^{-2} and $\sim 1 \times 10^5$ cm^{-1}, respectively [89], ELO by HVPE has been demonstrated [93]. Growth with ELO patterns with stripes oriented along the [0001] and [11$\bar{2}$0] directions have been investigated. As templates MBE-grown *m*-plane GaN layers on LAO(100) have been used. A reduction of the BSF density by 2 orders of magnitude was achieved in the wing regions of layers with ⟨11$\bar{2}$0⟩-oriented stripes. As in the case of *a*-plane ELO layers, the formation of BSFs is only suppressed when overgrowth occurs along the [0001] direction (Ga-face wing). For both stripe orientations, the TD density in the overgrowth regions was reduced down to $\sim 5 \times 10^6$ cm^{-2}. In the window regions, the TD density was found to be an order of magnitude lower in [0001]-stripe layers than that in [11$\bar{2}$0]-stripe layers, which was attributed to the lateral TD redirection from the windows to the wing. The results from spatially resolved room-temperature CL measurements were in full correlation with the defect density distribution [93]. The CL spectra for coalesced layers with the two stripe orientations are dominated by the FE emission, although a low-energy shoulder related to the common D1 emission is observed in the [0001]-stripe layers. The wing/window emission intensity ratio was estimated to be 2 : 1 in the [0001]-stripe layers and 5 : 1 in the [11$\bar{2}$0]-stripe layers. The difference was attributed to the lower TD density in the windows of the [0001]-stripe layers rather than to any significant difference in the structural quality of the films grown from each stripe orientation [93].

8.4.2 *m*-Plane GaN Layers Grown on *m*-Plane SiC

Although less lattice matched to GaN than LAO, *m*-plane SiC is a chemically and thermally more robust substrate for growth of *m*-plane GaN. Growth of *m*-plane GaN layers on *m*-plane 6H-SiC and 4H-SiC has been reported by plasma-assisted MBE [3, 94] and MOCVD [95]. In all cases, a thin AlN buffer layer was used to improve nucleation and to partly accommodate the lattice mismatch. Structural characterization of these layers typically showed the same densities of TDs and BSFs as in *a*-plane layers and *m*-plane layers grown on LAO. Only layers grown by MOCVD were found to exhibit slightly lower TDs density and smoother surface morphology compared with the *a*-plane GaN on *a*-plane 6H-SiC [95]. MOCVD layers with an improved structural quality have been obtained by SELO [96]. Cross-section TEM examination of layers grown on templates with different groove orientations showed that for an effective reduction of the structural defects, the [11$\bar{2}$0] orientation is the optimal one, the same as in the case of *m*-plane ELO layers on LAO [93]. While the average TD density in fully coalesced SELO layers decreases irrespective of groove orientation, a reduction of the BSF density is found only in the Ga-face overgrown regions over the terraces. On the whole, however, the densities of TDs and BSFs even in the low-defective regions in these layers remain higher compared with those in *a*-plane SELO layers on *r*-plane sapphire and the *m*-plane ELO layers on LAO.

Luminescence spectra of *m*-plane layers grown by MOCVD on 4H-SiC with and without applying SELO are compared in Figure 8.15. The SELO layers were grown on *m*-plane templates with 3 μm/3 μm (groove/terrace) pattern and grooves along the [0001] or [11$\bar{2}$0] directions [96]. The SELO layers are named with the angle between the grooves and [0001] direction: ELO-0°

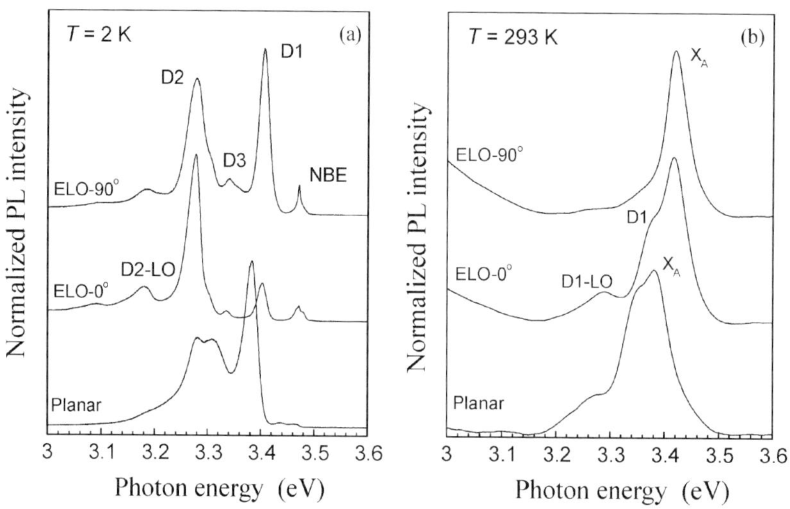

Figure 8.15 Low-temperature (a) and room-temperature (b) PL spectra of *m*-plane GaN layers grown by MOCVD on 4H-SiC. The spectra labeled with ELO-0° and ELO-90° are for samples grown using SELO with grooves oriented along the [0001] and [11$\bar{2}$0] directions, respectively. The spectra of a planar *m* plane are also shown for comparison.

for ⟨0001⟩-oriented grooves and ELO-90° for [11$\bar{2}$0]-oriented grooves. At low temperature, the planar *m*-plane layer exhibits only defect-related emissions peaked at 3.39, 3.31, and 3.28 eV (Figure 8.15a). The PL spectra of SELO layers are dominated by the emissions at 3.41 and 3.28 eV, but the NBE emission at 3.47 eV is also clearly resolved. As expected, the NBE is narrower and better pronounced in the ELO-90° layer. According to their energy positions, the defect-related emissions are labeled in analogy with those observed in *a*-plane GaN layers grown by MOCVD (Figure 8.3). The emission at 3.41 eV (3.39 eV in the planar layer) obviously corresponds to the D1 emission associated with BSFs of type I_1. The D2 emission at 3.28 eV, which in *a*-plane layers is tentatively attributed to the partial dislocations terminating BSFs, is also quite distinct; in ELO-0° layer, the D2 is even stronger than the D1 emission. This is somewhat surprising if both emissions are to be related with the BSFs. In fact, without detailed CL and TEM studies in these *m*-plane GaN layers, it is difficult to argue the same origin of the D2 emission as in *a*-plane layers. Reduced BSF density in ELO-90° layer is clearly evidenced by the room-temperature spectra shown in Figure 8.15b. While in the planar and the ELO-0° layers the D1 emission is still present, only the FE emission is observed in the ELO-90° layer.

8.5
Luminescence in GaN Layers Grown along Semipolar Directions

Growth along semipolar directions – that is in the directions inclined to the *c* axis, is an alternative means of bypassing the polarization-induced effects in group III nitrides. Semipolar layers have reduced polarization effects compared with *c*-plane GaN because the polarization vector is tilted with respect to the growth direction. Also, for specific strain states on specific semipolar planes, the net polarization in the growth direction vanishes [97, 98]. In semipolar growth, any plane with a nonzero *h*, *k*, or *i* index *and* a nonzero *l* index in the (*hkil*) Miller–Bravais indexing convention can serve as a growth surface. However, only growth on low index semipolar planes, for example {10$\bar{1}\bar{1}$}, {10$\bar{1}\bar{3}$}, and {11$\bar{2}$2} has been proven to result in planar layers with a reliably smooth surface.

Different substrates have been used for heteroepitaxial growth of semipolar GaN layers with different orientation. (10$\bar{1}\bar{1}$) and (10$\bar{1}\bar{3}$) layers were grown on (100)$MgAl_2O_4$ and (110)$MgAl_2O_4$, respectively, [6], while growth on *m*-plane sapphire resulted in (10$\bar{1}\bar{3}$) or (11$\bar{2}$2) layers depending on the nitridation condition [7]. Semipolar layers on tilted *m*-plane sapphire [99] on 7°-off-axis (001)Si [100] and on (30$\bar{3}$8) 4H-SiC [101] have also been reported. Although the quality of the layers varies depending on grown plane orientation and substrate, surface morphology and defect microstructure of semipolar layers normally resemble those in nonpolar layers. The typical densities of the TDs and BSFs in HVPE layers grown on $MgAl_2O_4$ and *m*-plane sapphire are $\sim 10^9 - 10^{10}$ cm^{-2}, and $\sim 10^5$ cm^{-1} [6, 7]. ELO technique by MOCVD has been applied for (11$\bar{2}$2) layers on *m*-plane sapphire [66, 102]. Different stripe orientations have been attempted but a successful reduction of the structural defects in the overgrown regions was achieved only for stripes oriented along the [$\bar{1}$100] direction that is perpendicular to *c*-axis projection. For this stripe orientation, SFs and partial TDs have been found to propagate only through the windows in a direction inclined to the layer surface [66]. The coalescence front where the extended defects stop was partly inclined and partly vertical. It has been predicted that for optimized growth conditions which would allow a perfectly vertical coalescence front, full filtration of inclined BSFs and TDs for thick enough overgrowth would take place [66]. Then the semipolar (11$\bar{2}$2) ELO layers would be perfectly defect-free on the top surface. It is worthy to note that that such a completely defect-free surface, in principle, cannot be achieved by one-step ELO in nonpolar *a*-plane and *m*-plane GaN.

Low-temperature PL spectra of (11$\bar{2}$2) layers grown by MOCVD are shown in Figure 8.16 [66]. The spectrum of the planar layer is very similar to that of MOCVD-grown *a*-plane and *m*-plane layers, the D1 and D2 emissions dominate while the NBE emission is practically absent. The emission in the ELO layer with stripes perpendicular to the *c*-axis projection (ELO-90°) is clearly improved reflecting the improved structural quality of the material. Although the D1 emission is still present, the spectrum is now dominated by the NBE emission. In addition, the D2 emission is much weaker compared with the one in the planar layer. Also shown in Figure 8.16 is the PL spectrum of the ELO layer with stripes inclined at 45° to the *c*-axis projection (ELO-45°). Compared with the planar layer, the NBE emission in the ELO-45° layer is slightly enhanced and the D2 emission is reduced. The BSF-related emission

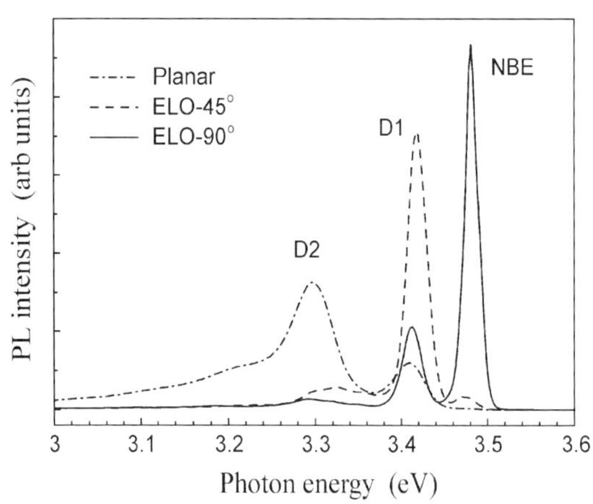

Figure 8.16 Low-temperature (7 K) PL spectra of semipolar (11$\bar{2}$2) GaN grown by MOCVD on *m*-plane sapphire. Spectra of a planar layer, an ELO layer with stripes inclined at 45° to the *c*-axis projection (ELO-45°) and an ELO layer with stripes perpendicular to the *c*-axis projection (ELO-90°) are compared.

D1 is, however, much stronger. The TEM study reveals that there is practically no reduction of the defect density for this stripe orientation. BSFs and TDs are observed everywhere above the mask and above the inclined facets.

8.6 Luminescence in GaN with Nonpolar Surfaces Sliced from Boules Grown along the *c* Axis

As seen from the previous sections, heteroepitaxial GaN layers grown along nonpolar or semipolar directions contain a high density of TDs and SFs. Reduction of the extended defects and improvement of structural quality of the layers have been achieved by applying different ELO-based techniques. However, this approach requires a complicated technological process, which involves *ex situ* processing steps and regrowth. In addition, the nonuniformity in the defect reduction limits the usable area for the device structures and confines the development of a large-scale production.

An alternative approach for producing templates with nonpolar or semipolar surfaces is growth of GaN boules in the conventional [0001] direction up to a significant thickness by HVPE that could then be sliced into bars or wafers with desired orientations. This type of GaN material was characterized by different techniques such as X-ray diffraction, TEM, and CL, and confirmed to be of high structural quality with TD density in the range of 10^5 cm^{-2} [103, 104].

The emission properties of such bulk GaN bars with *a*-plane and *m*-plane surfaces are very similar to those observed in polar *c*-plane GaN layers. Low-temperature PL spectra (Figure 8.17) in the NBE region show the FE emissions (X_A and X_B) and narrow lines of DBEs involving the O and Si donors (D^oX_A(O) and D^oX_A(Si)). The FWHM of 0.7 and 0.6 meV is found for DBEs in *a*-plane and *m*-plane bars, respectively. At lower photon energies, several TET lines related to DBE recombination leaving the donors in excited s- or p-like states are also seen. The 2s and 2p TETs related to the Si donor, as well as the 2s and 3s TETs related to the O donor are observed in the spectra. The broader peak, labeled as X_A:D^0(2s), is interpreted as a replica related to scattering of the X_A excitons at the Si donors, which is left in a 2s state [20]. Polarized PL spectra of *a*-plane and *m*-plane bars reveal the common optical selection rules in GaN and practically resemble the spectra measured from the edge of *c*-plane GaN layers [105]. For both samples, no defect-related emissions in the 3.29–3.42-eV region are observed implying a negligibly small density of SFs as it is generally expected for GaN grown along the [0001] direction.

Optical properties of GaN plates with a semipolar $(11\bar{2}2)$ surface sliced from thick HVPE layers as well as of homoepitaxial layers grown by MOCVD on such semipolar substrates have also been reported [106, 107]. As in the case of the nonpolar bars, low-temperature PL spectra exhibit narrow DBE lines and well-resolved X_A and X_B FEs. However, the polarized PL spectra measured in backscattering geometry show some peculiarities, the most prominent of

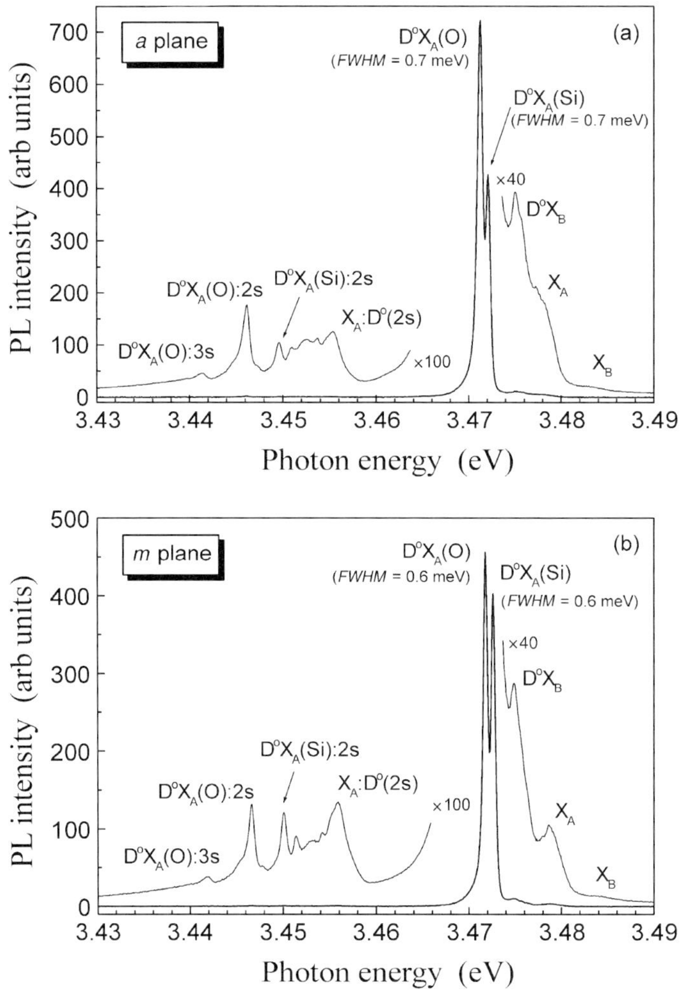

Figure 8.17 Low-temperature (2 K) PL spectra of nominally undoped bulk GaN bars with nonpolar *a*-plane GaN (a) and *m*-plane GaN (b) surface.

which is that the X_A exciton emission does not vanish for any polarization orientation in the $(11\bar{2}2)$ plane. The specific optical anisotropy in the FE emission in semipolar GaN has recently been explained in term of crystal orientation and the symmetry of the exciton states [108].

8.7 Summary

We have reviewed optical emission properties of GaN grown along nonpolar and semipolar directions. It has been shown that the low-temperature spectra of heteroepitaxial *a*-plane, *m*-plane, and semipolar layers are dominated by a number of emission lines in the 3.29–3.42-eV spectral region. These lines being not typical for GaN grown along the [0001] direction are associated with the extended structural defects, which are present with a high density in the nonpolar and semipolar layers. Although the details of the PL spectra vary in different types of layers and depend on both the substrate and the growth method, three common defect-related emission bands are found to occur in most of the layers. Detailed PL and CL studies combined with TEM structural characterizations allow associating the 3.40–3.42 eV emission with BSFs of type I_1, the 3.35–3.37 eV emission with BSFs of type I_2, and the 3.29–3.32 eV emission with partial dislocations or prismatic SFs terminating the BSFs.

Suppression of the defect-related emissions and enhancement of the NBE exciton emission has been found in layers produced by different ELO techniques. Such layers have an improved structural quality owing to reduced density of the TDs and BSF in the overgrowth regions if the ELO stripes are with optimal orientation, that is the orientation perpendicular to the *c* axis or to its projection. In addition, the ELO layers possess a smooth surface morphology and can serve as templates for subsequent growth of device structures.

We have also examined the emission in bulk GaN templates with nonpolar and semipolar surfaces sliced in appropriate directions from boules grown along the *c* axis by HVPE. The absence of structural defect-related emissions and the observation of sharp exciton lines in the low-temperature PL spectra reflect a high structural and optical quality of the material. It appears that the slicing of thick GaN grown along the *c* axis is so far the most promising approach for producing nonpolar and semipolar substrates to be used for light-emitting devices.

Acknowledgments

We would like to thank Dr B. Haskell, Dr P. Fini, Prof. J. Speck, and Prof. S. Nakamura (University of California, Santa Barbara), Prof. H. Amano and Prof. I. Akasaki (Meijo University), Dr S. Figge and Prof. D. Hommel (Bremen University), Dr E. Preble, Dr A. Hanser, and Dr K. Evans (Kyma Technologies Inc.) for kindly providing GaN samples with nonpolar surfaces. Dr T. Paskova is acknowledged for growing many of HVPE layers studied, performing CL measurements and for fruitful discussions. R. Schifano and T. Malinauskas are acknowledged for the assistance in a part of PL measurements. We are also indebted to Prof. F. Ponce for providing Figures 8.7 and 8.8, to Prof. O. Brandt for providing Figure 8.14, and to Dr Z. Bougrioua for sending the data shown in Figure 8.15. P. P. Paskov would like to acknowledge a partial support from the Swedish Research Council (VR) under contract 2005–5918.

References

1. Fiorentini, V., Bernardini, F., Della Sala, F. *et al.* (1999) *Physical Review A*, **60**, 8849.
2. Waltereit, P., Brandt, O., Trampert, A. *et al.* (2000) *Nature*, **406**, 865.
3. McLaurin, M., Mates, T.E., and Speck, J.S. (2005) *Applied Physics Letters*, **86**, 262104.
4. Craven, M.D., Lim, S.H., Wu, F. *et al.* (2002) *Applied Physics Letters*, **81**, 469.
5. Craven, M.D., Chakraborty, A., Imer, B. *et al.* (2003) *Physica Status Solidi C*, **0**, 2132.
6. Baker, T.J., Haskell, B.A., Wu, F. *et al.* (2005) *Japanese Journal of Applied Physics*, **44**, L920.
7. Baker, T.J., Haskell, B.A., Wu, F. *et al.* (2006) *Japanese Journal of Applied Physics*, **45**, L154.
8. Wang, H.M., Chen, C.Q., Gong, Z. *et al.* (2004) *Applied Physics Letters*, **84**, 499.
9. Monemar, B. (1999) *Material Science & Engineering B*, **59**, 122.
10. Monemar, B. (2001) *Journal of Physics. Condensed Matter*, **13**, 7011.
11. Reshchikov, M.A. and Morkoc, H. (2005) *Journal of Applied Physics*, **97**, 061301.
12. Paskov, P.P., Paskova, T., Holtz, P.O., and Monemar, B. (2001) *Physical Review A*, **64**, 115201.
13. Paskov, P.P., Paskova, T., Holtz, P.O., and Monemar, B. (2004) *Physical Review A*, **70**, 035210.
14. Kornitzer, K., Ebner, T., Thonke, K. *et al.* (1999) *Physical Review A*, **60**, 1471.
15. Thonke, K., Kornitzer, K., Grehl, M. *et al.* (2000) *IPAP Conference Series*, **1**, 587.
16. Kovalev, D., Averboukh, D., Volm, D. *et al.* (1996) *Physical Review A*, **54**, 2318.
17. Freitas J.A. Jr, Moore, W.J., Shanabrook, B.V., *et al.* (2002) *Physical Review A* **66**, 233311.
18. Gil, B., Bigenwald, P., Leroux, M. *et al.* (2007) *Physical Review A*, **75**, 085204.
19. Rodina, A.V., Dietrich, M., Göldner, A. *et al.* (2001) *Physical Review A*, **64**, 115204.
20. Paskov, P.P., Monemar, B., Toropov, A.A. *et al.* (2007) *Physica Status Solidi C*, **4**, 2601.
21. Stepniewski, R., Wysmolek, A., Potemski, M. *et al.* (2003) *Physical Review Letters*, **91**, 226404.
22. Korona, K.P., Doradzinski, R., Palczewska, M. *et al.* (2003) *Physica Status Solidi B*, **235**, 40.
23. Monemar, B., Paskov, P.P., Bergman, J.P. *et al.* (2006) *Physica Status Solidi B*, **243**, 1604.
24. Monemar, B., Paskov, P.P., Bergman, J.P. *et al.* (2007) *Physica Status Solidi B*, **244**, 1759.
25. Zhong, Y., Wong, K.S., Zhang, W., and Look, D.C. (2006) *Applied Physics Letters*, **88**, 022108.
26. Saarinen, K., Laine, T., Kuisma, S. *et al.* (1997) *Physical Review Letters*, **79**, 3030.
27. Sano, M. and Aoki, M. (1976) *Japanese Journal of Applied Physics*, **15**, 1943.
28. Yoshida, S., Misawa, S., and Gonda, S. (1983) *Applied Physics Letters*, **42**, 427.
29. Eddy, C.R. Jr, Moustakas, T.D., and Scanton, J. (1993) *Journal of Applied Physics* **73**, 448.
30. Sasaki, T. and Zembutsu, S. (1987) *Journal of Applied Physics*, **61**, 2533.
31. Sun, C.J. and Razeghi, M. (1993) *Applied Physics Letters*, **63**, 973.
32. Lei, T., Ludwig, K.F. Jr, and Moustakas, T.D. (1993) *Journal of Applied Physics* **74**, 4430.
33. Tripathy, S., Soni, R.K., Asahi, H. *et al.* (1999) *Journal of Applied Physics*, **85**, 8386.
34. Iwata, K., Asahi, H., Asami, K. *et al.* (1997) *Japanese Journal of Applied Physics*, **36**, L661.
35. Imura, M., Hoshino, A., Nakano, K. *et al.* (2005) *Japanese Journal of Applied Physics*, **44**, 6418.
36. Ni, X., Fu, Y., Moon, Y.T. *et al.* (2006) *Journal of Crystal Growth*, **290**, 166.

37. Haskell, B.A., Wu, F., Matsuda, S. *et al.* (2003) *Applied Physics Letters*, **83**, 1554.

38. Paskova, T., Paskov, P.P., Valcheva, E. *et al.* (2004) *Physica Status Solidi A*, **201**, 2265.

39. Ng, H.M. (2002) *Applied Physics Letters*, **80**, 4369.

40. Iyer, S., Smith, D.J., Bhattacharyya, A. *et al.* (2003) *Materials Research Society Symposium Proceedings*, **743**, L3.20.

41. Chen, Z., Lu, D.C., Liu, X. *et al.* (2003) *Journal of Applied Physics*, **93**, 316.

42. Yu, H.B., Chen, H., Li, D.S. *et al.* (2004) *Journal of Crystal Growth*, **263**, 94.

43. Liu, R., Bell, A., Ponce, F.A. *et al.* (2005) *Applied Physics Letters*, **86**, 021908.

44. Paskov, P.P., Schifano, R., Monemar, B. *et al.* (2005) *Journal of Applied Physics*, **98**, 093519.

45. Paskov, P.P., Schifano, R., Monemar, B. *et al.* (2006) *Superlattices and Microstructures*, **40**, 253.

46. Hussy, S., Meissner, E., Birkmann, B. *et al.* (2006) *Physica Status Solidi A*, **203**, 1676.

47. Zhu, T., Martin, D., Butte, R. *et al.* (2007) *Journal of Crystal Growth*, **300**, 186.

48. Gil, B. (1999) in *Gallium Nitride II*, (eds J.I. Pankove and T.D. Moustakas, Academic Press, San Diego, p. 209.

49. Liu, L. and Edgar, J.H. (2002) *Materials Science and Engineering Reports*, **37**, 61.

50. Ghosh, S., Misra, P., Grahn, H.T. *et al.* (2005) *Journal of Applied Physics*, **98**, 026105.

51. Gil, B. and Alemu, A. (1997) *Physical Review B*, **56**, 12446.

52. Paskov, P.P., Schifano, R., Paskova, T. *et al.* (2006) *Physica B*, **376–77**, 473.

53. Salviati, G., Albrecht, M., Zannotti-Fregonara, C. *et al.* (1999) *Physica Status Solidi A*, **171**, 325.

54. Bai, L., Dudley, M., Chen, L. *et al.* (2005) *Materials Research Society Symposium Proceedings*, **831**, E11.37.

55. Stampfl, C. and Van de Walle, C.G. (1998) *Physical Review A*, **57**, R15052.

56. Skromme, B.J., Chen, L., Mikhov, M.K. *et al.* (2004) *Materials Science Forum*, **457–60**, 1613.

57. Wu, F., Craven, M.D., Lim, S.H., and Speck, J.S. (2003) *Journal of Applied Physics*, **94**, 942.

58. Sumiya, M., Yoshimura, K., Ohtsuka, K., and Fuke, S. (2000) *Applied Physics Letters*, **76**, 2098.

59. Röder, C., Einfeldt, S., Figge, S. *et al.* (2006) *Journal of Applied Physics*, **100**, 103511.

60. Paskova, T., Darakchieva, V., Paskov, P.P. *et al.* (2005) *Journal of Crystal Growth*, **281**, 55.

61. Paskova, T., Kroeger, R., Paskov, P.P. *et al.* (2006) *Proceedings of SPIE*, **6121**, 47.

62. Paskov, P.P., Schifano, R., Malinauskas, T. *et al.* (2006) *Physica Status Solidi C*, **3**, 1499.

63. Beaumont, B., Vennegues, P., and Gibart, P. (2001) *Physica Status Solidi B*, **227**, 1.

64. Craven, M.D., Lim, S.H., Wu, F. *et al.* (2002) *Applied Physics Letters*, **81**, 1201.

65. Chen, C.Q., Yang, J.W., Wang, H.M. *et al.* (2003) *Japanese Journal of Applied Physics*, **42**, L640.

66. Ni, X., Özgur, U., Fu, Y. *et al.* (2006) *Applied Physics Letters*, **89**, 262105.

67. Bougrioua, Z., Laugt, M., Vennegues, P. *et al.* (2007) *Physica Status Solidi A*, **204**, 282.

68. Haskell, B.A., Wu, F., Craven, M.D. *et al.* (2003) *Applied Physics Letters*, **83**, 644.

69. Özgur, U., Ni, X., Fu, Y. *et al.* (2006) *Applied Physics Letters*, **89**, 262117.

70. Chichibu, S.F., Uedono, A., Onuma, T. *et al.* (2005) *Applied Physics Letters*, **86**, 021914.

71. Jursenas, S., Kuokstis, E., Miasojedovas, S. *et al.* (2004) *Applied Physics Letters*, **85**, 771.

72. Chen, C.Q., Zhang, J.P., Yang, J.W. *et al.* (2003) *Japanese Journal of Applied Physics*, **42**, L818.

73. Wang, T.C., Lu, T.C., Ko, T.S. *et al.* (2006) *Applied Physics Letters*, **89**, 251109.
74. Imer, B.M., Wu, F., DenBaars, S.P., and Speck, J.S. (2006) *Applied Physics Letters*, **88**, 061908.
75. Iida, D., Nagai, T., Kawashima, T. *et al.* (2007) *Materials Research Society Symposium Proceedings*, **955**, I4.2.
76. Linthicum, K., Gehrke, T., Thomson, D. *et al.* (1999) *Applied Physics Letters*, **75**, 196.
77. Wagner, B.P., Preble, E.A., Reitmeier, Z.J. *et al.* (2004) *Materials Research Society Symposium Proceedings*, **798**, Y10.34.
78. Li, J.W., Gong, Z., Chen, C.Q. *et al.* (2005) *Materials Research Society Symposium Proceedings*, **831**, E2.9.
79. Wagner, B.P., Reitmeier, Z.J., Park, J.S. *et al.* (2006) *Journal of Crystal Growth*, **290**, 504.
80. Hellman, E.S., Liliental-Weber, Z., and Buchanan, D.N.E. (1997) *MRS Internet Journal of Nitride Semiconductor Research*, **2**, 30.
81. Waltereit, P., Brandt, O., and Ploog, K.H. (1999) *Applied Physics Letters*, **75**, 2029.
82. Xu, K., Hu, J., Deng, P.Z. *et al.* (1999) *Physica Status Solidi A*, **176**, 589.
83. Sun, Y.J., Brandt, O., Jahn, U. *et al.* (2002) *Journal of Applied Physics*, **92**, 5714.
84. Liu, T.Y., Trampert, A., Sun, Y.J. *et al.* (2004) *Philosophical Magazine Letters*, **84**, 435.
85. Gerlach, J.M., Hofmann, A., Höche, T. *et al.* (2006) *Applied Physics Letters*, **88**, 011902.
86. Rivera, C., Misra, P., Pau, J.L. *et al.* (2007) *Journal of Applied Physics*, **101**, 053527.
87. Ghosh, S., Waltereit, P., Brandt, O. *et al.* (2002) *Physical Review A*, **65**, 075202.
88. Maruska, H.P., Hill, D.W., Chou, M.M.C. *et al.* (2003) *Opto-Electronics Review*, **11**, 7.
89. Haskell, B.A., Chakraborty, A., Wu, F. *et al.* (2005) *Journal of Electronic Materials*, **34**, 357.
90. Reshchikov, M.A., Teke, A., Maruska, H.P. *et al.* (2004) *Materials Research Society Symposium Proceedings*, **798**, Y5.53.
91. Jasinski, J., Liliental-Weber, Z., Maruska, H.P. *et al.* (2003) *Materials Research Society Symposium Proceedings*, **764**, C6.6.
92. Chen, C.Q., Gaevski, M.E., Sun, H.W. *et al.* (2002) *Applied Physics Letters*, **81**, 3194.
93. Haskell, B.A., Baker, T.J., McLaurin, M.B. *et al.* (2005) *Applied Physics Letters*, **86**, 111917.
94. Armitage, R., Horita, M., Suda, J., and Kimoto, T. (2007) *Journal of Applied Physics*, **101**, 033534.
95. Imer, B., Wu, F., Craven, M.D. *et al.* (2006) *Japanese Journal of Applied Physics*, **45**, 8644.
96. Kawashima, T., Nagai, T., Iida, D. *et al.* (2007) *Physica Status Solidi B*, **244**, 1848.
97. Takeuchi, T., Amano, H., and Akasaki, I. (2000) *Japanese Journal of Applied Physics*, **39**, 413.
98. Romanov, A.E., Baker, T.J., Nakamura, S., and Speck, J.S. (2006) *Journal of Applied Physics*, **100**, 023522.
99. Matsuoka, T. and Hagiwara, E. (2001) *Physica Status Solidi A*, **188**, 485.
100. Honda, Y., Kameshiro, N., Yamaguchi, M., and Sawaki, N. (2002) *Journal of Crystal Growth*, **242**, 82.
101. Kamiyama, S., Honshio, A., Kitano, T. *et al.* (2005) *Physica Status Solidi C*, **2**, 2121.
102. Ni, X., Özgur, U., Baski, A.A. *et al.* (2007) *Applied Physics Letters*, **90**, 182109.
103. Paskova, T., Kroeger, R., Figge, S. *et al.* (2006) *Applied Physics Letters*, **89**, 051914.
104. Paskova, T., Kroeger, R., Hommel, D. *et al.* (2007) *Physica Status Solidi C*, **4**, 2536.
105. Paskova, T., Paskov, P.P., Darakchieva, V. *et al.* (2007) *Materials*

Research Society Symposium Proceedings, **955**, I3.4.

106. Kojima, K., Ueda, M., Funato, M., and Kawakami, Y. (2007) *Physica Status Solidi B*, **244**, 1853.

107. Ueda, M., Kojima, K., Funato, M. *et al.* (2006) *Applied Physics Letters*, **89**, 211907.

108. Gil, B. (2007) *Applied Physics Letters*, **90**, 121903.

9
Optical Phonons in *a*-plane GaN under Anisotropic Strain

Vanya Darakchieva, Tanya Paskova, and Mathias Schubert

9.1
Introduction

The phonon spectrum is one of the fundamental characteristics of crystals. The behavior of phonon dispersion branches reflects specific features of the crystal structure and the interatomic interactions, and therefore gives the most comprehensive and detailed information about the dynamical properties of crystals. Phonons are straightforward signatures of bond lengths and chemical species, and phonon parameters can be used to obtain information on strain, alloy composition, free-carrier concentration, and mobility in individual layers and constituents of device structures. Phonon mode parameters can be also used to study compositional fluctuations, phase mixing, interface morphologies, as well as defects and impurities. To use the phonons as a gauge and tool to extract information on material properties and characteristics, it is mandatory to establish first the phonon mode behavior.

Phonon frequencies are often employed as a tool for strain assessment in mismatched semiconductor heterostructures [1]. This involves IR and Raman measurements of phonon frequency shifts with respect to unstrained material, and their conversion into strain or stress components via the phonon deformation potentials. The knowledge of precise values of the strain-free frequencies and the phonon deformation potentials is of key importance in such studies. Phonon deformation potentials can be obtained from control experiments, in which the frequencies of the phonons under consideration are calibrated versus well-defined applied strains (or stresses) on otherwise strain-free material. Alternatively, the phonon frequencies can be calibrated versus strain components in layers exhibiting different intrinsic strains. The latter requires determination of the intrinsic strain components.

The application of these principles for studying phonons and strain effects in GaN and related materials turned out to be challenging. The main obstacle is the lack of native substrates in the growth process and consequent difficulties

Nitrides with Nonpolar Surfaces. Growth, Properties, and Devices. Edited by Tanya Paskova

ISBN: 978-3-527-40768-2

in the preparation of strain-free material. Substantial efforts have been directed toward the fabrication of bulk GaN and AlN using growth from solution [2], sodium melt [3], or in supercritical ammonia [4] and sublimation epitaxy [5]. Alternatives, such as growth of thick GaN layers on sapphire by hydride vapor phase epitaxy (HVPE) single crystals followed by substrate separation has been also suggested [6, 7]. Although promising results have been achieved, the available quasi-bulk GaN single crystals still suffer from one or more drawbacks: small size, high density of point defects, relatively high residual carrier doping, relatively high dislocation density, and so on. Consequent difficulties in performing 'pure' control vibrational measurements have lead to a large scatter in the experimental phonon deformation potentials and strain-free frequencies of GaN [8–10] and AlN [9, 11, 12], while for InN, such information is mostly unavailable [13]. In addition, there is a lack of agreement between some phonon deformation potentials of GaN and AlN as determined in theory, [14] and by employing Raman scattering spectroscopy [8, 10, 11] that makes further clarification of these issues by independent techniques highly desirable.

In this respect, GaN films and bulk material with nonpolar orientations offer interesting opportunities. For instance, non-*c*-axis-oriented GaN allows access to the complete set of polar phonons by generalized infrared spectroscopic ellipsometry (GIRSE) [15], which provides an alternative and/or complementary tool to Raman scattering spectroscopy for the study of vibrational properties. Hence, non-*c*-axis GaN layers experiencing different strains will make possible the determination of phonon deformation potentials by GIRSE. However, this will require assessment of the strain components in the films. Nonpolar nitride growth is often realized on non-*c*-plane sapphire and SiC substrates. As a consequence, a peculiarity of nonpolar GaN heteroepitaxial layers is the presence of anisotropic strain, being the result of the anisotropy of the growth surfaces. Anisotropically strained group III nitrides are very appealing from a material physics point of view, since they provide a possibility for the study of new effects, although they impose some challenges on measurements and analyses. Theory predicts that some of the phonons of group III nitrides split under anisotropic strain [14, 16] since the GaN $[11\bar{2}0]$ and $[1\bar{1}00]$ directions are no longer equivalent in terms of strain as is in the case of isotropic strain. Very recently, the first experimental evidence of phonon splitting has been given and the *c* phonon deformation potentials, related to the strain anisotropy in the basal hexagonal plane have been determined [17, 18].

In this chapter, we review recent studies of optical phonons in relation to the anisotropic strain in GaN heteroepitaxial layers and quantum dots (QDs) with nonpolar orientations. For reasons of comparison, results on anisotropically strained *c*-plane GaN films grown on *a*-plane sapphire is also presented where appropriate. The application of GIRSE and Raman scattering to the studies of optical phonon frequencies of anisotropic wurtzite GaN is described to reveal the phonon mode behavior. The assessment of anisotropic strain components in *a*-plane GaN films and their evolution with thickness is presented. Special attention is paid to the effect of anisotropic strain on phonon frequencies and determination of the phonon deformation potentials related to the strain anisotropy in the basal plane. Further, the determination of the isotropic A_1(TO) and E_1(LO) phonon deformation potentials by GIRSE will be described, and a comparative analysis of the results with respect to recent theoretical studies and Raman experiments is made. The use of phonon frequencies to extract information on the internal electric fields present in *a*-plane GaN QDs is also reviewed.

9.2 Background

9.2.1 Structure of Wurtzite GaN

The strongly ionic GaN compound crystallizes in three polytypes: the wurtzite, zinc blende, and rock salt structures. At ambient conditions, the thermodynamically stable structure is the wurtzitic with space group C_{6v}^4 ($P6_3mc$). The crystal structure and the bond stacking of wurtzite GaN are shown in Figure 9.1a and b, respectively. Typical planes of epitaxial growth of GaN are indicated in Figure 9.1a. Unless specified otherwise, the crystal coordinate system is chosen as $x \parallel [11\bar{2}0]$, $y \parallel [1\bar{1}00]$, and $z \parallel [0001]$ as shown in Figure 9.1a. The wurtzite structure is characterized by two hexagonal lattice parameters, a and c, and by the internal cell parameter $u = (1/3)(a^2/c^2) + 1/4$, which is related to the relative length of the III–N bonds parallel to the c axis (i.e. along the [0001] direction, Figure 9.1). There is a large scatter in the

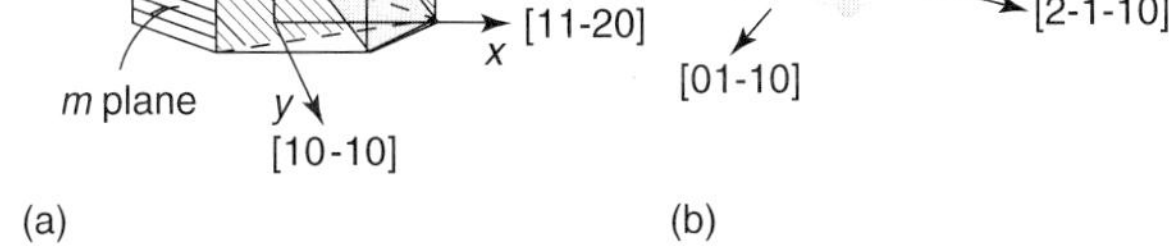

Figure 9.1 (a) Schematic presentation of the wurtzite crystal structure of GaN. Hatched areas indicate the most often used planes for epitaxial growth: *c*-plane (0001), *a*-plane ($11\bar{2}0$), *m*-plane ($1\bar{1}00$), and *r*-plane ($1\bar{1}02$). (b) Perspective view of the hexagonal unit cell in hexagonal Cartesian coordinate system. The bonds in ($11\bar{2}0$) plane are indicated by thick solid lines, the stacking sequence is shown, and the lattice parameters *a* and *c* are indicated.

reported lattice parameters of GaN, and consensus on the strain-free values have not been reached yet [19–21]. Recently, $c = 5.18521$ Å and $a = 3.18923$ Å have been obtained for state-of-the-art undoped HVPE bulk GaN material [22]. In general, the experimental c/a ratios for the wurtzite GaN lattice deviate from the ideal value of 1.633, suggesting a slight distortion of the hexagonal lattice [23]. This deviation is also influenced by structural defects, high doping, or incomplete strain relaxation in the GaN films.

9.2.2
Phonons in Wurtzite GaN

In Figure 9.2, the calculated phonon dispersion curves of hexagonal GaN are shown along several high-symmetry lines in the Brillouin zone together with the corresponding density of states [24]. Since the wurtzite unit cell contains four atoms, there are 12 vibrational modes producing complicated dispersions. The acoustic phonon branches are well separated from the optical ones. According to the factor group analysis, the optical and acoustic phonon modes of wurtzite GaN belong to the following irreducible representation at the Γ point of the Brillouin zone [24]:

$$\Gamma_{\mathrm{ac}} + \Gamma_{\mathrm{opt}} = 2(\Gamma_1 + \Gamma_4 + \Gamma_5 + \Gamma_6) = 2(A_1 + B_1 + E_1 + E_2)$$
$$\Gamma_{\mathrm{ac}} = \Gamma_1 + \Gamma_6 = A_1 + E_1$$
$$\Gamma_{\mathrm{opt}} = \Gamma_1 + 2\Gamma_4 + 2\Gamma_5 + \Gamma_6 = A_1 + 2B_1 + E_1 + 2E_2$$

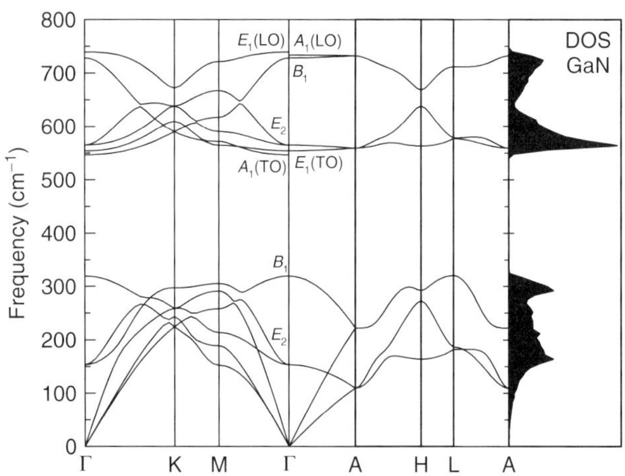

Figure 9.2 Phonon dispersions and densities of states for wurtzite GaN. (Reprinted figure with permission from Davydov, V.Y., Kitaev, Y.E., Goncharuk, I.N. *et al.* (1998) *Physical Review B*, **58**, 12899. Copyright 1998 by the American Physical Society.)

The A_1 and E_1 modes are both IR and Raman active, the twofold E_2 (E_2^{low} and E_2^{high}) are only Raman active, and the B_1 modes (B_1^{low} and B_1^{high}) are silent. Owing to the macroscopic electric field associated with the relative atomic displacement of the longitudinal phonons, the polar A_1 and E_1 modes are split into longitudinal optical (LO) and transverse optical (TO) components. This electric field serves to stiffen the force constants of the phonon and thereby rises the LO frequency over that of the respective TO mode. The comparatively large mode polarity in GaN is a consequence of the strong ionic character of the bonds between group III and N atoms.

The short-range interatomic forces, which exhibit the anisotropy of the force constants, cause the mode anisotropy resulting in different frequencies of the A_1 and E_1 modes. Since the electrostatic forces dominate the anisotropy in the short-range forces for group III nitrides, the TO–LO splitting is larger than the A_1-E_1 splitting. For the lattice vibrations with A_1 and E_1 symmetry, the atoms move parallel and perpendicular to the c axis, respectively. The atomic displacement patterns corresponding to the optical modes in group III nitride compounds are illustrated in Figure 9.3.

9.2.3 Lattice Deformation and Strain

The deformation in a crystal is described by two normalized quantities related to each other: *strain* and *stress*. The stress, σ, is defined as the internal force created to balance an external force, F, applied to an area of a crystal, A, normalized to A:

$$\sigma = \frac{F}{A} \tag{1}$$

The strain, ε, is defined as the deformation, Δu_0, per unit length, L_0:

$$\varepsilon = \frac{\Delta u_0}{L_0} = \frac{(L - L_0)}{L_0} \tag{2}$$

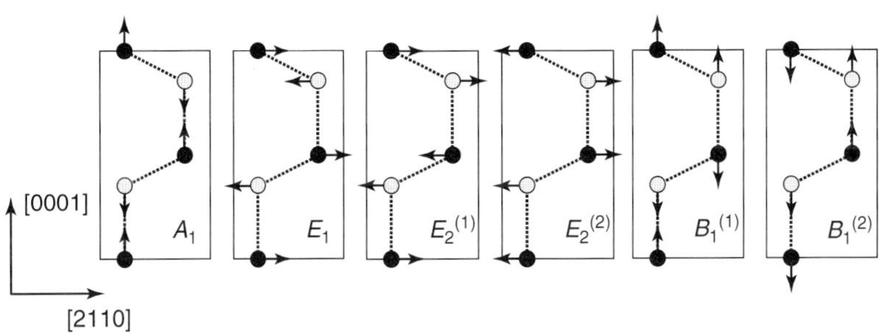

Figure 9.3 Displacement patterns for the six optical phonon modes in wurtzite group III nitrides.

The applied forces are vector quantities and can be resolved into three components: two within the plane they are acting on, and the third, normal to this plane. The force component acting in the normal direction is the normal force, while the components acting in the plane are called *shear forces*. The stress caused by these forces is a second-rank tensor:

$$\sigma_{ij} = \begin{pmatrix} \sigma_{xx} & \sigma_{xy} & \sigma_{xz} \\ \sigma_{xy} & \sigma_{yy} & \sigma_{yz} \\ \sigma_{xz} & \sigma_{yz} & \sigma_{zz} \end{pmatrix} \tag{3}$$

The stresses with mixed indices are the shear stresses where the second index defines the normal to the plane in which the stress acts, and the first index indicates its direction. Normal stresses act along the direction of the axis indicated by the repeated index. Positive values of stress indicate tensile stress, while negative values indicate a compressive stress state.

The strain is also a second-rank tensor:

$$\varepsilon_{ij} = \begin{pmatrix} \varepsilon_{xx} & \varepsilon_{xy} & \varepsilon_{xz} \\ \varepsilon_{xy} & \varepsilon_{yy} & \varepsilon_{yz} \\ \varepsilon_{xz} & \varepsilon_{yz} & \varepsilon_{zz} \end{pmatrix} \tag{4}$$

In the linear elasticity theory, the relation between the stress and strain tensors σ_{ij} and ε_{ij} for wurtzite structure semiconductors, such as GaN is given by:

$$\begin{pmatrix} \sigma_{xx} \\ \sigma_{yy} \\ \sigma_{zz} \\ \sigma_{yz} \\ \sigma_{xz} \\ \sigma_{yz} \end{pmatrix} = \begin{pmatrix} C_{11} & C_{12} & C_{13} & 0 & 0 & 0 \\ C_{12} & C_{11} & C_{13} & 0 & 0 & 0 \\ C_{13} & C_{13} & C_{33} & 0 & 0 & 0 \\ 0 & 0 & 0 & C_{44} & 0 & 0 \\ 0 & 0 & 0 & 0 & C_{44} & 0 \\ 0 & 0 & 0 & 0 & 0 & 1/2(C_{11} - C_{12}) \end{pmatrix} \begin{pmatrix} \varepsilon_{xx} \\ \varepsilon_{yy} \\ \varepsilon_{zz} \\ 2\varepsilon_{yz} \\ 2\varepsilon_{xz} \\ 2\varepsilon_{xy} \end{pmatrix} \tag{5}$$

where C_{ij} are the elastic stiffness constants and the x, y, and z axes are chosen along the GaN $[11\bar{2}0]$, $[1\bar{1}00]$, and [0001] directions, respectively (see Figure 9.1a).

9.2.3.1 Biaxial Isotropic Strain in GaN

The growth on foreign substrates typically leads to the presence of built-in strain in heteroepitaxial GaN layers owing to the difference in lattice parameters and thermal expansion coefficients between layers and substrates [19, 25–27]. Sapphire and SiC are among the most often used substrates, and typically growth is realized on the basal (0001) *c*-plane of sapphire and SiC. In such instances, nitride films grow along the *polar* [0001] direction. The sixfold symmetry of the basal planes of the wurtzite (nitrides, SiC) and rhombohedral (sapphire) crystal structures dictates their isotropy in the basal plane and hence, the thermal expansion coefficients, piezoelectric and elastic properties should be the same for any direction within the *c*-plane. As a consequence, the films grown on *c*-plane substrates experience *isotropic biaxial in-plane strain*. In such instances, the stress along the growth direction σ_{zz} of the *c*-plane films vanishes because the surface is free to expand or contract. It follows then from Equation 5 that the strain along the growth direction can be expressed by the two in-plane strain components as follows:

$$\varepsilon_{zz} = -\frac{C_{13}}{C_{33}}(\varepsilon_{xx} + \varepsilon_{yy}) \tag{6}$$

9.2.3.2 **Anisotropic Strain in GaN**

Whenever growth is realized on non-*c*-plane-oriented sapphire and SiC surfaces, the nitride films are under *anisotropic biaxial strain* (independently of the film orientation) as a consequence of the anisotropy of the growth surfaces. The growth of *nonpolar a*-plane GaN on *r*-plane sapphire can be taken as an exemplary case. The epitaxial relationships between the *a*-plane GaN layer and the *r*-plane sapphire substrate, illustrated in Figure 9.4, are GaN [0001]||sapphire [$\bar{1}$101] and GaN [$\bar{1}$100]||sapphire [11$\bar{2}$0] [17]. The difference in thermal expansion coefficients between sapphire and GaN [28] results in highly anisotropic thermal mismatch of −28% along the GaN [$\bar{1}$101] and <−1% along the GaN [0001], respectively. Further, different lattice mismatches along the two main in-plane directions of −16% along the GaN [$\bar{1}$101] and −1% along the GaN [0001] can be estimated. Therefore, *a*-plane GaN films grown on *r*-plane sapphire are expected to be under anisotropic in-plane compression ($0 < \varepsilon_{zz}, \varepsilon_{yy} < 0$).

For *a*-plane GaN films the growth is realized along the [11$\bar{2}$0] direction and consequently σ_{xx} vanishes. In this case, it follows from Equation 5 that

$$\varepsilon_{xx} = -\frac{C_{12}}{C_{11}}\varepsilon_{yy} - \frac{C_{13}}{C_{11}}\varepsilon_{zz} \tag{7}$$

Then, according to the elasticity theory (Equation 7), the strain along the growth direction is expected to be tensile ($\varepsilon_{xx} > 0$). Thus, the *a*-plane GaN

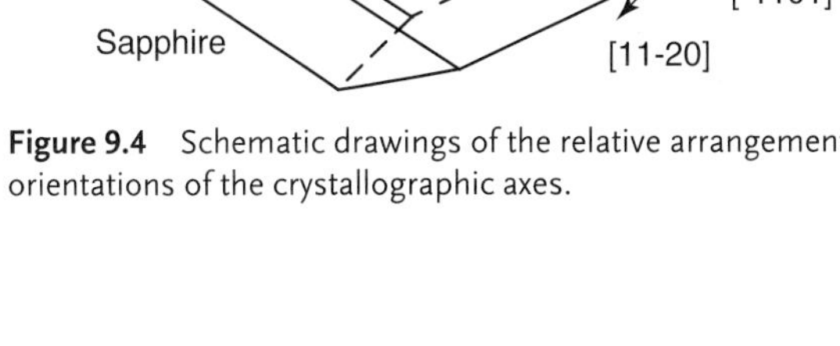

Figure 9.4 Schematic drawings of the relative arrangement of the unit cells and the orientations of the crystallographic axes.

films grown on *r*-plane sapphire are expected to experience *anisotropic* strain in the basal plane ($0 < \varepsilon_{xx} \neq \varepsilon_{yy} < 0$).[1)]

9.2.4
Phonon Deformation Potentials

The frequencies of the zone-center optical phonons in GaN are shifted or split with respect to the strain-free values. In the linear strain limit, these shifts and splittings are related to the strain tensor via [14, 16]:

$$\Delta\omega_j = a_j(\varepsilon_{xx} + \varepsilon_{yy}) + b_j\varepsilon_{zz} \tag{8}$$

for phonons polarized along the GaN *c* axis, that is, the A_1(TO) and A_1(LO) phonons, and

$$\Delta\omega_j = a_j(\varepsilon_{xx} + \varepsilon_{yy}) + b_j\varepsilon_{zz} \pm c_j[(\varepsilon_{xx} - \varepsilon_{yy})^2 + 4\varepsilon_{xy}^2]^{1/2} \tag{9}$$

for the nonpolar E_2 phonons, and the phonons polarized perpendicular to the GaN *c* axis, that is, the E_1(TO) and E_1(LO) phonons. The silent B_1 mode obeys a similar relation with strain, as the A_1 mode. The coefficients a_j, b_j, and c_j are the corresponding mode deformation potentials per unit strain. The combinations $2a_j + b_j$ and $a_j - b_j$ directly give the effects of hydrostatic strain or uniaxial strain parallel to the *c* axis. The effect of a trigonal strain is characterized by the deformation potential c_j. Trigonal strain may be caused by *anisotropic distortion of the basal plane* of wurtzite crystals, for example, in the case of nonpolar *a*-plane GaN grown on *r*-plane sapphire [30] or polar *c*-plane GaN grown on *a*-plane sapphire [29]. In such instances, GaN E_1 and E_2 phonon frequencies will be expected to split according to Equation 9.

Using Hooke's law (Equation 5),[2)] and defining the mode deformation potentials for a given stress, $\tilde{a}_j$, $\tilde{b}_j$, and $\tilde{c}_j$, we can obtain an alternative form of Equation 8 that relates the mode frequency shift with the stress tensor [14].

9.2.5
Raman Scattering Spectroscopy

Raman scattering in solids occurs as a result of modulation of the electronic polarizability induced by elementary excitations, such as phonons and

1) Note that in case of growth of polar GaN on non-*c*-plane sapphire substrate, for example *c*-plane GaN on *a*-plane sapphire, the strain in the GaN basal plane will also be anisotropic [17, 29].

2) The effect of the polarization fields is neglected.

plasmons, and has been extensively applied in the study of conventional c-plane GaN and related III nitrides (for a review see [31]).

Because of the momentum conservation rule during the light scattering process, first-order Raman scattering is caused by phonons at the Brillouin zone center. The Raman scattering efficiency S with polarization detection is given by:

$$S = |e_{\text{i}} \mathbf{R} e_{\text{s}}|^2$$

where e_{i} and e_{s} denote the polarization of the incident and scattered light, respectively, and R is the Raman tensor of the scattering process. The symmetries of the medium and of the vibrational modes involved in the scattering impose some restrictions on the Raman tensor. As a result of these symmetry requirements, the scattered radiation vanishes for certain choices of the polarizations of the scattered and incident light and scattering geometries. In the case of wurtzite structure, the Raman tensor has the following form for each mode. For the polar phonons, the polarization is indicated within parentheses [32]:

$$E_1(x): \begin{pmatrix} 0 & 0 & c \\ 0 & 0 & 0 \\ c & 0 & 0 \end{pmatrix}, \quad E_1(y): \begin{pmatrix} 0 & 0 & 0 \\ 0 & 0 & c \\ 0 & c & 0 \end{pmatrix},$$

$$E_2^L: \begin{pmatrix} 0 & d & 0 \\ d & 0 & 0 \\ 0 & 0 & 0 \end{pmatrix}, \quad E_2^H: \begin{pmatrix} d & 0 & 0 \\ 0 & -d & 0 \\ 0 & 0 & 0 \end{pmatrix},$$

$$A_1(z): \begin{pmatrix} a & 0 & 0 \\ 0 & a & 0 \\ 0 & 0 & b \end{pmatrix}. \tag{10}$$

The Raman tensor has only diagonal elements for the A_1 phonon mode and only off-diagonal components for the E_1 mode. For the E_2 mode, the nonvanishing components of the Raman tensor are diagonal and off-diagonal. This means that the A_1 modes could be observed when the incident and scattered light have parallel polarization, while the E_1 modes are observed only in cross-polarization geometry. These so-called Raman selection rules are used for determining Raman active phonons. The Raman selection rules for GaN are listed in Table 9.1. The respective Raman scattering configurations are illustrated in Figure 9.5.

As a result of the anisotropic nature of the wurtzite structure (uniaxial symmetry), the A_1 and E_1 modes exhibit an angular dependence. In other words, the mode frequency depends on the angle θ between the phonon propagation direction and the c axis. The angular dispersions of the optical modes in GaN at the Γ point of the Brillouin zone are shown in Figure 9.6 [33]. If the phonon propagation direction is along the x, y, or z axis (where z axis coincides with the hexagonal optical c axis), only pure longitudinal and pure transverse phonons with well-defined symmetry character are observed.

Table 9.1 Raman selection rules for pure and mixed quasi-transverse *Q*(TO) and quasi-longitudinal *Q*(LO) phonon modes of wurtzite group III nitrides. The *z* direction is parallel to the hexagonal *c* axis.

Scattering configuration	Allowed mode
$z(yy)\bar{z}$	E_2, A_1(LO)
$z(xy)\bar{z}$	E_2
$x(zz)\bar{x}$	A_1(TO)
$x(yz)\bar{x}$	E_1(TO)
$x(yy)\bar{x}$	A_1(TO), E_2
$x(yz)y$	E_1(TO), E_1(LO)
$x(zx)z$	Q(TO), Q(LO)
$x(yy)z$	E_2, Q(TO), Q(LO)

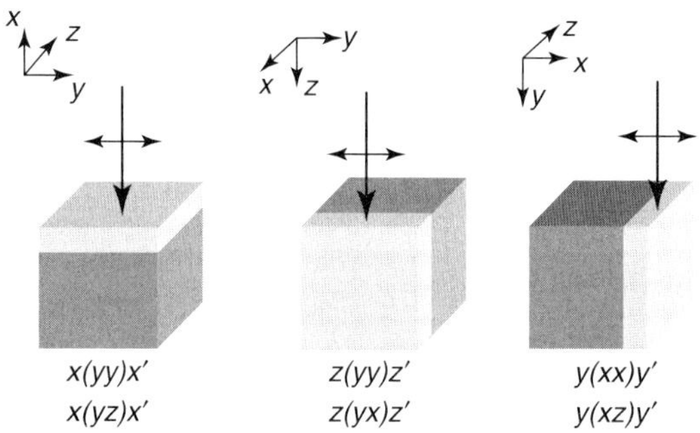

Figure 9.5 Schematic drawings of the Raman scattering configurations for an *a*-plane-oriented sample. The *z* direction is parallel to the hexagonal *c* axis.

Table 9.2 lists the various phonons that may be observed for each of the Cartesian propagation directions and each of the polarizations. It is seen from Table 9.2 that if a phonon is propagating along the *z* direction, one may observe three possible modes: a $TO^{\perp}$ mode polarized in the *x* direction (*xz* component of the polarizabilty tensor), another $TO^{\perp}$ polarized in *y* direction (*yz* component of the polarizabilty tensor), and a $LO^{\parallel}$ phonon polarized in the *z* direction (*zz* component of the polarizabilty tensor). By symmetry, the twofold $TO^{\perp}$ modes are degenerate. The twofold degenerate E_1(TO) mode at $\theta = 0$ splits into E_1(TO) and A_1(TO) for phonon propagation perpendicular to the *c* axis (see Figure 9.6). Thereby, one TO frequency remains constant, whereas the other one changes from E_1(TO) to A_1(TO) when the propagation direction is varied. The latter is referred as *quasi-TO phonon* because of its mixed E_1 and A_1 symmetry. The LO phonon has a pure A_1 character at $\theta = 0$ and pure E_1 character at $\theta = 90^{\circ}$ only. For intermediate phonon propagation directions these phonons no longer preserve a well-defined symmetry but have a mixed character and are termed as quasi-LO phonons. The frequencies of

Table 9.2 Types of phonons possible for various polarizations and propagation directions. The z direction is parallel to the nitride c axis. The notations $\perp$ and $\parallel$ refer to polarizations perpendicular to the c axis (E_1 symmetry) and parallel to the c axis (A_1 symmetry), respectively.

		Phonon polarization		
		$x(E_1)$	$y(E_1)$	$z(A_1)$
Phonon propagation	x	$LO^{\perp}$	$TO^{\perp}$	$TO^{\parallel}$
	y	$TO^{\perp}$	$LO^{\perp}$	$TO^{\parallel}$
	z	$TO^{\perp}$	$TO^{\perp}$	$LO^{\parallel}$

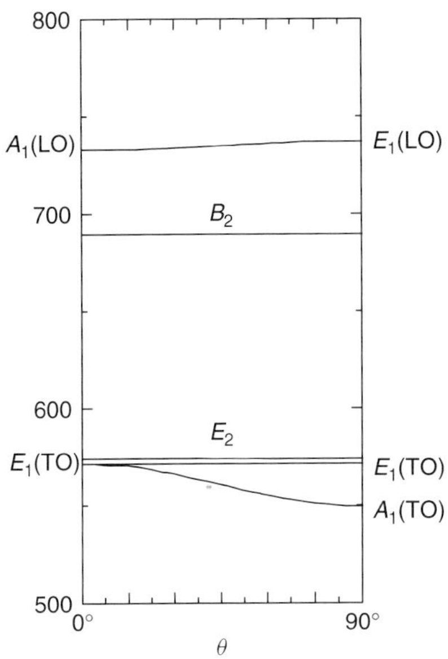

Figure 9.6 Angular dispersion of the optical phonon modes in wurtzite GaN at the Γ point of the Brillouin zone. (Reprinted (figure) with permission from Bungaro, C., Rapcewicz, K., and Bernholc, J. (2000) *Physical Review B*, **61**, 6720. Copyright 2000 by the American Physical Society.)

the quasi-LO and quasi-TO phonons are given by [32, 34]:

$$\omega_{\mathrm{LO}}^2 = \omega_{A_1(\mathrm{LO})}^2 \cos^2\theta + \omega_{E_1(\mathrm{LO})}^2 \sin^2\theta$$
$$\omega_{\mathrm{TO}}^2 = \omega_{A_1(\mathrm{LO})}^2 \sin^2\theta + \omega_{E_1(\mathrm{LO})}^2 \cos^2\theta \qquad (11)$$

9.2.6 Infrared Spectroscopic Ellipsometry

Ellipsometry, in general, determines the complex ratio ρ of linearly independent electric field components of polarized electromagnetic plane waves. More specifically, it measures the *change* of the polarization state of an electromagnetic plane wave upon interaction with a sample. Explicitly, ellipsometry determines the ratio:

$$\rho = \left(\frac{B_p}{B_s}\right) \Big/ \left(\frac{A_p}{A_s}\right) \qquad (12)$$

where A and B stand for the amplitudes of the incident and exciting waves, respectively. Traditionally, this ratio is measured in reflection or transmission configuration or light polarized parallel (p), or perpendicular (s) to the plane of incidence [35]. The plane of incidence with the respective notations is defined in Figure 9.7. The result of an ellipsometry measurement is usually presented by real-valued parameters Ψ and Δ, where $\tan\Psi$ is defined as the absolute value of the complex ratio, and Δ denotes the relative phase change of the p and s components of the electric field vector [35].

9.2.6.1 Standard Ellipsometry

In the standard ellipsometry situation Ψ and Δ do not depend on the polarization state of the incident plane waves, which are then defined as follows

$$\rho = \frac{r_p}{r_s} = \tan\Psi \exp(i\Delta) \qquad (13)$$

Here r_p and r_s denote the p- and s-polarized complex reflection coefficients.

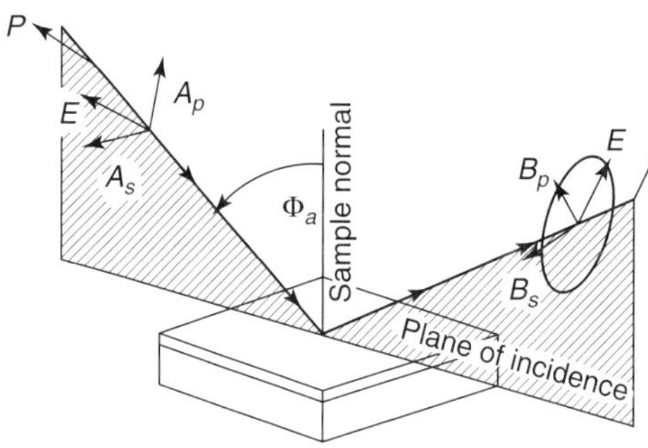

Figure 9.7 Definition of the plane of incidence (p plane) and the incidence angle Φ_a through the wave vectors of the incident and emerging (reflection set up) plane waves. A_p, A_s, B_p, and B_s denote the complex amplitudes of the p and s modes before and after reflection, respectively. P and A are the azimuth angles of the linear polarizer used, for example, in the standard arrangement of rotating-analyzer (polarizer) ellipsometer.

9.2.6.2 Generalized Ellipsometry

For a general situation of optical anisotropy and sample orientation, the simple definition of ρ in Equation 13 is no longer a complete description, and the generalized ellipsometry approach is needed. In such instances, Equation 13 depends on the polarization state of the incident plane wave. The generalized ellipsometry parameters are then described by three ratios of the polarized-light reflection coefficients among the four available complex-valued elements of the Jones reflection matrix **r** [15, 36]:

$$\mathbf{r} = \begin{pmatrix} r_{pp} & r_{sp} \\ r_{ps} & r_{ss} \end{pmatrix}.$$

In this notation, the first index denotes the incident polarization mode, and the second index refers to the outgoing polarization mode. The off-diagonal elements are nonzero for optical systems that convert p into s waves and vice versa. These ratios defining the generalized ellipsometry parameters Ψ_{ij} and Δ_{ij} are given by [15, 36]

$$\begin{aligned} \frac{r_{pp}}{r_{ss}} &\equiv R_{pp} = \tan\Psi_{pp}\exp(i\Delta_{pp}) \\ \frac{r_{ps}}{r_{pp}} &\equiv R_{ps} = \tan\Psi_{ps}\exp(i\Delta_{ps}) \\ \frac{r_{sp}}{r_{ss}} &\equiv R_{sp} = \tan\Psi_{sp}\exp(i\Delta_{sp}) \end{aligned} \tag{14}$$

The real-valued quantities Ψ_{pp}, Ψ_{ps}, Ψ_{sp}, Δ_{pp}, Δ_{ps}, Δ_{sp} now comprise the generalized ellipsometry data format. For small off-diagonal ratios, the on-diagonal ratio R_{pp} is very similar to the complex ratio determined by standard ellipsometry. In particular, if R_{ps} and R_{sp} vanish, R_{pp} is identical to the standard ellipsometric ratio ρ defined by Equation 13.

9.2.6.3 Ellipsometry Data Analysis

In principle, the dielectric function and thickness of each sample layer can be determined from an ellipsometry experiment by adjusting calculated data to measured data using a parameterized model dielectric function. The standard model for analyzing spectroscopic ellipsometry (SE) data consists of a sequence of parallel layers with abrupt interfaces and specially homogeneous dielectric functions, bound between the semi-infinite substrate and the ambient. Next, a regression analysis is performed, in this case, using Levenberg–Marquardt fitting algorithm, where the model parameters are varied until the calculated and experimental data match as closely as possible. Minimization is realized by using the maximum likelihood approach [15, 36]. Fitting parameterized model dielectric functions to experimental data, simultaneously for all spectral data points, provides a direct connection between measured data and physical parameters of interest.

9.2.6.3.1 Infrared Dielectric Model

In the infrared spectral region, the dielectric function $\varepsilon_j(\omega)$ is sensitive to phonon and plasmon contribution. Therefore, the infrared model dielectric function can be written as a sum of lattice, $\varepsilon_j^{\mathrm{L}}(\omega)$ and free-charge carrier, $\varepsilon_j^{\mathrm{FC}}(\omega)$ contributions:

$$\varepsilon_j^{\mathrm{L}}(\omega) = \varepsilon_j^{\mathrm{L}} + \varepsilon_j^{\mathrm{FC}} \tag{15}$$

where the subscript j refers to a specific parametrization with respect to the polarization state.

A product presentation for harmonic oscillator lineshapes with Lorentzian broadening is employed for the calculation of the lattice contribution ε^{L} to the dielectric function of uniaxial crystals, such as wurtzite-structure GaN [37]:

$$\varepsilon_j^{\mathrm{L}}(\omega) = \varepsilon_{\infty,j} \frac{\omega^2 + i\gamma_{\mathrm{LO},j}\omega - \omega_{\mathrm{LO},j}^2}{\omega^2 + i\gamma_{\mathrm{TO},j}\omega - \omega_{\mathrm{TO},j}^2} \tag{16}$$

The anisotropic dielectric functions are parameterized along Cartesian lattice directions [17]: $j = x = [11\bar{2}0]$, $j = y = [1\bar{1}00]$, and $j = z = [0001]$ (see Figure 9.1), where $\omega_{\mathrm{TO},j=x,y,z}$ denote the frequencies of the $E_1^x(\mathrm{TO})$, $E_1^y(\mathrm{TO})$, and $A_1(\mathrm{TO})$, while $\omega_{\mathrm{LO},j=x,y,z}$ denote the frequencies of the $E_1^x(\mathrm{LO})$, $E_1^y(\mathrm{LO})$, and $A_1(\mathrm{LO})$ modes. $\gamma_{\mathrm{TO},j}$, and $\gamma_{\mathrm{LO},j}$ are the corresponding mode-broadening parameters, and $\varepsilon_{\infty,x}$, $\varepsilon_{\infty,y}$ and $\varepsilon_{\infty,z}$ are the high-frequency limits for polarization perpendicular and parallel to the optical c axis. Note that in this presentation, the E_1 modes may be different owing to different strains along x and y directions, for instance caused by anisotropic basal plane distortion.

Free-charge carriers in semiconductors form collective excitation modes, the so-called plasma mode (plasmon). The plasma modes will couple to the LO lattice modes and form the so-called coupled LO plasmon–phonon (LPP) modes. Depending on the strength of the coupling, the free carriers thereby influence the dielectric function. A possible contribution from free carriers to the dielectric functions is also accounted for by virtue of the classical Drude model [38]:

$$\varepsilon_j^{(\mathrm{FC})}(\omega) = -\frac{(\omega_{p,j}^*)^2}{\omega(\omega + i\gamma_{p,j})} \tag{17}$$

with

$$\omega_{p,j}^* = e\sqrt{\frac{N}{\varepsilon_0 m_j}} \tag{18}$$

The unscreened plasma frequency ω_p^* depends on the free-charge-carrier concentration N and the effective conduction band mass m; ε_0 is the vacuum permittivity, and e is the electrical unit charge. The plasmon- broadening parameter γ_p is related to the optical free-charge-carrier mobility μ:

$$\gamma_{p,j} = \frac{e}{m_j \mu_j} \tag{19}$$

Note that only the quantities, $\frac{N}{m}$ and $\frac{N}{\mu}$ can be determined from the plasma frequency and broadening parameters. Hence, the effective mass and mobility parameters can be obtained only when the free-charge-carrier concentration N is known. Likewise, N can be determined when the m and/or μ values are known [36].

9.2.6.3.2 **Infrared Selection Rules and Anisotropic Tensor**

In noncubic solids, the phonon mode frequencies of the polar lattice vibrations depend, in general, on the phonon mode propagation direction. Likewise, directionally dependent free-charge-carrier scattering rates and the anisotropic inverse effective free-charge-carrier mass tensor will produce nonscalar free-charge-carrier contributions. The infrared dielectric function is then represented by a complex-valued second-rank tensor ε, which can be expressed in Cartesian coordinates (x,y,z) as:

$$\vec{D} = \tilde{\varepsilon}_0 \varepsilon \vec{E} = \begin{pmatrix} \varepsilon_{xx} & \varepsilon_{xy} & \varepsilon_{xz} \\ \varepsilon_{yx} & \varepsilon_{yy} & \varepsilon_{yz} \\ \varepsilon_{zx} & \varepsilon_{zy} & \varepsilon_{zz} \end{pmatrix} \vec{E} = \tilde{\varepsilon}_0 \left(\vec{P} + \vec{E} \right) \tag{20}$$

where the field-phasors displacement D, polarization P, and electric field E are given along the unit directions x, y, z [15].

For uniaxial materials (*hexagonal, tetragonal, trigonal*), such as for the wurtzite group-III nitrides, or for sapphire and SiC, the dielectric tensor is symmetric in all indices and has the following form:

$$\varepsilon = \begin{pmatrix} \varepsilon_x & 0 & 0 \\ 0 & \varepsilon_y & 0 \\ 0 & 0 & \varepsilon_z \end{pmatrix} \tag{21}$$

where the index j corresponds to electric field polarization parallel to the high-symmetry [0001] direction, $j = z \| [0001]$, or perpendicular to it: $j = x \| [11\bar{2}0]$ and $j = y \| [1\bar{1}00]$ as shown in Figure 9.1a. TO- and LO-phonon modes with E_1 (A_1) symmetry represent poles and zeroes within $\varepsilon_{x,y}(\varepsilon_z)$. For unstrained or biaxially strained uniaxial materials, such as bulk wurtzite GaN or c-plane GaN grown on c-plane sapphire, respectively, $\varepsilon_x = \varepsilon_y = \varepsilon_\perp$ and the standard ellipsometry approach is applicable. In the latter case, however, the ellipsometry data are insensitive to the parameters of the TO resonance with polarization parallel to the sample normal (0001), that is, to the A_1(TO) mode. Except for wavenumbers, where the real part of $\varepsilon_\|$ is between 0 and 1 (which is the case in a certain spectral range above the LO-phonon frequency), the incident light beam propagates almost parallel to the sample normal within the film and thus senses mostly $\varepsilon_\perp$. The lattice displacement pattern that belongs to the TO resonance in $\varepsilon_{\parallel}$ cannot be excited by the IR light beam in this configuration and the A_1(TO) mode remains inaccessible. However, specifically when grown on low-symmetry cuts of anisotropic substrates, for instance on *r*-plane sapphire, wurtzite-structure layers are likely to adopt a different growth orientation, namely *a*-plane (see Figure 9.4). For the latter case, the optical axis is parallel to the growth surface and the result of an optical measurement will depend on the orientation of the optical axis with respect to the plane of incidence, defined by the angle φ as well as with respect to incident polarization [39]. The important implication is that for an *a*-plane-oriented wurtzite sample the A_1(TO) mode parameters are now accessible via IR ellipsometry.

Further, the presence of anisotropic distortion of the basal plane of *a*-plane wurtzite layers, ($\varepsilon_{xx} \neq \varepsilon_{yy}$), will lift-off the degeneracy of the $\varepsilon_{\perp}$ and $\varepsilon_x \neq \varepsilon_y$. Therefore, the IR dielectric functions ε_x and ε_y provide access to the frequencies and broadening parameters of the TO and LO phonons with E_1 symmetry polarized along the $x = [11\bar{2}0]$ and $y = [1\bar{1}00]$ directions. In other words, the splitting of the TO and LO with E_1 symmetry that is predicted theoretically by Equation 9 [14, 16] can be obtained from the IR ellipsometry data analysis. Note, that polar *c*-plane GaN heteroepitaxial layers that experience anisotropic distortion of the basal plane, for instance when grown on *a*-plane sapphire [29] will also allow assessment of the E_1 phonon splitting [17, 18]. In this case, the optical measurement will depend on the orientation of the plane of incidence and incident polarization with respect to the two in-plane directions $x = [11\bar{2}0]$ and $y = [1\bar{1}00]$. The standard ellipsometry measurement for non-*c*-plane-oriented and anisotropically strained wurtzite crystals is inapplicable and the generalized ellipsometry approach is needed.

9.3
Anisotropic Strain in *a*-plane GaN

9.3.1
Assessment of Anisotropic Strain Components

It is convenient to chose the *x*, *y*, and *z* axes of the laboratory system along the GaN principal axes $[11\bar{2}0]$, $[1\bar{1}00]$, and [0001] directions, respectively (see Figure 9.1a), since in such a case, the shear stress and strain components ($i \neq j$, Equation 5) are zero [40]. Therefore, in such a case, knowledge of the normal strain components ($i = j$) is sufficient to fully describe the complex strain state in the GaN films that experience anisotropic basal distortions ($\varepsilon_{xx} \neq \varepsilon_{yy}$).

Assessment of the anisotropic strain components in *a*-plane GaN can be realized by performing high-resolution X-ray diffraction (HRXRD), using standard alignment procedures [41]. To measure both *a* and *c* lattice parameters of the GaN layers, different GaN symmetric and asymmetric reflections should

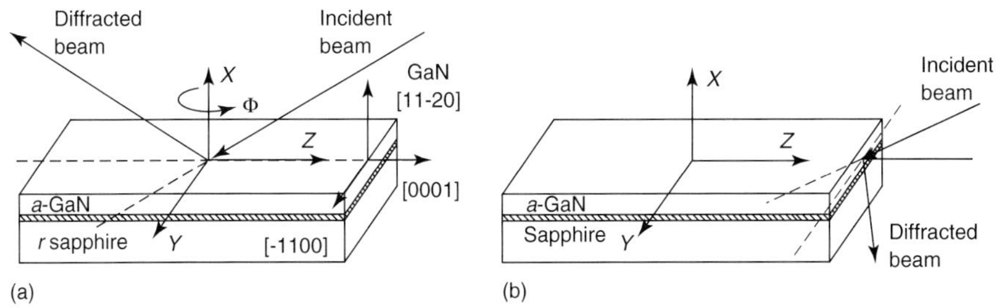

Figure 9.8 Schematic drawings of experimental geometries used in the measurements of the lattice parameters of the *a*-plane GaN films: (a) normal symmetric scans and (b) edge symmetric scans.

be measured. Therefore, the measurements have to be performed either in symmetric or edge symmetric geometries illustrated in Figure 9.8. A typical measurement procedure as suggested in [30] involves determination of the *a* lattice parameter along the growth direction of the *a*-plane GaN films, a_1 from the $11\bar{2}0$ and $22\bar{4}0$ reflections for four azimuth positions (ϕ) separated by 90° (see Figure 9.8a). The measurements of the *c* and a_2 lattice parameters are performed in edge symmetric geometry, preferably at the two azimuth angles 0° and 180° for the *c*, and 90° and 270° for the a_2 lattice parameters, respectively (see Figure 9.8b) [17, 30]. For the determination of the second lattice parameter, a_2 (see Figure 9.4) the $1\bar{1}00$ and $\bar{1}100$ peaks can be used, whereas the *c* lattice parameter can be determined from the 0002, 0004, and 0006 $2\theta - \omega$ diffraction spectra. Precise determination of the lattice parameters requires to account for the refractive index correction [42, 43]. Following this, the strain in the growth plane along the $[\bar{1}100]$ direction can be estimated using the interplanar distance $d_{1\bar{1}00}$ [30], while for the estimation of the rest of the strains, the values of the lattice parameters are typically used [17, 30]. The strain anisotropy can be estimated as the difference in the strains along the GaN $[11\bar{2}0]$ and $[1\bar{1}00]$ directions [17, 29].

9.3.2
Anisotropic Lattice Distortion

A-plane GaN films with different thickness between 1 μm and 67 μm were grown on *r*-plane sapphire by metal-organic vapor phase epitaxy (MOVPE) and HVPE. The *a*-plane MOVPE films were grown at University of Stuttgart and University of Bremen at 1030–1050 °C using a conventional two-step process employing GaN or AlN buffer layers [45, 46]. The *a*-plane HVPE films were grown at University of California, Santa Barbara, directly on sapphire in a three-zone horizontal reactor at 1040–1070 °C [47].

Figure 9.9 shows a contour plot of ε_{yy} as a function of the strain components, ε_{xx} and ε_{zz}, according to Equation 7 and using the stiffness constants from [44]

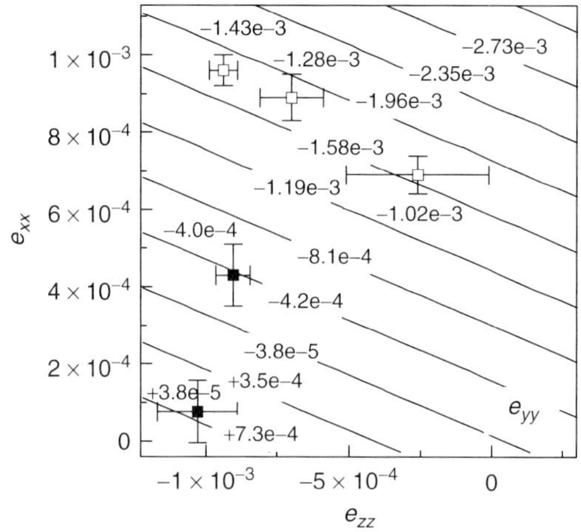

Figure 9.9 Contour plot of ε_{yy} in a-plane GaN films versus the strain components, ε_{xx} and ε_{zz}, according to Equation 7 and using the stiffness constants from [44]. The squares show the calculated values of ε_{yy} using the measured ε_{xx} and ε_{zz}, and the respective labels give the ε_{yy} values as determined by X-ray diffraction. The hollow squares represent the results for the HVPE films, whereas the filled squares represent the results for the MOVPE films.

for the a-plane GaN films [17]. The squares show the calculated frequencies using the measured ε_{xx} and ε_{zz}; each data point is labeled by the respective experimental value of ε_{yy} as determined by XRD [17].

The results in Figure 9.9 show that with one exception, all a-plane GaN films are compressively strained in the growth plane (both along the [0001] and [1$\bar{1}$00] directions) [17], as one can expect from the difference between the thermal expansion coefficients of GaN and sapphire [28]. The strain along the growth direction is found to be tensile (Figure 9.9) as the elasticity theory predicts in the case of biaxial stress. This anisotropic strain results in a complex distortion of the basal GaN plane, which is compressed along the [1$\bar{1}$00] and expanded along the [11$\bar{2}$0] directions as illustrated in Figure 9.10. The compression along the [1$\bar{1}$00] direction was found to be more significant than that along the [0001] direction for the thick HVPE-GaN films, where a certain degree of strain relaxation with increasing thickness may be expected. Similar findings were also reported for relatively thick MOVPE-GaN films [48]. On the other hand, the thin MOVPE films (Figure 9.9) show the opposite trend, being more compressed along the [0001] than along the [1$\bar{1}$00] directions.

One can also see from Figure 9.9 that the experimentally determined and theoretical values of ε_{yy} estimated via Equation 7 for the a-plane-oriented films agree fairly well. Note that generally the edge geometry used in the experimental determination of ε_{yy} and ε_{zz} leads to larger uncertainties (in particular for

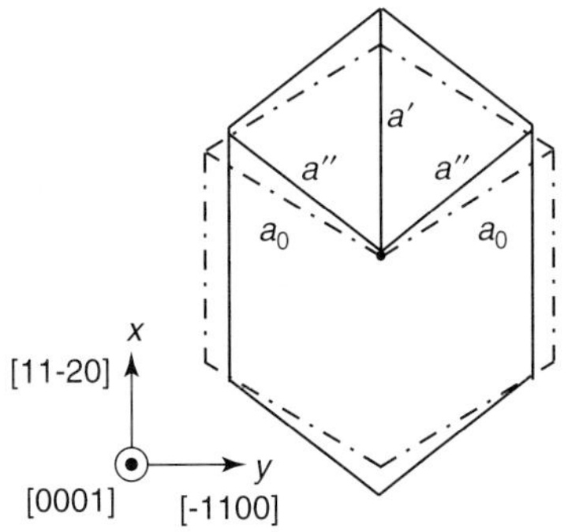

Figure 9.10 Schematic drawing of the distortion of the GaN basal plane of *a*-plane GaN films under anisotropic strain. The undistorted basal planes are indicated by dashed lines.

ε_{yy}) compared to the symmetric geometry used for the determination of ε_{xx}. Consequently, the experimentally and theoretically estimated ε_{yy} values may deviate up to 20% for some of the samples [17, 18]. In addition, for the thick HVPE samples, different volumes of the films were probed when measuring in normal and edge geometries (Figure 9.8). Consequently, the estimated out-of-plane and in-plane strains are averaged over different thickness volumes, with larger difference for the thicker samples. This might explain the deviations between experimental and elasticity-theory results for the HVPE samples (Figure 9.9, hollow squares). It is also concordant with the very good agreement between experimental and predicted ε_{yy} for the thin MOVPE sample, where the whole thickness was probed during normal and edge geometry measurements (Figure 9.9, filled square). To circumvent the effect of the uncertainty in ε_{yy} on the assessment of the vibrational properties in all related calculations, the ε_{yy} values as estimated by elasticity theory may be employed.

For the purpose of comparison, a short description of the case of polar *c*-plane GaN films grown on *a*-plane sapphire that experience anisotropic strain is given. *c*-Plane films with thickness of 2 μm were grown at Aixtron, AG in an Aixtron planetary reactor at 1170 °C using a low temperature nucleation GaN layer. In this case, it was found that the $2\theta - \omega$ peak of the HRXRD asymmetric scans appears at different positions, when measured at the six azimuthal positions, which indicates an anisotropic distortion of the basal plane [17, 29]. The values of the interplanar distances derived from these peak positions can be discriminated accordingly into two groups and consequently, the in-plane lattice parameter has two different values [29]. It was found that the two *a* lattice parameters are smaller than the strain-free value [49] as one can expect from the difference between the thermal expansion coefficients of GaN and sapphire [28]. Similarly, for the nonpolar *a*-plane films, the strain along the growth direction in the *c*-GaN films was found to be tensile: the *c* lattice parameters are larger than the strain-free value as the elasticity theory predicts in the case of biaxial strain. A comparison between the experimentally determined and theoretical values of ε_{zz} estimated by Equation 6 and using the stiffness constants from [44] gives excellent agreement for the *c*-plane-oriented films.

Further details about the data and measurement, and calculation procedures of the anisotropic strain components in both *a*- and *c*-plane GaN films can be found in [17–19, 29, 30, 41].

9.4
Phonons in GaN Layers under Anisotropic Strain

9.4.1
Strain-free Frequencies of GaN Phonons

Recent advances in the HVPE technique have made possible the fabrication of high-crystal quality GaN bulklike material with relatively low carrier concentration below 10^{17} cm^{-3} [7]. Bulk material with nonpolar surfaces was grown by HVPE at Kyma Technologies [50]. First, a homoepitaxial growth was conducted on free-standing GaN film also produced by HVPE and separated from the sapphire. Boules with 2″-inch diameter and thickness up to 7–8 mm were grown in [0001] direction. Then rectangular bars with sizes of 10×6 mm were sliced parallel to *a* and *m* plane (see Figure 9.1). Bulk *c*-plane GaN with thickness of 2 mm was fabricated by HVPE at Furukawa Co, Japan [51]. The free-carrier concentration is estimated to be below 10^{17} for the *a*- and *m*-plane GaN, and below 5×10^{16} cm^{-3} for the *c*-plane bulk GaN. Such materials, therefore represent the prime choice for establishing the bulk strain-free properties of GaN.

GaN phonon frequencies have been measured in these high-structural quality and low-defect density bulk GaN samples with different surface orientations [17, 22, 52]. Figure 9.11 shows unpolarized Raman spectra of the bulk GaN samples with *a*-, *m*-, and *c*-plane surface orientations. The measurements were performed in backscattering configurations in different geometries, and at multiple locations on the samples. The GaN A_1(TO), E_1(TO), E_2, A_1(LO), and quasi-LO (QLO) phonons were identified according to the Raman selection rules. Their frequencies were extracted from the Raman spectra using a standard best-match procedure with Lorentzian lineshapes. The frequency values averaged over different sample locations are listed in Table 9.3 together with the results reported in Ref. 9. The latter were estimated from hydrostatic pressure Raman experiments for free-standing 50-μm-thick GaN films and are often quoted in the literature as strain-free frequencies. The phonon frequencies determined for the *a*-, *m*-, and *c*-plane bulk GaN agree very well (coinciding with each other within the limits for experimental errors for most of the phonons) being somehow smaller for the *c*-plane GaN. An overall good agreement with the results from Ref. 9 can be also seen.

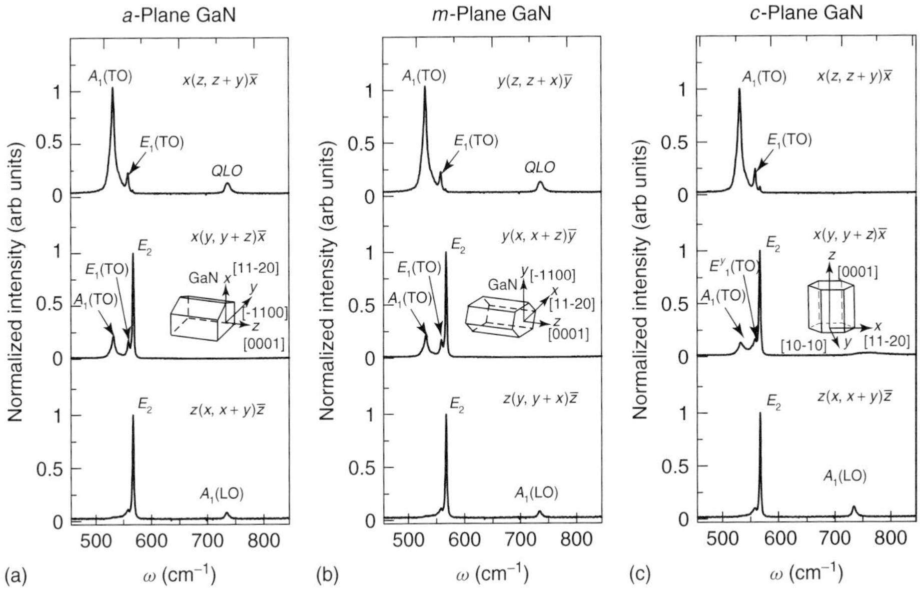

Figure 9.11 Raman spectra of bulk (a) *a*-plane GaN; (b) *m*-plane GaN; and (c) *c*-plane GaN in $x(z, z + y)\bar{x}$, $x(y, y + z)\bar{x}$, and $z(x, x + y)\bar{z}$ with x parallel to $[11\bar{2}0]$, y parallel to the $[1\bar{1}00]$, and z parallel to the [0001] directions.

Table 9.3 Phonon strain-free frequency in cm^{-1} as determined for bulk *a*-, *m*-, and *c*-plane GaN, and those reported in Ref. 9.

	A_1(TO)	E_1(TO)	E_2	A_1(LO)	QLO
a-GaN	531.1 ± 0.3	558.5 ± 0.3	567.2 ± 0.5	734.7 ± 0.1	741.7 ± 0.1
m-GaN	531.15 ± 0.4	558.3 ± 0.5	567.5 ± 0.3	734.5 ± 0.1	742.0 ± 0.1
c-GaN	530.6 ± 0.1	557.9 ± 0.1	567.0 ± 0.2	733.6 ± 0.1	740.3 ± 0.15
[9]	531.7 ± 0.3	558.2 ± 0.1	567.0 ± 0.1	736.5 ± 0.2	$742 + 0.2$

9.4.2 Infrared Anisotropy

9.4.2.1 Infrared Dielectric Tensor

Experimental and best-fit model calculated Ψ_{ij} spectra for a MOVPE *a*-plane GaN layer on *r*-plane sapphire [17, 18] are shown in Figure 9.12 as an example. The coupling between *p*- and *s*-polarized light, sensed by the off-diagonal elements is small at $\varphi = 0°$ and minimal at $\varphi = 90°$ (Figure 9.12b and c). A noticeable conversion of *p*-(*s*-) to *s*-(*p*-) polarized light related to the GaN film is observed in the GIRSE spectra for intermediate φ, particularly across frequency regions of resonant excitation of TO or LO phonons. It can be seen

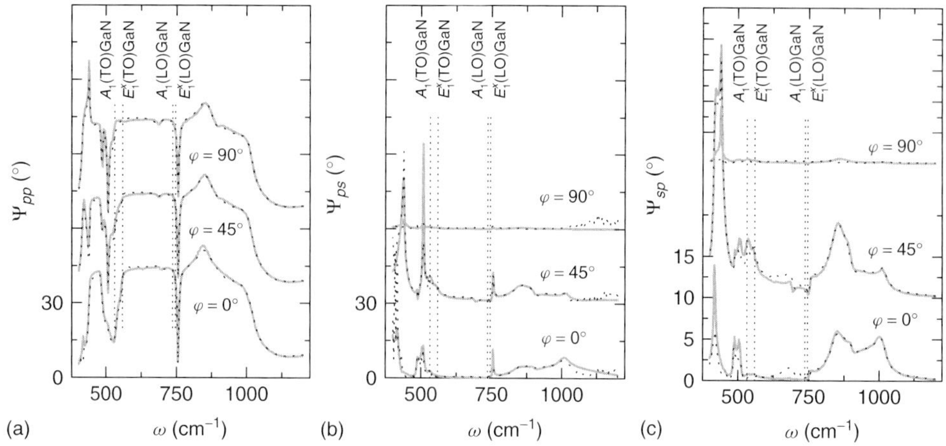

Figure 9.12 Experimental (dots) and calculated (lines) GIRSE spectra of one representative *a*-plane GaN on *r*-plane sapphire for different angles φ between the plane of incidence and the GaN [0001] direction: (a) Ψ_{pp}; (b) Ψ_{ps}; and (c) Ψ_{sp}. The frequencies of the GaN phonon modes of A_1 and E_1 symmetry are indicated. For the sake of clarity, the phonon frequencies are given only for polarization along the [11$\bar{2}$0] direction.

that the low-frequency edge of the GaN reststrahlen band in Ψ_{pp} moves from E_1(TO) to A_1(TO) when the azimuth angle, φ, between the plane of incidence and the GaN c axis changes between 0° and 90° (Figure 9.12a). The E_1(TO) direction and A_1(TO) are excited resonantly depending on φ, as the electric field parallel to the film interface senses lattice excitations with polarization perpendicular or parallel to the GaN c axis, respectively. This effect can be better visualized for thick *a*-plane layers, where contribution from the substrate to the GIRSE response becomes negligible, as illustrated in Figure 9.13 [39]. In this case the small features above the restrahlen band are due to incoherent back-reflected light, which interfere with the ellipsometric measurement.

The anisotropy of the dielectric function in the x–y plane is revealed upon model layer calculations. Figure 9.14 shows the real (ε_1) and the imaginary parts of the dielectric functions (ε_2), and the imaginary part of the dielectric loss function ($-1/\varepsilon$) as obtained from the model lineshape analysis. The z direction response is along the GaN c axis, whereas the x and y directions are parallel to the GaN [11$\bar{2}$0] and [$\bar{1}$100] directions, respectively. The different strains along the axes x and y result in different phonon mode shifts and cause spectrally narrow differences between the imaginary parts within the phonon mode resonance region (dichroism). Causality requirements then produce birefringence over a much larger spectral range than imposed by the phonon mode shifts, which also affects the imaginary parts of the dielectric loss functions. Despite the fact that the spectral resolution of the ellipsometer system employed is larger than the transverse optical phonon mode frequency

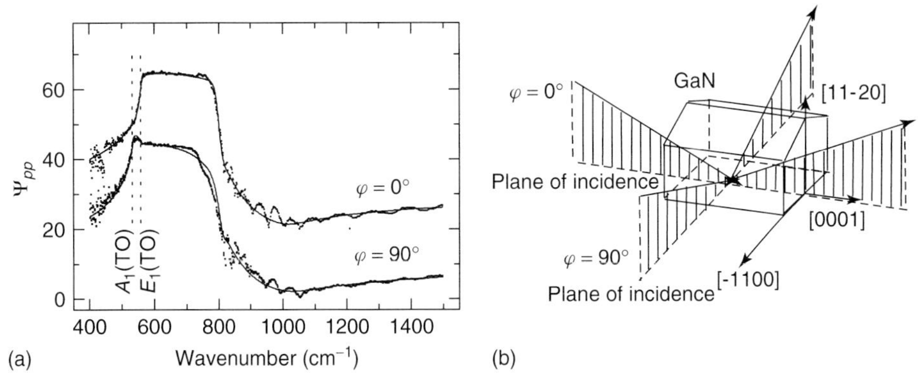

Figure 9.13 Experimental (dots) and calculated (lines) Ψ_{pp} spectra of a 12-μm-thick *a*-plane GaN on *r*-plane sapphire for different angles φ between the plane of incidence and the GaN [0001] direction (a) and a schematic drawing of sample orientation with respect to the plane of incidence (b). The frequencies of the GaN phonon modes of A_1 and E_1 symmetry are indicated. For the sake of clarity, the phonon frequencies are given only for polarization along the [11$\bar{2}$0] direction.

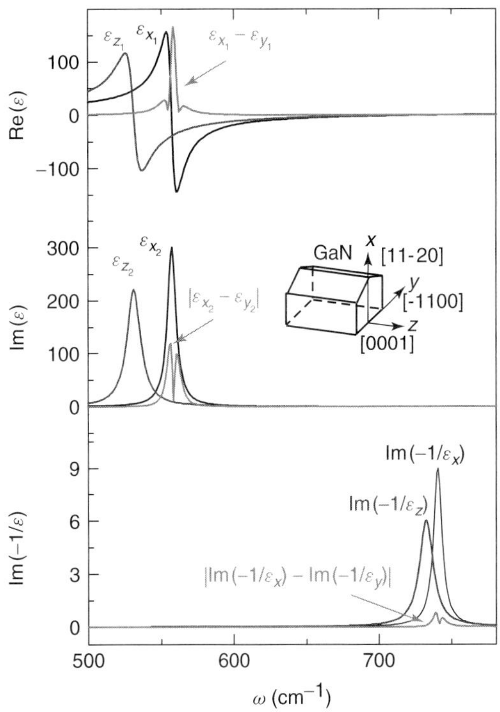

Figure 9.14 Real and imaginary parts of the dielectric function ε and imaginary part of the dielectric loss functions $-1/\varepsilon$ for in- (*x*) and out-of-plane (*z*) directions of one representative *a*-plane GaN film grown on *r*-plane sapphire surface. For clarity, the functions for directions *y* are given through the differences between those for directions (*x*) parallel to the GaN [11$\bar{2}$0] and (*y*) parallel to the GaN [1$\bar{1}$00] directions (cf. Color Plate XXX).

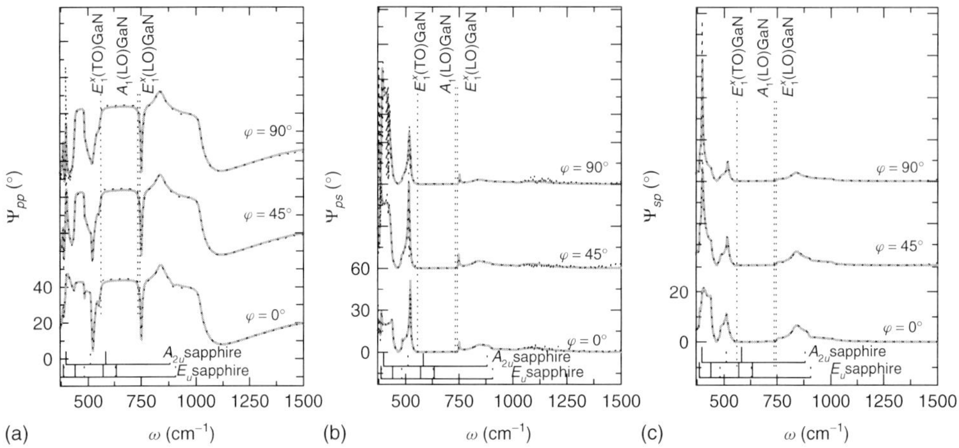

Figure 9.15 Experimental (dots) and calculated (solid lines) GIRSE spectra of a polar *c*-plane GaN film on *a*-plane sapphire for different angles φ between the plane of incidence and the GaN [11$\bar{2}$0] direction: (a) Ψ_{pp}; (b) Ψ_{ps}; and (c) Ψ_{sp}. The strain-free frequencies of the GaN phonon modes of A_1 and E_1 symmetry are indicated by dashed lines. For the sake of clarity, the phonon frequencies are given only for polarization along the [11$\bar{2}$0] direction. The A_{2u} and E_u phonon modes of sapphire are indicated by brackets.

shifts, the phonon mode resonance shifts can be located precisely by the model lineshape analysis of the GIRSE data [17]. Further details about the data analysis can be found in [17, 36, 53] and references therein.

For the purpose of comparison, GIRSE experimental and best-fit model calculated Ψ_{ij} spectra are also shown for an anisotropically strained polar *c*-plane GaN layer grown on *a*-plane sapphire in Figure 9.15 [17, 18]. It is seen that a distinct spectral feature in Ψ_{pp} appears near the GaN E_1(TO) resonance, which changes in a subtle manner upon rotation of the sample by φ. Spectral features are related to the LO modes, and the A_1(LO) and E_1^j(LO) phonon mode parameters were obtained from the lineshape analysis. Basal-plane anisotropy of the dielectric function is revealed upon layer model calculations in agreement with the findings for *a*-plane films. However, for polar *c*-plane-oriented films the GIRSE data are not sensitive to the TO resonance frequency with polarization vector parallel to the sample normal [37] and consequently, the A_1(TO) phonon mode resonance could not be found. Therefore, the complete anisotropy of the GaN dielectric function could be revealed only for anisotropically strained *a*-plane GaN films (Figure 9.14).

9.4.3
Phonon Splitting

The frequencies and the broadening parameters of the E_1 phonons polarized in the x–y plane parallel to the GaN [11$\bar{2}$0] and [$\bar{1}$100] directions, extracted from

Table 9.4 Best-fit values and standard deviations of the E_1^j (TO), E_1^j (LO), and E_2^j phonon frequencies, ω, and broadening parameters, γ, for polarizations along the GaN [11$\bar{2}$0] ($j = x$) and [$\bar{1}$100] ($j = y$) directions for an *a*-plane GaN film. For the E_1 modes, *j* denotes the phonon polarization, while for the E_2 modes, *j* denotes the polarization of the incident light in the parallel and cross-polarized configurations, respectively.

	ω (cm^{-1})		γ (cm^{-1})	
	j = [11$\bar{2}$0]	j = [1$\bar{1}$00]	j = [11$\bar{2}$0]	j = [1$\bar{1}$00]
E_1^j(TO)	557.2 ± 0.4	559.4 ± 0.7	7.1 ± 0.2	7.4 ± 0.2
E_1^j(LO)	740.3 ± 0.1	740.7 ± 0.2	7.1 ± 0.2	7.4 ± 0.2
E_2^j	567.3 ± 0.1	568.81 ± 0.1	1.7 ± 0.1	2.2 ± 0.1

the best-match modeling to the experimental GIRSE data (Figure 9.12), are summarized in Table 9.4. The phonon mode parameters for both polarizations were determined with high precision, much lower than the spectral resolution. The use of appropriate lineshape to describe the IR response and its cause is mandatory to allow resolvability of the phonon parameters when the resolution is larger than the error bars of the respective parameter. It has been shown previously that the Lorentzian-broadened harmonic oscillator describes to a high degree of closeness the polar phonon contribution to the IR response for GaN [37]. The accurate location of the center of energy of the Lorentzian-broadened harmonic oscillator is possible as its broadening is sufficiently larger than the spectral resolution.

As seen from Table 9.4, a small difference in the frequency values of the E_1 phonons was determined for polarizations along the GaN [11$\bar{2}$0] and [$\bar{1}$100] directions based on the model layer calculations. The splitting of the E_1(TO) and E_1(LO) is demonstrated by the difference between the dielectric function and the dielectric loss function along the [11$\bar{2}$0] and [$\bar{1}$100] directions (Figure 9.14). The E_1^x(TO) and E_1^y(TO) appear as strong peaks in the imaginary part of ε_x and ε_y, respectively, while the E_1^x(LO) and E_1^y(LO) appear in the respective dielectric loss function along the two directions, $\mathrm{Im}(-1/\varepsilon_x)$ and $\mathrm{Im}(-1/\varepsilon_y)$, respectively. Phonon splittings have also been observed for polar *c*-plane films grown on *a*-plane sapphire (Figure 9.14) [17, 18]. The finding that E_1(TO) and E_1(LO) phonons possess different frequencies for the two basal polarizations when the basal plane is anisotropically distorted (see Figure 9.10) agrees with the theoretical predictions that the frequencies of the E_1 phonons in GaN are split when $\varepsilon_{xx} \neq \varepsilon_{yy}$ (see Equation 9 [14, 16]).

It can be also seen from Table 9.4 that the broadening parameters of the E_1 phonons polarized along the [11$\bar{2}$0] and [1$\bar{1}$00] directions also show a subtle difference. The broadening parameters of the E_1^x phonons are smaller compared to the E_1^y phonons. It has been suggested that this can be related to a smaller tensile strain (in absolute magnitude) along the [11$\bar{2}$0] direction compared to the compressive strain (in absolute magnitude) along [1$\bar{1}$00] direction (see Figures 9.4 and 9.9) [17, 18]. A more pronounced difference in the broadening parameters between the two polarizations of about 0.6 cm^{-1} is observed in the case of the anisotropically strained *c*-plane films [17, 18]. The E_1^x(TO) and E_1^x(LO), in this case, exhibit larger broadening, which could be related to the larger compressive in-plane strain along the GaN [11$\bar{2}$0] direction as compared to the [1$\bar{1}$00] in agreement with the measured strain in the *c*-plane films [29]. The high-frequency dielectric constant along the [11$\bar{2}$0] and [1$\bar{1}$00] directions follow from the best-fit to the GIRSE data of *c*-plane films with 5.18 ± 0.01 and 5.15 ± 0.01, respectively [17]. This subtle anisotropy of the dielectric constant in the x–y plane may likely be ascribed to strain-related effects also. The in-plane dielectric constants are both smaller than $\varepsilon_{\infty,z}$, determined to be 5.28 ± 0.01 along the [0001] direction, in agreement with previous theoretical and experimental findings for GaN [14, 37].

The splitting of the E_1(TO) phonons can be also observed in the Raman spectra of the *a*-plane films. Figure 9.16 shows Raman spectra from an *a*-plane film in two backscattering configurations: $x(y, y+z)\bar{x}$ and $y(x, x+z)\bar{y}$ with x parallel to [11$\bar{2}$0], y parallel to the [1$\bar{1}$00], and z parallel to the [0001] directions. In the two configurations there was no polarization detection, which allows the simultaneous detection of the A_1(TO), E_1(TO), and E_2 phonon modes. For the phonon propagation direction along the x axis (see Tables 9.1 and 9.2) A_1(TO) (yy component of the polarizabilty tensor), E_1^y(TO) (yz component of the polarizabilty tensor), and E_2 (yy component of the polarizabilty tensor) phonon modes are observed. While, when the light propagates along the y axis, A_1(TO) (xx component of the polarizabilty tensor), E_1^x(TO) (xz component of the polarizabilty tensor), and E_2 (xx component of the polarizabilty tensor) are observed. It is seen from Figure 9.16 that for the two configurations, E_1^x(TO) and E_1^y(TO) appear at different frequency positions in agreement with the GIRSE results.

A similar splitting is also observed for the E_2 modes (Figure 9.16) in agreement with theoretical predictions for $\varepsilon_{xx} \neq \varepsilon_{yy}$ (see Equation 9). Experimental values of the E_2 frequency and broadening parameters have been extracted from the Raman spectra using a standard best-match procedure with Lorentzian lineshapes, also shown in Figure 9.16, and the extracted phonon parameters are listed in Table 9.4. At the same time, the A_1(TO) phonon does not split (Figure 9.16); this is in agreement with theory that predicts only shift of the A_1 phonons with respect to the strain-free frequency (see Equation 8) [14, 16].

Splitting of the E_2 phonon have been also observed for anisotropically strained *c*-plane films on *a*-plane sapphire [17, 18]. A subtle difference of 0.5 cm^{-1} in the E_2 frequency was determined when measuring in $z(xx)\bar{z}$ and z(yx)$\bar{z}$ configurations. This is in agreement with the theoretical prediction

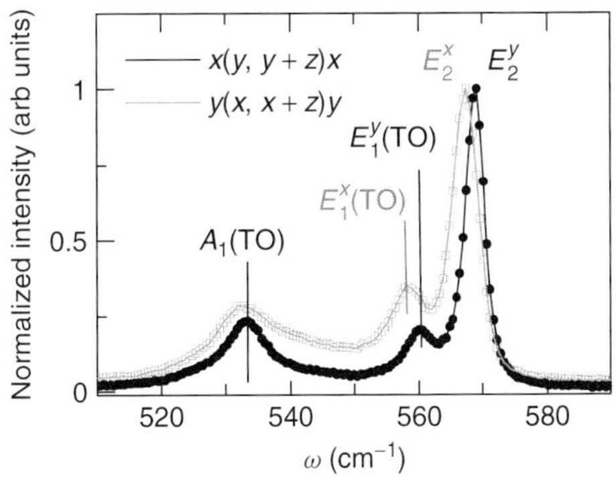

Figure 9.16 Raman spectra in $x(y, y+z)\bar{x}$ and $y(x, x+z)\bar{y}$ scattering geometries from an *a*-plane GaN film on *r*-plane sapphire (cf. Color Plate XXXI).

[14, 16] for the E_2 behavior under basal anisotropic strain (Equation 9) and with the findings for the splitting of the phonons in the *a*-plane films.

9.4.4
Phonon Deformation Potentials

9.4.4.1 Anisotropic Phonon Deformation Potentials

The shear strain component ε_{xy} for the anisotropically strained *a*-plane GaN films is zero when the axes of the laboratory system are chosen along the crystal principal axes (see Section 9.3). It follows from Equation 9 then that the splittings of the E_1 and E_2 phonons depend only on the magnitude of the in-plane strain anisotropy, $\varepsilon_{xx} - \varepsilon_{yy}$. Consequently, the difference between the frequencies of the $E_{1,2}$ phonons for polarizations along the GaN $[11\bar{2}0]$ and $[\bar{1}100]$ directions is given by:

$$\omega(E_{1,2})_{[11\bar{2}0]} - \omega(E_{1,2})_{[1\bar{1}00]} = 2c(E_{1,2})\left|\varepsilon_{xx} - \varepsilon_{yy}\right|$$

The *c* phonon deformation potentials for the E_1(TO), E_1(LO), and E_2 phonons can then be estimated using the experimental results for the strain components and the phonon frequency splittings [17, 18]. The results obtained for anisotropically strained nonpolar *a*-plane and polar *c*-plane GaN films are listed in Table 9.5.

Table 9.5 Anisotropic phonon deformation potentials, $c_{E_1(\mathrm{TO})}$, $c_{E_1(\mathrm{LO})}$, and c_{E_2} in cm^{-1}, and the respective standard deviations.

	$c_{E_1(\mathrm{TO})}$	c_{E_2}	$c_{E_1(\mathrm{LO})}$
a-Plane GaN	369 ± 37	369 ± 36	–
c-Plane GaN	379 ± 43	379 ± 107	678 ± 49

Excellent agreement between the values of the $c_{E_1(\mathrm{LO})}$, and c_{E_2} was determined for *a*-plane and *c*-plane GaN layers, respectively. The difference between the *c* deformation potentials of the E_1(TO)(379 cm^{-1}) and E_1(LO)(682 cm^{-1}) modes is relatively large, which can be related to the large bond ionicity in GaN causing substantial mode polarity. In principle, the high bond strength of GaN is expected to result in relatively large values of the *c* phonon deformation potentials as compared to II–VI materials with the same crystal symmetry. A comparison with CdS, for example, reveals that the GaN *c* phonon deformation potentials [17, 18] are significantly larger than in CdS [16], concordant with the argument above.

9.4.4.2 A_1(TO) and E_1(LO) Phonon Deformation Potentials

A correlation between phonon mode frequencies and strain components in films with different strains can, in principle, allow determination of the isotropic deformation potentials, *a* and *b* (see Equations 9 and 8). *a*-Plane GaN films offer a unique opportunity to obtain the A_1(TO) phonon deformation potentials by GIRSE. This is particularly important in view of the existing discrepancy in the literature between the values of the A_1(TO) phonon deformation potentials determined by Raman scattering [54] and theory [14]. Further, the $a_{E_1(\mathrm{LO})}$ and $b_{E_1(\mathrm{LO})}$ have not been experimentally determined yet.

a-Plane films with different thicknesses that exhibit different degree of strain (see Figure 9.9) have been investigated by GIRSE [17, 18]. The A_1(TO) frequency extracted from the GIRSE analysis changes in a complex manner with film thickness following the change in ε_{xx} and ε_{zz} (Figure 9.17). To obtain the deformation potentials and strain-free frequency, it is necessary to perform a regression analysis of the experimental frequencies and strains. However, Equation 8 cannot be directly used as the three strain components are not independent. Substituting Equation 7 in Equation 8, we obtain

$$\Delta\omega_{A_1(\mathrm{TO})} = a_{A_1(\mathrm{TO})}\left(1 - \frac{C_{11}}{C_{12}}\right)\varepsilon_{xx} - \left(a_{A_1(\mathrm{TO})}\frac{C_{13}}{C_{12}} - b_{A_1(\mathrm{TO})}\right)\varepsilon_{zz} \qquad (23)$$

Then, using the experimental results obtained for the A_1 (TO) frequency and the strain components ε_{xx}, and ε_{zz}, a regression analysis of Equation 23 has been performed. The obtained values of the phonon deformation potentials and the strain-free frequency of the GaN A_1(TO) mode and their standard deviations are listed in Table 9.6 together with previous theoretical and experimental results. The contour plot in Figure 9.17 demonstrates the dependence of the A_1(TO) frequency on the in-plane strain components, ε_{xx} and ε_{zz}, according to Equation 23 and using the results about the phonon deformation potentials and strain-free frequency as described in [17]. The squares show the calculated frequencies using the measured ε_{xx} and ε_{zz}, and the respective labels give the frequency values as determined from the GIRSE analysis.

It has been suggested that the relatively large uncertainties of the deformation potentials (Table 9.6) stem mostly from the regression procedure itself [55].

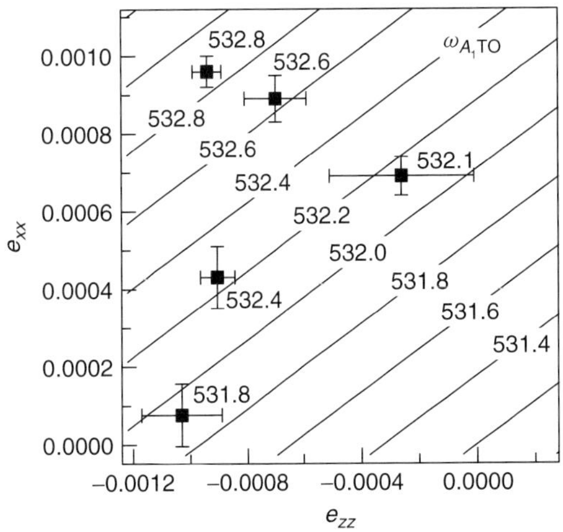

Figure 9.17 Contour plot of A_1 (TO) frequency versus the strain components, ε_{xx} and ε_{zz}, according to Equation 23 and using the results about the phonon deformation potentials ($a = -664$ cm^{-1} and $b = -1182$ cm^{-1}) and strain-free frequency ($\omega_0 = 531.2$ cm^{-1}) [17]. The squares show the calculated frequencies using the measured ε_{xx} and ε_{zz}, and the respective labels give the frequency values as determined by GIRSE analysis.

Table 9.6 A_1 (TO) and E_1 (*LO*) strain-free frequency, ω^0 (in cm^{-1}), and deformation potentials, *a* and *b* (in cm^{-1}), and the respective standard deviations.

	A_1 (TO)			**E_1 (LO)**		
	$\omega^0_{A_1(TO)}$	$a_{A_1(TO)}$	$b_{A_1(TO)}$	$\omega^0_{E_1(LO)}$	$a_{E_1(LO)}$	$b_{E_1(LO)}$
GIRSE [17]	531.2 ± 0.1	-664 ± 129	-1182 ± 282	739.9 ± 0.1	-789 ± 80	-1169 ± 532
Raman experimental	531.8[a]	−630[a]	−1290[a]	742[b]	–	–
theory	540[c]	−640[c]	−695[c]	757[c]	−775[c]	−703[c]

a Ref. 8.
b Ref. 9.
c Ref. 14.

The regression does not account for the errors of the experimental strains, and therefore may limit the accuracy with which the deformation potentials are determined (see Figure 9.17 and Equation 23). The large uncertainties in the stiffness constants and the narrow frequency range could also contribute to large standard deviations of the deformation potentials. We note, however, that experimentally obtained phonon deformation potentials of GaN and AlN typically suffer from uncertainties as large as 20–23% [10, 11].

The E_1(LO) phonon deformation potentials a and b have also been determined by performing a regression analysis of the following equation [17]:

$$\Delta\omega_{E_1(\mathrm{LO})} = a_{E_1(\mathrm{LO})}\left(1 - \frac{C_{11}}{C_{12}}\right)\varepsilon_{xx} - \left(a_{E_1(\mathrm{LO})}\frac{C_{13}}{C_{12}} - b_{E_1(\mathrm{LO})}\right)\varepsilon_{zz}$$
$$\pm c_{E_1(\mathrm{LO})}\left|\left(1 - \frac{C_{11}}{C_{12}}\right)\varepsilon_{xx} - \frac{C_{13}}{C_{12}}\varepsilon_{zz}\right| \quad (24)$$

obtained by substituting Equation 7 in Equation 9. The value of the $c_{E_1(\mathrm{LO})}$ from Table 9.5 has been used as an input parameter. The obtained results are listed in Table 9.6. The $a_{E_1(\mathrm{LO})}$ is determined with good accuracy. However, the larger uncertainties of the phonon frequencies and insufficient number of data sensitive to the phonon anisotropy in the basal plane resulted, in this case, in considerably larger error in the second deformation potential, $b_{E_1(\mathrm{LO})}$.

It is seen from Table 9.6 that there is a very good agreement between the strain-free frequency values obtained by GIRSE and Raman scattering, whereas the theoretical values slightly deviate. We note that strain-free frequencies of $530.7 \pm 0.1\ \mathrm{cm}^{-1}$ for the A_1(TO) and $740.3 \pm 0.15\ \mathrm{cm}^{-1}$ for the E_1(LO) phonons in a high-quality bulk GaN, which are in a very good agreement with the GIRSE results [17, 18], have recently been determined by Raman experiments [17, 52]. The $a_{A_1(\mathrm{TO})}$ obtained by GIRSE in [17, 18] agrees well with the values previously determined by Raman scattering and theory [8, 14]. On the other hand, the $b_{A_1(\mathrm{TO})}$ [17, 18] is significantly larger than the theoretical estimation [14], confirming the experimental results of Davydov *et al.* [8]. There is a very good agreement between the $a_{E_1(\mathrm{LO})}$ obtained by GIRSE in [17] and the theoretical estimation [14]. Similarly, the second deformation potential of the E_1(LO) was found to be significantly larger than the theoretically estimated $b_{A_1(\mathrm{TO})}$ (Table 9.6). However, owing to the large error, it is difficult to draw firm conclusions in this case. The reasons for the observed difference between the theoretical estimations and experimental determinations are not well understood, but we note that a similar discrepancy between theory and experiment was found to be a general tendency for other phonon modes of GaN and AlN as well [8, 10–12]. In this respect, it is important to emphasize that in the GIRSE determination of the A_1(TO) and E_1(LO) deformation potentials, only the experimental results about the strains and A_1(TO) frequencies were used without the need of additional hydrostatic experiments (usually performed on a different set of samples) or calibration of strain by other optical experiments. The latter are often used in the experimental determination of phonon deformation potentials of III nitrides and are considered as possible sources of errors [14]. The uncertainties in the published stiffness constants of GaN, another source of possible errors in experimentally determined phonon deformation potentials, cannot alone account for the observed difference between theory and experiment. On the other hand, the choice of strain-free lattice parameters may significantly affect the experimental phonon deformation potentials and therefore, most of the experimentally determined phonon deformation potentials of III nitrides may be biased, suffering from the lack of unification of strain-free lattice parameters of GaN (AlN, InN).

9.4.5
Quantum Dots

The growth of *a*-plane GaN QDs on nonpolar sapphire and SiC has recently been demonstrated in an attempt to overcome the impact of structural defects on emission properties minimizing, at the same time, the detrimental effects of polarization fields [56]. In contrast to quantum wells and superlattices, an electric field is expected to be present along the in-plane [0001] direction of *a*-plane QDs owing to the existence of interfaces across the polar axis. The shift of the phonon frequencies with respect to their strain-free values can be used to estimate the piezoelectric contribution to the internal electric field in the dots. This has been demonstrated in recent Raman studies on *a*-plane GaN QDs [57, 58]. Figure 9.18, as taken from [57], shows micro-Raman spectra of a sample consisting of 18 periods *a*-plane GaN QDs grown on *a*-plane 6H-SiC with 8.8-nm-thick AlN spacers. A_1(TO), E_1(TO) and E_2 phonon modes originating from both the QDs and the spacer have been identified in agreement with the Raman selection rules. Garro *et al.* determined shifts of $34.5 \pm 3(-4.7 \pm 1)$, $28 \pm 3(-5 \pm 1)$, and $30.2 \pm 3(-4.6 \pm 1)$ cm^{-1} for the GaN (AlN) A_1(TO), E_2, and E_1(TO) modes, respectively using as strain-free phonon frequencies the

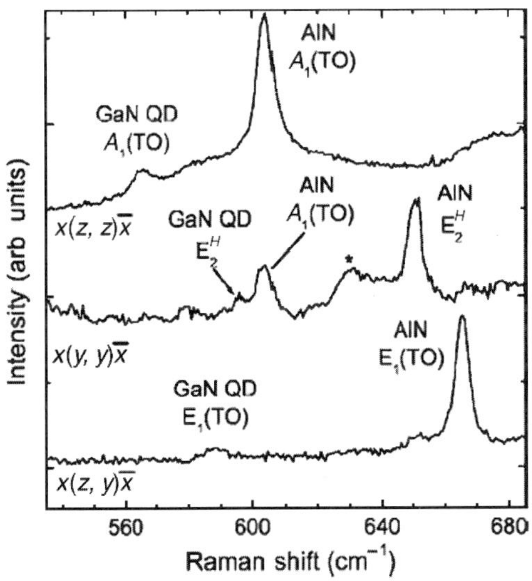

Figure 9.18 Backscattering micro-Raman spectra from *a*-plane QD sample, obtained for three polarization configurations under 2.41 eV (514.5 nm) excitation. The *z* direction coincides with the *c* axis of the wurtzite structure. The asterisk refers to features attributed to the substrate. (Reprinted figure with permission from Garro, N., Cros, A., Budagosky, J.A. *et al.* (2005) *Applied Physics Letters*, **87**, 011101. Copyright 2005 by the American Institute of Physics.)

values reported in [31]. Further, the authors estimated the strain components of the AlN spacer of $\varepsilon_{zz} = 0.015$ and $\varepsilon_{xx} + \varepsilon_{yy} = -0.009$, which fit with a 10% accuracy the measured strain-induced frequency shifts. On the other hand, the estimated strain components for the GaN QDs, $\varepsilon_{zz} = -0.023$ and $\varepsilon_{xx} + \varepsilon_{yy} = -0.008$ were derived only from the A_1(TO) and E_2 phonon shifts, and give a deviation of 30% for the E_1(TO) phonon shift [56]. As a further step, Garro *et al.* estimated the average piezoelectric polarization in the dots and the barriers using the strain components obtained from the Raman shifts [57]. They concluded that in contrast to *c*-plane QDs the contribution of the piezoelectric polarization compensates the effect of the spontaneous polarization [57, 58]. The polarization charge in the *a*-plane QD is further reduced by a factor of two owing to geometrical consideration, and therefore reducing even more, compared to *c*-plane QD, the electrostatic field in the *a*-plane QD.

The inconsistency in the strain components in the *a*-plane QD samples reported in [57, 58] indicates a significant impact of the assumptions made in their estimation from the phonon frequencies. We note that in all estimations of the strain components, the anisotropic phonon deformation potentials $c_{E_1(\mathrm{TO})}$ and c_{E_2} were assumed to be zero [57]. The results of the anisotropically strained GaN films presented in Sections 9.4.2 and 9.4.3 show clear splittings of the E_1 and E_2 indicating nonzero anisotropic deformation potentials for these phonon modes (Tables 9.4 and 9.5). We point out, however, that the values of the $c_{E_1(\mathrm{TO})}$ and c_{E_2} in Ref. 17 may be also affected by the assumptions made in the strain component determination by XRD (for instance the deviation from hexagonal symmetry is assumed to be small). The studies of the vibrational properties of nitride materials with nonpolar surface orientations are scarce and clearly, further experimental and theoretical investigations are needed to clarify these issues.

9.5
Summary and Outlook

GaN layers with vicinal surfaces, such as *a*-plane surface orientations, reveal all information on their IR active phonon modes upon application of GIRSE. Valuable information on the complete dielectric tensor anisotropy and A_1(TO) phonons in GaN, that is otherwise obscured for *c*-plane-oriented films, is retrieved. The capability of the GIRSE technique to detect spectrally narrow dichroism in nonpolar GaN films under anisotropic strain, thereby allowing to precisely locate the phonon mode resonances for different polarizations, has been demonstrated. As a result, splittings of the GaN E_1(TO) and E_1(LO) phonon modes under anisotropic strain have been identified. Further splitting of the GaN E_2 phonons measured by Raman spectroscopy has been also demonstrated.

It is shown that the combination of high-resolution X-ray diffraction, applied to assess strain components, with the GIRSE and Raman scattering spectroscopy represents a valuable approach for the determination of the anisotropic phonon mode properties and their interrelation with lattice deformations. More specifically, a correlation of phonon splittings with anisotropic strain components allows the determination of the *anisotropic* c phonon deformation potential of the GaN E_1(TO), E_1(LO), and E_2 phonons. Further, the *isotropic* a and b phonon deformation potentials of the GaN A_1(TO) and E_1(LO) phonon modes are determined by GIRSE, thereby shedding light on the existing discrepancy between the GaN A_1(TO) phonon deformation potentials determined by theory and Raman experiments, and providing experimental reference for the E_1(LO) phonon deformation potentials. The acquired knowledge of the phonon deformation potentials and strain-free phonon frequencies of GaN is indispensable for the use of phonon frequencies to extract information on strain components and related polarization fields in layers, low-dimensional, and complex device structures. However, the information on phonon -deformation potentials of GaN is scarce, in particular for the anisotropic phonon deformation potentials and further refinement of their values is needed. Moreover, new theoretical efforts are called for, to calculate the anisotropic phonon deformation potentials in group III nitrides.

The GIRSE, Raman, and HRXRD analyzes presented here are models for comprehensive and systematic studies of the IR optical properties and can be applied to other nonpolar group III nitride materials, such as AlN, InN, and the ternary alloys, and different material systems, such as wurtzite-structure wide band-gap oxides, II–VI semiconducting materials, and so on.

Acknowledgments

We would like to thank Dr J. Off, Prof. F. Scholz, Dr S. Figge, Prof. D. Hommel, Dr B. Haskell, Prof. S. Nakamura, Dr P. Fini, Prof. J. Speck, Prof. A. Usui, Dr E. Preble, Dr A. Hanser, Dr K. Evans, and Dr M. Heuken for kindly providing us with the GaN samples. Prof. H. Arwin and Prof. S. Lourdudoss are acknowledged for access to the IRSE and Raman equipment, and Dr P. Paskov and Prof. B. Monemar for fruitful discussions. We thank M. Saenger, Dr C. Bundesmann and U. Techner for help with some of the Raman measurements. We would like to thank Prof. V. Davydov for kindly providing the original of Figure 9.2. V. Darakchieva would like to acknowledge support from the Swedish Research Council (VR) under contract 2005-5054 and the University of Nebraska-Lincoln. M. Schubert acknowledges DFG research grant SCHUH 1338/3-1, support by the J. A. Woollam Foundation, and starting funds from UNL.

References

1. Anastassakis, E. and Cardona, M. (1998) *Semiconductor and Semimetals*, Vol. 55. (eds T. Suski and W. Paul), Academic Press, London.
2. Porowski, S. and Grzegory, I. (1997) *Journal of Crystal Growth*, **178**, 174.
3. Aoki, M., Yamane, H., Shimada, M. *et al.* (2000) *Journal of Crystal Growth*, **218**, 7.
4. Ketchum, D.R. and Kolis, J.W. (2001) *Journal of Crystal Growth*, **222**, 431.
5. Schlesser, R. and Sitar, Z. (2002) *Journal of Crystal Growth*, **234**, 349.
6. Kelly, M.K., Ambacher, O., Dalheimer, B. *et al.* (1996) *Applied Physics Letters*, **69**, 1749.
7. Paskova, T. and Monemar, B. (2002) in *Low-Dimensional Nitride Semiconductors*, (B. Gil), Oxford University press, p. 79.
8. Davydov, V.Y., Averkiev, N.S., Goncharuk, I.N. *et al.* (1997) *Journal of Applied Physics*, **82**, 5097.
9. Goñi, A.R., Siegle, H., Syassen, K. *et al.* (2001) *Physical Review B*, **64**, 035205.
10. Demangeot, F., Frandon, J., Baules, P. *et al.* (2004) *Physical Review B*, **69**, 155215.
11. Gleize, J., Renucci, M.A., Frandon, J. *et al.* (2003) *Journal Of Applied Physics*, **93**, 2065.
12. Darakchieva, V., Paskov, P.P., Paskova, T. *et al.* (2002) *Applied Physics Letters*, **80**, 2302.
13. Darakchieva, V., Paskov, P.P., Valcheva, E. *et al.* (2004) *Applied Physics Letters*, **84**, 3636.
14. Wagner, J.-M. and Bechstedt, F. (2002) *Physical Review B*, **66**, 115202.
15. Schubert, M. (2004) in *Handbook of Ellipsometry*, (eds H. Thompkins and E.A. Irene), William Andrew Publishing, Highland Mills.
16. Briggs, R.J. and Ramdas, A.K. (1976) *Physical Review B*, **13**, 5518.
17. Darakchieva, V., Paskova, T., Schubert, M. *et al.* (2007) *Physical Review B*, **75**, 195217.
18. Darakchieva, V., Paskova, T., Schubert, M. *et al.* (2007) *Journal of Crystal Growth*, **300**, 233.
19. Darakchieva, V., Paskova, T., Paskov, P.P. *et al.* (2005) *Journal of Applied Physics*, **97**, 013537.
20. Leszczynski, M., Teisseyre, H., Suski, T. *et al.* (1996) *Applied Physics Letters*, **69**, 73.
21. de Walle, C.G.V. (2003) *Physical Review B*, **68**, 165209.
22. Darakchieva, V. Monemar, B. and Usui, A. (2007) *Applied Physics Letters*, **91**, 031911.
23. Leszczynski, M. (1998) INSPEC, London.
24. Davydov, V.Y., Kitaev, Y.E., Goncharuk, I.N. *et al.* (1998) *Physical Review B*, **58**, 12899.
25. Amano, H., Hiramatsu, K., and Akasaki, I. (1988) *Japanese Journal of Applied Physics*, **27**, L1384.
26. Siegle, V., Hoffmann, A., Eckey, L. *et al.* (1997) *Applied Physics Letters*, **71**, 2490.
27. Jasinski, J., Swider, W., Liliental-Weber, Z. *et al.* (2001) *Applied Physics Letters*, **78**, 2297.
28. Leszczynski, M., Suski, T., Teisseyre, H. *et al.* (1994) *Journal of Applied Physics*, **76**, 4909.
29. Darakchieva, V., Paskov, P.P., Paskova, T. *et al.* (2003) *Applied Physics Letters*, **82**, 703.
30. Paskova, T., Darakchieva, V., Paskov, P.P. *et al.* (2005) *Journal of Crystal Growth*, **281**, 55.
31. Harima, H. (2002) *Journal of Physics Condensed Matter*, **14**, R967.
32. Loudon, R. (1964) *Advances in Physics*, **13**, 423.
33. Bungaro, C., Rapcewicz, K., and Bernholc, J. (2000) *Physical Review B*, **61**, 6720.
34. Arguello, C.A., Rousseau, D.L., and Porto, S.P.S. (1969) *Physical Review*, **181**, 1351.
35. Azzam, R.M.A. and Bashara, N.M. (1984) *Ellipsometry and Polarized Light*, North-Holland, Amsterdam.
36. Schubert, M. (2004) *Infrared Ellipsometry on Semiconductor Layer Structures: Phonons, Plasmons and Polaritons*, Vol. 209. Springer, New York.

37. Kasic, A., Schubert, M., Einfeld, S. *et al.* (2000) *Physical Review B*, **62**, 7365.
38. Schubert, M. (2006) *Annalen der Physik*, **15**, 480.
39. Darakchieva, V., Paskova, T., Paskov, P.P. *et al.* (2006) *Physica Status Solidi B*, **243**, 1594.
40. Romanov, A.E., Baker, T.J., Nakamura, S., and Speck, J.S. (2006) *Journal of Applied Physics*, **100**, 023522.
41. Darakchieva, V., Birch, J., Schubert, M. *et al.* (2004) *Physical Review B*, **70**, 045411.
42. Bond, W.L. (1960) *Acta Crystallographica*, **13**, 814.
43. Fewster, P.F. and Andrew, N.L. (1995) *Journal of Applied Crystallography*, **28**, 451.
44. Yamaguchi, M., Yagi, T., Azuhata, T. *et al.* (1997) *Journal of Physics. Condensed Matter*, **9**, 241.
45. Off, J. (2001) Ph.D. Thesis, Universität Stuttgart.
46. Figge, S., Böttcher, T., Einfeldt, S., and Hommel, D. (2000) *Journal of Crystal Growth*, **221**, 262.
47. Haskell, B.A., Wu, F., Matsuda, S. *et al.* (2003) *Applied Physics Letters*, **83**, 1554.
48. Tsuda, M., Furukawa, H., Honshio, A. *et al.* (2006) *Physica Status Solidi B*, **243**, 1524.
49. Angerer, H., Brunner, D., Freudenberg, F. *et al.* (1997) *Applied Physics Letters*, **71**, 1504.
50. Paskova, T., Kroeger, R., Figge, S. *et al.* (2006) *Applied Physics Letters*, **90**, 051914.
51. Usui, A., Sunakawa, H., and Yamaguchi, A.A. (1997) *Japanese Journal of Applied Physics*, **36**, L899.
52. Paskova, T., Paskov, P.P., Darakchieva, V. *et al.* (2007) *Materials Research Society Symposium Proceedings*, **955E**, I3.4.
53. Schubert, M., Tiwald, T.E., and Herzinger, C.M. (2000) *Physical Review B*, **61**, 8187.
54. Davydov, V.Y., Averkiev, N.S., Goncharyk, I.N. *et al.* (1997) *Journal of Applied Physics*, **82**, 5097.
55. Kendall, M.G. and Stuart, A. (1961) *The Advanced Theory of Statistics*, Vol. 2. Charles Griffin and Company Ltd, London.
56. Founta, S., Rol, F., Bellet-Amalric, E. *et al.* (2005) *Applied Physics Letters*, **86**, 171901.
57. Garro, N., Cros, A., Budagosky, J.A. *et al.* (2005) *Applied Physics Letters*, **87**, 011101.
58. Cros, A., Budagosky, J.A., García-Cristóbal, A. *et al.* (2006) *Physica Status Solidi B*, **243**, 1499.

10
Defects Formed in Nonpolar GaN Grown on SiC and Al_2O_3 and their Reduction in Pendeo-epitaxial and Laterally Overgrown GaN Layers

Zuzanna Liliental-Weber and Dmitri Nikdaj Zakharov

10.1
Introduction

III-nitride semiconductors (InN-GaN-AlN) have a direct band gap ranging from 0.7 eV for InN [1, 2] to 3.4 eV for GaN to 6.2 eV for AlN [3, 4]. Such materials have become very attractive for applications in light emitting diodes (LEDs) and laser diodes (LDs), the wavelength of emission ranging from red to ultraviolet (UV) [5–7]. However, the lack of large diameter wafers of native substrates for growth of wurtzite GaN (a = 3.189 Å, c = 5.185 Å) and its alloys forces the use of heteroepitaxial growth using foreign substrates. The most commonly used substrates are (0001) Al_2O_3, 6H-SiC, and (111) Si. Of these the most frequently used is Al_2O_3 with about 14% lattice mismatch to GaN. This mismatch results in a high density of misfit and threading dislocations (TDs) in the GaN layers, typically $10^9 - 10^{10}$ cm^{-2}. In the case of 6H-SiC the mismatch is only 3.5%, but typical dislocation densities are still high at ($10^8 \div 10^9$ cm^{-2}). For the Si substrate, a 17% lattice mismatch with GaN, in addition to the difference in thermal expansion coefficients, leads to tensile stress resulting in cracking of the grown layer and high density of TDs. It has been predicted that dislocations may act as nonradiative recombination centers [8, 9] and that scattering at charged dislocations could be a limiting factor of carrier mobility in a two-dimensional electron gas [10]. Experiments using cathodoluminescence (CL) studies have proved that these TDs reduce efficiency and lifetime of optoelectronic devices by creating energy levels inside the band gap [11, 12]. These centers have more impact on short wavelength LEDs than on longer wavelength devices, because the carrier migration distances are believed to be much smaller in the latter case [13, 14]. According to one theoretical calculation [15], the dislocation density must be reduced at least to $\sim 10^7$ cm^{-2} to provide internal quantum efficiency close to unity. In devices based on heterostructures containing III-nitride materials, the polar *c* axis is perpendicular to the sample surface and to the interfaces

Nitrides with Nonpolar Surfaces. Growth, Properties, and Devices. Edited by Tanya Paskova

ISBN: 978-3-527-40768-2

of the active layers. Since Ga and N atoms are arranged on top of each other along the polar [0001] crystallographic direction, this leads to nonzero spontaneous polarization. In addition, stress applied along the same axis can also give rise to piezoelectric polarization in heterostructures (GaN–AlGaN or GaN–InGaN). The total polarization, which is the sum of spontaneous and piezoelectric polarizations, is aligned along the [0001] direction of the wurtzite III-nitride heterostructures. This polarization produces electrostatic fields that reduce electron-hole recombination efficiencies. The change in the total polarization across the heterolayers results in high interface charge densities and spatial separation of the electron and hole wave functions, red shifting of the photoluminescence peak, and a decrease in the peak intensity [16–18]. These deleterious effects can be eliminated if GaN-based epilayers are grown in nonpolar orientations, such as [$11\bar{2}0$] and [$1\bar{1}00$]. Recently, nonpolar *m*-plane and *a*-plane GaN/AlGaN multiple quantum well structures have been grown on ($1\bar{1}00$) *m*-plane $LiAlO_2$ [19], ($1\bar{1}02$) *r*-plane Al_2O_3 [20, 21], 4H-SiC [22], and 6H-SiC [23] substrates. These layers show a lack of red-shift and improved photoluminescence intensity [19, 24, 25].

Since optoelectronic properties of GaN-based devices strongly depend on the structural quality of the heterostructure layers, the first part of this chapter will discuss the defects formed in GaN *a*-plane layers grown on nonpolar ($11\bar{2}0$) 4H-SiC. In the second part of this chapter, pendeo-epitaxial and laterally overgrown epitaxial layers will be characterized and possible defect reduction using these methods will be discussed. Transmission electron microscopy (TEM) methods will be used to characterize these defects.

10.2 Defects Formed in Nonpolar ($11\bar{2}0$) *a*-plane GaN Grown on ($11\bar{2}0$) 4H-SiC

10.2.1 Growth Procedure

Epitaxial AlN buffer layers, on which an epitaxial GaN film was deposited, were grown *in situ* in a cold-wall, vertical, pancake style metal organic vapor phase epitaxy (MOVPE) system on a 4H-SiC ($11\bar{2}0$) substrate. The layer sequence in such samples is shown on (Figure 10.1). Before the growth, the SiC substrate was etched in a 10% HF solution to remove the native oxide. The AlN buffers and GaN layers were grown at 20 Torr and 1100 and 1020 °C, respectively. Trimethylaluminum (TMA), triethylgallium (TEG), and ammonia were used as the precursors; the hydrogen was used both as the carrier and the dilutant. The V/III ratios used during the growth of AlN and GaN were 2400 and 1300, respectively. More details are described in [26].

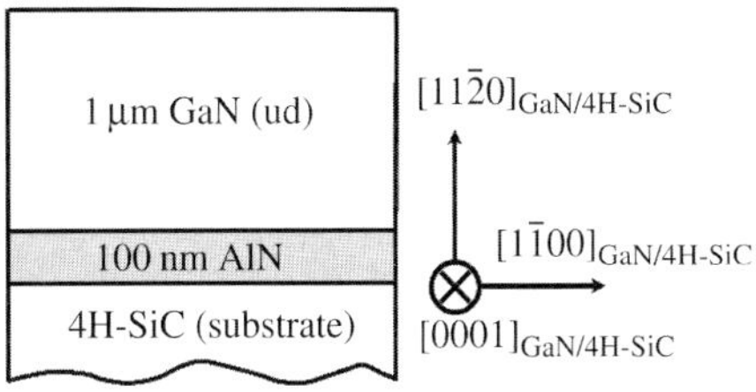

Figure 10.1 Schematic of the layer arrangement in the studied samples.

10.2.2 Defect Characterization

To get 3D information about defects present in a GaN layer grown on a nonpolar ($\bar{2}$110) *a*-plane 4H-SiC surface, one needs to understand how the particular planes are arranged in the unit cell and how the atoms are arranged on these planes in comparison to the *c* plane popularly used for the growth of III-Nitrides. Figure 10.2 shows a schematic of a GaN unit cell, where *c*, *a*, *m* and *r* planes are marked.

In addition, Figure 10.3a–d shows the atomic arrangement on the *c* and *a* planes for both 4H-SiC and GaN. One may notice that the *c* planes (Figure 10.3a and b) are polar, since in both cases (SiC or GaN) they are occupied only by one kind of atom. This means that such planes can have different growth polarity, for example, Si or C for 4H-SiC (or 6H-SiC). Atom distribution on the ($\bar{2}$110) *a* plane is different (Figure 10.3c and d) (e.g. for growth of GaN on 4H-SiC). An equal number of Ga and N (or Si and C for SiC) must be present on these planes. In these circumstances, growth of the layers proceeds along the [$\bar{2}$110] direction and *c* planes are along this direction. It is, therefore expected that different types of defects may be formed for the growth

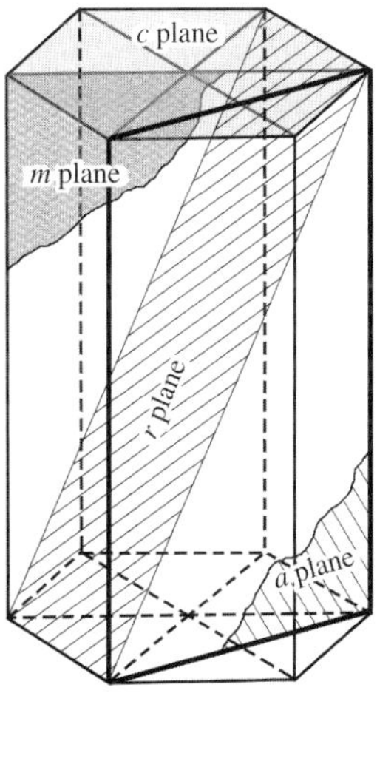

Figure 10.2 The wurtzite unit cell of GaN, with *a*, *c*, *m*- and *r* planes indicated.

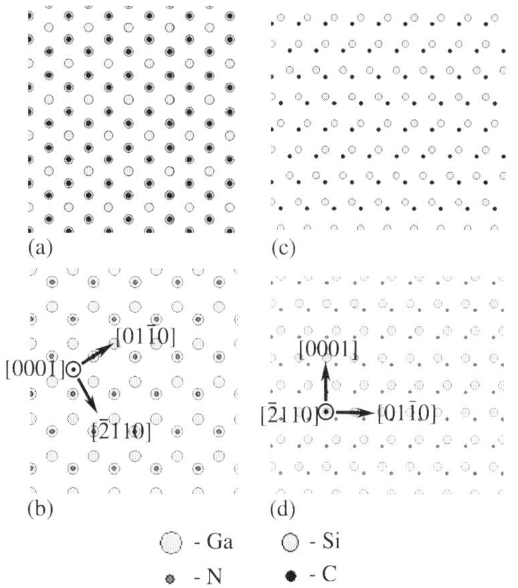

Figure 10.3 Schematic of the atomic arrangement on different crystallographic planes: (a) (0001) 4H-SiC; (b) (0001)GaN; (c) ($\bar{2}$110) 4H-SiC; (d) ($\bar{2}$110) GaN (cf. Color Plate XXXI).

in the nonpolar direction compared to growth along the polar direction. To determine the nature of defects formed in layers grown in a nonpolar direction, cross-sectional and plan-view samples were investigated by conventional and high-resolution TEM along the [0001], [1$\bar{1}$00], [$\bar{2}$110], [1$\bar{2}$10], and [11$\bar{2}$0] zone axes. For the (11$\bar{2}$0) GaN layer grown on (11$\bar{2}$0) 4H-SiC substrate, the orientation relationship obtained from the diffraction patterns from cross-sectional samples can be described as $(11\bar{2}0)_{GaN} \| (11\bar{2}0)_{AlN} \| (11\bar{2}0)_{4H\text{-}SiC}$ and $[1\bar{1}00]_{GaN} \| [1\bar{1}00]_{AlN} \| [1\bar{1}00]_{4H\text{-}SiC}$. Schematics of these interfaces in two perpendicular orientations in cross-section samples are shown in Figure 10.4a and b.

TEM images show a high density of basal stacking faults (BSFs) distributed on *c* planes along the growth direction (Figure 10.5a). In majority of cases they propagate to the sample surface. These defects are visible in the image (Figure 10.5a) as thin lines perpendicular to the 4H-SiC/AlN and AlN/GaN interfaces. Once these BSFs are formed at the substrate/buffer layer interface (Figure 10.5b), they propagate to the GaN layer owing to the similarity of the AlN and GaN crystal structures. Some BSFs then propagate to the GaN layer surface and some are terminated inside the layer. Partial dislocations terminate the faults ending inside the layers.

Three types of BSFs are known to exist in a wurtzite structure: intrinsic I_1 and I_2 types and an extrinsic *E* type [27]. The wurtzite structure of GaN consists of two hexagonal close-packed sublattices: Ga and N shifted by 0.377*c* along [0001]. In both sublattices (0002) planes are closely packed. Stacking these basal (0002) planes in the ...*Aa Bb Aa Bb Aa Bb*... sequence forms the wurtzite structure, in which the atoms are tetrahedrally coordinated. Capital letters

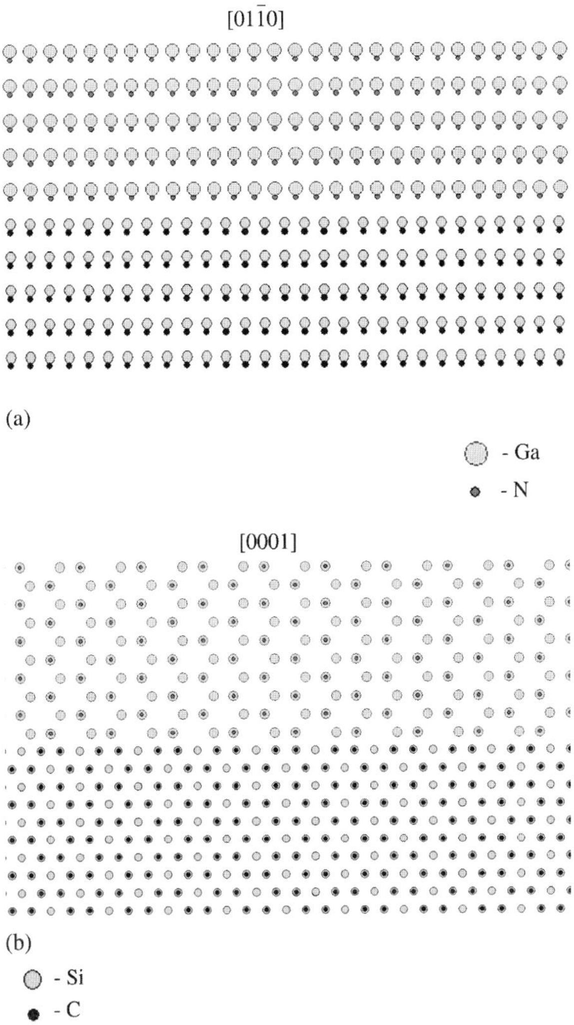

Figure 10.4 Schematic of the atomic arrangement in GaN/4H-SiC cross-sectional samples in (a) $[01\bar{1}0]_{GaN}$ and (b) $[0001]_{GaN}$ orientations (cf. Color Plate XXXII).

correspond to Ga layers and the small letters to N layers. For simplicity, the hexagonal stacking sequence will henceforth be referred to as . . .ABABAB. . ., where capital letters represent Ga–N atomic bilayers. From a crystallography point of view, the basal faults are equivalent to local transitions from the hexagonal to the cubic structure with a few (0001) atomic planes. In hexagonal GaN, the bonds have mirror symmetry, whereas in the cubic structure, the bonds are rotated through 60° with respect to the nearest neighbors. The local change of crystal symmetry is from 2H to 3C. Each Ga–N bilayer of the cubic

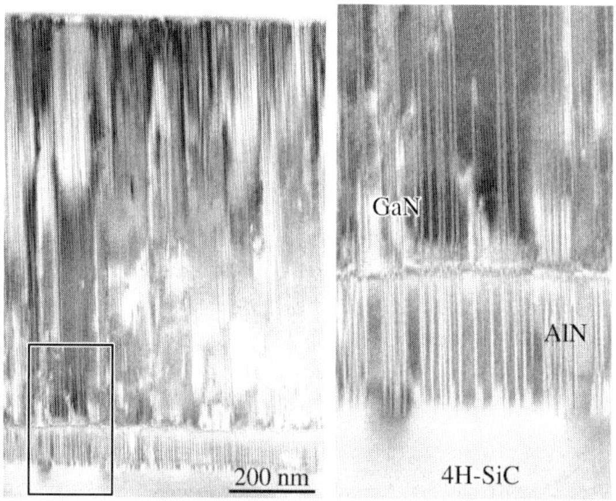

(a) (b)

Figure 10.5 (a) Stacking faults in GaN and AlN. Note that majority of the faults originating at the AlN/SiC interface ((b) the enlarged area from the box indicated in (a)) propagate into the GaN layer, and continue to the sample surface. Only some of these faults are terminated by partial dislocations within the layer.

GaN might be situated in one of three possible positions assigned as A, B, or C, with the ideal stacking sequence being . . .ABCABC. . . , while in hexagonal GaN each bilayer has only two possible positions A and B, with a perfect sequence . . .ABABAB. . . . Therefore, BSFs locally change the bond arrangement and introduce a number of atomic planes with the zinc-blende structure.

There are several ways for BSFs to end inside the GaN layer, depending on the type of BSF. The I_1-type BSF changes the usual . . .ABABAB. . . wurtzite stacking sequence into . . .ABABCBC. . . . Such a BSF, if it ends inside a crystal, has to be bounded by a sessile Frank–Shockley dislocation with Burgers vector $\mathbf{b} = 1/6\langle 20\bar{2}3\rangle$, or, as will be shown later, it can be terminated by prismatic stacking faults (PSFs).

The I_2 BSFs have an . . .ABABCACA. . . sequence and are formed by a $1/3\langle 1\bar{1}00\rangle$ shear of one part of the crystal with respect to the other or by dissociation of a perfect dislocation with Burger's vector $\boldsymbol{b} = 1/3\langle 11\bar{2}0\rangle$ into two Shockley partials, with $\mathbf{b} = 1/3\langle 1\bar{1}00\rangle$ at both ends. The *E*-type BSF with a stacking sequence . . .ABABCABAB. . . is formed owing to the insertion of an extra basal plane, in most cases, by precipitation of point defects on a basal plane, and it is bound by a Frank partial with $\mathbf{b} = 1/2[0001]$. This fault is always surrounded by a dislocation loop. If $1/3\langle 1\bar{1}00\rangle$ shear occurs on the basal plane, then this fault can be converted into an I_1 fault [28].

The energy of BSFs, γ, is proportional to the number of cubic bilayers in the particular fault. As described I_1-, I_2-, and *E*-type defects consist of one, two, and three cubic bilayers, respectively, and their energy to a first approximation is $\gamma_E \sim 3/2\ \gamma_{I2} \sim 3\gamma_{I1}$. Although the I_1-type BSF has the lowest energy, its formation requires removal of a basal bilayer, in addition to a basal shear of $1/3\langle 1\bar{1}00\rangle$. This means that the I_1 BSF is likely to be formed during the growth process. Deformation caused by stress during or after growth can cause formation of the I_2-type BSF by movement of a glissile Shockley partial dislocation in the basal plane.

The best way to determine density of stacking faults in a sample grown on the nonpolar (11$\bar{2}$0) plane is to observe these defects in plan-view configuration. In this orientation, BSFs are seen as thin lines parallel to the *c* plane of GaN when two-beam conditions with $\mathbf{g} = 1\bar{1}00$ are applied (Figure 10.6a). The density of stacking faults was estimated to be about $\sim 1.6 \times 10^6$ cm^{-1}. Owing to such a high density of these faults, diffraction patterns show streaks along the [0001] crystallographic direction. Figure 10.6b, taken with $\mathbf{g} = 0002$ from the same area of the sample, shows partial dislocations terminating these faults. Detailed analysis using well defined two-beam imaging conditions, for which one type of dislocation is out of contrast, while the other is in contrast (Figure 10.6b) showed the presence of two types of partial dislocations, one with a Burger's vector with component parallel to [0001] and the second that has no component of Burgers vector along [0001]. The first type is at the

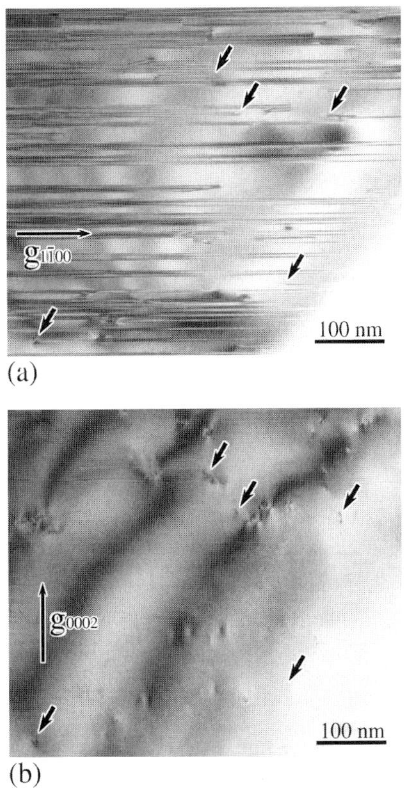

Figure 10.6 Plan-view micrographs of *a*-plane GaN grown on *a*-plane of 4H-SiC recorded with two different diffraction conditions:
(a) $\mathbf{g} = 1\bar{1}00$ – stacking faults formed on *c* planes are in contrast;
(b) $\mathbf{g} = 0002$ – partial dislocations with nonzero [0001] component; terminating stacking faults are visible.

boundary of an I_1 BSF and the second at an I_2 BSF. The densities of these dislocations are $\sim 4.0 \times 10^{10}$ cm^{-2} and $\sim 0.4 \times 10^{10}$ cm^{-2}, respectively. These are sessile Frank–Shockley dislocations and Shockley partial dislocations with $\mathbf{b} = 1/6\langle 20\bar{2}3\rangle$ and $\mathbf{b} = 1/3\langle 1\bar{1}00\rangle$ respectively, bounding I_1 and I_2 BSFs [29].

High-resolution electron microscopy (HREM) of these faults in a plan-view sample confirms the presence of I_1 and I_2 types of BSFs (Figure 10.7a and b). The insets in these two images show the configuration of Ga and N atomic planes. We have not observed *E*-type BSFs in our samples. In addition to the well known I_1 and I_2 types of BSFs, an I_3-type BSF has been observed (Figure 10.7c). This type of defect is characterized by the ...ABABCBABA... sequence, when one of an 'A' bilayer is 'switched' to the 'C' position. This type of fault can be treated as two I_1 BSFs forming a nanotwin of cubic structure. A schematic of such a fault is shown in the inset of Figure 10.7c. It was predicted theoretically that I_3 has low formation energy second to that of I_1 BSF [30].

The density of partial dislocations allows us also to estimate that the density of I_1-type BSFs is one order of magnitude larger than the density of I_2-type faults. However, occasionally it is possible that two I_1 faults with different Burgers vectors meet each other, resulting in the appearance of one Shockley partial dislocation instead of two Frank–Shockley dislocations. This special case is illustrated in Figure 10.8.

Besides stacking faults formed on basal *c*-planes, there are also faults formed on prismatic $\{\bar{1}2\bar{1}0\}$ planes. A high-resolution image (Figure 10.9a) of a cross-sectional sample, along the [0001] zone axis, shows PSFs formed on $(\bar{2}110)$ and $(1\bar{2}10)$ planes. When imaged in bright field with $\mathbf{g} = 11\bar{2}0$ close to the [0001] zone axis, these faults are seen as thin lines intersecting the sample surface at a 60° angle.

A zigzag-like propagation from substrate toward the GaN layer surface is exhibited by PSFs. It is believed that such zigzag-like behavior decreases the total energy of the PSFs. Theoretical calculations predicted relatively high formation energy for PSFs of 72 meV Å^{-2} [31], which is about 30 times higher than that for I_2 BSFs [29]. The distribution and density of these PSFs in the sample can be estimated from both cross-sectional and plan-view sample configurations, when imaged close to or exactly at $\langle 11\bar{2}0\rangle$ directions (Figure 10.10, PSFs are marked with arrows). One can see from this image that both PSF boundaries on the *c* planes are surrounded by BSFs. It is clear that BSFs can not only be terminated by partial dislocations, but also by planar defects on prismatic planes; therefore, sometimes their presence can be beneficial. Figure 10.11 illustrates schematically a possible three-dimensional configuration of a PSF bounded by two BSFs. The density of PSFs from cross-sectional and plan-view images was estimated to be about 0.7×10^2 cm^{-1}.

To determine the atomic structure of a PSF, a number of image simulations were performed near the Scherzer defocus for various sample thicknesses using MacTempas software [32]. On the basis of our experimental images, it

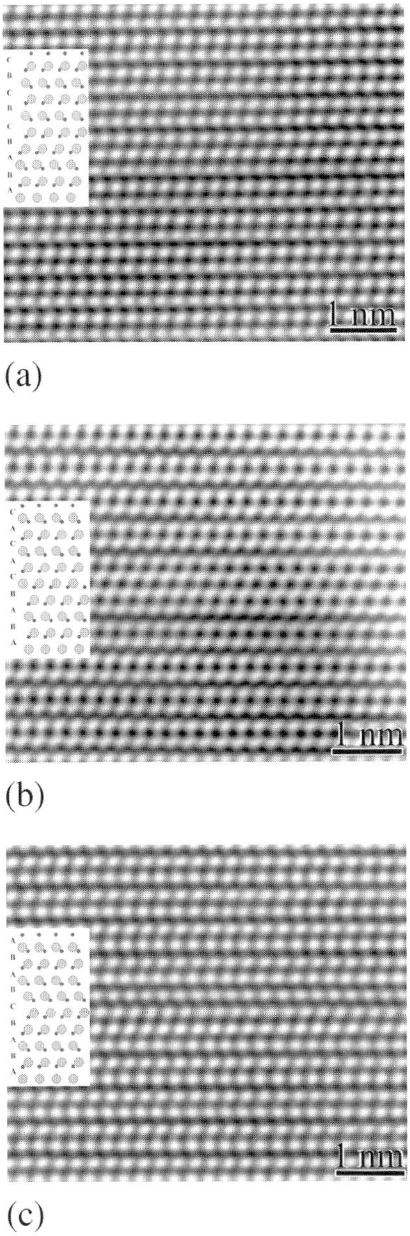

Figure 10.7 High-resolution images of basal stacking faults with indicated stacking: (a) I_1 fault with the sequence . . .ABABCBCB. . . ; (b) I_2 fault with the stacking sequence . . .ABABCACA. . . and (c) I_3 fault with the stacking sequence . . .ABABCABAB. . . .

was found that the size of PSFs in the [0001] direction does not exceed 6–7 nm. Therefore, simulated images were calculated for different thicknesses up to 7 nm. Drum's atomic model [33] (Figure 10.9c) was used for image simulations of PSFs. Calculated and experimental images were compared to obtain the

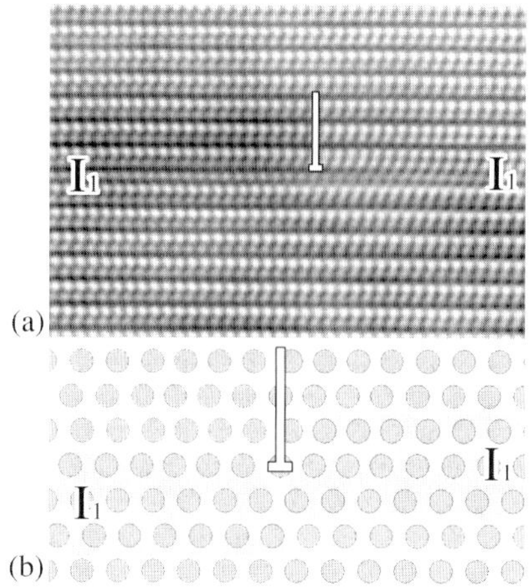

Figure 10.8 (a) HREM image shows the intersection of two I_1-type basal stacking faults. The intersection gives rise to Shockley partial dislocation (marked); (b) Schematic representation of (a); only Ga atoms are shown for simplicity.

best matching. The simulated image, using the defocus value of −45 nm and sample thickness of 4.7 nm, shown in Figure 10.9d has the closest match to the experimental HREM image (Figure 10.9b).

A HREM image of a PSF, which is formed on $(1\bar{2}10)$ or $(\bar{2}110)$, is shown in Figure 10.12a. One can see the contrast change due to the presence of an inclined PSF between two I_1-type BSFs. The contrast inside the dashed box area (enlarged in Figure 10.12b) has a hexagonal-like appearance. One could arbitrarily assign . . .ABAB. . . for the perfect hexagonal stacking sequence to atomic columns on the left and the right sides of the HREM image (see the assigned letters on both sides of Figure 10.12a). One can notice that the stacking sequence on the left side is changed by the first I_1-type BSF from . . .ABAB. . . to . . .ACAC. . . and by the second I_1 from . . .ACAC. . . to . . .CBCB. . . , whereas on the right side the sequence stays as . . .ABAB. . . . This is an indication of a 1/2[0001] shift between the left and the right parts of a crystal separated by a PSF. The displacement vector **R** for a PSF is equal to $1/2\langle 10\bar{1}1\rangle$. Since displacement vectors for BSFs and PSFs not being equal, stair-rod dislocations have to be formed at BSF/PSF intersections (marked 's' and '−s' in Figure 10.11), to compensate for the difference in the displacement vectors.

Figure 10.13a shows a projection on the $(11\bar{2}0)$ plane of a super cell used to simulate an image of a PSF formed on $(1\bar{2}10)$ or $(\bar{2}110)$ planes. One can see that atoms in the central part of the image are overlaid in a hexagonal-like way because of the presence of a shift between the upper and the lower parts of the crystal separated by a PSF inclined at a 60° angle with respect to the $[11\bar{2}0]$ direction. The simulated image in Figure 10.13b calculated for a defocus value of −57.5 nm and sample thickness of 3 nm gives the best fit with the experimental HREM image shown in Figure 10.12b.

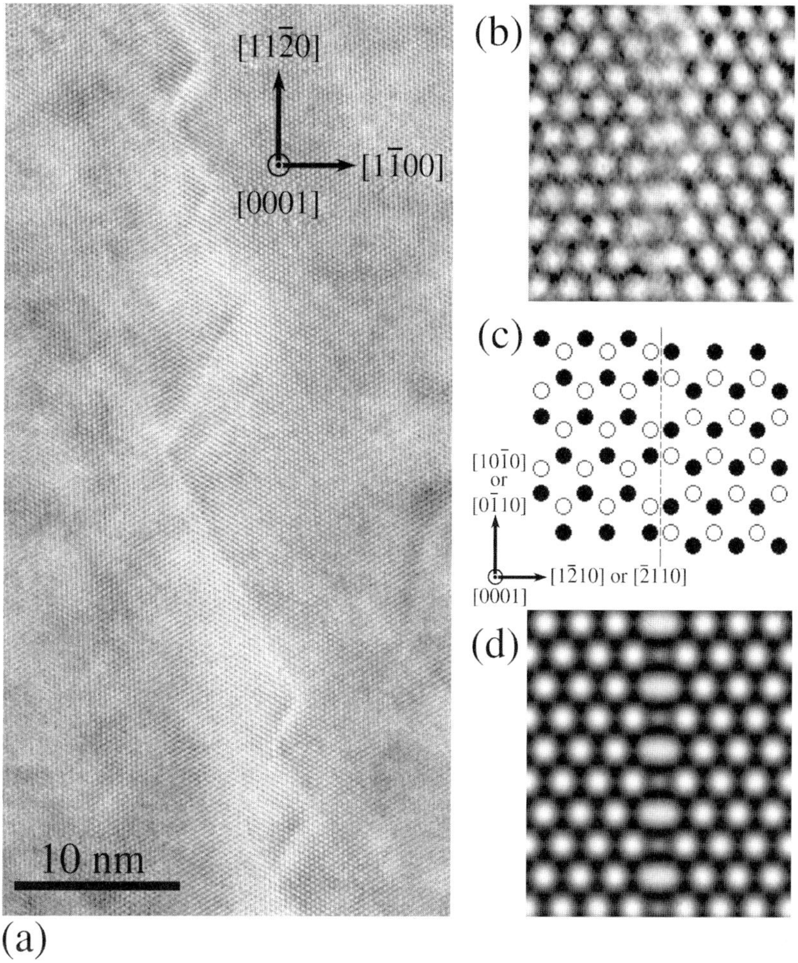

Figure 10.9 (a) High magnification of prismatic stacking faults (PSF) formed on ($\bar{2}$110) or (1$\bar{2}$10) planes showing zigzag-like configuration; (b) An experimental high-resolution image of a PSF formed on ($\bar{2}$110) or (1$\bar{2}$10) planes and recorded along $[0001]_{GaN}$ zone axis; (c) Drum's model of such a fault; (d) Simulated image based on Drum's model, showing a good agreement with the experiment.

10.2.3
Why Planar Defects are Formed in the Layers Grown on Nonpolar Surfaces

There are two ways to form a PSF in GaN; one by a shift of one part of a GaN crystal over the other part by $\mathbf{R} = 1/2[1\bar{1}01]$ or, alternatively, by removing a (1$\bar{2}$10) plane and then applying a 1/6[1$\bar{2}$10] shift. In our samples, GaN is grown on a thin AlN buffer (∼200 nm), which undergoes the Stranski–Krastanov [34] growth mode. Therefore, GaN grains nucleate at steps or pits in the AlN buffer

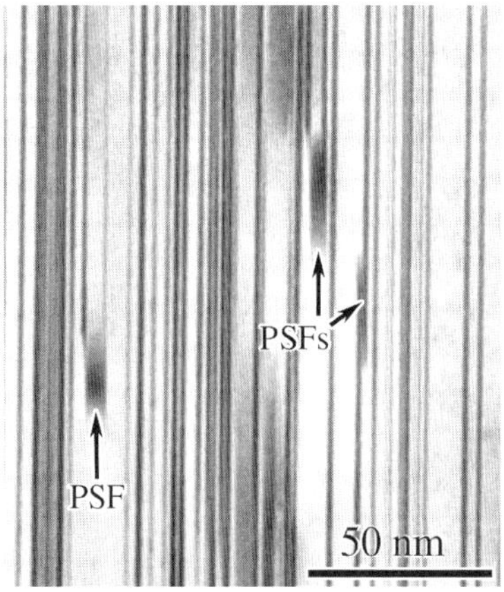

Figure 10.10 Bright-field image ($\mathbf{g} = \bar{1}010$) of inclined prismatic stacking faults, each bounded by two basal stacking faults.

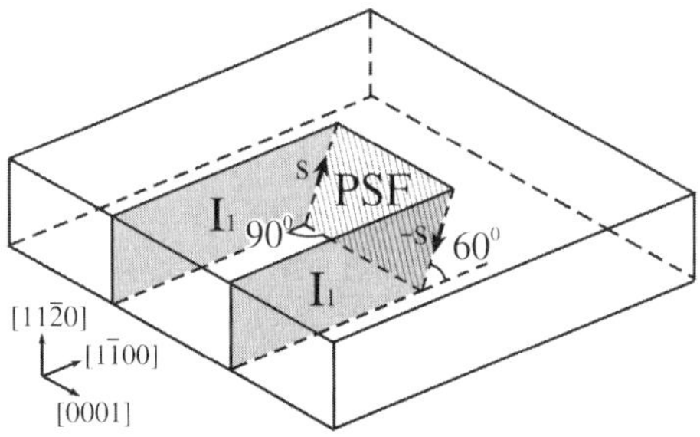

Figure 10.11 Termination of two I_1 basal stacking faults (BSFs) by PSF (schematic representation), s and −s represent 1/6[10$\bar{1}$0] and 1/6 [$\bar{1}$010] stair-rod dislocations.

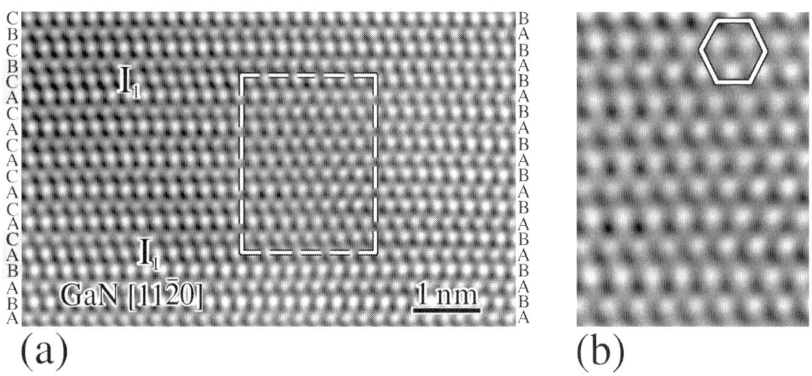

Figure 10.12 (a) High-resolution image (plan-view configuration) of an inclined PSF (inside outlined rectanglar) projected onto (11$\bar{2}$0) plane, showing hexagonal-like contrast; stacking sequences on the right and left sides of the image show the 1/2[0001] shift induced by this fault. The first I_1 fault changes the stacking sequence from . . .ABAB. . . to . . .ACAC. . . . This sequence is preserved until the second I_1 basal fault converts it to . . .BCBC. . . ; (b) The magnified image from the dashed box shown in (a).

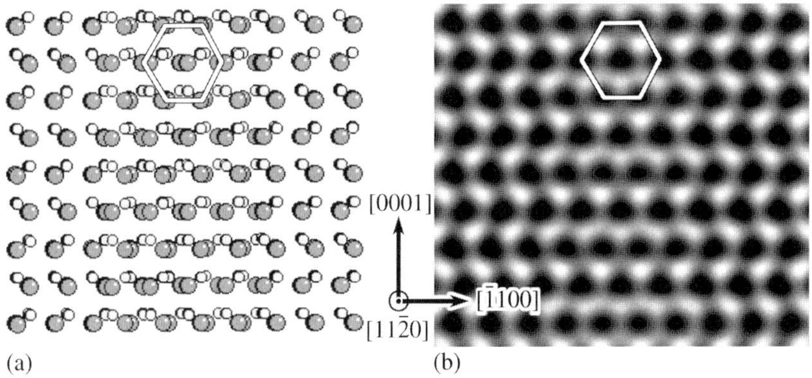

Figure 10.13 (a) Drum's atomic model of a PSF formed on ($\bar{2}$110) or (1$\bar{2}$10) planes projected onto the (11$\bar{2}$0) plane of GaN; (b) Simulated high-resolution image of an inclined PSF based on the model shown in (a).

layer, which is the most energetically favorable places for nucleation. It is expected therefore, that locally the thickness of GaN can be different, and lateral growth needs to take place until the islands finally merge. At this stage, PSFs are likely to appear, to accommodate lattice mismatch between the GaN and AlN structures. Moreover, recently it has been found [20] that the roughness of a SiC substrate is only about 7 nm, much smaller than that of a 200-nm thick AlN buffer layer deposited on a SiC substrate. This roughness is expected to propagate through the AlN and it would be in agreement with our observation that PSFs should be formed in GaN to accommodate the stress. Later PSFs tend to fold into pairs of I_1 BSFs, since the formation energy of the latter is much smaller. Even further energy minimization is possible owing to PSF folding in a (1$\bar{2}$10)/($\bar{2}$110) set of planes resulting in an almost vertical zigzag-like PSF propagation toward the sample surface.

Growth-related defects such as I_1 BSFs, which are surrounded by sessile – Shockley partial dislocations, are probably responsible for accommodation of azimuthal rotation of adjacent GaN grains and compensation of lattice mismatch in the [0001] direction between the substrate and the GaN layer. The appearance of I_2 BSFs surrounded by glissile Shockley partial dislocations is probably a result of local stress relaxation on the *c* plane from lattice mismatch and different thermal expansion coefficients between the GaN and the AlN/4H-SiC. If there is no effective way to neutralize stress, then formation of cracks is expected. Such cracks are visible in cross-section samples imaged by TEM. These cracks appear always at some distance from the AlN buffer layer, usually about 100 nm. Once formed, they continue to propagate to the sample surface. The measured average distance between cracks was about 2 μm.

Another reason for BSFs formation could be a lack of epitaxial relationship between a substrate and a grown layer. Since the majority of BSFs are

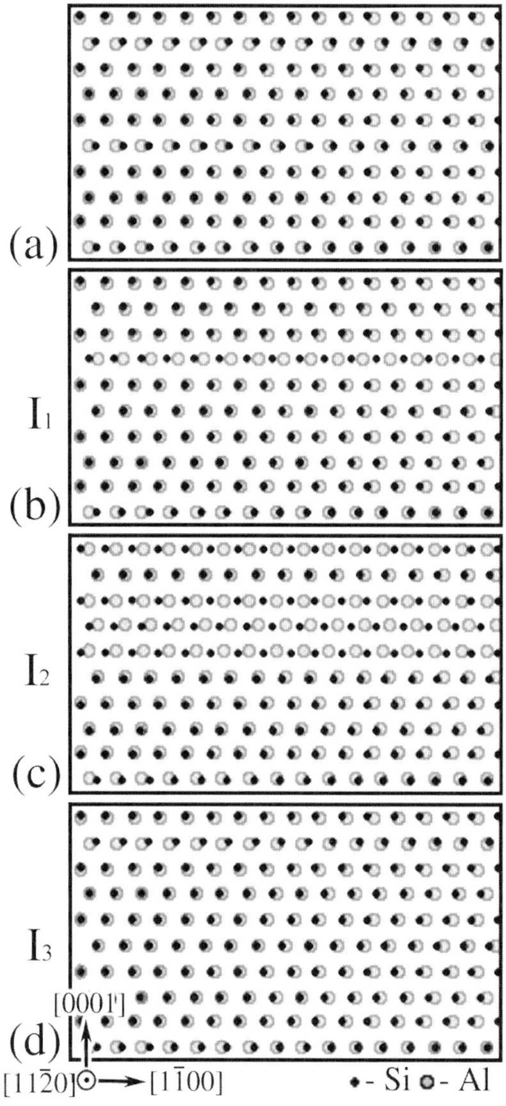

Figure 10.14 Epitaxial relationships between 4H-SiC substrate and AlN buffer layer. Different basal stacking faults are considered to be present in the AlN layer: (a) without stacking fault; (b) with I_1 stacking fault; (c) with I_2 stacking fault and (d) with I_3 stacking fault.

formed at the AlN/4H-SiC interface, the epitaxial relationship between the 4H-SiC and the AlN structure projected on the (11$\bar{2}$0) plane was considered (Figure 10.14a). On this figure, small and large circles represent Si and Al atoms, respectively. Carbon and nitrogen atoms are not shown to sustain image clarity. On each schematic image, the first atom in a second row (lower left corner) of a 4H-SiC projection was aligned with the similar atom on the AlN projection. The lattice mismatch between 4H-SiC and AlN in [0001] and [$1\bar{1}00$] directions is 0.9% and −1.28%, respectively. This means that an aligned projected pattern of the epitaxial relationship will be repeating approximately at each 21 nm in the [$1\bar{1}00$] direction. Figure 10.14a illustrates superposition of ideal 4H-SiC and AlN structures. Each fourth layer of the AlN has a disturbed epitaxial relationship, whereas the first three layers have perfect alignment. Introduction of an I_1 BSF to the AlN structure will make a perfect epitaxy of five layers instead of three (Figure 10.14b).

If the I_2 BSF was introduced instead of I_1, it would create a perfect epitaxy of only four layers (Figure 10.14c). Similarly, the introduction of I_3 BSF should extend the perfect epitaxy to 7 layers (Figure 10.14d). This would suggest that I_3-type BSF has the maximum ability to accommodate the AlN structure to the 4H-SiC, despite the fact that its formation energy is slightly higher than I_1 (I_3 has the second lowest formation energy after I_1). However, to make the most ideal coincidence between the substrate and the buffer layer, the I_3 BSF has to be introduced at each eighth layer. All BSFs formed in the AlN buffer layer will propagate further into the GaN layer because of structural similarity with AlN. Lattice mismatch between AlN and GaN layers should also result in the formation of extra BSFs at the AlN/GaN interface. However, we were not able to reveal any regularity or periodicity in the formation of BSFs. The reason for this can be irregularities on the SiC substrate and AlN buffer layer, such as surface steps and/or small azimuthal rotation of adjacent grains. Therefore, it is not absolutely clear what stabilizes I_3-type BSFs. Observation of termination of these defects would help understand their nature.

10.3 Defect Reduction

10.3.1 Growth of Thick Layers

There are several approaches for dislocation density reduction [35]. It was shown [36, 37, 38] that the number of dislocations decreases with layer thickness, which results from dislocation reactions. The experimental behavior of dislocation density versus layer thickness ‘h’ was found to be proportional to $1/h^n$, where $n = 0.66$ [39]. This means that to obtain dislocation density in GaN in the mid 10^6 cm^{-2}, the thickness of the layer should be in submillimeter range. However, cracking can appear in thick layers due to difference in thermal expansion coefficients between the substrate and the layer. Experimental [40] and theoretical calculations [41] show that screw and edge TDs in GaN are parallel to the [0001] direction and only mixed type of dislocations are inclined to the [0001] direction. When dislocations are parallel to each other, they can interact only if the distance between them is within the range of an image force of a particular dislocation, for example, not more than a few tens of nanometers (e.g. high density of dislocations). It is clear from earlier work that rapid decrease of dislocation density with an increase of a layer thickness is observed only in layers with a high dislocation density. When the distance between the dislocations is larger (e.g. lower dislocation density), much smaller interaction between the dislocations can occur for the same layer thickness, and the rate of decrease of dislocation density becomes smaller. The observation also shows that edge-type TDs do not play a significant part in TD density reduction. Therefore, by growing a thicker layer, a reduction in screw and mixed dislocations can be expected, and this should saturate at some thickness, after which no further reduction is possible. Once this saturation is reached, the growth of a thicker layer will only lead to formation of cracks. For dislocations to interact, their line directions should be inclined, promoting interaction between different types of dislocations.

10.3.2 Lateral Overgrowth

Other methods to reduce dislocation density include lateral epitaxial overgrowth (LEO) [42–45] and pendeo-epitaxy (PE) [45–48]. The LEO technique has been successfully applied to improve structural quality in Si [49], GaAs [50], and InP [51] thin films. In general, growth occurs in several steps in these techniques. First, several hundred nanometers of AlN or GaN buffer layer is grown on top of a substrate, before deposition of GaN a few micrometers thick. Second, parallel stripes of a dielectric SiO_2 or SiN_x mask are deposited, using a photolithographic technique. Then, regrowth of GaN takes place only on the uncovered stripes of GaN regions called *seed* regions. After the thickness of GaN reaches the upper surface of the deposited dielectric height, the GaN stripes start to overgrow the dielectric laterally, forming 'wing' regions. If the growth process is stopped before 'wing' coalescence occurs, an uncoalesced GaN layer is formed. It is anticipated that the overgrown 'wings' will have drastically reduced threading-dislocation density, because TDs mainly propagate along the growth direction. In this case, growth is lateral and line directions of dislocations in the overgrown area will be arranged parallel to the substrate. This would mean that these dislocations, which were formed in the seed area close to the seed edge, would bend and propagate in the overgrown area, parallel to the substrate (Figure 10.15). It turned out that the bending angle of a dislocation depends on dislocation type [52]. Bending angles for edge, mixed, and screw type dislocations were determined based on calculated dislocation energy minima. According to the calculations, vertical propagation is energetically more favorable for screw dislocations. This might explain the

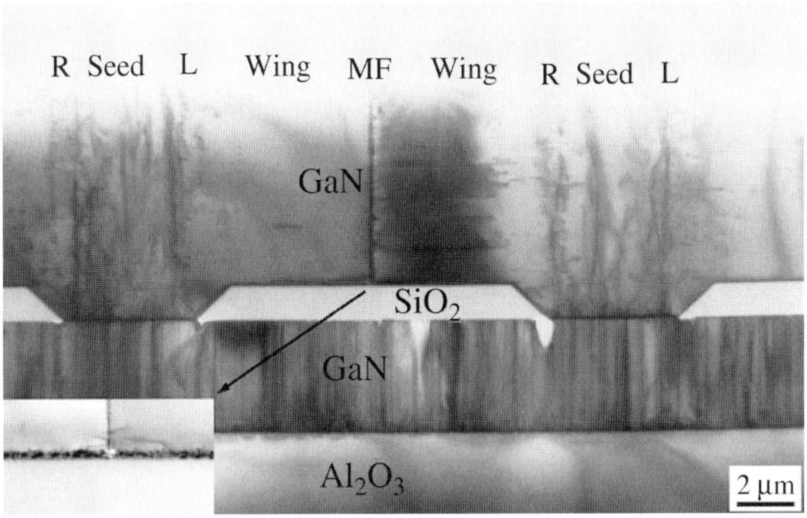

Figure 10.15 Cross-section TEM micrograph of the laterally overgrown GaN layer showing overgrown seed and wing areas: L-left interface at the seed/wing area, R-right interface of the wing/seed area, MF-meeting front of two wings areas with high density of dislocations. Note high density of threading dislocations in the seed area propagating to the sample surface. Different contrast on two sides of the MF indicates tilt/twist across this interface. The inset shows the interface roughness and formation of small voids at the SiO_2 interface.

fact that some dislocations in the wing areas reach the sample surface instead of propagating to the wing side (Figure 10.16). When two wings coalesce, these bent dislocations will interact with each other and a final dislocation (a reaction product of two or more dislocations) will occur at the meeting front of the two wings (Figure 10.16). Experimental observations show a high dislocation density at the meeting front of two wings [44, 45, 53].

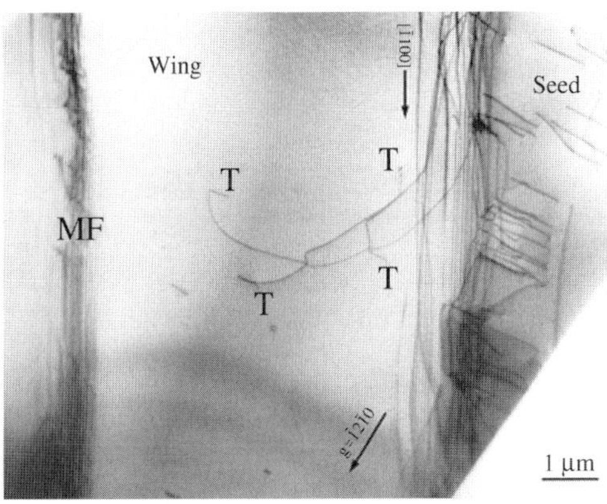

Figure 10.16 Two-beam bright-field image from a plan-view sample showing sets of dislocations at the seed/wing and wing/wing boundaries and two half-loop dislocations extending into the wing region. The half-loop dislocations end in threading segments T.

In most cases some misorientation (tilt/twist) is observed for two adjacent wings and additional new dislocations can be formed at the meeting front to accommodate tilt and twist. If such overgrown areas are used for future devices, then the meeting front of two wings needs to be omitted. Such nonuniformity of dislocation distribution in the overgrown layer is a limiting factor for device application.

To determine local misorientation, large angle convergent beam electron diffraction (LACBED) can be used [54]. In this mode, an image of the studied area and a diffraction pattern (Kikuchi lines) are superimposed (Figure 10.17), therefore, it becomes easy to measure local misorientation [55–56]. Miller indices can be assigned to each Kikuchi line, thereby the position of a particular crystallographic plane in relation to a studied boundary will be known, and tilt and twist across such boundary can be derived depending on the considered Kikuchi line and its relation to the boundary. By measuring a distance between a particular pair of Kikuchi lines, a relation between this distance and the Bragg angle can be determined. The length of shift of a Kikuchi line across the boundary, like the one marked by double arrow on Figure 10.17, will give a tilt angle on two sides of the boundary since this line is oblique to the boundary.

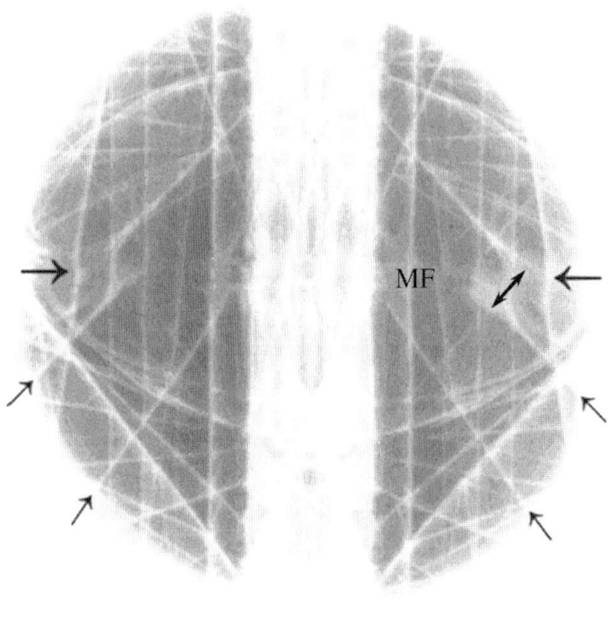

Figure 10.17 LACBED pattern showing the crystal rotation across the meeting front (→ MF ←) of two wings. A tilt across the boundary of about 0.5° is revealed by displacements of the obliquely inclined Kikuchi lines marked by arrows. The double arrow indicates a shift of the Kikuchi line (same indices) across the boundary.

10.3.3 Pendeo-epitaxial Layers

In pendeo-epitaxial samples [46–48] GaN posts (also called *seed* regions) are produced by etching away unmasked GaN layer areas. PE takes place on vertical faces of the 'seed' regions forming 'wing' regions hanging above a bare substrate (a schematic of such growth is shown in Figure 10.18). After a certain growth time, adjacent 'wings' meet each other and form a continuous GaN layer with reduced dislocation density in the 'wing' regions [56–58]. However, there is no TD density reduction in the 'seed' regions, owing to vertical growth in these areas and propagation of dislocations formed at the interface to the sample surface (Figure 10.19). Therefore, similar to laterally overgrown layers, device fabrication is limited only to 'wing' areas with the exception of the meeting fronts (MF) of two wings. Very often, a higher density of impurities covers the growth front of wings and two fronts are prevented from complete coalescence, resulting in voids forming at the meeting front [57, 59]. These observations apply for both lateral overgrowth and PE [58].

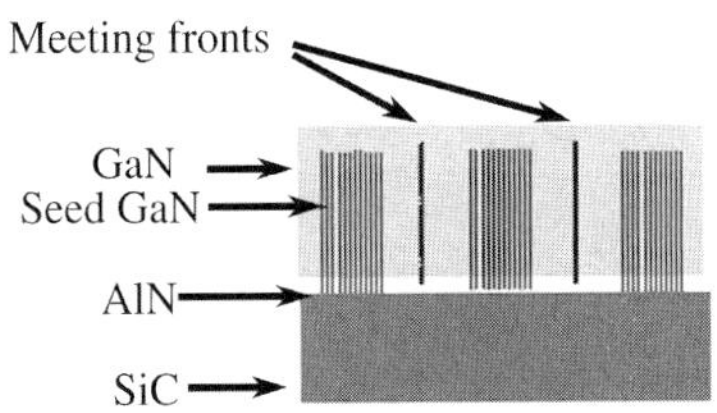

Figure 10.18 Schematic diagrams of the pendeo-epitaxial layers. Vertical lines indicate defects in the seed area propagating to the sample surface.

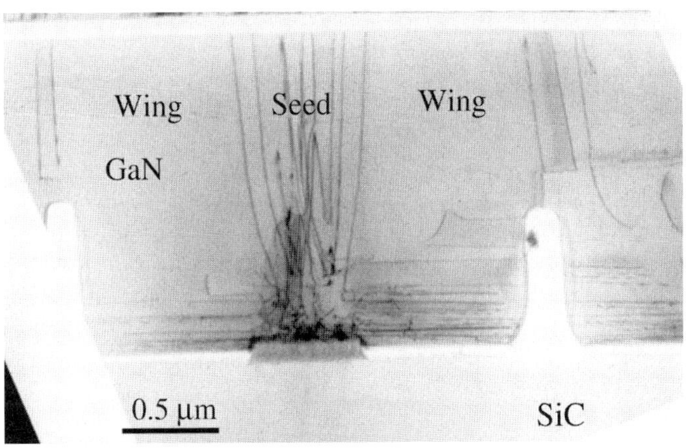

Figure 10.19 Cross-section micrograph from the pendeo-epitaxial GaN layer showing defect distribution over the seed and in the wings. Note the uniform contrast indicating that the misorientation in this sample is small.

Lateral overgrowth and pendeo-epitaxial methods result in lowering of dislocation densities ($10^6 \div 10^7$ cm^{-2}) at least by two or three orders of magnitude in the overgrown areas [42–48, 53, 55–58, 60]. However, MFs where two adjacent 'wings' meet each other have high dislocation density, comparable to, or even higher than that observed in conventionally grown GaN layers on SiC or Al_2O_3. Nevertheless, implementation of LEO or PE techniques drastically improves the nitride laser diode performance [61].

10.4
Application of Pendeo-epitaxy for the Layers Grown on Nonpolar Substrates

As discussed at the beginning of this chapter, the growth of GaN along *c* axis results in a red-shift of the photoluminescence peak and decrease in the peak intensity [62–64], which can be avoided by growth of GaN-based epilayers in nonpolar orientations (*a* plane or *m* plane) [65–67]. We showed in the first part of this chapter, that such growth leads to a high defect density (mainly stacking faults and dislocations). Once the BSFs nucleate at substrate/buffer layer interface, they propagate to the epilayer surface making the approach of growing thicker layers for reducing the defect density useless. Therefore, LEO and Pendeo techniques need to be used to make these layers applicable for devices, since defects play an important role in the performance of GaN-based device structures.

In the samples overgrown on nonpolar substrates, the *c* plane is aligned along the growth direction of the overgrown layer (wings). Therefore, wings grow in two opposite polar directions. Consistent with earlier studies, the growth rate in opposite polar directions [0001] and [$000\bar{1}$] is different. On the basis of this knowledge, one can expect different widths for the two wings [55], as is shown on Figure 10.20a. This figure shows a bright-field image of an uncoalesced *a*-plane PE layer in cross section. In total, the layers grew to a thickness of 2.7 μm. One can also notice the difference in width of the 'wings' on the left and right sides of the 'seed' area.

Convergent beam electron diffraction (CBED) was applied to study growth polarity of this uncoalesced sample. A set of experimental CBED patterns (only one is shown in Figure 10.20b) were taken for different sample thicknesses and compared to the patterns simulated for the same thicknesses. The analysis showed that wings grown with Ga-polarity are much wider (Figure 10.20b) and the width ratio of the wing grown with Ga-polarity was 17 times larger than wings grown with N-polarity (the example is shown on Figure 10.20a). These findings agree with previous experiments for epitaxial growth of GaN, which show that GaN samples grown by metal organic chemical vapor deposition (MOCVD) grow faster for Ga-growth polarity [68]. Samples grown with N-polarity usually grow at a much lower rate. To reach a comparable growth rate, different growth parameters are required for Ga- and N-polarities. In our overgrown samples it is impossible to use different growth temperatures

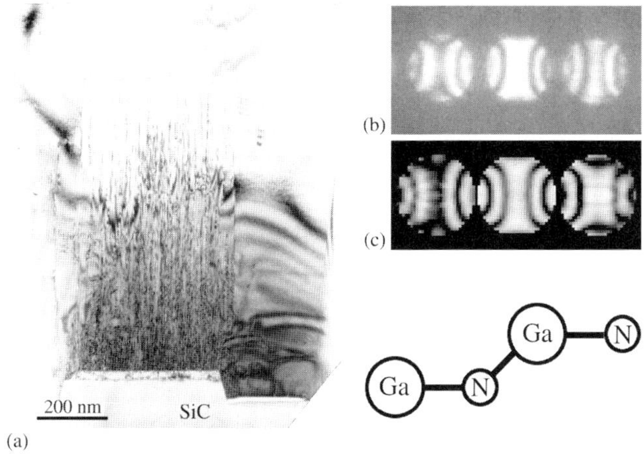

Figure 10.20 TEM micrograph of a cross-section uncoalesced pendeo-epitaxial GaN layer grown on *a*-plane 4H SiC. High density of defects (mainly stacking faults (SFs) and dislocations) is observed in the seed area; (b) CBED pattern taken in the wider wing and (c) simulated CBED pattern for the sample thickness $t = 240$ nm indicating that the wider wing is growing with Ga-polarity as indicated schematically. Bending of dislocations occurs in the Ga-wing in the area close to the substrate.

or different nitrogen overpressure for the opposite wings since the two wings grow simultaneously. Therefore, the resulting structures have wider Ga-polarity wings.

In the PE layers, growth of the wings started from the slightly inclined sidewalls of the lithographically prepared seeds (SiC + AlN + GaN) and partial side growth started also from the inclined SiC. For a wing grown with Ga-polarity, one can notice a gap between the overgrown wings and the substrate, confirming lateral growth without interaction with the substrate, as required for pendeo growth [46–48]. However, when the wing reaches a width of 500 nm, it starts to touch the substrate suggesting possible tilt (Figure 10.21). Three CBED patterns were taken from different areas of the sample: a wing area grown with Ga-polarity (Figure 10.22a), the central part (seed area) (Figure 10.22b), and a wing grown with N-polarity (Figure 10.22c). To compare these CBED patterns the perfect arrangement of intensity rings for the $[1\bar{1}00]$ pattern was set on the wing grown with Ga-polarity. Then the sample was shifted, with the same electron beam and a goniometer setting, and subsequent patterns were obtained from other areas (the central part and N-wing). The pattern from the central area is diffuse, but the pattern taken on N-wing was clear, showing a smaller number of defects. Analyzing these two patterns, one can notice a tilt in $[11\bar{2}0]$ direction and twist along [0002] direction for the N-wing compared to the Ga-wing, suggesting a small tilt and twist between these two wings. One can also notice that the CBED pattern from the seed area is very diffuse. This is owing to the high density of dislocations and stacking faults in this area of the sample. Only a diffuse pattern can be observed owing to a strong interaction between the strain field of the defects and the electron beam. The presence of a single dislocation can cause a splitting in a particular Kikuchi line. This can be observed in Figure 10.17. From the multiplication of the number of nodes and the Miller indices of a Kikuchi line, a Burgers vector of such dislocation can be determined [69]. Since the number of dislocations in this seed area is large, and the strain field of these dislocations interact with each other, their Burger's vector cannot be determined using this method.

Misorientation between two wings was also confirmed by LACBED studies (Figure 10.23). In this figure, LACBED patterns were taken from both wings (Figure 10.23a and c) and the seed area (Figure 10.23b) for the same beam tilt and a goniometer setting. Since the LACBED pattern is a superposition of the image and Kikuchi lines, the top surface of the uncoalesced area (both

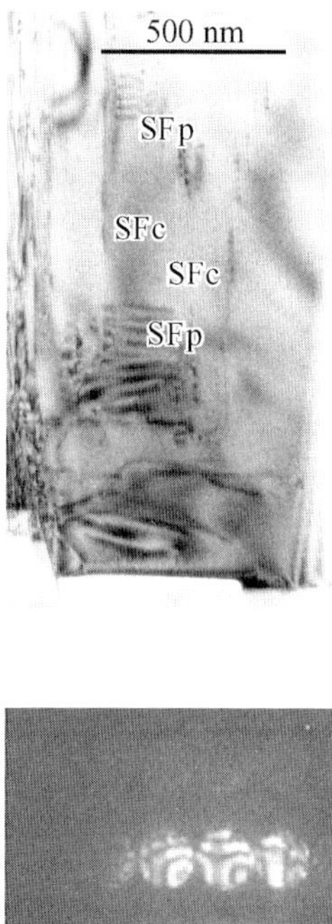

Figure 10.21 The micrograph showing Ga-polarity wing and a small part of the seed area with high defect density. Note a gap between the wing and the substrate (white contrast). This gap decreases after the wing reaches the width of about 500 nm. At this point, the wing starts to touch the substrate and indicates possible tilting of the wing. Note also a bending of the dislocations, which were close to the edge of the seed.

Figure 10.22 (a) CBED pattern from the wing grown with Ga-polarity; (b) the seed area and (c) the wing grown with N-polarity. Note a substantial deviation (tilt and twist) in (b) and (c) compared to (a) area.

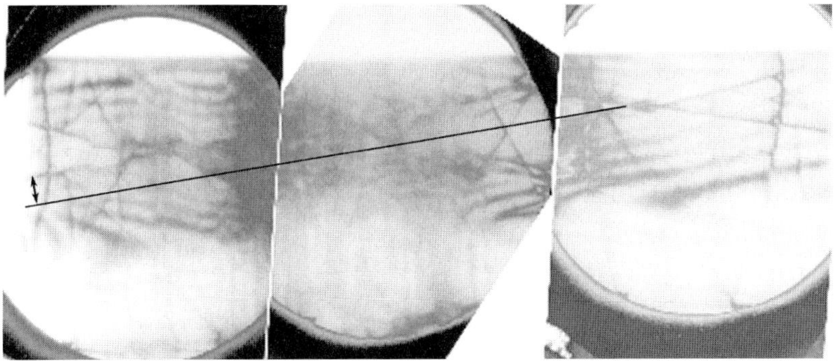

Figure 10.23 (a) LACBED pattern from the wing grown with Ga-polarity; (b) the seed area and (c) the wing grown with N-polarity. Note that Kikuchi lines disappear in the seed area as a result of interaction with defects. A double arrow in (c) indicates the deviation of the same Kikuchi lines between two wings grown with opposite polarities.

wings grown with the opposite polarity on two sides of the seed) can be easily recognized. If the surface of the overgrown area from each of the three areas is aligned, it can be noticed that the Kikuchi lines are shifted, especially when comparing the wing grown with Ga-polarity (Figure 10.23a) and the wing grown with N-polarity (Figure 10.23c). Since the sample was tilted to a particular orientation, where low index Kikuchi lines can be recognized and used as an internal standard, the dependence between the distance between the same Kikuchi lines and the tilt/twist angle can be determined. By measuring the shift (such as marked by the arrow in Figure 10.23c) of the Kikuchi lines (with the same Miller indices) between both wings, one can determine a tilt of 0.25° as well as some twist (0.09°) between wings grown with Ga- and N-polarity.

Since the two wings in these PE layers are growing with opposite polarities, it is not easy to make them to coalesce. Usually, it requires a longer growth period (i.e. larger layer thickness). When coalescence is complete, one can see that this occurs mainly owing to faster growth of the wings with Ga-polarity because this wing is much broader than the N-wing (Figure 10.24a). In many cases, a small gap between two wings can be present, and it is difficult to distinguish between uncoalesced wings and a still existing small gap between wings [70]. The advantage of the presence of such a small gap is the lack of dislocations with a vertical line direction at these MFs.

A high density of SFs formed on the basal plane of hexagonal GaN was observed in cross-sectional TEM images from 'seed' areas. These BSFs are visible in a bright-field image obtained under two-beam conditions with $\mathbf{g} = 1\bar{1}00$ as thin lines arranged perpendicular to the c-direction of the wurtzite GaN (Figure 10.24b), similar to that observed during direct growth and described in the first part of this chapter. The measured density of SFs is $\sim 1.3 \times 10^6\ \mathrm{cm}^{-1}$.

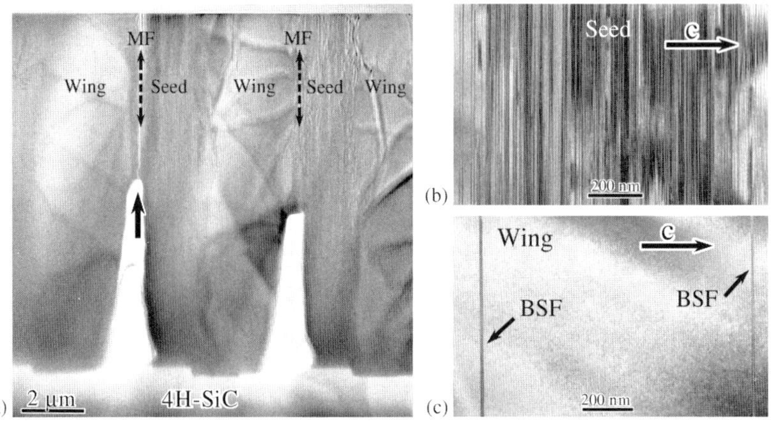

Figure 10.24 (a) Cross-section TEM micrograph of a coalesced *a*-plane pendeo-epitaxial GaN layer grown on $(11\bar{2}0)$ 4H-SiC substrate. The seeds, wings and MFs of two wings are indicated. Note that wings to the left of the MF are much wider (17 times) than wings on the right. Often cracks (one is indicated by the arrow) are formed at the MFs. Note the formation of large voids; (b) High density of defects formed in the seeds is seen on higher magnification; (c) Basal stacking faults present in one of the wings.

Figure 10.24c represents a bright-field image of a 'wing' area obtained with the same two-beam condition as Figure 10.24b. A drastic decrease in the density of BSFs in the 'wing' areas in comparison to the 'seed' areas can be noticed. Arrows on Figure 10.24c mark two BSFs. The average density of BSFs measured over a number of 'wings' is $\sim 1.2 \times 10^4\ cm^{-1}$.

10.5
Application of Lateral Overgrowth for *a*-plane GaN Layers Grown on the *r* plane of Al_2O_3

For lateral overgrowth of $(11\bar{2}0)$ *a*-plane GaN layers grown on $(1\bar{1}02)$ *r* plane of Al_2O_3, a 1.5 μm thick *a*-plane GaN was deposited to be used as subsequent LEO template, after the low-temperature growth of a GaN nucleation layer. Following this, about 100-nm thick SiO_2 was grown on *a*-GaN template using plasma-enhanced chemical vapor deposition. Using conventional photolithography a striped mask pattern was transformed onto the SiO_2 and oriented along the $[1\bar{1}00]$ direction of the GaN, consisting of 4-μm wide openings and 10–20-μm-wide SiO_2 stripes, to cause the lateral growth fronts along c^+ and c^- directions. The patterned sample was subsequently reloaded into the growth chamber for the overgrowth at 1050 °C for about 3 h. The GaN layer was grown to a thickness of 1 μm (Figure 10.25a). Formation of

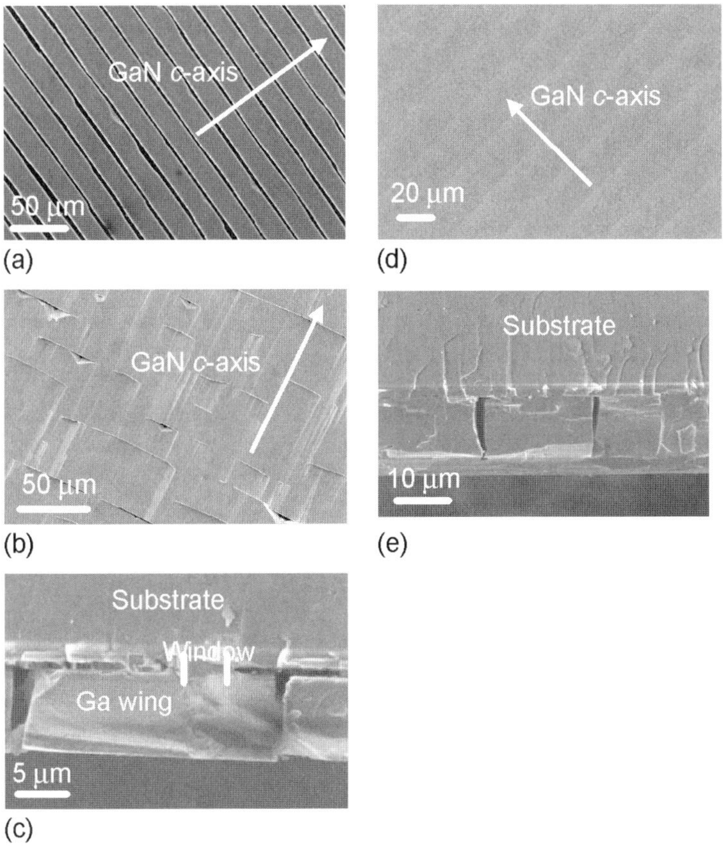

Figure 10.25 SEM images of the laterally overgrown GaN layer using one-step growth at 1050 °C (a) plan-view image of the layers after short time of growth (0.5 hour). There is no coalescence between the wings; (b) Plan-view of the layers grown 3 h revealing noticeable cracking; (c) The layers from (b) in cross section confirming presence of cracks and uneven sample surface; (d) SEM image of a plan-view sample grown for 2 h at 1000 °C. This sample was overgrown at 1050 °C for another 3 h (two-step growth) and is shown in cross section in (e). Note overgrowing of cracks and a smooth sample surface. The arrows indicate *c* axis of GaN.

voids was observed at the MFs of the two wings (Figure 10.26). In some areas these voids were overgrown, and a continuous layer was formed when the total layer thickness reached 0.8 µm, but there were areas where a wing with Ga-polarity grew faster along *c* plane and some cracks were formed (Figure 10.25b). This can be observed in cross section on an optical micrograph (Figure 10.25c). It was noticed that the two sides of the void were not parallel to each other. The side of the void on the N-polarity wing was practically vertical to the substrate, but some steps were clearly visible in the Ga-polarity wing. These steps were associated with the presence of BSFs.

Figure 10.26 Cross-section TEM micrograph showing a coalescence of the N- and Ga-polarity wings and high density of defects in the seeds present in the laterally overgrown GaN layer which was grown at 1050 °C (one-step growth). Bending contours are due to bending of thin TEM sample. Note some defects (mainly stacking faults) present in the Ga-wing (vertical lines).

Stacking faults formed on basal *c* planes were observed in both wings (Figure 10.27), and their density decreased more than two orders of magnitude compared to the density of these faults observed in the seed area. Interaction of BSFs with a PSF leads very often to a decrease in the density of BSFs in the upper part of the layer (Figure 10.28). It was also noticed that the ratio of the wing width grown with Ga-polarity to that grown with N-polarity became smaller (5 : 1) in comparison to that observed in the PE layers grown on SiC (17 : 1). In the LEO layers, much wider wings were formed for the growth with N-polarity in comparison to the same wings in the PE layers. When two wings (Ga- and N-polarity) coalesced in the LEO layers, dislocations were formed at the MF (Figure 10.29). Since different diffraction conditions have to be applied for the disappearance of their contrast, one needs to conclude that their Burgers vector was also different. It is assumed that one set of dislocations was created to accommodate twist and another to accommodate tilt between these two wings.

LACBED studies were applied to observe plane bending in the N-wing close to the void wall (Figure 10.30a). In this area $\bar{2}110$ Kikuchi lines are almost parallel to the substrate, but strong inclination of these lines is observed above the tip of the void (Figure 10.30b). Higher order of $\bar{2}110$ Kikuchi lines, and their position at the meeting front (from a coalescence point on the top of a void to the sample surface) of these two wings, clearly show a decrease in this displacement toward the sample surface (Figure 10.30c). This would suggest that an increase of the total thickness of the layers might decrease the existing displacement of two wings.

The same LACBED method using the set of systematic 0002 and $11\bar{2}0$ Kikuchi lines was used to measure tilt and twist between two wings (Figure 10.31). The 0002 Kikuchi lines were distorted only at the meeting front (Figure 10.31a). The $\bar{2}110$ Kikuchi lines stayed parallel to the substrate in the area adjacent to the SiO_2 mask and further along the void walls. However,

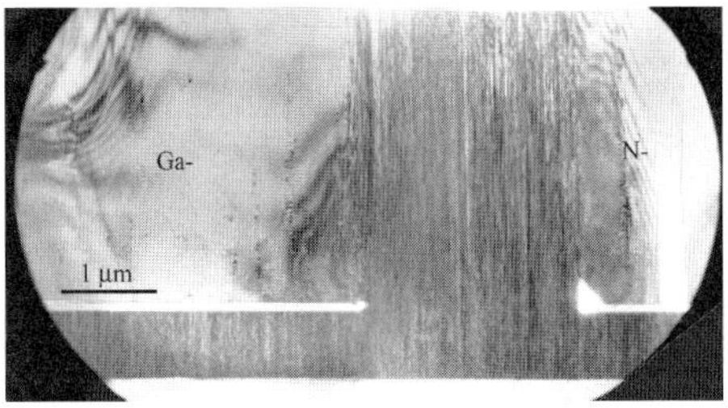

Figure 10.27 TEM cross-section micrograph showing drastic decrease of defects in the wings compared to the seed area.

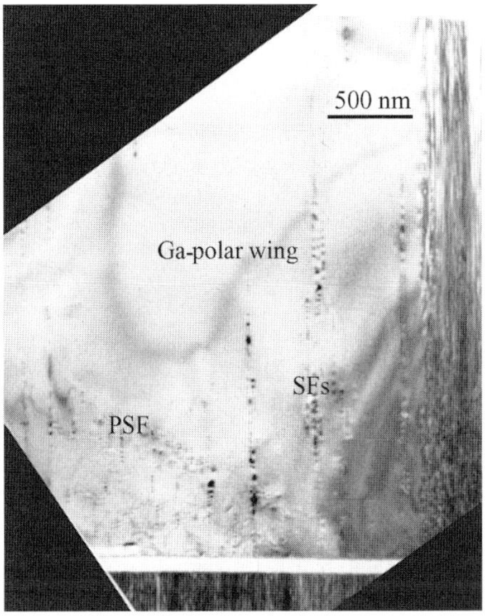

Figure 10.28 A TEM micrograph showing BSFs and PSFs present in the Ga-wing. Some BSFs are terminated by PSFs.

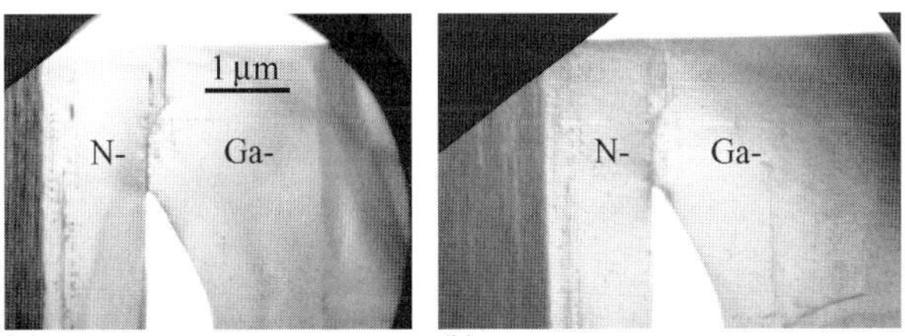

(a) (b)

Figure 10.29 Cross-section micrograph showing formation of dislocations at the meeting front of two wings. These two images were recorded with two different diffraction conditions. Note that a dislocation formed above the void is in contrast in both cases, but the dislocations in the upper part of the MF are in contrast in (a) but out of contrast in (b). This suggests that two different sets of dislocations were formed in this area to accommodate tilt and twist between the wings.

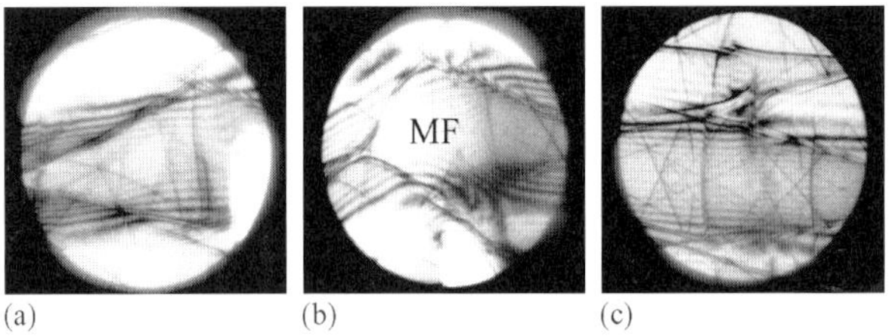

Figure 10.30 LACBED pattern at the meeting front of two wings; $(\bar{2}110)$ Kikuchi lines: (a) left side of the void (N-wing); (b) at MF slightly above the void. Note upward bending of Kikuchi lines; (c) at MF close to the surface where only slight bending is observed.

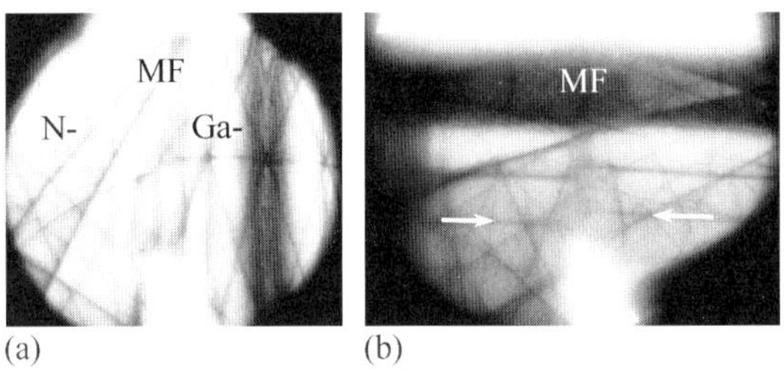

Figure 10.31 (a) LACBED patterns with (0002) Kikuchi lines parallel to the meeting front (MF) between two wings (b) A set of $(\underline{2}110)$ Kikuchi lines perpendicular to the MF (b). The arrows indicate the shift of $(\underline{6}330)$ Kikuchi lines across the boundary.

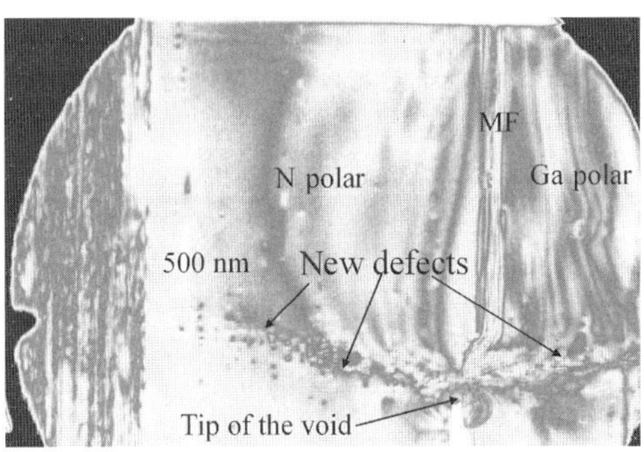

Figure 10.32 Formation of a new boundary on the tip of the void appearing in the two-step grown samples.

above the void all oblique and $\bar{2}110$ lines perpendicular to the boundary were shifted across the boundary. By measuring this shift, 0.3° tilt and 0.1° twist was observed. Rotations about the boundary normal were found to be smaller than tilts and vary from one region to another.

To decrease tilt/twist between the wings grown with opposite polarities, a two-step growth procedure was applied [71]. After the overgrowth reached a thickness of 2 μm at 1000 °C (about 2 h), the growth temperature was raised to 1050 °C for another 3 h. This allowed growth of the remaining part of the layer with a higher rate, to a total thickness of 5 μm. In this case, coalescence was practically observed at each MF of two wings grown with the opposite growth polarity. The measured tilt/twist was slightly smaller than that measured for the one-stage growth, and the width ratio of a Ga-polar wing to an adjacent N-polar wing was about 3 : 1. Unfortunately, when these wings coalesce, a grain boundary (Figure 10.32) is formed on a prismatic plane. Such boundaries sometime eliminate BSFs propagating from the substrate, but often they can be a source of new defects propagating to the surface.

10.6 Summary

This chapter consists of two major parts. In the first part, defects formed in *a*-plane grown GaN are characterized, and in the second, the results of defect reduction due to lateral overgrowth and pendeo-epitaxy are described. TEM was used for defect characterization for cross-sectional and plan-view *a*-plane GaN grown on *a*-plane 4H-SiC with AlN buffer layer, and also on *r*-plane Al_2O_3. The main defects in these layers are basal and prismatic stacking faults, in addition to perfect dislocations. Three types of low-energy stacking faults were described: I_1, I_2 and I_3, and their density in GaN totaled about 1.6×10^6 cm^{-1}. Both I_1 and I_2 types of stacking faults were terminated by Frank–Shockley and glissile Shockley partial dislocations with Burgers vectors $\mathbf{b} = 1/6\langle 2\bar{2}03\rangle$ and $\mathbf{b} = 1/3\langle 1\bar{1}00\rangle$ and their densities were 4.0×10^{10} cm^{-2} and 0.4×10^{10} cm^{-2}, respectively. It was also described how BSFs were terminated by PSFs. Since defects act as nonradiative recombination centers, their low density is desired for device application. The effects of LEO and pendeo-epitaxial samples on defect reduction in samples grown in nonpolar orientation show that it is much harder to eliminate SFs in these layers in comparison to dislocations, the main defects observed in *c* plane grown GaN [55]. However, comparison of the density of defects in the seed and wing areas showed a drastic decrease in the density of defects in the wing area, making them useful for device applications. It was also explained in this chapter that coalescence of two wings in the overgrown nonpolar samples is much more difficult since the two wings grow in opposite polar directions. Usually, some voids are formed at the MF between two wings. Formation of these voids can be influenced by impurities collected on the growth fronts of the wings, especially in the areas close to the substrate. In most cases, these voids are overgrown. Some tilt and twist between the two wings is observed. Since the growth rate in the N- and Ga-polar directions are different, and growth of these wings is taking place simultaneously, the resulting structures have much wider wings in the

Ga-polar direction. However, by adjusting the growth parameters (two-step growth), the Ga-polarity to N-polarity wing ratio can be as low as 3 : 1 and the sample surface is flat, as desired for application of this material in devices.

Acknowledgments

This research was supported by the Air Force Office of Scientific Research under the order No. OGMORD 52K44134 through the U.S Department of Energy under Cu. DE-ACOZ-05CH11231. The use of the facilities at the National Center for Electron Microscopy at Lawrence Berkeley National Laboratory is greatly appreciated.

References

1. Davydov, V.Y., Klochikhin, A.A., Seisyan, R.P. *et al.* (2002) *Physica Status Solidi B*, **229**, R1.
2. Wu, J., Walukiewicz, W., Yu, K.M. *et al.* (2002) *Applied Physics Letters*, **80**, 3967.
3. Yim, W.M., Stofko, E.J., Zanzucchi, P.J. *et al.* (1973) *Journal of Applied Physics* **44**, 292.
4. Perry, P.B. and Rutz, R.F. (1978) *Applied Physics Letters*, **33**, 319.
5. Nakamura, S. (1998) *Science*, **281**, 956.
6. Kahn, M.A., Yang, J.W., Simin, G. *et al.* (1999) *Applied Physics Letters*, **75**, 2806.
7. Nishida, T. and Kobayashi, N. (2001) *Physica Status Solidi A*, **188**, 113.
8. Nortrup, J.E. (2001) *Applied Physics Letters*, **78**, 2288.
9. Nothrup, J.E. (2002) *Physical Review B66*, 045204.
10. Jena, D., Gossard, A.C., and Mishra, U.K. (2000) *Applied Physics Letters*, **76**, 1707.
11. Zhonghai, Y., Johnson, M.A.L., Brown, J.D. *et al.* (1999) *MRS Internet Journal of Nitride Semiconductor Research*, **4S1**, G4.3.
12. Cherns, D., Henley, S.J., and Ponce, F.A. (2001) *Applied Physics Letters*, **78**, 2691.
13. Stephan, T. (2002) *Physica Status Solidi A*, **194**, 568.
14. Amano, H. (2003) *Physica Status Solidi*, **195**, 491.
15. Karpov, S.Y. and Makarov, Y.N. (2002) *Applied Physics Letters*, **81**, 4721.
16. Bernardini, F. and Fiorentini, V. (2001) *Physica Status Solidi A*, **188**, 113.
17. Jin Seo, I., Kollmer, H., Off, J. *et al.* (1998) *Physical Review B*, **57**, R9435.
18. Deguchi, T., Sekiguchi, K., Nakamura, A. *et al.* (1999) *Japanese Journal of Applied Physics Part 2-Letters*, **38**, L914.
19. Waltereit, P., Brandt, O., Trampert, A. *et al.* (2000) *Nature*, **406**, 865.
20. Craven, M.D., Lim, S.H., Wu, F. *et al.* (2002) *Applied Physics Letters*, **81**, 469.
21. Haskell, B.A., Wu, F., Matsuda, S. *et al.* (2003) *Applied Physics Letters*, **83**, 1554.
22. Wagner, B.P., Preble, E.A., Reitmeier, Z.J. *et al.* (2004) *Materials Research Society Symposium Proceedings*, **798**, Y10.34.
23. Craven, M.D., Wu, F., Chakraborty, A. *et al.* (2004) *Applied Physics Letters*, **84**, 1281.
24. Kuokstis, E., Chen, C.Q., Gaevski, M.E. *et al.*(2002) *Applied Physics Letters*, **81**, 4130.
25. Ng, H.M. (2002) *Applied Physics Letters*, **80**, 4369.
26. Wagner, B.P., Reitmer, Z.J., Park, J.S. *et al.* (2006) *Journal of Crystal Growth*, **290**, 504.

27. Hull, D. and Bacon, D.J. (1984) *Introduction to Dislocations*, Pergamon Press, Oxford.
28. Blank, H., Delavignett, P., Gevers, R., and Amelinckx, A. (1964) *Physica Status Solidi*, **7**, 747.
29. Zakharov, D.N., Liliental-Weber, Z., Wagner, B. *et al.* (2005) *Physical Review B*, **71**, 235334.
30. Stampfl, C. and Van de Walle, C.G. (1998) *Physical Review B*, **57**, R15052.
31. Northrup, J.E. (1998) *Applied Physics Letters*, **72**, 2316.
32. Programs are available from http://www.totalresolution.com.
33. Drum, C.M. (1965) *Philosophical Magazine*, **11**, 313.
34. Daudin, B., Widmann, F., Feuillet, G. *et al.* (1997) *Materials Science and Engineering B*, **50**, 8.
35. Jasinski, J. (2005) *Physica Status Solidi C*, **2**(3), 994.
36. Albrecht, M., Nikita, I.P., Nikolaev, A.E. *et al.* (1999) *Physica Status Solidi A*,**176**, 456.
37. Mathis, S.K., Romanov, A.E., Chen, L.T.F. *et al.* (2000) *Physica Status Solidi A*, **179**, 125.
38. Jasinski, J. and Liliental-weber, Z. (2002) *Journal of Electronic Materials*, **31**, 429.
39. Speck, J.S., Brewer, M.A., Beltz, G. *et al.* (1996) *Journal of Applied Physics*, **80**, 3808.
40. Golan, Y., Wu, X.H., Speck, J.S. *et al.* (1998) *Applied Physics Letters*, **73**, 3090.
41. Sheldon, P., Jones, K.M., Al-Jassim, M.M., and Yacobi, B.G. (1998) *Journal of Applied Physics*, **63**, 5609.
42. Marchand, H., Ibbetson, J.P., Fini, P.T. *et al.* (1998) *Journal of Electronic Materials*, **3**, 3.
43. Sakai, A., Sunakawa, H., and Usui, A. (1997) *Applied Physics Letters*, **71**, 2259.
44. Liliental-Weber, Z., Benamara, M., Swider, W. *et al.* (1998) *MRS Internet Journal of Nitride Semiconductor Research*, **4S1**, G4.6.
45. Liliental-Weber, Z. and Cherns, D. (2001) *Journal of Applied Physics*, **89**, 7833.
46. Zheleva, T., Smith, S., Thomson, D. *et al.* (1999) *Journal of Electronic Materials*, **28**(4), L5.
47. Zheleva, T., Smith, S., Thomson, D. *et al.* (1999) *Journal of Electronic Materials*, **4S1**, G49.
48. Davis, R.F., Gehrke, T., Linthicum, K.J. *et al.* (2001) *Journal of Crystal Growth*, **225**, 134.
49. Rathman, D.D., Silversmith, D.J., and Burns, J.A. (1982) *Journal of Electrochemical Society*, **129**, 2303.
50. Tsaur, B.Y., McClelland, R.W., Fan, J.C.C. *et al.* (1982) *Applied Physics Letters*, **41**, 347.
51. Vohl, P., Bozler, C.O., McClelland, R.W. *et al.* (1982) *Journal of Crystal Growth*, **56**, 410.
52. GradeŠak, S., Stadelmann, P., Wagner, V., and Ilegems, M. (2004) *Applied Physics Letters*, **85**, 4648.
53. Marchand, H., Wu, X.H., Ibbetson, J.P. *et al.* (1998) *Applied Physics Letters*, **73**, 747.
54. Tanaka, M. and Terauchi, M. (1985) *Convergent-Beam Electron Diffraction*, JEOL LTD.
55. Liliental-Weber, Z., Zakharov, D., Wagner, B., and Davis, R.F. (2006) *Proceedings of SPIE-The International Society for Optical Engineering*, vol. 6121, 612101.
56. Cherns, D. and Liliental-Weber, Z. (2001) *Materials Research Society Symposium Proceedings*, **639**, G.5.2.1.
57. Liliental-Weber, Z., Jasinski, J., Cherns, D. *et al.* (2002) *Materials Research Society Symposium Proceedings*, **693**, 309.
58. Liliental-Weber, Z., Jasinski, J., and Washburn, J. (2002) *Journal of Crystal Growth*, **246**, 259.
59. Liu, Q.K.K., Hoffmann, A., Siegle, H. *et al.* (1999) *Applied Physics Letters*, **74**, 3122.
60. Beaumont, B., Vennegues, Ph., and Gibart, P. (2001) *Physica Status Solidi B*, **227**, 1–43.
61. Ni, X., Fy, Y., Moon, Y.T. *et al.* (2006) *Journal of Crystal Growth*, **290**, 166.
62. Bernardini, F. and Fiorentini, V. (1998) *Physical Review B. Condensed Matter*, **57**, R9427.

63. Deguchi, T., Sekiguchi, K., Nakamura, A. *et al.* (1999) *Japanese Journal of Applied Physics Part 2-Letters*, **38**, L914.
64. Seo, I.J., Kollmer, H., Off, J. *et al.* (1998) *Physical Review B. Condensed Matter*, **57**, R9435.
65. Waltereit, P., Brandt, O., Trampert, A. *et al.* (2000) *Nature*, **406**, 865.
66. Fini, P., Marchand, H., Ibbetson, J.P. *et al.* (1999) *Materials Research Society Symposium Proceedings*, **572**, 315.
67. Kuokstis, E., Chen, C.Q., Gaevski, M.E. *et al.* (2002) *Applied Physics Letters*, **81**, 4130.
68. Piquette, E.C., Bridger, P.M., Bandic, Z.Z., and McGill, T.C. (1999) *Journal of Vacuum Science and Technology*, **17**, 1241.
69. Cherns, D. and Preston, A.R. (1998) *Journal of Electron Microscopy Technique*, **13**, 111.
70. Zakharov, D., Liliental-Weber, Z., Wagner, B. *et al.* (2004) *Materials Research Society Symposium Proceedings*, **798**, 747.
71. Ni, X., Ozgur, U., Fu, Y., Biykii, N., Morkoc, H. and Liliental-Weber, Z. (2006) Materials Research Society Symposium Proceedings, vol. 955-E, 0955-I07-51.

11
Defects and Interfacial Structure of *a*-plane GaN on *r*-plane Sapphire

Roland Kröger

11.1
Introduction

Crystal imperfections, such as point defects, line defects, planar and volume defects are limiting factors for the applicability of semiconductors in microelectronics. The impact of these defects on band structure, carrier mobility, and optical properties is even found to create principal obstacles for the applicability of certain materials. One example is the II-VI material system for which a prominent obstacle is the creation of extended defects during device operation, connected with the diffusion of doping-related point defects [1], causing lifetime shortening and making the device unreliable.

In contrast, the quality of group III nitride compounds has been improved tremendously in the past years, so that they became applicable for reliable optoelectronic devices such as light-emitting diodes or laser diodes. However, the strong spontaneous polarization field caused by the charge separation between the Ga and the N for polar *c*-plane oriented growth leads to a significant reduction of the quantum efficiency of such devices owing to a reduced overlap of the electron and hole wave-functions in the quantum well region [2]. This motivates the research of approaches for the growth of nitride-based films having a nonpolar growth plane. For example, the growth along the $[11\bar{2}0]$ axis results in films with zero spontaneous polarization field in growth direction [3, 4] and can be realized on sapphire substrates, which are currently the least expensive substrates available in the market.

Although first deposition experiments of nonpolar GaN films on *r*-plane sapphire were performed as early as in 1987 [5], the defect structure in wurtzite GaN and related compounds has been mainly studied in the past focusing on films oriented along the hexagonal [0001] axis (*c*-axis). Growth techniques such as molecular beam epitaxy (MBE) [1], metal organic vapor phase epitaxy (MOVPE) [6], and hydride vapor phase epitaxy (HVPE) [7, 8] are established for polar as well as nonpolar GaN growth. The structural quality of *a*-plane GaN

Nitrides with Nonpolar Surfaces. Growth, Properties, and Devices. Edited by Tanya Paskova

ISBN: 978-3-527-40768-2

is yet far from being comparable with state-of-the-art *c*-plane films, although high-quality *a*-plane GaN films could be realized by HVPE growth of *c*-plane GaN, which were subsequently sliced in *a*-direction [9]. Heteroepitaxially grown *a*-plane nitride films basically show different defect patterns compared with those found in *c*-plane films. In order to elucidate this point, the present chapter focuses on defects identified in *a*-plane GaN on *r*-plane sapphire and discusses the role of anisotropy of the lattice mismatch between film and substrate. The peculiarities of GaN grown with the nonpolar *a*-plane as growth surface on *r*-plane sapphire are briefly discussed in the following section.

11.1.1 Conventions

For the heteroepitaxial growth of *a*-plane GaN, sapphire is a suitable substrate candidate if it is cut in such a way that the top surface exposes the $(1\bar{1}02)$ plane [10]. Wurtzite GaN has a hexagonal symmetry and belongs to the space group P63*mc*, whereas the corundum structure α-Al_2O_3 (sapphire) is trigonal belonging to the space group R3*c*. The difference in crystal structure and lattice parameters is the cause for an anisotropic lattice mismatch and thus creates an important source for defects formed in the GaN film. The wurtzite structure of the GaN as well as the trigonal structure of the sapphire is conveniently classified by the Miller–Bravais notation, where the lattice planes and the plane normal vectors have the same indexing. To describe GaN growing perpendicular to the $\{11\bar{2}0\}$ planes, it will be labeled as '*a*-plane GaN'. GaN grown perpendicular to the $\{1\bar{1}00\}$ planes is called '*m-plane GaN*' and '*c-plane GaN*' assigns the case for the growth along the $\langle 0001 \rangle$ directions.

The wurtzite unit cell is shown in Figure 11.1a. It consists of two anion and cation hexagonal sublattices, with one sublattice displaced by a vector 1/3[0001]

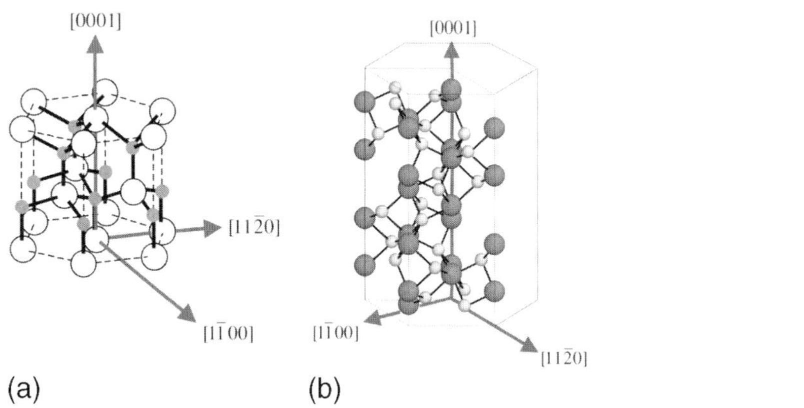

Figure 11.1 (a) The wurtzite structure and crystallographic directions in Miller–Bravais notation. (b) The trigonal structure of α-Al_2O_3.

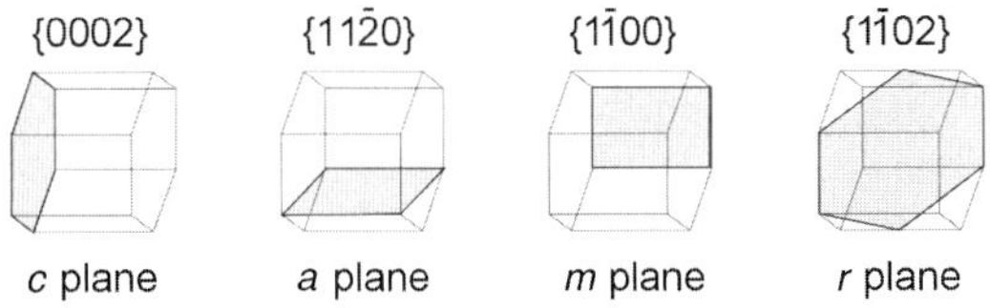

Figure 11.2 Orientations and planes in the hexagonal and trigonal crystal lattice. The respective planes are shaded.

relative to the other sublattice. In Figure 11.1b, the trigonal crystal structure of sapphire is shown. The symmetries of the atoms in the unit cell is such that the directions and planes can also be characterized using the Miller–Bravais notation for the hexagonal symmetries. Different prominent crystal planes in the hexagonal and trigonal crystal structures are shown in Figure 11.2. The basal, or c-planes are given by the {0002} facets, the a-planes by the {11$\bar{2}$0} facets, the m-planes by the {1$\bar{1}$00} facets, and the r-plane by the {1$\bar{1}$02} facets of the hexagonal unit cell.

11.1.2 Lattice Mismatch

The lattice parameters and thermal expansion coefficients at room temperature of GaN and sapphire are shown in Table 11.1 (on the basis of data given in Refs. 11 and 12). For the case of a-plane GaN grown on r-plane sapphire, the translational periods of the respective lattice planes [10] in $[1\bar{1}00]_{\mathrm{GaN}}$ direction is given by $\Delta_{[1\bar{1}00]\mathrm{GaN}} = \sqrt{3}a_{\mathrm{GaN}}$ for GaN and by $\Delta_{[0001]\mathrm{Al_2O_3}} = a_{\mathrm{Al_2O_3}}$ for the sapphire, and in the $[0001]_{\mathrm{GaN}}$ direction $\Delta_{[0001]\mathrm{GaN}} = c_{\mathrm{GaN}}$ for GaN and $\Delta_{[0\bar{1}10]\mathrm{Al_2O_3}} = \sqrt{3a^2_{\mathrm{Al_2O_3}} + c^2_{\mathrm{Al_2O_3}}}$ for the sapphire with a_{GaN}, $a_{\mathrm{Al_2O_3}}$, c_{GaN}, and $c_{\mathrm{Al_2O3}}$ being the corresponding a- and c-lattice parameters. For the $[1\bar{1}00]_{\mathrm{GaN}}$ direction, the respective lattice mismatch is

$$f_{[1\bar{1}00]} = \frac{\Delta_{[1\bar{1}00]\mathrm{GaN}} - \Delta_{[0001]_{\mathrm{Al_2O_3}}}}{\Delta_{[0001]_{\mathrm{Al_2O_3}}}} \quad (1)$$

Table 11.1 Lattice parameters and thermal expansion coefficients at room temperature (perpendicular and parallel to the c-plane) for GaN and sapphire [11, 12].

Material	a (Å)	c (Å)	α_{par} $(10^{-6}\,°\mathrm{C}^{-1})$	α_{perp} $(10^{-6}\,°\mathrm{C}^{-1})$
GaN	3.189	5.185	4.3	8.1
α-Al_2O_3	4.758	12.991	3.17	5.59

The mismatch in $[0001]_{GaN}$ direction is given by

$$f_{[0001]} = \frac{\Delta_{[0001]_{GaN}} - \Delta_{[0\bar{1}10]_{Al_2O_3}}}{\Delta_{[0\bar{1}10]_{Al_2O_3}}} \quad (2)$$

Inserting the lattice parameter values given in Table 11.1 into Equations 1 and 2 results in $f_{[1-100]} = 16.1\%$ and $f_{[0001]} = 1.1\%$. It can be expected that the strong difference in the relative lattice mismatch between *a*-plane GaN and *r*-plane sapphire plays a significant role in the defect formation process during film growth.

An important additional difference between the two materials lies in the thermal expansion coefficients. The room-temperature values for the linear thermal expansion coefficients perpendicular to and in the basal plane for GaN are about 25–30% higher for GaN compared with sapphire [12], which will lead to a stronger expansion of the GaN lattice with respect to the sapphire lattice during deposition, which in general takes place in the temperature regime of 600–1000 °C.

Contrary to the growth of *c*-plane GaN on *c*-plane sapphire, the lattice mismatch between the substrate and the film in the case of *a*-plane GaN growth on *r*-plane sapphire is not spatially isotropic but depends on the crystal direction and respective lattice plane distances. Figure 11.3 shows the projected surface of the *r*-plane sapphire and the surface unit cell associated to it. On top of this, 1.5-unit cells of *a*-plane GaN are superimposed. This representation visualizes the relatively small lattice mismatch in $[1\bar{1}00]$ direction of about

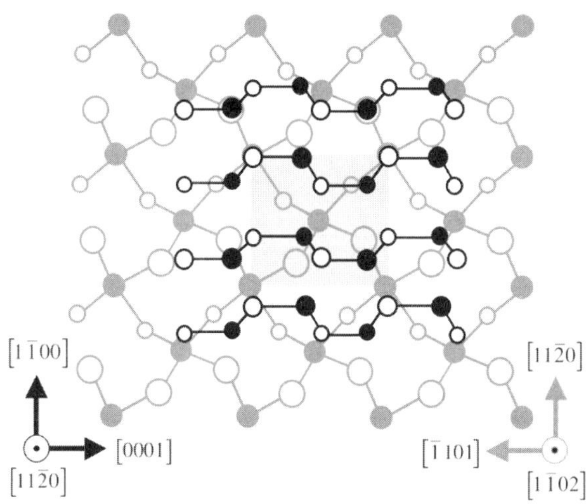

Figure 11.3 Superposition of *a*-plane GaN and *r*-plane α-Al_2O_3: gray filled and empty circles assign Al and O, respectively, and the black filled and empty circles assign Ga and N, respectively. The shaded rectangle indicates the surface unit cell of the *r*-plane sapphire.

1.1% and the comparably large mismatch of 16.1% in [0001] direction. The resulting interfacial structure is discussed in Section 11.3.

11.2 Defects in *a*-plane GaN on *r*-plane Sapphire

The origins of the defect formation in heteroepitaxial growth are connected with the lattice misfit between film and substrate, the different thermal expansion coefficients of the film and the substrate as well as the growth conditions, especially gas flow rates, pressure, and substrate temperature [13]. In *c*-plane GaN a variety of structural defects have been identified and characterized comprising zero-dimensional point defects and impurities [14, 15], one-dimensional line defects [16], such as dislocations, two-dimensional planar defects [17], such as basal and prismatic stacking faults (PSFs), and three-dimensional volume defects, such as inclusions and inversion domains or surface pits [18, 19]. Structural defects can create obstacles for the applicability of the nitride layers in electronic devices. Understanding the defect formation processes and finding ways to avoid defect formation were crucial steps for the development of high-quality *c*-plane GaN. Thus, there is strong interest to obtain similar information on nonpolar nitride films and to compare the defect formation mechanisms with those found for *c*-plane GaN. The structural defects characterized in *a*-plane GaN films grown so far show the special challenges that growers face when trying to improve the crystal quality of the layers. Therefore, this section focuses on the types of defects found in *a*-plane GaN grown on *r*-plane sapphire, namely, point defects, dislocations, and stacking faults, as well as voids and surface pits.

11.2.1 Point Defects and Impurities

The native point defects and impurities are strongly influenced by the growth conditions, that is substrate temperature, flow rates, and partial pressures of precursors during growth. Because these parameters depend on the growth technique, it is expected that the incorporation of point defects, as well as impurities depend on the deposition method. Point defects and impurities affect the optical and electrical properties of the GaN in several ways. They can limit the device performance and lifetime by the formation of deep level traps in the energy gap [20] or they can promote the formation of extended defects such as dislocation loops [21], both reducing the performance and lifetime of devices. As for the optical properties, Ga vacancies were found to limit the nonradiative photoluminescence lifetime significantly [22]. The growth conditions for MOVPE and HVPE GaN differ strongly [13]. Therefore, the study of point defect and impurity concentrations in the films is an important issue. Suitable methods for the determination of point defect character and distribution and for the impurity measurements are positron annihilation spectroscopy (PAS) and time-of-flight secondary ion mass spectrometry (ToF-SIMS), respectively.

11.2.1.1 Point Defects

PAS has already been successfully applied to *c*-plane GaN as well as *a*-plane GaN [22–24]. PAS uses the strong interaction between positrons and neutral or negatively charged vacancies, which act as positron traps. By measuring the positron–electron momentum and the positron lifetime and by observing the narrowing of the Doppler-broadened 511-keV annihilation peak, information can be obtained regarding the concentration of vacancies as well as their charge state. Recently, it was found that the predominant vacancy type in *a*-plane GaN could be identified as V_{Ga}-O_N pairs [25].

The measurement of vacancy distribution [26] was performed along the growth *a*-axis and evaluation of the spectra was carried out using the conventional line-shape parameters S and W [27], determined as fractions of positrons annihilating with low-momentum electrons ($|p_L| < 2.1 \times 10^{-3} m_0c$) and high-momentum electrons ($11 \times 10^{-3} m_0c < |p_L| < 29 \times 10^{-3} m_0c$), respectively. To study the impact of the different growth techniques on the point defect concentration, a representative *a*-plane GaN/*r*-plane sapphire samples grown by HVPE (18-μm thick) on a MOVPE (1–2-μm thick) template layer was investigated. V_{Ga}-O_N pairs created the major type of native point defects also in this sample. Using a typical positron trapping coefficient of $\mu_V = 3 \times 10^{15}\ s^{-1}$ for the Ga vacancy, the concentration of the V_{Ga}-O_N could be determined from the experimental data (see Figure 11.4). The estimated values of the vacancy-complex concentrations were $[V_{Ga}\text{-}O_N] = 1 \times 10^{18}\ cm^{-3}$ in the 1.2-μm-thick metal organic chemical vapor deposition (MOCVD) GaN template and $[V_{Ga}\text{-}O_N] = 2 \times 10^{17}\ cm^{-3}$ in the 5–25-μm-thick HVPE layers. It is remarkable that

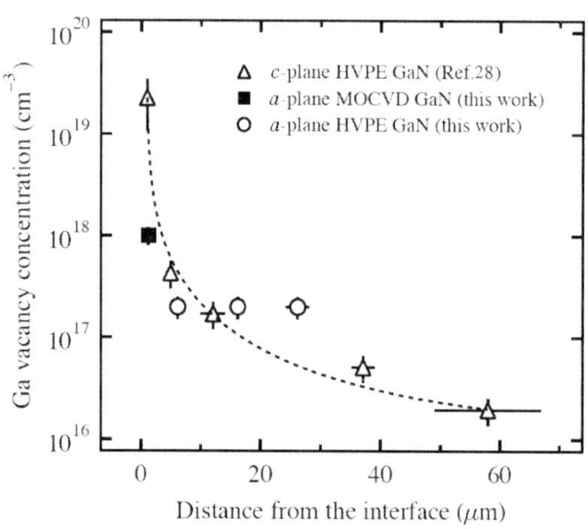

Figure 11.4 Ga vacancy concentration as determined by positron annihilation spectroscopy comparing the results for *a*-plane GaN and *c*-plane GaN films with different sample thickness values [26, 28].

the behavior of the Ga vacancy concentration as a function of distance from the GaN–sapphire interface differs significantly from that in the *c*-plane HVPE GaN in the *a*-plane HVPE GaN layer.

While in the *c*-plane GaN films, the vacancy concentration decreases strongly with increasing film thickness in the *a*-plane GaN. The concentration remains almost on the same level showing about 1 order of magnitude higher vacancy concentration for the MOVPE GaN compared with the HVPE GaN. A possible explanation for this observation could be that the MOCVD growth is clearly more N-rich than the HVPE growth, and hence from kinetics point of view the incorporation of the O_N should be more efficient in the latter in an otherwise defect-free crystal. Following this argumentation, the V_{Ga} formation should be more efficient in MOCVD-grown GaN compared to the HVPE growth [26, 29].

11.2.1.2 **Impurities**

ToF-SIMS was applied to perform line-profile measurements of the element distribution [26]. The qualitative sputter profile is given in Figure 11.5a. It shows Ga_2N, GaO, Cl, and Al signals as a function of sputter depth and shows an approximately homogenous distribution of impurities in both the HVPE and MOVPE sublayers. The Ga_2N and Al signal slopes mark the GaN–sapphire interface.

In order to monitor the qualitative O distribution in the GaN layers and to screen out the effect of O in the sapphire, the GaO signal is plotted in Figure 11.5a indicating a significantly higher O concentration in the HVPE compared with that of the MOVPE GaN. The increased O and Cl signals

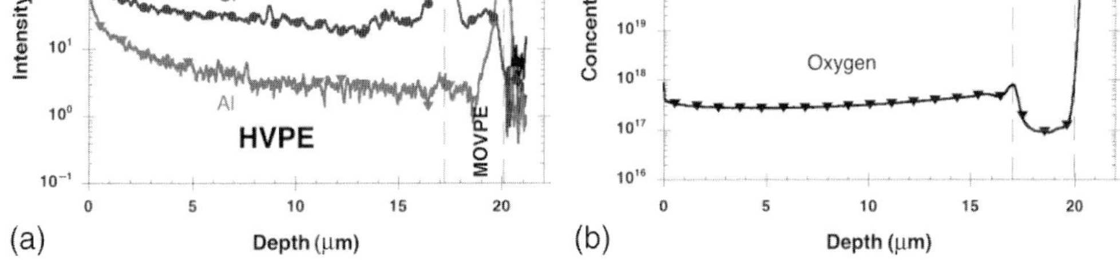

Figure 11.5 ToF-SIMS profiles: (a) intensity profiles for Al, GaO, Cl and Ga_2N and (b) O concentration profile. Note that owing to the different crystallographic structure of the sapphire compared with GaN, the O concentration value in the sapphire is prone to large errors.

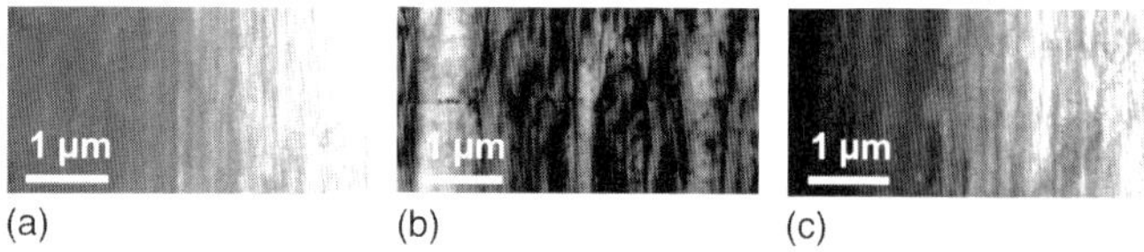

Figure 11.6 Bright-field TEM images taken along the [0001] zone axis (a) from the MOVPE area, (b) the HVPE–MOVPE interface, and (c) the HVPE area [28].

at the HVPE/MOVPE interface are most likely caused by a nonoptimized surface pretreatment of the MOVPE GaN prior to the HVPE deposition. The O concentration, shown in Figure 11.5b, remains almost constant in each of the sublayers with a significant contrast between them.

The quantitative analysis allows an estimation of the absolute O concentration of approximately 1×10^{17} and 3×10^{17} cm^{-3} in the MOCVD GaN and HVPE GaN, respectively. The higher oxygen concentration in the HVPE material as compared with the MOCVD material agrees comparably well with previously observed trends in the residual impurity concentrations in *c*-plane GaN [30, 31]. Most noticeable here is the constant behavior of the O concentration (as well as of the other impurities like H and C, not shown here) along the thickness, which is in strong contrast to the nonuniform distribution of the O concentration observed in conventional GaN grown in *c* direction, which ranges over a few orders of magnitude along a thickness of several micrometers.

Figure 11.6a–c shows $[1\bar{1}00]$ cross-section views of a sample slightly tilted away from the zone axis, so that all structural defects become visible. Figure 11.6a shows an area inside the MOVPE GaN, Figure 11.6b an area at the HVPE/MOVPE interface, and Figure 11.6c shows an area well inside the HVPE GaN layer and about 5 μm apart from the HVPE/MOVPE interface. The interface between the MOVPE and the HVPE GaN shows a dark contrast, which is likely to be connected with the impurity accumulation in agreement with the findings by SIMS. Moreover, the transmission electron microscopy (TEM) results show that the density of extended defects, such as dislocations and stacking faults [in the 10^{10} cm^{-2} range for dislocations and in the 10^5 cm^{-1} range for basal plane stacking faults (BSFs)] remains constant throughout the whole layer and is not affected by the change of growth technique. This is inline with the constant vacancy and oxygen impurity profiles found by PAS and SIMS giving rise to the conclusion that the impurity concentration is coupled with the presence of extended defects such as dislocations and stacking faults, which will be discussed in the next sections.

11.2.2 Dislocations

The impact of dislocations on the optical and electrical properties, especially of *c*-plane GaN, has been investigated in numerous works (see e.g. [32, 33]). They can act as nonradiative recombination centers diminishing the emission from light-emitting devices as well as forming current short-cut paths reducing the lifetime of the device. Furthermore, it was found that dislocations affect the wet-etching during device processing [34]. The dislocations are characterized by their line direction, as well as by their Burgers vectors (BVs), representing the atomic displacement with respect to the perfect lattice, which causes the formation of the dislocation. Figure 11.7 shows the characteristic atomic arrangement of the wurtzite structure and the possible perfect and partial dislocations indicated by dark gray and light gray arrows, respectively. Typical dislocation densities found in heteroepitaxially grown *a*-plane GaN are in the range of 10^{10} cm^{-2}, which could be reduced somewhat using the lateral epitaxial overgrowth technique [8].

There exists a strong difference in the predominant types of dislocations found in *c*- and *a*-plane GaN grown on sapphire. In general, the *c*-plane GaN

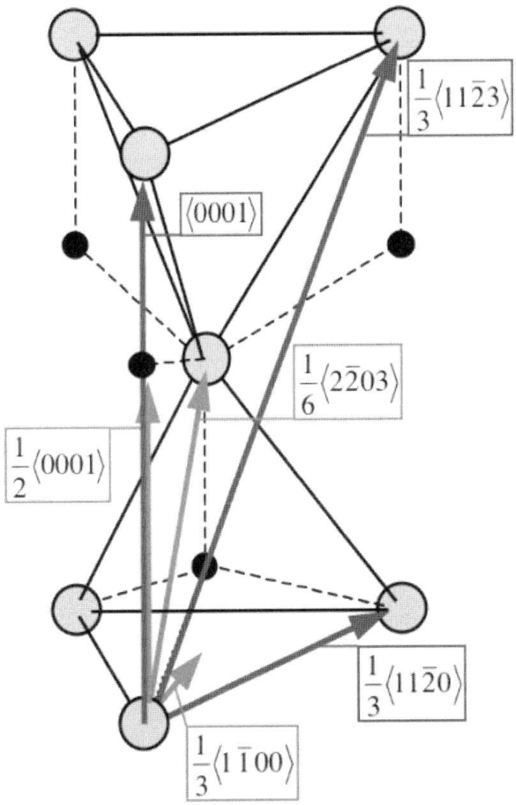

Figure 11.7 Schematic drawing of possible Burgers vectors in the wurtzite structure. Light and dark gray arrows indicate partial and perfect dislocations, respectively. The dashed lines assign the atomic bonds.

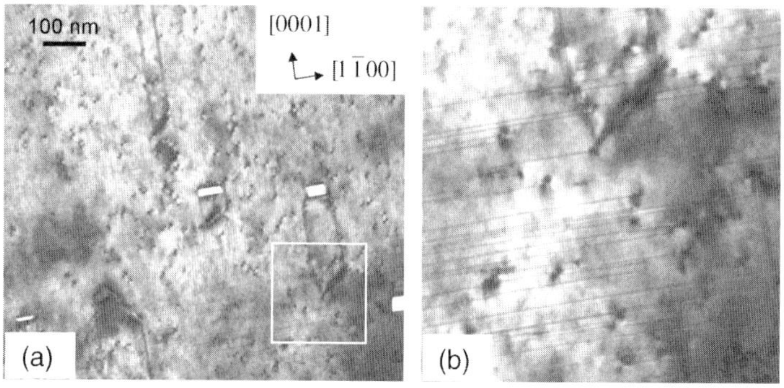

Figure 11.8 (a) A bright-field TEM image of HVPE *a*-plane GaN film grown on *r*-plane sapphire taken along the [11$\bar{2}$0] zone axis, slightly tilted away from the zone axis showing dislocations, basal plane, and prismatic stacking faults as well as rectangular voids; (b) enlarged view of the area indicated by a white rectangle in (a).

films show threading dislocations of screw, edge, or mixed type owing to the lattice mismatch between the film and the sapphire substrate [35]. Beside screw dislocations, with BVs $\mathbf{b}_s = \langle 0001 \rangle$, edge dislocations of type $\mathbf{b}_e = 1/3\langle 11\bar{2}0 \rangle$ and mixed dislocations of type $\mathbf{b}_{a+c} = 1/3\langle 11\bar{2}3 \rangle$ were observed in these films [35]. The edge dislocations are caused by the small twist between the different grains nucleating on the sapphire, whereas the screw and mixed dislocations are, in general, related to the growth parameters.

Partial dislocations connected with stacking faults are rarely found in high-quality *c*-plane GaN and are usually associated with poor nucleation conditions, thereby promoting the formation of the sphalerite phase besides the wurtzite phase [36]. For the *a*-plane GaN, the situation is different as is illustrated by the plan-view bright-field transmission electron microscopy (BFTEM) image shown in Figure 11.8a and b. The image was taken close to the [11$\bar{2}$0] zone axis orientation of the sample with a slight tilt toward the 0001 pole. Under these conditions, a high density of BSFs can be observed. These BSFs are terminated by partial dislocations as will be discussed in Section 11.2.2, which give rise to a strong bright and dark contrast owing to the strain fields associated with them (see Figure 11.8b). Moreover, PSFs can be seen, which are connected with rectangular voids. The observed stacking faults and voids will be discussed in Sections 11.2.3 and 11.2.4, respectively.

The interface of the MOVPE layer with the sapphire substrate was investigated in cross-section geometry using BFTEM imaging conditions in the [0001] and [1$\bar{1}$00] zone axis orientations with respect to the GaN film, slightly tilted away from the exact zone axes orientation to emphasize the defect-related contrast. Figure 11.9a shows a BFTEM image along the [0001] direction showing threading dislocations, as well as 60° stacking faults emerging at the interface. Furthermore, relaxation-related contrast is visible on the sapphire side of the interface. In addition, facetted voids were occasionally observed. In

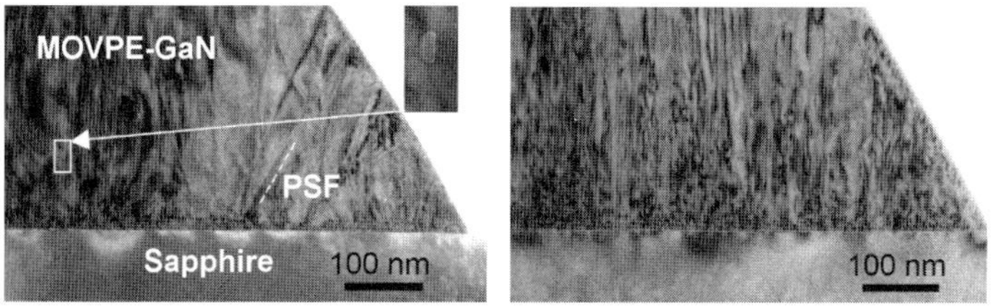

Figure 11.9 Cross section of bright-field TEM images of an *a*-plane GaN film grown by MOVPE on *r*-plane sapphire taken along the (a) $[0001]_{GaN}$ and (b) $[1\bar{1}00]_{GaN}$ zone axes.

the $[1\bar{1}00]$ direction (Figure 11.9b), the BSFs become visible and it can be seen that they also emerge directly at the interface, thus giving rise to the conclusion that the formation of BSFs is connected with the early growth stage, namely, the nucleation of the *a*-plane GaN on top of the sapphire substrate.

11.2.2.1 Dislocation Types in *a*-GaN

In order to characterize line defects in *a*-plane GaN, it is useful to discuss the possible dislocation configurations for this orientation in more detail. Figure 11.10 shows the possible configurations of BVs for dislocations in the hexagonal close-packed crystal system with line orientations in growth direction $[11\bar{2}0]$ and the respective glide planes (with the exception of screw dislocations, which do not have a glide plane). For simplification, the schematic in Figure 11.10 shows the ABAB stacking along the *c* direction with the second atom type being omitted. An atomic displacement from the A into the B layer will result in a sessile partial dislocation since the coordination of the atoms will change upon such a displacement out of the {0002} glide planes. Since the majority of experimentally found dislocations are aligned in $[11\bar{2}0]$ direction, only the configurations for this line direction are discussed here. In Figure 11.10a and b, the BVs of partial and perfect dislocations are represented by black arrows. For the possible partial dislocations, the respective BVs are $\mathbf{b}_p = 1/3\langle 1\bar{1}00\rangle$, $1/2\langle 0001\rangle$, and $1/6\langle 2{-}203\rangle$. The first are Shockley partial dislocations, while the second are Frank dislocations and the third type corresponds to Frank–Shockley (FS) partial dislocations. From the schematic picture it becomes clear that the screw dislocation is of type $\mathbf{b}_s = 1/3[11\bar{2}0]$ and the pure edge dislocation is of type $\mathbf{b}_e = [0001]$ in opposition to the case of the *c*-plane GaN, whereas the perfect mixed dislocations are $\mathbf{b}_s = 1/3\langle 11\bar{2}3\rangle$ and $\mathbf{b}_m = 1/3\langle -12{-}10\rangle$ (except $\mathbf{b}_s = 1/3[11\bar{2}0]$, which is screw type in this case).

To study the character of the dislocations present in the *a*-plane GaN films, HVPE-grown films were analyzed in cross-section, as well as in plan-view geometry by conventional and high-resolution transmission electron microscopy (HRTEM). The dislocation density was determined to be $(1{-}2) \times 10^{10}\ \mathrm{cm}^{-2}$ and almost all the dislocations were found to bound BSFs as

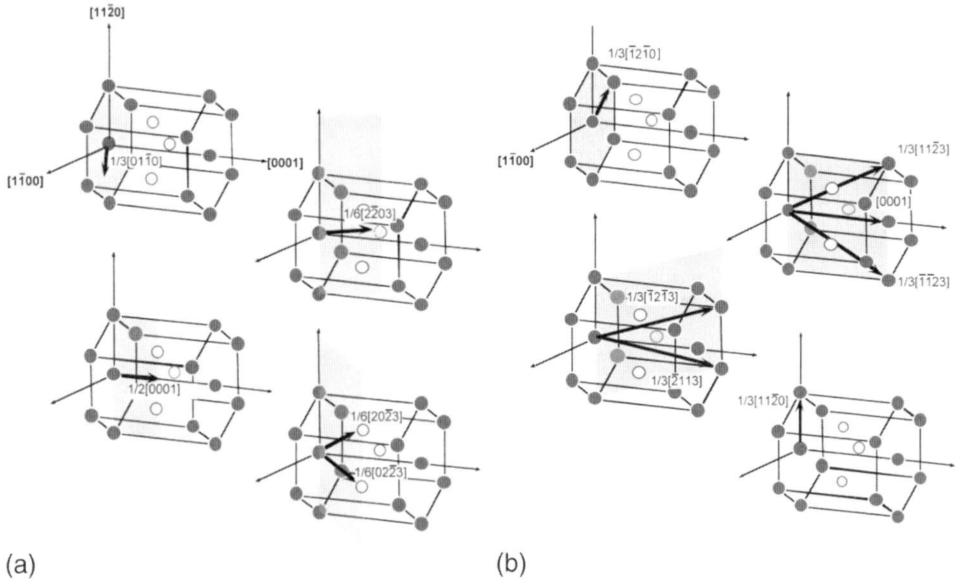

Figure 11.10 Burgers vectors and glide planes in the hexagonal close-packed structure for a dislocation line direction [11$\bar{2}$0]: (a) partial dislocations and (b) perfect dislocations. The shaded planes are the respective glide planes. Thick black arrows indicate the possible Burgers vector configurations.

could be seen in Figure 11.8b. As discussed in Section 11.2.3, these dislocations bound I_1-type stacking faults. The partial separation was estimated to be in the range of $d = 200$ to 500 nm and the stacking fault density was determined to be about 10^5 cm^{-1}. This means that, in average, the stacking faults were separated by about $\Delta = 100$ nm. Assuming that each stacking fault is bounded by two partials, the observed stacking fault density would result in a density of partials, $\sigma_{part} = 2 \times d^{-1} \times \tilde{\Delta}^{l} = (0.4\text{–}1) \times 10^{10}$ cm^{-2}. This value agrees relatively well with the total density of observed threading dislocations indicating that the vast majority of dislocations are indeed partials.

11.2.2.2 Dislocation Formation Energy

The line energy required for the formation of a dislocation depends on the magnitude of its BV b and on the distance r. It can be calculated on the basis of linear elasticity theory according to [37],

$$E_{\text{disl}} = E_{\text{core}} + Kb^2 \ln\left(\frac{r}{r_0}\right) \tag{3}$$

where r_0 is the core radius and E_{core} is the dislocation core energy, which depends on the actual atomic configuration and cannot be determined by linear elasticity theory as the linearity of elastic properties breaks down close to the dislocation core. Atomistic calculations [38] show that for a pure edge dislocation in *c*-plane GaN, the line energy E_{core} values are comparable to the stress field related part of the dislocation energy close to the dislocation core. However, for distances larger than the critical radius (which is usually in the order of 1 nm) the stress-related contribution increases rapidly, dominating the energy related with the dislocation. The prelogarithmic factor $K_{iso}b^2$ in Equation 3 for the isotropic case can be calculated from [37]:

$$K_{\text{iso}}(\phi)b^2 = b^2\mu\left(\frac{\sin^2\phi}{1-\nu} + \cos^2\phi\right) \tag{4}$$

with the shear modulus $\mu = C_{44}$, the Poisson ratio $\nu = C_{12}/2(C_{44} + C_{12})$ (C_{ij}: elastic stiffness matrix components) and the angle between the BV and the dislocation line direction ϕ. Hence, the energy factor can be divided into an edge component (with $\phi = \pi/2$) and a screw component (with $\phi = 0$). However, this assumption is valid only for dislocations running perpendicular to the *c* plane in the wurtzite structure because the strain fields are isotropic in the basal plane. For the *a*-plane GaN with dislocations running perpendicular or at an angle different from zero, for example, in $[11\bar{2}0]$ direction, this assumption is not valid anymore and anisotropy theory has to be applied.

11.2.2.3 Impact of Anisotropy

The energy factor for a dislocation in the anisotropic case is given by [39],

$$K_{\text{aniso}}(\phi)b^2 = \sum_{i,j}^{3} K_{ij}(\phi)b_i b_j \tag{5}$$

where the b_i are the BV components in the three spatial directions given in the dislocation coordinate system where z is always chosen to be parallel to the dislocation line. The K_{ij} in Equation 5 depends on the angle ϕ and can, in general, only be calculated numerically, although for the case of hexagonal crystals, analytical solutions for the elastic properties of straight dislocations for certain symmetry planes are available [40, 41]. However, it becomes clear that K_{aniso} cannot be easily separated into screw and edge components like for the isotropic case. Yoo and Loh [42] found that in general, significant differences exist for the dislocation strain fields between the isotropic and anisotropic case. For example, the anisotropy can give rise to interaction forces between dislocations of the same or different types, which are absent under the strictly isotropic approximation [42]. As already mentioned, the anisotropy of the crystals has to be taken into account if the probability of the formation of a dislocation and its stability are aimed at. The elastic stiffness parameters of GaN are commonly given in the crystal reference system. For *a*-plane GaN, the C_{ij} elasticity matrix has to be transformed to take into account the crystal rotation around an axis lying in the basal plane (here the [1$\bar{1}$00] axis). The energy and elastic properties of the straight dislocations in anisotropic media such as the displacement field and the stresses associated with them can be determined on the basis of the Stroh-theory for anisotropic crystals [39], for example by following the approach of Teodosiu *et al.* [43]. Table 11.2 shows the possible BV and glide plane configurations for *a*-plane GaN, as well as the resulting strain field related energy factor Kb^2 in comparison with the energy factors obtained for the isotropic assumption. The comparison makes clear that, in general, the dislocation energy is larger for the anisotropic case and the differences between the anisotropic and isotropic values become most significant for the mixed perfect 1/3[−2110] and 1/3[1$\bar{2}$10] dislocations, as well as for the Shockley partial dislocation with energy values up to twice as large as for the isotropic case.

Another important property of the dislocations is the hydrostatic pressure imposed by the presence of the defect. The hydrostatic pressure field around the dislocation can be calculated from the diagonal elements σ_{ii} of the stress tensor matrix according to

$$p = -\frac{1}{3}\sum_{i}^{3}\sigma_{ii} \tag{6}$$

Figure 11.11 shows the resulting hydrostatic pressure fields for a FS partial dislocation 1/6[2−203], running in the [11$\bar{2}$0] direction (which was found as major dislocation type in the films) in comparison with the isotropic approximation. The effect of the anisotropy becomes clearly visible in this figure by a strong elongation of the pressure field in the ⟨1$\bar{1}$00⟩ directions,

Table 11.2 Energies of the different possible dislocations in *a*-plane GaN calculated on the basis of anisotropic and the isotropic elasticity theory. Different angles ϕ lead to different energy factors for BVs having the same strength.

$\bar{b}$	Type	Glide plane(s)	$\Phi°$	Kb^2 (eV Å^{-1})	
				Anisotropic	Isotropic
⟨0001⟩	Edge	(1$\bar{1}$00)	90	10.98	8.90
$\frac{1}{3}$⟨11$\bar{2}$0⟩	Screw	–	0–180	0.80	0.74
	Mixed	(0001)	–	2.10	0.93
$\frac{1}{3}$⟨11$\bar{2}$3⟩	Perfect	{1$\bar{1}$01}	58.4–121.6	1.19–1.33	0.92–0.97
$\frac{1}{2}$⟨0001⟩	Frank	(0001)	90	2.74	2.23
$\frac{1}{3}$⟨1$\bar{1}$00⟩	Shockley	(1$\bar{1}$00)	30–150	1.33–2.00	0.972
$\frac{1}{6}$⟨2$\bar{2}$03⟩	Frank–Shockley	{11$\bar{2}$2}	60–105	0.30–0.35	0.23–0.24

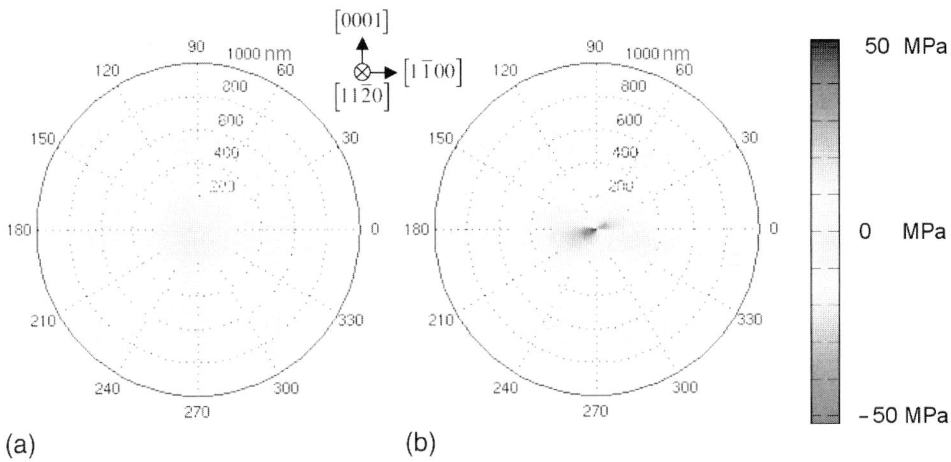

Figure 11.11 Polar plots of the hydrostatic pressure field of a 1/6[2–203] Frank–Shockley partial dislocation in *a*-plane GaN (a) assuming the isotropic case, (b) taking anisotropy into account (cf. Color Plate XXXIII).

which could be responsible for the asymmetry of the rocking curve shape of the 0002 reflection measured by X-ray diffraction in *a*-plane GaN films [44]. A significant difference can also be observed regarding the magnitude of the pressure, with values up to 50 MPa close to the dislocation core and is by almost half an order of magnitude larger in the anisotropic case compared to the isotropic calculation.

These findings underline the strong impact of the crystal anisotropy on the dislocation-related stress in the films, which has to be considered for examples for the interpretation of X-ray diffraction patterns performed on these layers. It can be expected that the strong anisotropy of the lattice mismatch results in an anisotropic formation of defects for stress accommodation, thus leading to anisotropic strain fields.

11.2.3
Planar Defects

In the wurtzite structure, the basal plane is the most prominent plane for stacking fault formation due to the low energy required to insert a faulted stacking sequence in this plane [37]. Different BSF types are conceivable depending on the type of stacking sequence. Besides the BSFs, PSFs are also possible on second-order glide planes such as the $\{10\bar{1}0\}$ or $\{10\bar{1}1\}$ planes. Stacking faults can have a significant impact on the electronic and optical properties of GaN. For example, BSFs are found to emit in UV spectral range [45], which affects the performance of GaN-based optoelectronic devices.

11.2.3.1 **Basal Plane Stacking Faults**

Three types of BSFs have been identified in the wurtzite structure: the intrinsic I_1-type with stacking sequence ABABCBCBC, the intrinsic I_2-type with stacking sequence ABABCACA, and the extrinsic E type ABABCABAB. These BSF configurations are depicted in Figure 11.12a–c. Zakharov *et al.* [46] reported the existence of another type of BSF, namely, a double I_1 sequence stacking fault in *a*-plane GaN grown on SiC.

Stampfl and Van de Walle [47] have determined stacking fault energies from density-functional pseudopotential calculations for wurtzite GaN, and identified the I_1 type as the lowest energy BSF with formation energy of about 1.81×10^{-2} J m^{-2}. Typical BSF densities found in *a*-plane GaN films grown on *r*-plane sapphire are well in the range of 10^5 cm^{-1} almost independent of the used substrate material [46, 48–50] and despite some improvement that could be achieved by using Si_xN_{1-x} micromasking to reduce the defect density down to 3×10^5 cm^{-1} [48].

Regarding PSFs in *a*-plane GaN, one type has been identified lying on $\{1\bar{2}10\}$ planes inclined at an angle of 60° with respect to the $(11\bar{2}0)$ growth plane [46]. The lateral extension of these PSFs is only a few nanometers and they are terminated by BSFs of I_1-type bounded by $1/6[10\bar{1}0]$ and $1/6[\bar{1}010]$ stair-rod dislocations.

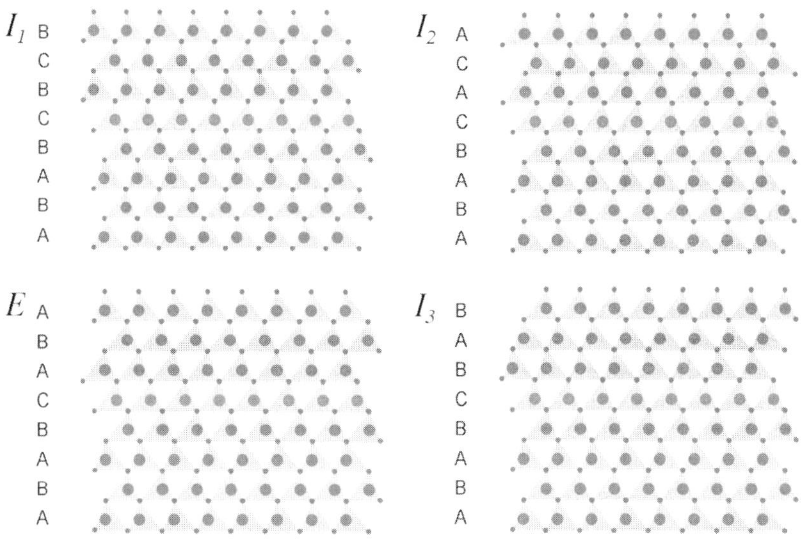

Figure 11.12 Types of intrinsic (*I*) and extrinsic (*E*) BSF types in the wurtzite structure showing the change of stacking sequence of the tetrahedrons drawn around the N-atoms: I_1, I_2, *E*, and I_3 (cf. Color Plate XXXIII).

11.2.3.2 Analysis of Basal Plane Stacking Faults

In Figure 11.13a, a BFTEM plan-view image of an HVPE *a*-plane GaN film directly grown on *r*-plane sapphire is shown recorded in the area containing a void [51]. The image was taken with the sample tilted by 4.5° towards the [1$\bar{1}$00] direction showing the high density of predominantly partial dislocations. Figure 11.13b shows a BFTEM image taken close to the [11$\bar{2}$0] zone axis tilted by 4° towards the [0001] direction showing a high density of BSFs (in the range of 10^5–10^6 cm^{-1}), and almost exclusively having a dislocation line orientation in the growth direction, namely the [11$\bar{2}$0] orientation. The BSFs were found to be, in general, of I_1 type as was determined from HRTEM with the characteristic stacking sequence AB<u>ABC</u>BCB (see Figure 11.14). Figure 11.13c shows the selected area diffraction (SAD) pattern in the [11$\bar{2}$0] zone axis orientation with streak-like contrast caused by the high density of BSFs.

The fact that most of the BSFs are found to be of I_1 type (the lowest energy BSF) is an indication for the near-equilibrium conditions during growth. FS type partial dislocations with BVs 1/6⟨2–203⟩ terminate I_1 type BSFs as was confirmed by g − 3g weak-beam analysis in cross-section geometry along the

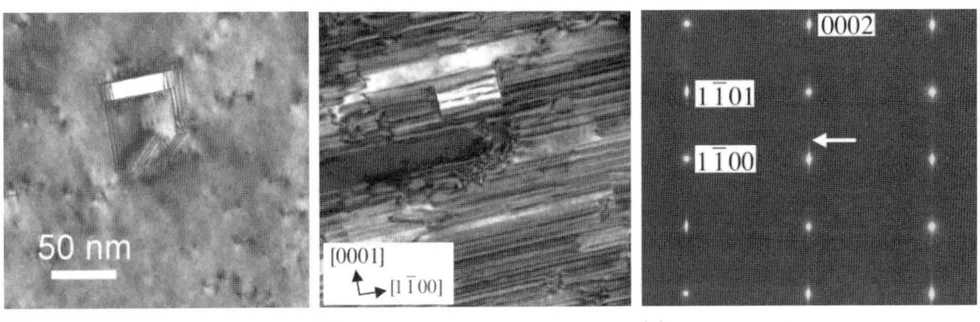

Figure 11.13 Plan-view bright-field TEM images of HVPE *a*-plane GaN: (a) tilted by 4° towards the 1$\bar{1}$00 pole and (b) tilted by 4.5° towards the 0002 pole. (c) A selected area diffraction pattern showing streaks originating from the BSFs (white arrow).

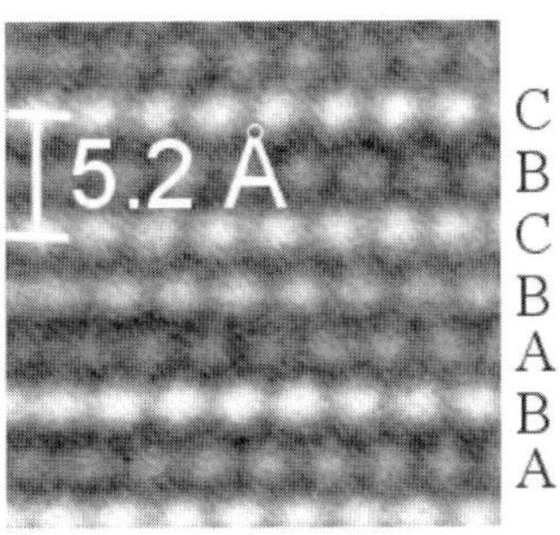

Figure 11.14 A HRTEM image of the stacking sequence of a BSF, characterizing the BSF as I_1-type stacking fault.

[0001] direction (not shown here). The length of the BSFs was found to vary between 0.2 and 1.5 μm.

The energy γ of a stacking fault bounded by two partial dislocations with the (0002) slip plane and a separation distance d after a dissociation reaction can be calculated for anisotropic crystals by using [52]

$$\gamma = \frac{b^2}{24\pi d}[(3\sin^2\phi - \cos^2\phi)K_1 + (3\cos^2\phi - \sin^2\phi)K_3] \tag{7}$$

In this equation, b is the magnitude of the BV of the partial dislocations, ϕ is the angle between BV and dislocation line direction, whereas the values for K_1 and K_3 can be calculated from the equations given by Steeds [52]. For the basal glide plane, the resulting values for K_1 and K_3 are 146.5 and 120.1 GPa, respectively. Using these values and the stacking fault formation energy for I_1-type stacking faults given by Stampfl and van de Walle [47], a separation distance d of 27.5 nm results, which is much lower than the experimentally determined distances. The latter indicates that the main driving force for the BSF formation is unlikely to be caused by a dislocation split reaction but is rather connected with the nucleation step of the film formation and the strong anisotropy of the lattice mismatch between a-plane GaN and r-plane sapphire.

11.2.3.3 Prismatic and Pyramidal Stacking Faults

Prismatic and pyramidal stacking faults (PrSF and PySF, respectively) were also observed in a-plane GaN films grown directly on r-plane sapphire lying on the $\{1\bar{1}00\}$ and $\{1\bar{1}02\}$ planes, respectively (see Figure 11.13a and b). These PrSF and PySF segments frequently occurred in the arrangement of closed domains created by a stacking fault folding process. Folded ribbons of stacking faults were also identified in a-plane GaN grown by epitaxial lateral overgrowth (ELO), which is shown in Figure 11.15a and b. In the 0002 dark-field TEM images (Figure 11.15c), the stacking fault related fringe contrast appeared when the sample was tilted toward the $1\bar{1}00$ pole. This indicates that the displacement vectors $\mathbf{R}$ of the different stacking fault segments do not lie in the reflecting planes, that is, the {0002} planes, as the visibility of the stacking fault under these conditions means that the product of diffraction vector and displacement vector is $\mathbf{g}_{0002} \times \mathbf{R} \neq 0$.

The process of stacking fault folding involves stair-rod dislocations to mediate between the different prismatic and pyramidal crystal planes because the displacement vectors of the stacking faults on the $\{1\bar{1}00\}$ and $\{1\bar{1}02\}$ planes are not the same. An HRTEM plan-view image of the tip of a prismatic/pyramidal stacking fault domain found in an HVPE film is shown in Figure 11.16. The arrows indicate the position of stair-rod dislocations. It is noteworthy that the tip involves the interaction of an I_1-type BSF, representative for all such domains found in the film. The tip position coincides with a terminated BSF and consists of two PySF, each of them subsequently folding onto the m plane.

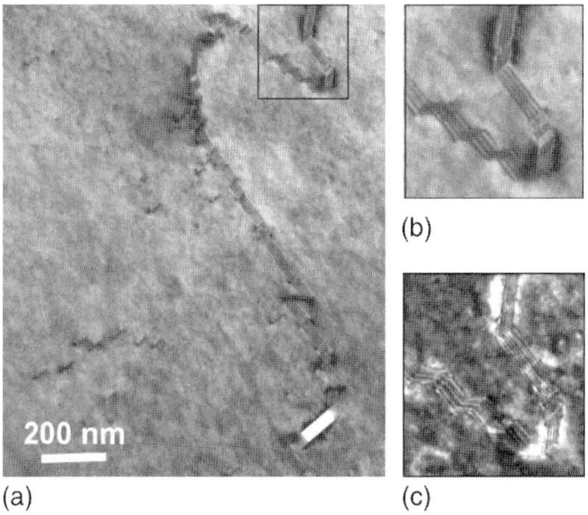

Figure 11.15 (a) BFTEM plan-view image (sample tilted by 3° towards the $1\bar{1}00$ pole) of an HVPE film grown by epitaxial lateral overgrowth. A folded stacking fault ribbon is visible which extends over tens of μm inside the film being connecting rectangular voids. (b) A magnified view of the area indicated by the rectangle. (c) 0002 dark-field image of the same area.

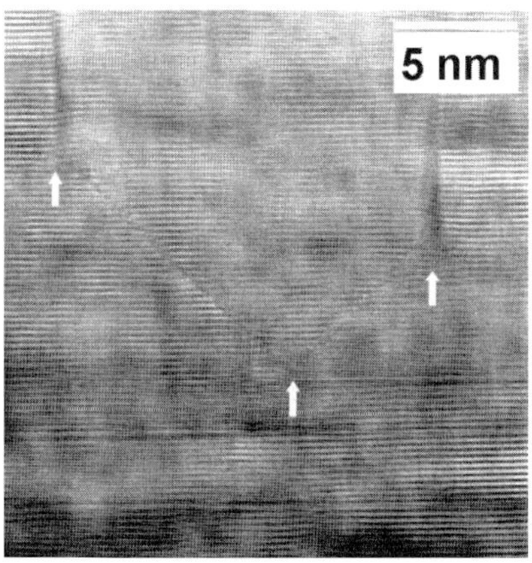

Figure 11.16 A plan-view HRTEM image, showing the tip of a folded stacking fault. White arrows assign the positions of stair-rod dislocations.

A magnified part of a PrSF on the *m*-plane is shown in Figure 11.17 with a BSF crossing the PrSF boundary. The arrows mark the position of the BSF inside (left) and outside (right) of the stacking fault domain with a significant relative shift in *c* direction. This shift indicates that the domain was formed after the BSF.

A schematic overview of the different planar defects found in *a*-plane GaN grown on *r*-plane sapphire is given in Figure 11.18. Beside the I_1 BSFs (A) bounded by FS partial dislocations folded stacking fault domains were

Figure 11.17 A HRTEM image of a prismatic stacking fault. The arrows assign a BSF crossing the PrSF being shifted in *c* direction.

found containing *m*- and *r*-plane facets (B) as well as 60° PSFs bounded by two I_1 stacking faults (C). The fact that prismatic and pyramidal stacking faults form indicates the presence of strong external stresses during growth, which is likely to be connected with wafer bending owing to the lattice mismatch and different thermal expansion coefficients of the GaN and the sapphire.

Remarkably, all reports on *a*-plane GaN give similar stacking fault and threading dislocation densities independent on the respective lattice mismatch (even varying between 1 and 16.1% depending on the chosen substrate). Also for the growth of *m*-plane GaN on γ-$LiAlO_2$ [53], similar values for the defect types and densities were made. These findings indicate that the defect formation mechanism during the initial growth state is most likely connected with the low formation energy of the I_1 stacking faults. Its formation is triggered even at low lattice mismatches during growth. This is consistent with the reports by Liu *et al.* [53] who found that the intrinsic stacking fault formation is connected with the film growth rather than the relaxation. However, as it will be shown in Section 11.3, the film relaxation introduces shear stress to the film parallel to the basal plane, which is likely to influence the stacking fault formation process.

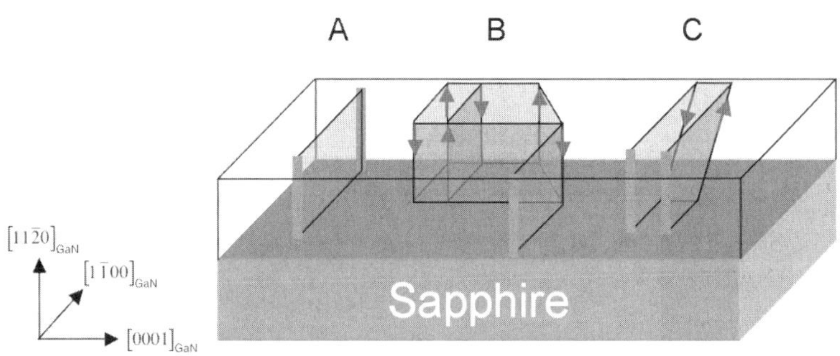

Figure 11.18 Sketches of stacking faults found in *a*-plane GaN. Gray arrows indicate stair-rod dislocations mediating between stacking faults on different planes and gray slabs assign terminating partial dislocations. (a) Basal plane stacking faults, (b) stacking fault domain with enclosed void and (c) prismatic stacking fault bounding two BSFs.

11.2.4 Volume Defects

Extended three-dimensional defects such as inversion domains, inclusions, or voids have been found in *c*-plane GaN grown on *c*-plane sapphire. Some of these defects are related to doping, like the pyramid-shaped inversion domains formed in highly Mg-doped films [54–56]. The same is valid for nanopipes in *c*-plane GaN on *c*-plane sapphire identified by atomic force microscopy or TEM [57]. These hexagonal-shaped elongated voids are caused by bundles of open-core screw dislocations, which form during the early stage of growth and coalesce. In *a*-plane GaN, two types of voids were observed, which are discussed in detail in this section.

11.2.4.1 Nanopipes

Figure 11.19a and b shows cross-section BFTEM images taken along the [0001] axis and showing the presence of pipe-like voids, which could also be visualized in scanning electron microscopy (SEM) images after the preparation of a cross section by focused ion beam (not shown here). The nanopipes originate at the GaN–sapphire interface (Figure 11.19b), possess $\{0001\}$ and $\{1\bar{1}00\}$ sidewalls (as can be seen in the plan-view image shown in Figure 11.12), and extend up to several micrometers into the film. They terminate by the formation of two $\{01\bar{1}0\}$ facets. The density of these voids amounts to about 10^{-7} cm^{-2}. As these nanopipes are found to be connected with folded PrSF and PySF ribbons (see previous section), their formation is likely to be caused by large strain fields during growth.

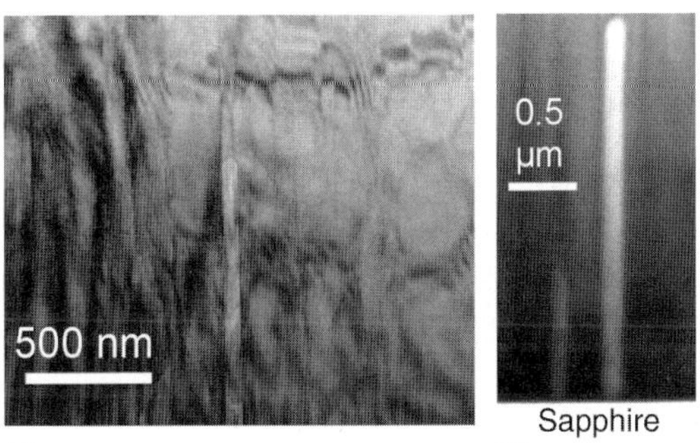

Figure 11.19 (a) A bright-field TEM cross-section views of a nanopipe along the [0001] direction formed in HVPE GaN grown directly on sapphire. (b) Similar images taken in a thicker part of the specimen show that the pipes originate at the GaN–sapphire interface.

11.2.4.2 Voids

It was shown in Section 11.2 that not only rectangular nanopipes running in $[11\bar{2}0]$ direction could be observed (as is the case for HVPE GaN grown directly on sapphire) but also voids being orientated in the [0001] direction were found in MOVPE GaN films grown on sapphire. Figure 11.20 shows a BFTEM image of an area close to the GaN–sapphire interface. The density of the voids is found to be in the range of 10^7 cm^{-2} and they possess $\{1\bar{1}00\}$ sidewall facets where these voids occurred. The origin of the voids is yet unclear and their origin appears to differ from that of the nanopipes observed in HVPE *a*-plane GaN directly grown on sapphire, which were discussed in the previous section. However, etching experiments with HCl (10% for 1 minute at 100 °C) show the formation of hexagonal etch pits in the GaN layer, indicating the presence of screw dislocations in the [0001] direction close to the interface (see Figure 2.17c). Thus, the formation mechanism of these voids is most likely to be comparable to the one observed in *c*-oriented GaN, namely, via open-core screw dislocations [57].

11.2.5 Surface Defects

Dominant topological defects are surface pits (shown in Figure 11.21). They vary in size between 200 nm and 1 μm and show a hexagonal $\{1\bar{1}01\}$ faceting. The pits are often found to form groups, and their density shows a spatial variation in the range of 10^6 cm^{-2}. The origin of the pits is not yet clarified. Wu *et al.* [58] suggest that the pits are pyramidal V-defects, which are distinguished from the V-shaped defects found in *c*-plane GaN [59] by the fact that they are not caused by screw dislocations. They assume that the pits form already during nucleation of the film. According to this interpretation, the $\{1\bar{1}01\}$ side facets of the pits grow much slower than the $(11\bar{2}0)$ growth plane and therefore do not vanish throughout the growth even of very thick films.

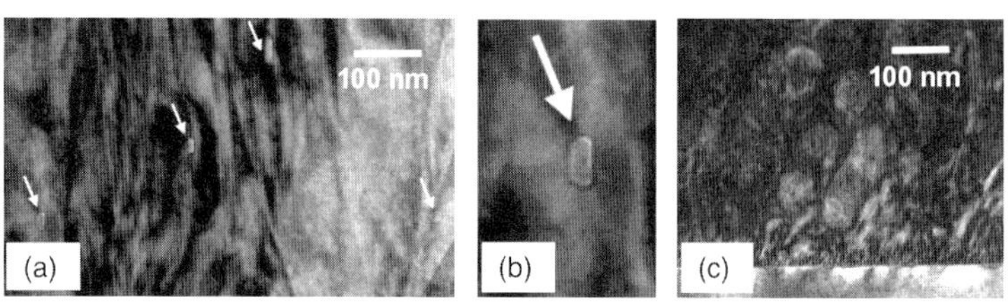

Figure 11.20 (a) BFTEM image of an MOVPE *a*-plane GaN film taken along the [0001] orientation close to the GaN–sapphire interface. White arrows mark facetted void. (b) A magnified view of a representative void. (c) Facetted etch pits in the GaN film formed by etching of the sample with hydrochloric acid.

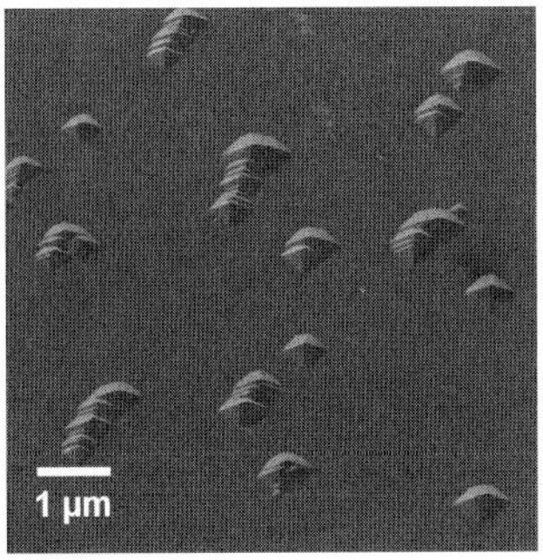

Figure 11.21 SEM image of surface pits found in HVPE GaN grown directly on *r*-plane sapphire viewed under an angle of 52°.

11.3 Interfacial Structure

Most of the structural defects in *a*-plane GaN grown on *r*-plane sapphire such as the basal plane and PrSF–PySF, and the corresponding dislocations originate at the substrate–film interface. Thus, it is important to obtain a thorough understanding of the interfacial structure, which in general affects the defect formation during the epitaxial growth of the film. To follow this aim, three aspects are of interest: the epitaxial relationships between film and substrate, the atomic arrangement at the interface, and the character of misfit-related dislocations formed as a result of the lattice mismatch between the sapphire and the GaN.

Conventional TEM, comprising dark-field TEM, bright-field TEM and SAD, was applied to determine the relative crystallographic orientations of the film and substrate, and HRTEM was used to obtain information regarding the atomic structure of the interface and the misfit-related dislocations at the interface. The following epitaxial relationships for the growth of *a*-plane GaN on *r*-plane sapphire were found [10, 49]: $[1\bar{1}00]_{GaN} || [11\bar{2}0]_{Al_2O_3}$, and $[0001]_{GaN} || [1\bar{1}01]_{Al_2O_3}$. Vennéguès and Bougrioua [60] characterized the absolute epitaxial relationship of *a*-plane AlN on *r*-plane sapphire and confirmed the suggested epitaxial relationships for AlN and found the Al face of the AlN film to grow parallel to the $[1\bar{1}01]_{Al_2O_3}$ direction.

11.3.1 Epitaxial Relationship

The epitaxial relationships between the film and the substrate can be determined by analysis of SAD patterns in cross-section taken from an area comprising the two materials using an SAD aperture of appropriate size. Figure 11.22a and b shows the cross-section BFTEM images in $[1\bar{1}00]_{GaN}$ and the $[0001]_{GaN}$ orientation. The SAD patterns taken along the $[1\bar{1}00]_{GaN}$ and the $[0001]_{GaN}$ zone axis orientation are given in Figure 11.22c and d, respectively.

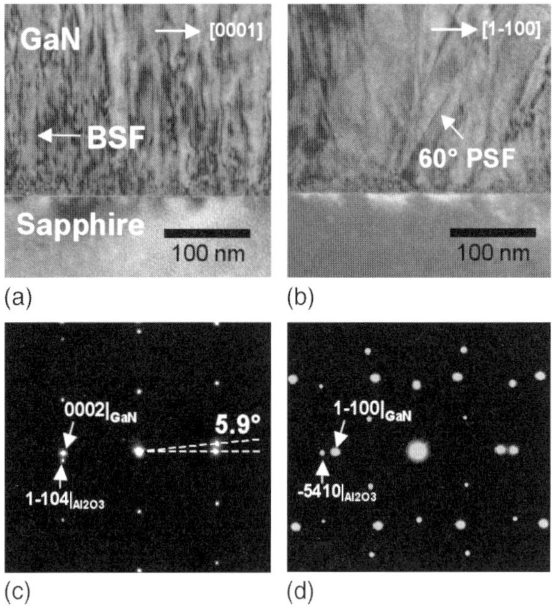

Figure 11.22 Bright-field TEM images taken along the $[1\bar{1}00]_{GaN}$ zone axis (a) and along the $[0001]_{GaN}$ zone axis (b). Interfacial SAD pattern along the $[1\bar{1}00]_{GaN}$ zone axis (c) and along the $[0001]_{GaN}$ zone axis (d).

The $[0001]_{GaN}$ direction is aligned along the $[11\bar{2}0]_{Al_2O_3}$ direction of the sapphire and the SAD pattern shows an inclination of 5.9° of the lattice planes with respect to the $\{0002\}_{GaN}$ lattice planes and the small mismatch of 1.2% gives rise to the almost identical distance of the respective reflections from the nondiffracted beam. The SAD pattern taken along the $[0001]_{GaN}$ direction illustrates that the corresponding orientation of the sapphire is $[01\bar{1}0]_{Al_2O_3}$, and the relative positions of the corresponding reflections reveal the complete relaxation of the 16% mismatch between the $\{1\bar{1}00\}_{GaN}$ and the $\{1\bar{1}04\}$ lattice planes.

11.3.2
Interfacial Structure

In order to investigate the atomistic structure of the interface and the relaxation-related dislocations, HRTEM analysis was applied. These studies aim to provide information on the relative atomic column positions of the two materials. Since the contrast in HRTEM images depends on defocus, lens aberrations, and foil thickness, [61] it is not possible to directly derive atomic column positions from the images. However, the atomic positions with respect to the phase contrast can be obtained by a comparison between the experimental HRTEM images and simulated images. The latter can be obtained using Bloch-wave simulation techniques [62] for GaN and sapphire in the respective zone axis orientations applying the known spherical aberration C_s of the microscope ($C_s = 0.5$ mm) and calculating thickness-defocus maps. These maps are subsequently compared with the experimental images and the best match for thickness-defocus values are searched for both the GaN and the sapphire, simultaneously.

11.3.2.1 HRTEM along the $[1\bar{1}00]_{GaN}$ Zone Axis

Figure 11.23a and b shows the interface in the $[1\bar{1}00]_{GaN}$ zone axis orientation and a magnified part of the defect-free area between the dislocations, respectively. The interface with the substrate was found to be relatively sharp, which allows a detailed study of the atomic structure by HRTEM. Some remaining dot contrast can be seen on the GaN side of the interface. This might be connected with delocalization effects owing to the finite spherical aberration of the microscope. A comparison between the experimental image taken at Scherzer-defocus with Bloch-wave simulated images for different specimen thickness values gave best agreement for a thickness of 15 ± 2 nm (see Figure 11.23c) for the GaN layer as well as for the sapphire. Because of the resolution limit of the microscope, only the $\{0002\}_{GaN}$ lattice planes could be resolved in GaN, leading to a fringe contrast pattern, which is also reproduced by the simulated images. Clearly visible is the 5.9° inclination of the $\{11\bar{2}0\}_{Al_2O_3}$ planes with respect to the $\{0002\}_{GaN}$ planes, and the actual atomic column positions of Ga and N in the GaN lattice match with the oxygen columns in the sapphire lattice (indicated by a dashed line in Figure 11.23c.

Only few misfit dislocations (MFDs) could be observed because of the small lattice mismatch of about 1.2% between the GaN and the sapphire in the $[1\bar{1}00]_{GaN}$ orientation. As only the $\{0002\}_{GaN}$ lattice fringes could be resolved, it was not possible to perform a Burgers circuit analysis. However, the magnified image of an MFD displayed in Figure 11.23d shows an extra half plane inserted on the sapphire side, giving rise to a BV component in the $(1\bar{1}00)$ plane of $\mathbf{b}_{(1-100)} = 1/3\langle 0001 \rangle$. One should bear in mind that the BV is not exactly defined for interfacial MFDs because the BV assigns the atomic displacement in the defect containing crystal relative to the defect-free crystal. This is exactly valid only for a single crystal containing a defect. Since the Burgers circuit around an MFD comprises areas of two different materials, the definition of the BV does not hold for this case. However, it is still useful to apply this concept to obtain information on the accommodation process of the film with respect to the substrate. Hence, the BV, or its components in the planes perpendicular to the imaging direction provide information on the type of atomic displacements in the film/substrate region and allows a discussion of possible strain relief mechanisms.

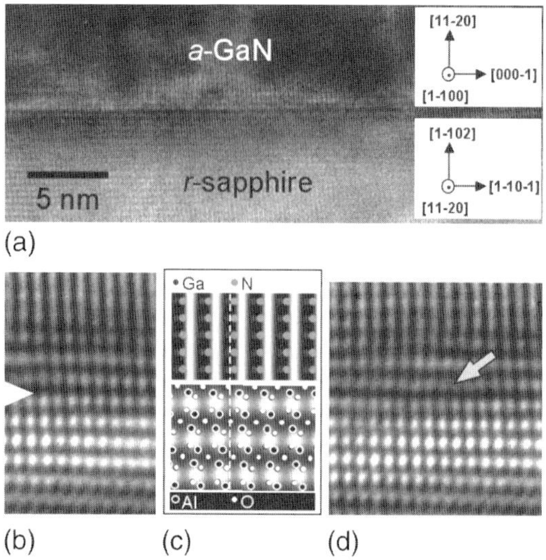

Figure 11.23 Cross-section HRTEM image along the $[1\bar{1}00]_{GaN}$ zone axis: (a) a-GaN–r-sapphire interface showing the clearly distinguishable transition between substrate and film and the registry between the $[0002]_{GaN}$ planes and the $\{11\bar{2}0\}_{Al_2O_3}$ planes. (b) A magnified view of a defect-free area at the interface (the white triangle marks the interface). (c) Best-fit Bloch-wave simulated images of GaN and sapphire for the respective orientations showing the relative atomic column positions. (d) Magnified view of an area containing a misfit dislocation.

11.3.2.2 **HRTEM along the $[0001]_{GaN}$ Zone Axis**

A similar HRTEM analysis was performed in the $[0001]_{GaN}$ zone axis orientation. For this orientation also, a comparably sharp interface between the sapphire and the GaN film was observed (Figure 11.24a). The expected hexagonal atomic arrangement in the GaN can be seen, as well as the corresponding sapphire structure along the $[\bar{1}101]_{Al_2O_3}$ direction demonstrating the registry of the $\{11\bar{2}0\}_{Al_2O_3}$ lattice fringes of the sapphire with the $[1\bar{1}00]_{GaN}$ lattice fringes of the GaN (see Figure 11.24b).

The comparison between the experimental images and Bloch-wave simulations resulted in the best agreement for a specimen thickness of 9 ± 2 nm (Figure 11.24c). The remaining deviation between the simulation and the experimental image on the sapphire side can be attributed to interfacial strain. The atomic columns of the GaN are centered between the bright dots visible in the experimental image aligned with the Al columns of the sapphire (dashed line in Figure 11.24c). As the lattice mismatch along the $[1\bar{1}00]_{GaN}$ direction is about 16%, the distance between the MFDs is expected to be significantly smaller compared to that in the perpendicular $[0001]_{GaN}$ direction if full relaxation is assumed. The Burgers circuit analysis, shown in Figure 11.24d determines the Burger vector component of the MFD in the $\{0002\}$ planes to be $\mathbf{b}_{(0002)} = 1/3\langle -2110 \rangle$.

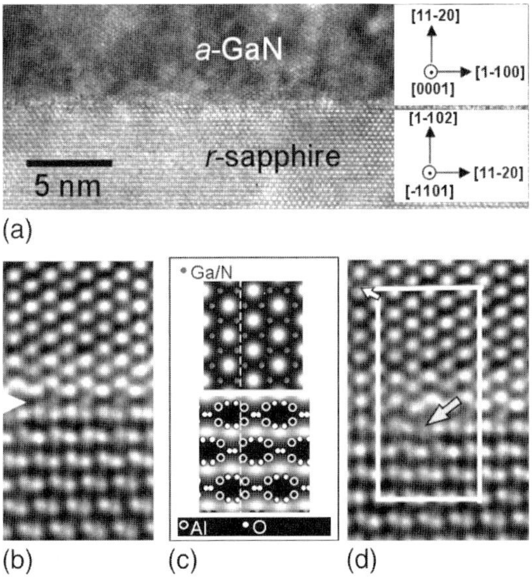

Figure 11.24 Cross-section HRTEM image taken along the $[0001]_{GaN}$ zone axis: (a) a-GaN–r-sapphire interface showing the registry between the $\{1\bar{1}00\}_{GaN}$ planes and the $\{1\bar{1}04\}_{Al_2O_3}$ planes. (b) A magnified view of a defect-free area at the interface (the white triangle marks the interface). (c) Best-fit Bloch-wave simulated images of GaN and sapphire for the respective orientations showing the respective atomic column positions. (d) A magnified view of an area containing a misfit dislocation and the respective Burgers circuit.

11.3.3
Model of Interfacial Structure

From the experimental HRTEM images discussed in the previous sections, some conclusions may be drawn regarding the possible atomic arrangement of the *r*-plane sapphire with respect to the *a*-plane GaN. Figure 11.25a and b shows possible interfacial arrangements in the $[0001]_{GaN}$ and $[1\bar{1}00]_{GaN}$ directions for which the Al–N and Ga–O bonds have been adjusted in such a way that they have minimum lengths. The low mismatch in $[1\bar{1}00]_{GaN}$ direction leads to an almost complete match between the O atoms of the sapphire and the Ga atoms of the nitride film. Contrary to this, the larger lattice mismatch in $[0001]_{GaN}$ direction results in a rather poor match between the Al and O columns of the sapphire with respect to the N and Ga columns of the film. This poor match results in shear forces parallel to the *c* planes, which are likely to increase the probability of BSF formation in the presence of stress fields, for example, those associated with the introduction of MFDs upon relaxation.

For a total relaxation, the relation between MFD distance d_{MF} and the respective lattice plane distance d as well as the lattice mismatch f is given

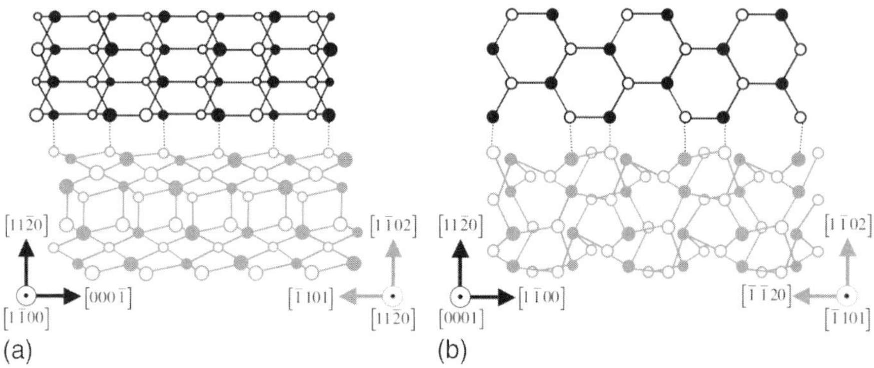

Figure 11.25 Possible interfacial configuration in: (a) $[1\bar{1}00]_{GaN}$ and (b) $[0001]_{GaN}$ zone axis orientation. The gray filled and empty circles represent Al and O atoms, respectively, and the black filled and empty circles represent Ga and N, respectively. The circle sizes represent the different layers to which the respective atoms belong.

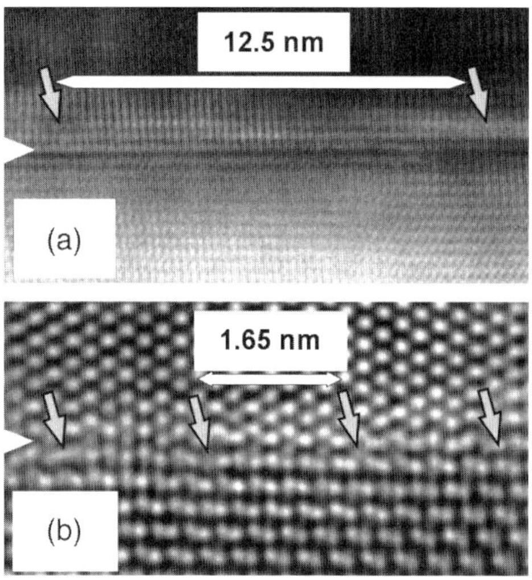

Figure 11.26 Magnified HRTEM images of the GaN–sapphire interface taken along: (a) the $[1\bar{1}00]_{GaN}$ zone axis and (b) the $[0001]_{GaN}$ zone axis. Gray arrows mark the presence of misfit dislocations.

by $d_{MF} = d/f$. Hence, for the $[1\bar{1}00]_{GaN}$ direction, with a lattice spacing of $d_{0002} = 0.26$ nm of the {0002} GaN planes, an average distance between the MFDs of about 23.5 nm would be expected. Because the investigated TEM sample had a lateral dimension – to be thin enough for HRTEM analysis – in the range of 100 nm, it was not possible to obtain statistically reliable data for this orientation. Nevertheless, MFD distances as low as 12.5 nm were found (Figure 11.26a). This does not necessarily imply overrelaxation but could also be caused by a low mobility of the dislocations along the interface during growth, so that an equilibrium distance might not be reached. For the $[0001]_{GaN}$

zone axis orientation, the average MFD distance was found to be about (1.65 ± 0.10) nm (Figure 11.26b) in very good agreement with the expected distance of 1.7 nm for full relaxation in this orientation. The dislocations were almost equally distanced, indicating that they were sufficiently mobile during the early stage of growth. Given that the average BSF distance is in the range of 15 to 40 nm, it is unlikely that the MFDs are directly related to the BSFs. On the basis of microstructural investigations of *m*-plane GaN grown on γ-$LiAlO_2$, Liu *et al.* [53] claim that the stacking faults do not form because of strain relaxation but are rather caused by the presence of substrate surface steps. According to their argumentation, the BSFs form if neighboring domains with different surface configurations, separated by these steps, coalesce and the resulting mismatch is accommodated by the insertion of a stacking fault. However, although the origin of BSF formation cannot be fully clarified yet on the basis of the experimental findings, it is likely that the formation of MFDs influences their formation as a result of stress associated with them.

11.4 Summary

The growth of *a*-plane GaN films on *r*-plane sapphire by HVPE or MOVPE is challenged by the formation of numerous growth-related defects, namely, point defects and impurities, dislocations, stacking faults, and voids. An analysis was presented for all of these defect types using various experimental techniques such as positron annihilation spectroscopy, secondary ion mass spectrometry, and transmission electron microscopy. Significant differences between the native point defect and impurity distribution were found between *a*-plane and *c*-plane GaN films grown on *a*-plane and *r*-plane sapphire, respectively. This was explained by the strong coupling of their formation with the extended defect pattern in *a*-plane GaN films. The extended defects were found to be caused by the misfit between film and substrate, which is anisotropic in the growth plane. The important role of anisotropy for dislocations and stacking faults was pointed out. Only low-energy I_1-type BSFs could be identified, which are rather owing to the film nucleation than the split of perfect dislocations. Besides the BSFs, prSF and pySF were identified forming folded stacking fault ribbons, which in some cases lead to the creation of closed domains, comprising nanopipe voids, which originate at the film–substrate interface. An analysis of the epitaxial relationships, as well as the interfacial structure of the film and substrate was presented. Furthermore, the anisotropically distributed interfacial MFDs were characterized. Especially, the extreme likeliness of BSF formation for the heteroepitaxial growth of *a*-plane GaN – which is most probably caused by the low formation energy for these stacking faults – remains one of the major obstacles for the applicability of the films for optoelectronic devices.

Acknowledgments

Support and discussions by and with T. Paskova, K. Müller, A. Rosenauer, S. Figge, D. Hommel, A. Rosenauer, R. Kersting, B. Monemar, F. Tuomisto, and P. Fini are gratefully acknowledged. The research was in part carried out with the support of the Deutsche Forschungsgemeinschaft under the project number FOR-506/KR-2195.

References

1. Shirakawa, T. (2002) *Materials Science Engineering B*, **91**, 470.
2. Ng, H.M. (2002) *Applied Physics Letters*, **80**, 4369.
3. Koida, T.,Chichibu, S.F., Craven, M.D. *et al.* (2004) *Applied Physics Letters*, **84**, 3768.
4. Ng, H.M., Bell, A., Ponce, F.A., and Chu, S.N.G. (2003) *Applied Physics Letters*, **83**, 653.
5. Sasaki, T. and Zembutsu, S. (1987) *Journal of Applied Physics*, **61**, 2533.
6. Li, D.S., Chen, H., Yu, H.B. *et al.* (2004) *Journal of Crystal Growth*, **265**, 107.
7. Paskova, T.,Darakchieva, V., Paskov, P.P. *et al.* (2005) *Journal of Crystal Growth*, **281**, 55.
8. Haskell, B.A., Wu, F., Craven, M.D. *et al.* (2003) *Applied Physics Letters*, **83**, 644.
9. Paskova, T., Kröger, R., Figge, S. *et al.* (2006) *Applied Physics Letters*, **89**, 051914.
10. Kung, P., Sun, C.J., Saxler, A. *et al.* (1994) *Journal of Applied Physics*, **76**, 236.
11. Bougrov, V., Levinshtein, M.E., Rumyantsev, S.L., and Zubrilov, A. (2001) Gallium Nitride (GaN), in *Properties of Advanced SemiconductorMaterials GaN, AlN, InN, BN, SiC, SiGe*, eds M.E. Levinshtein, S.L. Rumyantsev, and M.S. Shur, John Wiley & Sons, Ltd, New York.
12. Qian, W., Skowronski, M., and Rohrer, G.R. (1996) *Materials Research Society Symposium Proceedings*, **423**, 475.
13. Gil, B. (ed.) (1998) *Group III Nitride Semiconductor Compounds*, Clarendon Press, Oxford.
14. Van de Walle, C.G., Neugebauer, J., and Stampfl, C. (1999) Impurities and native defects in GaN and related compounds, in *Properties, Processing and Applications of Gallium Nitride and Related Semiconductors*, eds J.H. Edgar, S. Strite, I. Akasaki, *et al.* INSPEC.
15. Boguslawski, P. and Bernholc, J. (1997) *Physical Review A*, **56**, 9496.
16. Cherns, D. (2000) *Journal of Physics. Condensed Matter*, **12**, 10205.
17. Ruterana, P. and Nouet, G. (2001) *Physica Status Solidi B*, **227**, 177.
18. Romano, L.T., Northrup, J.E., and O'Keefe, M.A. (1996) *Applied Physics Letters*, **69**, 2394.
19. Pecz, B., Makkai, Z., di Forte-Poisson, M.A. *et al.* (2001) *Applied Physics Letters*, **78**, 1529.
20. Hierro, A., Hansen, M., Boeckl, J.J. *et al.* (2001) *Physica Status Solidi B*, **228**, 937.
21. Kröger, R., Kruse, C., Roder, C. *et al.* (2006) *Physica Status Solidi B*, **243**(7), 1533.
22. Chichibu, S.F., Uedono, A., Onuma, T. *et al.* (2005) *Applied Physics Letters*, **86**, 021914.
23. Saarinen, K. I., Laine, T., Kuisma, S. *et al.* (1997) *Physical Review Letters*, **79**, 3030.
24. Saarinen K. (2000) Characterization of native point defects in GaN by positron annihilation spectroscopy, in *III-V Nitride Semiconductors: Electrical, Structural and Defects*

Properties, ed. by M.O. Manasreh, Elsevier, Amsterdam, p. 109.
25. Oila, J., Kivioja, J., Ranki, V. *et al.* (2003) *Applied Physics Letters*, **82**, 3433.
26. Tuomisto, F., Saarinen, K., Lucznik, B. *et al.* (2005) *Applied Physics Letters*, **86**, 031915.
27. Saarinen, K., Hautojärvi, P., and Corbel, C. (1998) Poistron Annihilation Spectroscopy of Defects in Semiconductors, in *Identification of Defects in Semiconductors*, ed. by M. Stavola, Academic Press, New York, p. 209.
28. Tuomisto, F., Paskova, T., Kröger, R. *et al.* (2007) *Applied Physics Letters*, **90**, 121915.
29. Gogova, D., Siche, D., Fornari, R., etal (2006) *Semicond. Sci. Technol.*, **21**, 702.
30. Paskova, T., Paskov, P.P., Darakchieva, V. *et al.* (2002) *Physica Status Solidi C*, **0**, 209.
31. Wickenden, A.E., Koleske, D.D., Henry, R.L. *et al.* (2004) *Journal of Crystal Growth*, **260**, 54.
32. Reshchikov, M.A. and Morkoç, H. (2005) *Journal of Applied Physics*, **97**, 061301.
33. Elsner, J., Jones, R., Sitch, P.K. *et al.* (1997) *Materials Science Forum*, **258**, 1203.
34. Gao, Y., Craven, M.D., Speck, J.S. *et al.* (2004) *Applied Physics Letters*, **84**, 3322.
35. Ponce, F.A. (1998) Structural defects and materials performance of the III-V nitrides, in *Group III Nitride Semiconductor Compounds*, (B. Gil), Oxford University Press, Oxford, p. 123.
36. Liliental-Weber, Z., Sohn, H., Newman, N., and Washburn, J. (1995) *Journal of Vacuum Science and Technology B*, **13**, 1578.
37. Hirth, J.P. and Lothe, J. (1982) *Theory of Dislocations*, 2nd ed, Wiley, New York, p. 102.
38. Belabbas, I., Chen, J., Akli Belkhir, M. *et al.* (2006) *Physica Status Solidi C*, **3**, 1733.
39. Stroh, A.N. (1958) *Philosophical Magazine*, **3**, 625.
40. Teutonico, L.J. (1970) *Materials Science and Engineering*, **6**, 27.
41. Savin, M.M., Chernov, V.M., and Strokova, A.M. (1976) *Physica Status Solidi A*, **35**, 747.
42. Yoo, M.H. and Loh, B.T.M. (1972) *Journal of Applied Physics*, **43**, 1373.
43. Teodosiu, C., Vicolea, V., and Paven, H. (1975) *Physica Status Solidi A*, **27**, 191.
44. Wang, H., Chen, C., Gong, Z. *et al.* (2004) *Applied Physics Letters*, **84**, 499.
45. Liu, R., Bell, A., Ponce, F.A. *et al.* (2005) *Applied Physics Letters*, **86**, 021908.
46. Zakharov, D.N., Liliental-Weber, Z., Wagner, B. *et al.* (2005) *Physical Review A*, **71**, 235334.
47. Stampfl, C. and van de Walle, C. (1998) *Physical Review A*, **57**, R15052.
48. Chakraborty, A., Kim, K.C., Wu, F. *et al.* (2006) *Applied Physics Letters*, **89**, 041903.
49. Craven, M.D., Lim, S.H., Wu, F. *et al.* (2002) *Applied Physics Letters*, **81**, 1201.
50. Waltereit, P.,Brandt, O., Ramsteiner, M. *et al.* (2000) *Journal of Crystal Growth*, **218**, 143.
51. Kröger, R., Paskova, T., Rosenauer, A. *et al.* (2007) Proceedings of the 28th International Conference on the Physics of Semiconductors (ICPS), July, 23–28 2006, Vienna, Austria, ICPS 2006, AIP Conference Proceedings, **893**, 341 (2007).
52. Steeds, J.W. (1973) *Introduction to Anisotropic Elasticity Theory of Dislocations*, Clarendon Press, Oxford.
53. Liu, T.Y., Trampert, A., Sun, Y.J. *et al.* (2004) *Philosophical Magazine Letters*, **84**, 435.
54. Vennégués, M., Leroux, S.Dalmaasso, M. *et al.* (2003) *Physical Review A* **68**, 235214.
55. Liliental-Weber, Z., Jasinski, J., Benamara, M. *et al.* (2001) *Physica Status Solidi B*, **228**, 345.
56. Figge, S., Kröger, R., Böttcher, T., *et al.* (2002) *Applied Physics Letters*, **81**, 4748.
57. Qian, W., Skowronski, M., Doverspike, K. *et al.* (1995) *Journal of Crystal Growth*, **151**, 396.
58. Wu, F., Craven, M.D., Lim, S.-H, and Speck, J.S. (2003) *Journal of Applied Physics*, **94**, 942.

59. Bright, A.N., Sharma, N., and Humphreys, C.J. (2001) *Journal of Electron Microscopy*, **50**, 489.

60. Vennéguès, P. and Bougrioua, Z. (2006) *Applied Physics Letters*, **89**, 111915.

61. Reimers, L. (1997) *Transmission Electron Microscopy*, 4th ed. Springer-Verlag, Berlin Heidelberg New York.

62. Stadelmann, P. (2006) http://cimewww.epfl.ch/people/stadelmann/jemsWebSite/ jems.html.

Further Reading

Kröger, R., Paskova, T., Figge, S. *et al.* (2007) *Applied Physics Letters*, **90**, 081918.

PART III
Nonpolar Heterostructures and Devices

Nitrides with Nonpolar Surfaces. Growth, Properties, and Devices. Edited by Tanya Paskova

ISBN: 978-3-527-40768-2

12
Nonpolar Nitride Heterostructures and Devices grown by MOCVD

Arpan Chakraborty, Shigefusa Chichibu, and Umesh Mishra

12.1
Introduction

Current state-of-the-art wurtzite (Al,Ga,In)N heterostructure optoelectronic devices grown along the polar *c* direction are characterized by the presence of polarization-induced electric fields. These internal fields result in the spatial separation of the electron and hole wavefunctions within quantum wells, and give rise to a red shift of peak emission wavelengths owing to the quantum confined Stark effect, a blue shift of peak emission wavelengths with increasing injection current, and reduced radiative recombination efficiency because of limited wavefunction overlap. In contrast, growth of light emitters along nonpolar directions (e.g. growth along the *a*- and *m*-directions) is an effective means to circumvent the adverse effects of polarization. In recent years, several groups have confirmed the absence of polarization effects in nonpolar III-nitride-based heterostructures. A nonpolar *m*-plane GaN/AlGaN multiple-quantum well (MQW) structure free of polarization effects was first demonstrated by Waltereit *et al.* in the year 2000 [1]. The MQW structure was grown by plasma-assisted molecular beam epitaxy (MBE) using lithium aluminate substrates. Since this first demonstration, free-standing *m*-plane GaN substrates grown by hydride vapor phase epitaxy (HVPE) were used for subsequent epitaxial GaN/AlGaN MQW growths by both MBE [2] and metalorganic chemical vapor deposition (MOCVD) [3]. Growth and characterization of *m*-plane InGaN/GaN MQW by MBE was reported by Sun *et al.* [4–7]. In addition to the *m*-plane, research efforts have investigated *a*-plane GaN/AlGaN MQW structures grown on *r*-plane sapphire substrates by MBE [8] and MOCVD [9–10]. Optical characterization of these structures has shown that nonpolar quantum wells are not affected by polarization-induced electric fields.

Nitrides with Nonpolar Surfaces. Growth, Properties, and Devices. Edited by Tanya Paskova

ISBN: 978-3-527-40768-2

12.2 Nonpolar InGaN/GaN Multiple-quantum Wells

In spite of the above reports on the growth and characterization of nonpolar quantum wells (QWs), there have been very few reports on the growth and properties of *a*- and *m*-plane InGaN/GaN MQWs grown by MOCVD. Moreover, in comparison to *c*-plane InGaN/GaN MQWs, very little is known about the structural and optical properties of nonpolar InGaN/GaN MQWs. The purpose of this section of the chapter is to describe the structural and optical characterization of nonpolar *a*-plane and *m*-plane InGaN/GaN MQW structures grown by MOCVD. In particular, the presence of defects and its effect on the luminescence properties of MQWs grown along different planes have been investigated and compared.

12.2.1 Growth and Properties of *a*-Plane InGaN/GaN MQWs

In this section, the properties of nonpolar $(11\bar{2}0)$ *a*-plane InGaN/GaN MQWs grown by MOCVD on both reduced-defect density lateral epitaxial overgrown lateral epitaxial overgrowth (LEO) *a*-plane GaN and high-defect density planar *a*-plane GaN are discussed. In particular, In incorporation, surface morphology, and photoluminescence (PL) characteristics of the MQWs on the low extended defect-density wing regions of the laterally overgrown region were compared with the properties of the MQWs grown on the defective window regions as well as on the high-defect density planar *a*-plane GaN template.

12.2.1.1 Experiment

The InGaN/GaN MQW samples studied here were regrown simultaneously on both 10-μm thick, fully coalesced, reduced-dislocation density *a*-plane LEO GaN templates fabricated by HVPE, and on MOCVD-grown 2-μm-thick planar *a*-plane GaN templates. For comparison, a 2-μm-thick planar *c*-plane GaN template was also coloaded during the regrowths. Details of the HVPE-based LEO GaN template process may be found elsewhere [11]. The mask for this particular LEO template consisted of parallel 15-μm-wide SiO_2 stripes separated by 5-μm-wide window openings oriented parallel to the GaN $\langle 1\bar{1}00 \rangle$ direction. Transmission electron microscopy (TEM) on similar samples showed that the threading dislocation (TD) and basal plane stacking fault (SF) densities in the overgrown GaN were below $\sim 5 \times 10^6$ cm^{-2} and 3×10^3 cm^{-1}, respectively. The TD and SF densities in the window region of the LEO GaN template were $\sim 9 \times 10^9$ cm^{-2} and 4×10^5 cm^{-1}, respectively. Previously conducted TEM investigations on planar *a*-plane GaN template showed TD and SF densities on the order of $\sim 2.6 \times 10^{10}$ cm^{-2} and 3.8×10^5 cm^{-1}, respectively [12]. The InGaN/GaN MQW structures were grown in a vertical close-spaced showerhead MOCVD reactor using trimethylgallium (TMGa), trimethylindium (TMIn), disilane (DiSi), and ammonia (NH_3) as precursors. First, a 1-μm-thick Si-doped GaN layer was deposited on the templates followed by 12 periods of InGaN/GaN MQWs with Si-doped GaN barriers. Following the growth, symmetric high-resolution x-ray diffraction (HRXRD) $\omega-2\theta$ scans were recorded with a Philips MRD X'Pert PRO triple axis diffractometer to determine the structural parameters of the MQWs. Atomic force microscopy (AFM) performed on a digital instrument D3000 operating in tapping mode was used to image the surface morphology. The optical characterization was carried out at room temperature using a UV-microphotoluminescence (μ-PL) system in confocal geometry. A reflecting microscope objective (52X, numerical aperture (NA) 0.65) was used to excite the sample and collect the signal at the same time. For excitation, the 325 nm line of a He-Cd laser (continuous wave (cw), 17 mW) was used. The final spot diameter was ~3 μm. The μ-PL signal was analyzed using a 0.55-m spectrometer and a charge-coupled device charge-coupled device (CCD) camera. Wide-area cw -PL measurements were carried out using the 325-nm line of a He–Cd laser with an excitation power density of ~20 W cm^{-2}. Spatially integrated or resolved cathodoluminescence (CL) was excited by a cw electron beam (*e* beam) with or without the scanning, and dispersed by a 20-cm focal-length grating monochromator equipped with a scanning electron microscope (SEM). The acceleration energy and current of the *e* beam were typically 5 kV and 500 pA, respectively. Monochromatic and spatially-resolved CL images were taken under a fixed wavelength at 300 K.

12.2.1.2 Results

Figure 12.1 shows a comparison of the $\omega-2\theta$ scans for concurrently grown InGaN/GaN MQWs on LEO *a*-plane GaN and planar *a*-plane GaN template showing a significant difference in the In composition between the LEO and the planar *a*-plane film. The structural parameters of the quantum wells were obtained by analyzing the angular positions of the satellite peaks, and also on the basis of growth times of the wells and the barriers. For the MQWs grown on the LEO template, the average In composition in the wells was determined to be 4.4% with the well and barrier thicknesses being 27 and 87 Å, respectively. For the MQWs grown on the planar *a*-plane GaN, the average In composition of the wells was 12.5%, and the well and barrier thicknesses were 30 and 90 Å, respectively. The In composition in the MQWs grown on the *c*-plane GaN sample was so high that the sample had a yellow color, and reliable structural parameters could not be extracted from the X-ray scan.

The In incorporation in reduced-defect *a*-plane GaN and planar *c*-plane GaN as a function of the growth temperature is plotted in Figure 12.2. It can be seen that the In incorporation efficiency in *a*-plane GaN is much smaller compared with that in *c*- plane for all temperatures. This suggests that reduced-defect *a*-plane GaN is a low In incorporating plane.

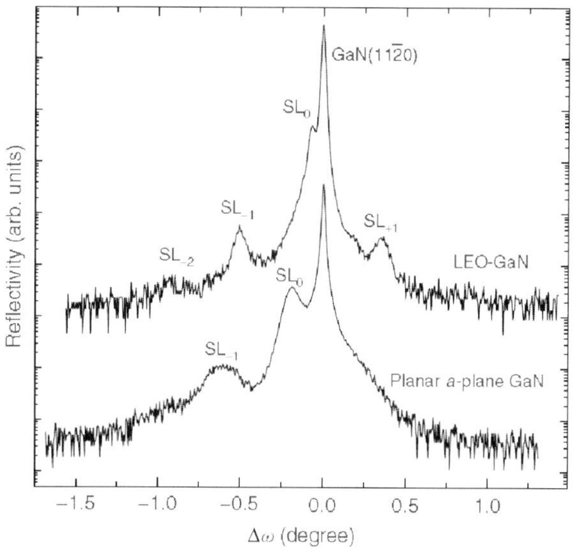

Figure 12.1 HRXRD ω–2θ scans for the InGaN/GaN MQWs grown in the same growth run on (a) LEO *a*-plane GaN and (b) planar *a*-plane GaN template.

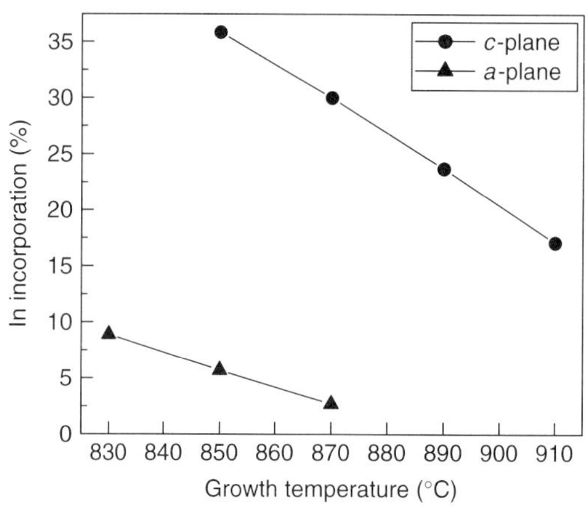

Figure 12.2 Comparison of In incorporation in *a*-plane and *c*-plane GaN at different temperatures.

Figure 12.3 shows a 20 μm × 20 μm AFM micrograph of the surface of the InGaN/GaN MQWs (terminated by an InGaN well) grown on a LEO *a*-plane GaN template. Smooth and rough areas were clearly visible in the micrograph. The virtually defect-free Ga-face wings, growing along the [0001] direction, exhibited a smooth surface morphology with minimal surface pitting. For the N-face wings, growing along the [000$\bar{1}$] direction, however,

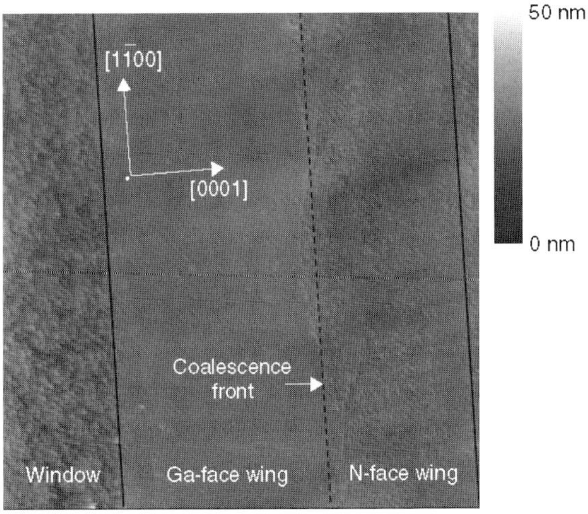

Figure 12.3 20 μm × 20 μm AFM micrograph showing the surface morphology in the Ga-face wing, N-face wing, and the window regions of the LEO sample.

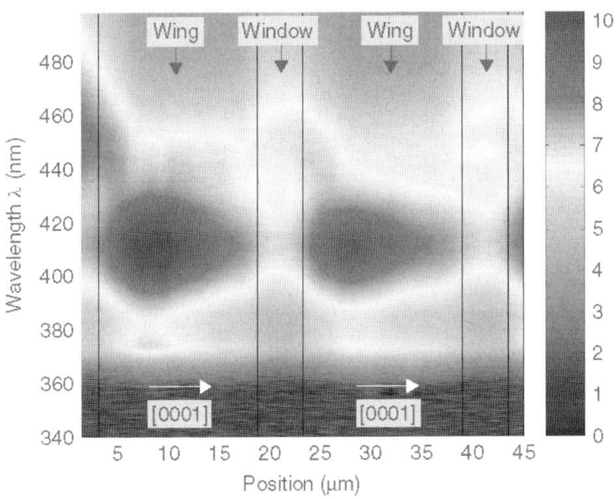

Figure 12.4 Contour plot of the μ-PL intensity on the wing and window regions of the LEO sample as a function of the wavelength (the intensity is plotted in a logarithmic scale) (cf. Color Plate XXXIV).

surface defects (pits) formed leading to an increase in surface roughness. The highest density of surface-pits was observed in the dislocated window region, resulting in an extremely rough surface morphology. The observation of surface defects is closely related to the presence of microstructural defects in the material [11].

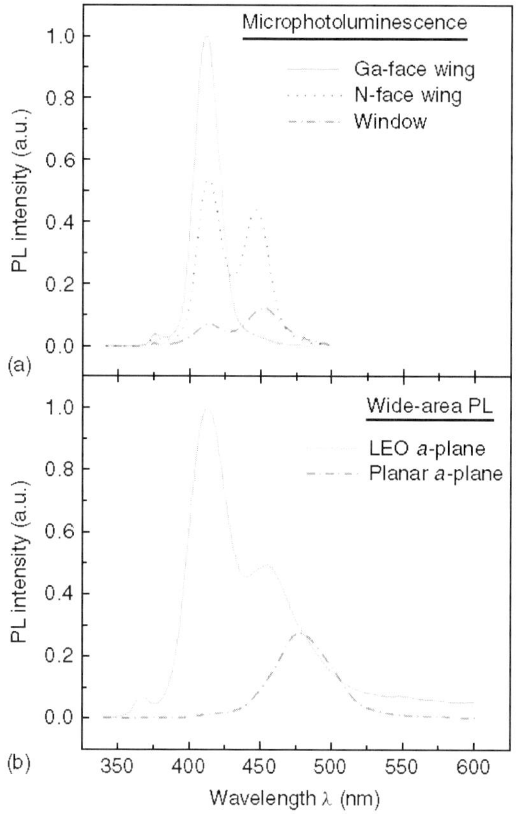

Figure 12.5 (a) μ-PL measurements on the Ga-face wing, N-face wing, and the window region of the LEO sample and (b) wide-area PL measurement on the LEO sample and the planar *a*-plane sample.

Figure 12.4 shows a contour plot of the μ-PL measurement performed on the MQWs grown on the LEO GaN template. The scan direction was perpendicular to the stripe direction (i.e. along [0001]). Two adjacent stripes were spatially scanned at intervals of 1 μm. At every position, the PL emission intensity was plotted for the spectral range $340 < \lambda < 500$ nm. Strong QW emission around 410 nm was observed from the reduced-dislocation density wings. In contrast, the emission from the defective window regions was an order of magnitude lower in intensity. Also, a second emission peak around 450 nm was observed from the windows. Figure 12.5a compares the PL spectra taken from regions on the Ga-face wing, the N-face wing, and the window of the LEO sample. Whereas only one strong emission peak at 413 nm with a full-width-at-half-maximum (FWHM) of 22.7 nm was seen on the Ga-face wing; two emission peaks at 413 and 446 nm were recorded on the N-face wing. Both peaks were about half as intense as the emission from the nearly defect-free Ga-face wing. Two weak emission peaks at ~410 and 453 nm were recorded from the defective windows as well. The appearance of the longer wavelength emission peaks in the N-face and window regions were correlated with enhanced In incorporation at surface defects observed by AFM.

Wide-area PL measurements were performed on the same LEO sample as was examined with μ-PL (Figure 12.5b). The spectra were characterized by a strong emission peak at 413 nm and the presence of a broad shoulder at ~450 nm. As shown in the μ-PL measurements, the shoulder can be attributed to the emission from the N-face wing and the window regions of the LEO sample. In contrast, wide-area PL measurement on the planar *a*-plane InGaN/GaN MQW sample exhibited only one peak at 478 nm with about one-fourth less intense emission compared to that from the LEO sample. The lower PL intensity and the absence of a shorter wavelength emission peak from the planar sample are the result of the significantly higher defect density in the relatively thin *a*-plane GaN template.

The longer wavelength PL emission from the planar *a*-plane sample agreed with the higher In content, determined from the HRXRD analysis. The results showed that the In incorporation and hence the PL emission wavelength was considerably higher in defective material. The presence of surface defects with facets other than the $(11\bar{2}0)$ orientation led to the increased In incorporation in the defective areas. Similar behavior was observed in the presence of the so-called 'V-defects' in *c*-plane InGaN/GaN MQWs by Wu *et al.* [13]. However, the reasons for lower In incorporation efficiency in *a*-plane GaN compared with *c*-plane GaN are unclear.

During SEM observation of the surface, in addition to the majority of regions that exhibited fully coalesced flat wings, uncoalesced and irregularly shaped wings were observed on occasion indicating that the LEO growth conditions were not fully optimized yet. A plan-view SEM image, which covers the regions of coalesced, uncoalesced, and abnormally grown wings of $(11\bar{2}0)$ 5.5-nm-thick $In_{0.06}Ga_{0.94}N/GaN$ MQWs is shown in Figure 12.6. The wing regions consisted of Ga- and N-face wings, which have (0001) and $(000\bar{1})$ vertical sidewalls, respectively. The Ga-face wings were six times wider than that of the N-face ones along the ⟨0001⟩ direction, reflecting the growth rate anisotropy. Some Ga-face wings exhibited paler parallelograms, and these anomalous regions have $\{20\bar{2}1\}$ facets, since their sides are 28° tilted from the underlying ⟨0001⟩ direction.

PL spectra at 8 and 300 K are shown in Figure 12.6b. The spectra exhibited PL peaks (bands) at 3.25, 3.14, 2.92, and 2.60 eV, which are labeled as UV, V, P, and B, respectively. Note that peak energies of these emissions did not change remarkably with the decrease in Λ, confirming that polarization fields were absent. A wide-area ($70 \times 50\ \mu m^2$) scanning CL spectrum at 300 K exhibited the corresponding peaks, as shown in Figure 12.6c. Spot-excitation CL spectra at 300 K are shown in bottom four traces in Figure 12.6c. Monochromatic CL images taken at the peak energies of UV, V, P, and B peaks are shown in

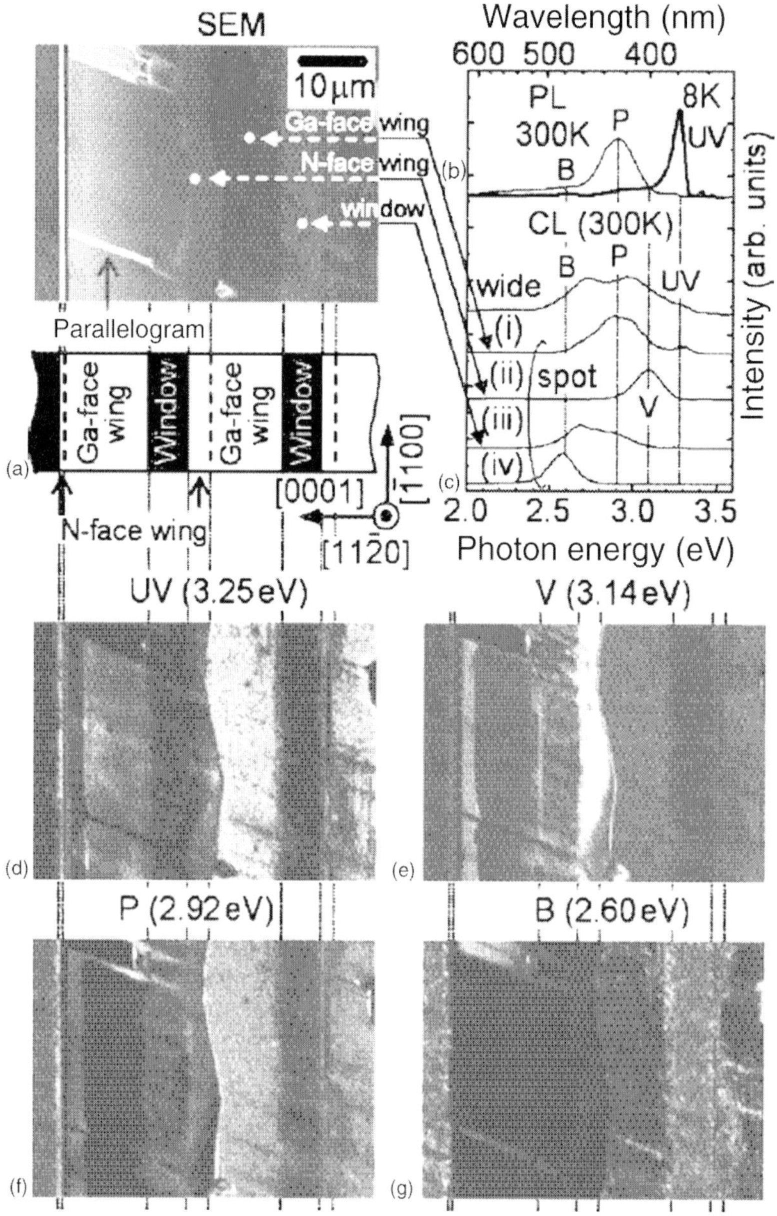

Figure 12.6 (a) Plan-view SEM image of (11$\bar{2}$0) 5.5-nm-thick $In_{0.06}$ $Ga_{0.94}N/GaN$ MQWs on LEO GaN; (b) PL spectra at 8 and 300 K; (c) a wide-area (70 × 50 μm^2) scanning CL spectrum and spot-excitation CL spectra in (i) coalesced Ga-face wing, (ii) N-face wing, (iii) window, and (iv) the parallelogram region in the uncoalesced Ga-face wing at 300 K. Monochromatic scanning CL images taken at (d) UV (3.25 eV); (e) V (3.14 eV); (f) P (2.92 eV); (g) B (2.60 eV) peaks.

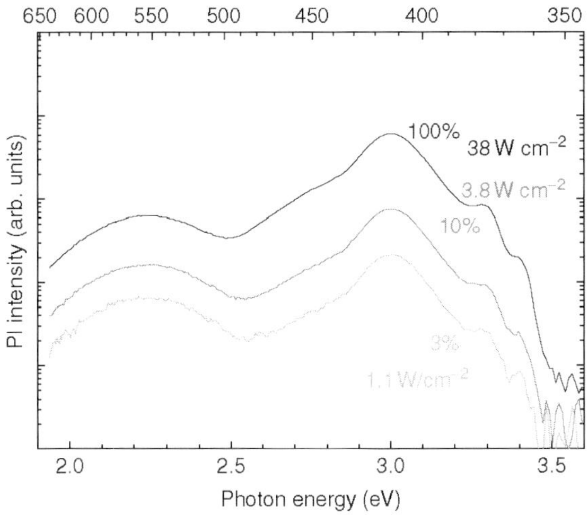

Figure 12.7 Excitation dependent PL measurements on an *a*-plane InGaN/GaN MQW sample.

Figure 12.6d–g, respectively. From these results, local *x* values of the QWs were recognized to vary depending on the base structure, although the macroscopic luminescence spectrum was dominated by the UV peak at 3.29 eV at 8 K and P peak at 2.92 eV at 300 K. For example, the dominant peak P was observed from the broadest Ga-face wing and window regions, while the V peak at 3.14 eV was found in the N-face wing and the window regions. Note that the window region exhibited an additional peak, B, at 2.60 eV (Figure 12.6g), indicating that defective regions have large inhomogeneity in *x*. The incorporation efficiency of In is considered to depend on the defect density and growth orientations, since the parallelograms on the Ga-face wing and the window regions exhibited the B peak (2.60 eV), while Ga-face and N-face wings exhibited P, V, and UV peaks. The former two regions have either $\{20\bar{2}1\}$ facet or spiral step edges initiated by TDs, which would not form in the LEO wings. The monochromatic CL mapping image of the UV peak (3.25 eV) exhibited a sharp contrast between the LEO wings and the window (Figure 12.6d) indicating that its intensity is strongly influenced by the TD (or nonradiative defect) density, as is the case with the near band edge (NBE) emissions in GaN and $In_xGa_{1-x}N$ alloys of very small *x* values. However, such contrast between the window and wings was less significant for the V and P peaks, as shown in Figure 12.6e and f, respectively. This result may reflect the defect-resistant nature of the localized excitons owing to the InN-rich disordered quantum nanoclusters reported for polar (0001) $In_xGa_{1-x}N$/GaN GaN QWs.

Excitation power density dependence PL measurements were carried out for an $In_{0.03}$ $Ga_{0.97}N$ (2.7 nm)/GaN (6.0 nm) MQW. Photoluminescence was excited by the 325.0 nm line of a cw He-Cd laser (38 W cm^{-2}). Neutral density filters were used to reduce the laser power. The excitation power density was changed by almost two orders of magnitude. However, no change in the PL emission wavelength was observed. This suggests the absence of polarization-induced electric field in the quantum wells.

12.2.1.3 Summary

In conclusion, nonpolar *a*-plane InGaN/GaN MQWs were grown on reduced-defect density LEO GaN templates as well as on high-defect density planar *a*-plane GaN templates. μ-PL measurements of the MQWs grown on the LEO GaN template showed an emission that was ten times stronger in the reduced-defect density wing compared with that from the defective window region. Spatially resolved μ-PL and CL measurements also showed the presence of additional longer wavelength emission peaks in the defective regions of the LEO sample. Evidence for enhanced indium incorporation in quantum wells grown on defective planar *a*-plane GaN in comparison to reduced-defect density LEO *a*-plane GaN was also seen in wide-area PL and HRXRD measurements. The enhanced indium incorporation was linked to the presence of surface defects on the sample. Excitation dependent PL measurements showed the absence of polarization-induced electric fields in the MQW.

12.2.2 Growth and Properties of *m*-Plane InGaN/GaN MQWs

In this section, the structural and optical properties of nonpolar $(1\bar{1}00)$ *m*-plane InGaN/GaN MQWs, grown simultaneously on reduced-defect LEO *m*-plane GaN and high-defect density planar *m*-plane GaN by MOCVD have been investigated. In particular, the effect of defects on In incorporation in *m*-plane has been studied and compared with that in *a*-plane MQW.

12.2.2.1 Experiment

The InGaN/GaN MQW samples studied here were grown simultaneously on both 10-μm thick, fully coalesced, reduced-defect density $(1\bar{1}00)$ *m*-plane LEO GaN templates grown by HVPE and on MOCVD-grown 1.2-μm-thick planar *m*-plane GaN templates. The LEO process utilized 140-nm-thick SiO_2 layers deposited on 0.5–1.0-μm MBE-grown *m*-plane GaN templates on *m*-plane 6H-SiC. The mask for this particular LEO template used in this study consisted of parallel 8-μm-wide SiO_2 stripes separated by 2-μm- wide window openings oriented along the $\langle 11\bar{2}0\rangle$ GaN direction. Details of the HVPE-based LEO GaN template process may be found elsewhere [14]. The root mean square (rms) roughness of the defective window regions was ~1 nm over 2 μm × 2 μm areas, while that of the low-defect density wing regions was ~0.5 nm over 5 μm × 5 μm areas. TEM on similar samples [12] showed that the TD and SF densities in the Ga-face wing were below $\sim 5 \times 10^6$ cm^{-2} and

3×10^3 cm^{-1}, respectively. The TD and SF densities in the N-face wing and window region of the LEO GaN template were $\sim 9 \times 10^9$ cm^{-2} and $\sim 10^5$ cm^{-1}, respectively.

The planar *m*-plane GaN templates were grown by MOCVD on *m*-plane 6H-SiC substrate using a 200-nm AlN nucleation layer. The rms roughness of the as-grown templates, measured by AFM, was $\sim$1.4 nm over 25 μm^2 area. TEM investigations on the planar *m*-plane GaN template showed TD and SF densities on the order of $\sim 2 \times 10^{10}$ cm^{-2} and $\sim 8 \times 10^5$ cm^{-1}, respectively [15]. The InGaN/GaN MQW structures were grown using TMGa, TMIn, DiSi, and NH_3 as precursors. First, a 1-μm-thick Si-doped GaN layer was deposited on the templates, which was followed by 12 periods of InGaN/GaN MQWs with undoped InGaN wells and Si-doped GaN barriers. Growth temperature for the MQWs was 880 °C and a reactor pressure of 100 torr was used. Ammonia to (TMGa + TMIn) ratio (V/III ratio) in excess of 30 000 was used for the MQW growths. Nitrogen was used as the carrier gas and a total gas flow of 12 slpm was used. Following the growth, the MQW structures were investigated by HRXRD and ω–2θ scans were recorded with a Philips MRD X'Pert PRO triple axis diffractometer with CuK$_{\alpha 1}$ radiation. Cross-sectional transmission electron microscope (XTEM) with an FEI Tecnai G2 Sphera microscope operating at 200 kV was used to image the MQW microstructure. The surface morphology was evaluated by AFM with a digital instrument D3100 operating in tapping mode. Spatially integrated or resolved CL was excited by a cw electron beam with or without the beam scanning. The acceleration voltage and current were typically 5 kV and 500 pA, respectively. The beam diameter was approximately 8.5 nm and the typical carrier concentration was 1.5×10^{18} cm^{-3} when the carrier lifetime was 1 ns. Monochromatic CL mapping images were taken under a fixed wavelength with slow beam scanning at 300 K. The spatial resolution was principally determined by the carrier diffusion length, which was shorter than $\sim$50–60 nm [16, 17]. All measurements were performed at room temperature.

12.2.2.2 **Results**

Figure 12.8 shows the HRXRD ω–2θ scans for concurrently grown InGaN/GaN MQWs on LEO *m*-plane GaN and planar *m*-plane GaN template. The structural parameters of the MQWs were determined by analyzing the position of the superlattice peaks of the X-ray scan, assuming that the superlattice was fully strained to the underlying GaN layer [18]. For the MQWs grown on the *m*-plane LEO template, the average In composition in the wells was determined to be 8% with the well and barrier thicknesses being 31 and 69 Å, respectively. For the MQWs grown on the planar *m*-plane GaN template, the average In composition of the wells was 8.4%, and the well and barrier thicknesses were 30 and 65 Å, respectively. Similar In incorporation in the two samples suggests that defects in *m*-plane GaN did not enhance In incorporation. This observation is in contrast to the higher incorporation of

In in defective *a*-plane GaN films compared with reduced-defect LEO *a*-plane GaN films [19].

Figure 12.9a shows a 10 μm × 10 μm image AFM micrograph of the surface of the InGaN/GaN MQWs grown on the LEO *m*-plane GaN template. Two

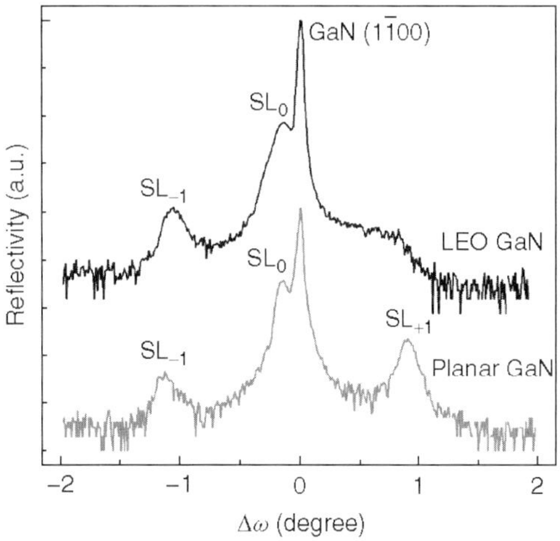

Figure 12.8 HRXRD ω–2θ scans for the InGaN/GaN MQWs grown on concurrently loaded LEO *m*-plane GaN and planar *m*-plane GaN template showing similar $\Delta\omega$ for the zeroth order superlattice peak, signifying similar average In composition.

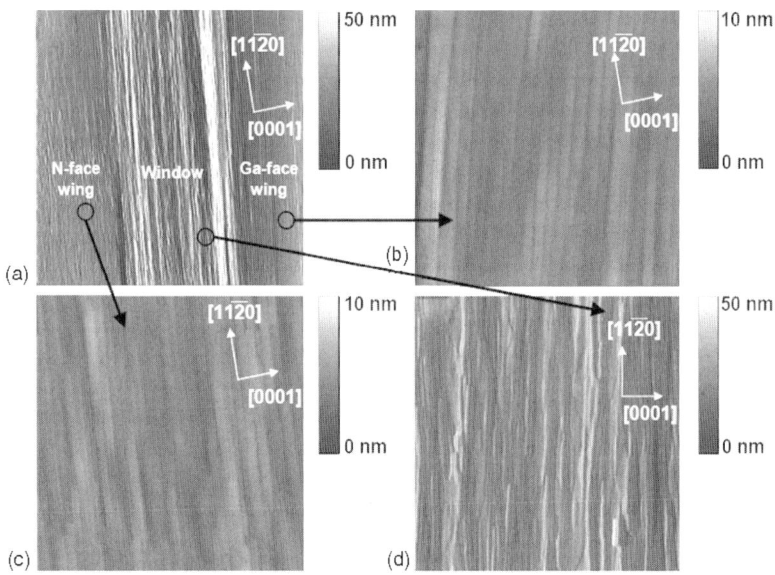

Figure 12.9 (a) 10 μm × 10 μm AFM micrograph showing the surface morphology of the LEO *m*-plane GaN after the InGaN/GaN MQW regrowth; (b) 2.5 μm × 2.5 μm AFM micrograph of the smooth Ga-face wing region; (c) 2.5 μm × 2.5 μm AFM micrograph of the rough N-face wing region; (d) 2.5 μm × 2.5 μm AFM micrograph of the rough and corrugated window region.

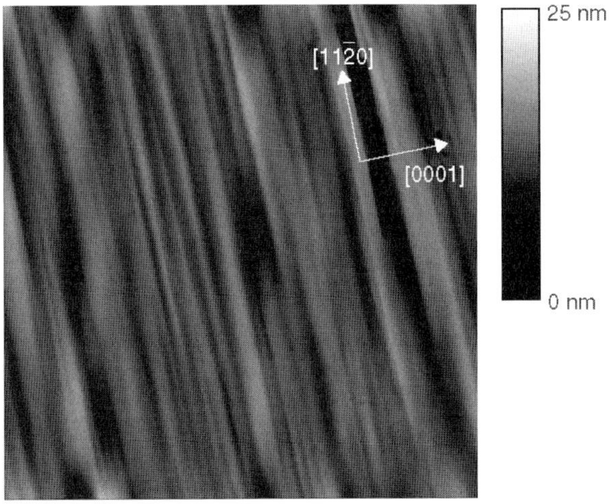

Figure 12.10 5 μm × 5 μm AFM micrograph showing the surface morphology of the InGaN/GaN MQW sample on planar *m*-plane GaN template.

distinct morphologies were apparent in the image. The Ga-face wings, growing along the [0001] direction, exhibited a much smoother, step-flow growth morphology as is shown in detail in Figure 12.9b. This step-flow growth morphology has only been observed on low-defect density *m*-plane GaN surfaces [20]. Figure 12.9c shows an enlarged micrograph of an N-face wing. In contrast to the Ga-terminated wings, the N-face wings growing along the [000$\bar{1}$] direction and the window regions exhibited the commonly observed 'slate' like morphology characteristic of *m*-plane GaN surfaces [21]. Unlike Ga-face wings, N-face wings were characterized by the presence of basal plane SFs that intersected the free surface along the ⟨11$\bar{2}$0⟩ direction. The approximate dimension of 0.1–0.2 μm of the slate feature in the [0001] direction is consistent with the typical spacings between SFs (as determined by TEM measurements). The highest density of surface-pits and SF-related morphological features was observed in the dislocated window region, resulting in an extremely rough surface morphology as shown in Figure 12.9d. Figure 12.10 shows an AFM image of the planar *m*-plane InGaN/GaN MQW sample. The surface mimicked the underlying template layer and was characterized by the presence of basal plane SF-related striations oriented along the ⟨11$\bar{2}$0⟩ direction. The rms roughness was ~1.4 nm over 25 μm^2 area. The observation of surface defects in the AFM micrographs is closely related to the presence of microstructural defects in the material [11, 14].

Figure 12.11 shows an XTEM image of the InGaN/GaN MQW sample grown on the LEO *m*-plane sample. On a nanometer scale, sharp and abrupt interfaces were observed. Measurements of the XTEM image indicated that the individual well and barrier thicknesses were in agreement with the thickness determined from the HRXRD ω–2θ scan. However, in lower-magnification

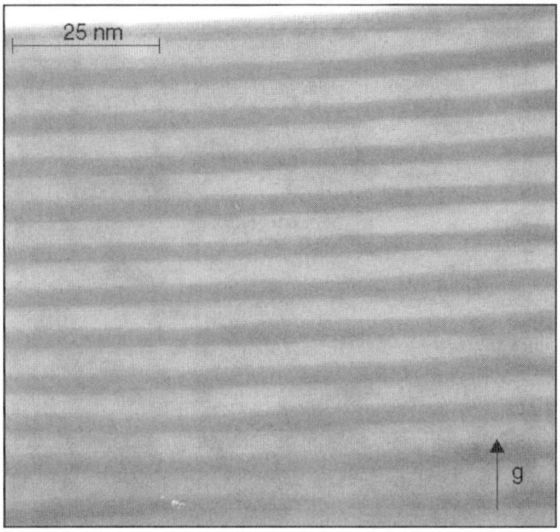

Figure 12.11 XTEM image of the 12-period InGaN/GaN MQW showing the sharp interfaces obtained by the MOCVD regrowth.

XTEM images (not shown), the QWs show lateral undulation with a wavelength and amplitude comparable with the surface height profile determined by AFM. We tentatively attribute the dampening of the higher order satellite peaks in the HRXRD $\omega-2\theta$ scan to the undulations in the QWs.

Spatially-resolved CL measurements were carried out on the *m*-plane $In_xGa_{1-x}N/GaN$ MQWs grown on LEO GaN to observe the relations between PL peak energies and exact locations. A plan-view SEM image, shown in Figure 12.12a, resembled the AFM image. A wide-area ($60 \times 45\ \mu m^2$) scanning CL spectrum at 300 K exhibited emission peaks at 3.24, 3.08, 2.93, and 2.83 eV, which are labeled UV, V, P, and B, respectively, as shown in Figure 12.12b. Spot-excitation CL spectra obtained from the Ga- and N-polar wings and the window are also shown. Monochromatic CL images taken at the peak energies of UV, V, P, and B peaks are shown in Figure 12.12c–f. From these results, the Ga-polar wings are found to exhibit shorter wavelength UV, V, and P peaks, while the N-polar wings and windows exhibit longer wavelength P and B peaks. Similar to the latter defective areas, PL spectrum of the MQW on planar-GaN exhibited P and B peaks. Therefore, local x in the QWs is recognized to vary with the base structure – x increases with the density of SFs and TDs, and consequently, striated, tilted, and uneven surfaces have higher In incorporation efficiency. The result is similar to that reported for polar *c*-plane InGaN QWs [16, 17, 22]. Clear dark lines and spots, which originate from SFs and TDs, and a sharp contrast between the Ga-polar wings and other areas were observed in the CL image of UV peak (Figure 12.12c). The result indicates that UV peak intensity is sensitive to nonradiative recombination centers normally associated with cation vacancies [23]). It should be pointed

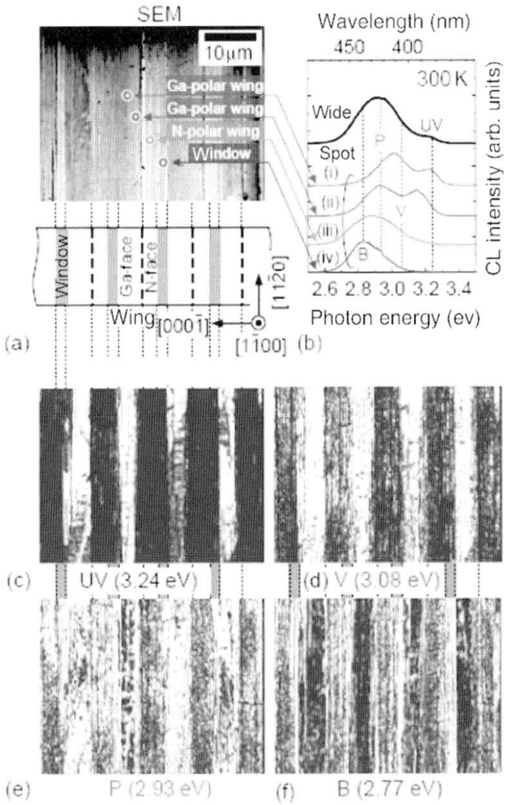

Figure 12.12 (a) Plan-view SEM image of the 12-period m-plane $In_{0.08}Ga_{0.92}N$ (3.1 nm)/GaN (6.9 nm) MQW on LEO GaN; (b) wide-area (60 × 45 μm^2) CL spectrum and spot-excitation CL spectra of the (i, ii) Ga-polar wing, (iii) N-polar wing, and (iv) window regions at 300 K. Monochromatic CL images taken at (c) 3.24 eV (UV); (d) 3.08 eV (V); (e) 2.93 eV (P); (f) 2.77 eV (B).

out that B peak is observed mainly in defective N-polar wings and windows (Figure 12.12f).

PL spectra of the MQWs on planar GaN and LEO GaN are shown as a function of temperature T in Figure 12.13a and b, respectively. The value of the equivalent internal quantum efficiency (η_{int}) at 300 K increased with the decrease in the peak energy from 1.1 (V peak) to 29% (B peak) for the MQW on planar-GaN. Corresponding values were 5.3 and 36%, respectively for the MQW on LEO GaN. The fivefold improvement in η_{int} for V peak is consistent with the decrease in TD and SF densities, which consequently gives rise to the decrease in the nonradiative recombination center concentration [23]. The MQW on LEO GaN exhibited an additional UV peak at 3.24 eV. Remarkably, high η_{int} values of B peak being 29% for the MQW on defective planar-GaN and 36% for the MQW on defective areas of LEO GaN suggest that B peak

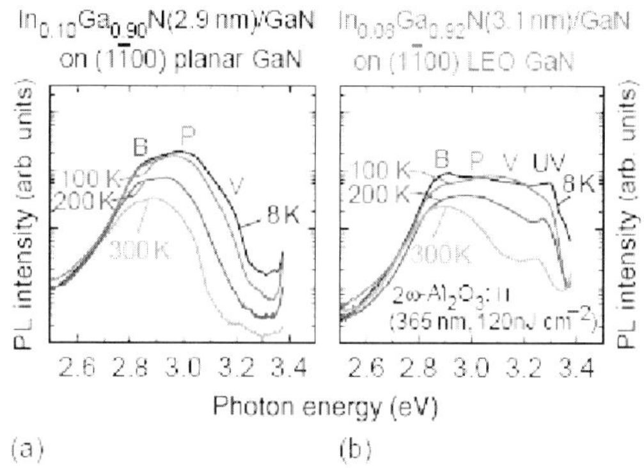

Figure 12.13 PL spectra of (a) the 12-period *m*-plane $In_{0.10}Ga_{0.90}N$ (2.9 nm)/GaN (6.5 nm) MQW grown on planar-GaN and (b) the 12-period *m*-plane $In_{0.08}Ga_{0.92}N$ (3.1 nm)/GaN (6.9 nm) MQW grown on LEO GaN as a function of T.

intensity is somewhat defect-insensitive [24] because of the beneficial effect of exciton localization into the atomic zig-zag chains [25] or atomic condensates of −In−N−.

12.2.2.3 **Summary**

In conclusion, 12-periods nonpolar *m*-plane InGaN/GaN MQWs structures were regrown by MOCVD on reduced-defect density LEO *m*-plane GaN templates as well as on high-defect density planar *m*-plane GaN templates. AFM measurements showed smooth steplike morphology in the Ga-face wing regions and presence of defects (mostly SFs) in the N-face wing and window regions. Cross-sectional transmission electron microscopy measurement showed sharp heterointerfaces, and the well and barrier thicknesses measured from the XTEM images agreed well with the values obtained from HRXRD. Spot-excitation CL measurement showed at least six times stronger emission from the Ga-face wing region compared with the adjacent window region. However, the presence of surface defects resulted in spatially inhomogeneous emission wavelength.

12.3 Nonpolar Light-emitting Diodes

12.3.1 Introduction and Background

In 2003, Chen *et al.* reported 363-nm ultraviolet LED using nonpolar *a*-plane AlGaN/GaN MQWs [25]. In 2004, Chitnis *et al.* reported visible LED using nonpolar *a*-plane InGaN/GaN MQWs [26]. Till 2004, there were no reports on optical power performance of nonpolar light emitters. The lack of low-defect density templates or substrates for the growth of LEDs was probably the reason for not realizing devices with decent optical power. The HVPE-grown reduced-defect nonpolar GaN templates, grown at UCSB by Haskell *et al.* [11], enabled the fabrication of InGaN/GaN MQWs-based visible LEDs with milliwatt output power. In 2004, the first visible *a*-plane LED with milliwatt-range output power was reported [27] and in 2005, *m*-plane LED was reported [28]. The first pulsed-power performance of *m*-plane LED lamps was reported in 2006 [29].

The purpose of this section of the chapter is to describe the growth, structural, electrical, and optical properties of nonpolar *a*-plane and *m*-plane LEDs using InGaN/GaN MQW structures grown by MOCVD.

12.3.2
Growth and Characterization of *a*-Plane InGaN/GaN LEDs

In this section, fabrication and electrical and luminescence characteristics of nonpolar *a*-plane violet, blue, and cyan InGaN/GaN LEDs using reduced-defect LEO *a*-plane GaN template have been discussed. In particular, the effect of drive current on output power, emission wavelength, and linewidth in the electroluminescence (EL) emission spectra were investigated. The current-voltage $(I-V)$ characteristics of the diodes have also been studied.

12.3.2.1 Experiment

The LED structures were regrown by MOCVD on ~10-μm-thick reduced-dislocation density *a*-plane GaN LEO template fabricated by HVPE. Details of the HVPE-based LEO process may be found in Ref. 11. The mask for the LEO consisted of parallel 8-μm-wide SiO_2 stripes separated by 2-μm-wide window openings oriented parallel to the GaN $\langle 1\bar{1}00 \rangle$ direction. The ratio of lateral growth rates of the Ga- and N-face {0001} wings was ~6 : 1, resulting in approximately 6.5-μm-wide defect-free overgrown areas between the windows and coalescence fronts. TEM of comparable samples has shown that TD and basal plane SF densities in the overgrown regions were lower than $\sim 5 \times 10^6\ \text{cm}^{-2}$ and $3 \times 10^3\ \text{cm}^{-1}$, respectively [11].

The regrowth consisted of a 2.2-μm Si-doped *n*-GaN base layer with an electron concentration of $2 \times 10^{18}\ \text{cm}^{-3}$ followed by the active region, which consisted of a 5 period MQW stack with 16-nm Si-doped GaN barriers and 4-nm $In_xGa_{1-x}N$ QWs. A 16-nm undoped GaN barrier separated the active region from the 0.3-μm Mg-doped *p*-type GaN layer with a hole concentration of $6 \times 10^{17}\ \text{cm}^{-3}$. The structure (shown in Figure 12.14) was capped with a 40-nm p^+-GaN contact layer.

Cross-sectional transmission electron microscope with an FEI Tecnai G2 Sphera microscope operating at 200 kV was used to image the MQW microstructure. The as-grown samples were investigated by PL measurements.

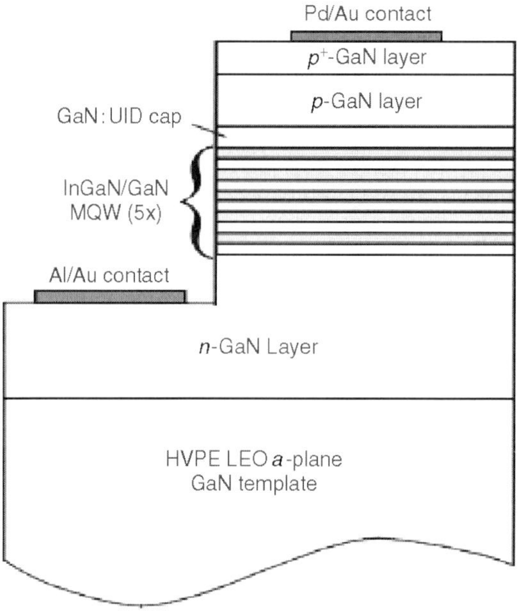

Figure 12.14 Schematic cross-section of the *a*-plane light-emitting diode structure.

$300 \times 300\,\mu m^2$ diode mesas were defined by chlorine-based reactive ion etching (RIE). Pd/Au (3/200 nm) and Al/Au (30/200 nm) were used as *p*-GaN and *n*-GaN contacts, respectively. The electrical and luminescence characteristics of the diode were measured by on-wafer probing of the devices. The I–V measurements were performed with a Hewlett-Packard 4155B semiconductor parameter analyzer. Relative optical power measurements under direct current (DC) conditions were obtained from the backside emission through the sapphire substrate onto a calibrated broad area Si photodiode. The emission spectrum and the optical power emission of the LEDs were measured as a function of drive currents. All measurements were carried out at room temperature.

12.3.2.2 **Results**

The XTEM micrographs from the reduced-defect wing region of the LED are shown in Figure 12.15. On a nanometer scale, sharp and abrupt interfaces were observed between the QWs and the barriers (Figure 12.15a). However, in lower-magnification XTEM images (Figure 12.15b), the QWs showed lateral undulation, and extended defects were observed. Unlike 'V-defects', which are commonly observed in *c*-plane InGaN/GaN MQWs, [13] the defects were asymmetric. These 'asymmetric V-defects' were characterized by the presence of an inclined non-$\{11\bar{2}0\}$-oriented facet and a vertical N-face $(000\bar{1})$

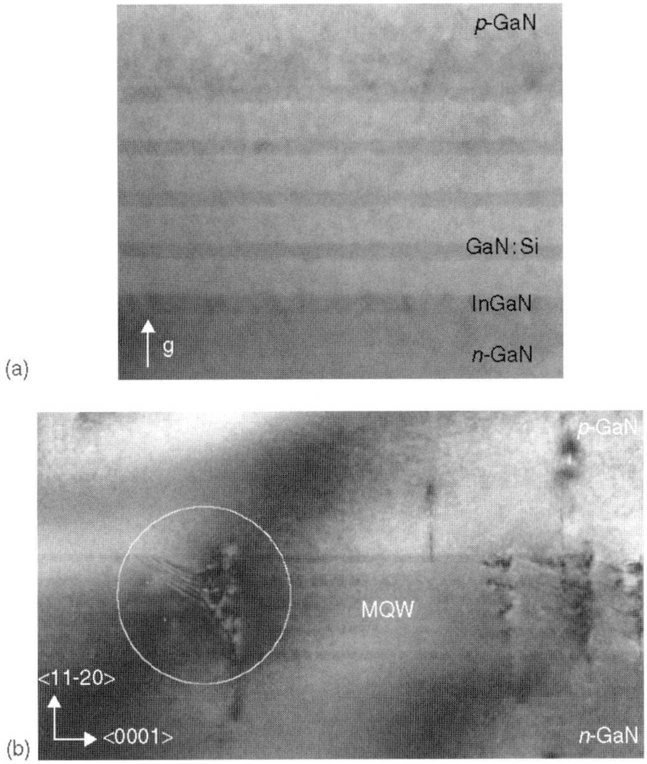

Figure 12.15 (a) High-magnification XTEM image showing the sharpness and abruptness of quantum-well interfaces; (b) low magnification image showing the generation of defects in MQWs. The encircled region shows an 'asymmetric V-defect'.

sidewall. The most stable of the inclined facets made an angle of $\sim 33^\circ$ with the $\langle 0001 \rangle$ direction and it was most likely to be either $\{20\bar{2}1\}$ or $\{11\bar{2}2\}$ facets. $\{20\bar{2}1\}$ facet was observed by Onuma *et al.* in the defective regions of *a*-plane InGaN/GaN MQWs [30] and $\{11\bar{2}2\}$ facets were seen in defective *a*-plane LEO samples by Haskell *et al.* [11]. The inclined and vertical facets are polar in nature. Like *c*-plane 'V-defects', the 'asymmetric V-defects' also originated from TDs. Furthermore, it was clearly seen from Figure 12.15b that the 'asymmetric V-defects' were generated during the growth of the MQW, which is most likely because of the instability of low-temperature InGaN growth with a reduced growth rate on the *a*-plane. Chakraborty *et al.* and Onuma *et al.* have shown that these facets facilitate higher In incorporation, and inhomogeneity in the local In content (x) is dependent on the defect density [19, 30]. The defect density increases from the Ga-face wing region to the N-face region and is largest in the window region.

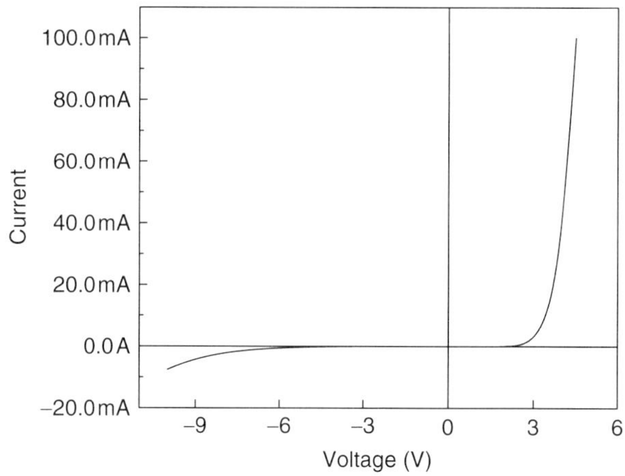

Figure 12.16 Current–voltage (*I*–*V*) characteristic of the violet *a*-plane LED.

The *I*–*V* curve (Figure 12.16) of the diode exhibited a forward voltage of 3.3 V with a low series resistance of 7.8 Ω. Nonpolar *a*-plane GaN *p-n* junction diodes grown under identical conditions on planar *a*-plane GaN templates exhibited similar forward voltage but had higher series resistances on the order of 30 Ω [31]. The lower series resistance in these LEDs can be attributed to the higher conductivity in the defect-free overgrown region of the LEO GaN template.

The dependence of the output power on the DC drive current was then measured. The output power increased sublinearly as the drive current was increased from 10 mA until it began to roll off at a current level close to 250 mA (Figure 12.17). The saturation of the output power can be attributed to heating

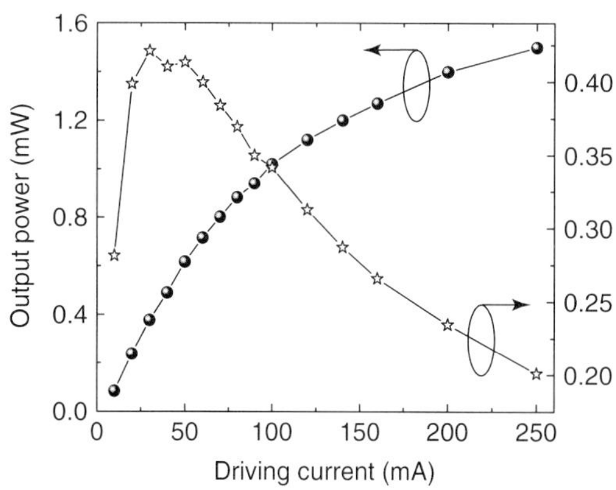

Figure 12.17 Variation of on-wafer output power and external-quantum efficiency of the violet *a*-plane LED with drive current.

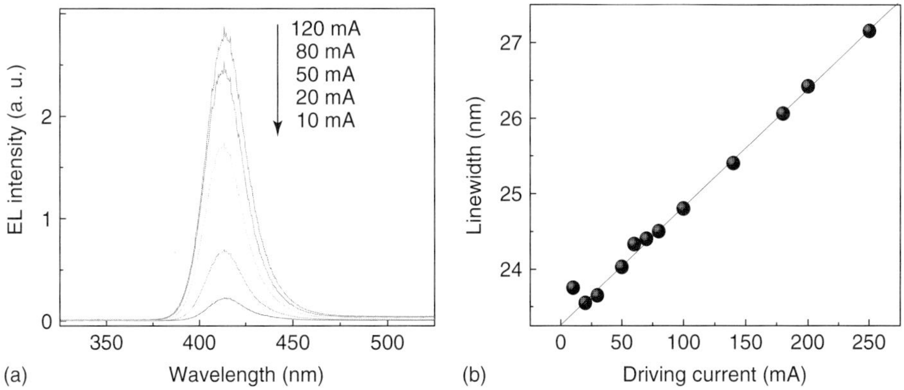

Figure 12.18 (a) EL spectra for different drive currents and (b) the variation of EL linewidth as a function of the drive current for the violet *a*-plane LED.

effects [32], thereby causing a reduction in the quantum efficiency. The output power at 20 mA forward current was 240 μW, corresponding to an external-quantum efficiency (EQE) of 0.4%. Direct current output power, as high as 1.5 mW, was measured for a drive current of 250 mA. The EQE increased as the drive current was increased, attaining a maximum of 0.42% at 30 mA, and then decreased rapidly as the forward current was increased beyond 30 mA. The comparably low EQE for these LEDs can be attributed partially to the poor reflectivity of the *p*-contact and partially to the 'dark' defective window regions of the LEO that do not emit light. An attempt of using transparent *p*-contact was made but severe current crowding was observed in the LEDs, most likely owing to the low conductivity of the *n*-type GaN layer.

The EL spectra of the devices were studied as a function of the DC drive current (Figure 12.18a). Emission spectra were measured at drive currents ranging from 10 to 250 mA. The devices emitted in the violet spectral range at 414 nm for all drive currents with minimal linewidth broadening (Figure 12.18b). The PL spectra on the as-grown sample (shown in Figure 12.19) showed a strong QW emission at 412 nm with a linewidth of 25 nm. The longer wavelength emission in the PL spectra is most likely because of defect-related deep-level emission from GaN layers. The absence of blue shift in the emission peak with increasing drive currents is in contrast to the commonly observed phenomenon of blue shift in *c*-plane LEDs working at this wavelength range and similar drive current range [33]. We attribute the absence of blue-shifted emission to the absence of polarization-induced electric field in the nonpolar *a*-plane QWs. The linewidth increased almost linearly with the driving current starting from a minimum of 23.5 nm at 20 mA to 27.5 nm at 250 mA. This minimal linewidth broadening with the increase in drive current suggests that the device heating was low in this current regime.

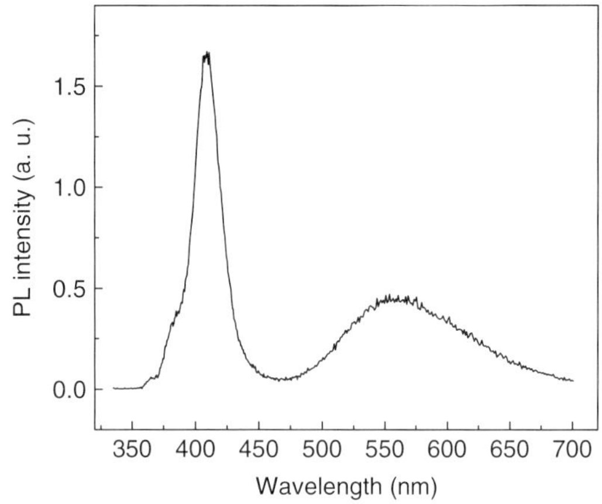

Figure 12.19 PL spectra for the violet *a*-plane LED.

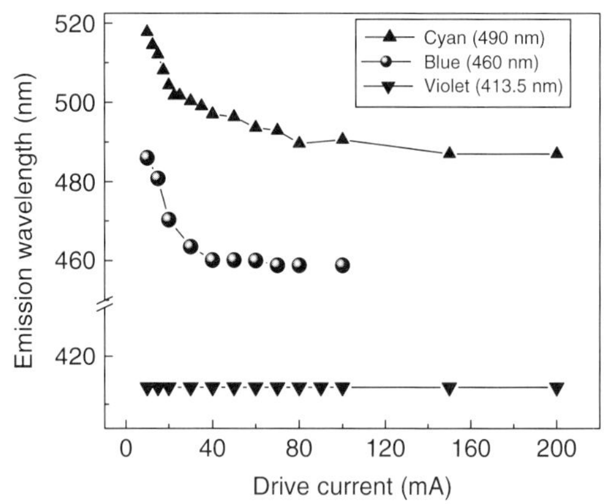

Figure 12.20 Comparison of the wavelength shift of *a*-plane LEDs emitting at different wavelengths of the visible spectrum.

12.3.2.3 **Wavelength Shift in Nonpolar *a*-Plane LEDs**

The EL spectra of the LEDs were measured at drive currents ranging from 10 to 200 mA. The wavelength shift for nonpolar *a*-plane violet, blue, and cyan LEDs is shown in Figure 12.20. It can be seen that the emission wavelength of the violet LED emitting at 413.5 nm does not shift with the increase in the drive current. However, the blue and the cyan LED emitting at 460 and 490 nm, respectively are characterized by current-dependent emission peak for small drive currents. However, for large drive currents, the emission peak saturates. An initial blue shift (~25–27 nm) was observed for the emission wavelength of the LEDs when the drive current was increased from 10 to 80 mA. For the cyan LED, the wavelength decreased from 517 to 490 nm, whereas for the blue LED, the wavelength increased from 486 to 460 nm. Above 80 mA, no appreciable shift in the emission wavelength was observed. We speculate that the initial shift in the emission wavelength with the increase in drive current is owing to the band-filling effect as a result of carrier localization in the 'asymmetric V-defects'. However, at higher drive currents, the emission wavelength is determined by the QW recombination in the defect-free regions.

12.3.3 Growth and Characterization of *m*-plane In GaN/GaN LEDs

In this section, the growth and fabrication of *m*-plane InGaN/GaN-based blue LEDs and packaged lamps, grown on free-standing *m*-plane GaN substrates, have been discussed. In particular, the cw and pulsed performance of the *m*-plane light emitters has been investigated.

12.3.3.1 Experiments

The *m*-plane LED structure described here was grown by MOCVD on 250-μm thick free-standing *m*-plane GaN substrates. The GaN substrates were grown in a conventional three-zone horizontal HVPE reactor at substrate temperatures of 860–890 °C. The substrates were optically transparent and mirrorlike, with gentle 50–200 nm peak-high undulations over millimeter length scales. AFM revealed a finely striated surface morphology with local rms roughness as low as 0.636 nm over 25 μm^2 areas. A SF density of $\sim 1 \times 10^5$ cm^{-1} and TD density of $\sim 4 \times 10^9$ cm^{-2} were determined from plan-view and cross-sectional TEM. Details of the HVPE-based free-standing *m*-plane GaN growth process may be found elsewhere [18].

The MOCVD-grown device structure, shown in Figure 12.21, consisted of a 2.2 μm Si-doped *n*-type GaN base layer with an electron concentration of 2×10^{18} cm^{-3}, followed by the active region, which consisted of a five period MQW stack with 16 nm Si-doped GaN barriers and 4-nm InGaN quantum wells. A 16-nm undoped GaN barrier separated the active region from the 0.3-μm Mg-doped *p*-type GaN layer with a hole concentration of 6×10^{17} cm^{-3}. A 40 nm p^+-GaN layer was used to cap the structure for improved electrical contact to the *p*-GaN layer. Two different LED structures, LED1 and LED2, were grown with the MQW growth temperatures of 870 and 890 °C, respectively.

The as-grown samples were investigated by optical microscopy and room temperature PL measurements. 300×300 μm^2 square diode were defined by chlorine-based RIE. Electron beam deposited Pd/Au (20/200 nm) and Al/Au (30/300 nm) were used as *p*-GaN and *n*-GaN contacts, respectively. For packaging, the wafer was diced into discrete dies for attachment to submounts. Conventional die-bonding and wire-bonding techniques were then applied, and the individual LEDs were packaged into conventional epoxy-encapsulated lamps. The electrical and luminescence characteristics of the diode were

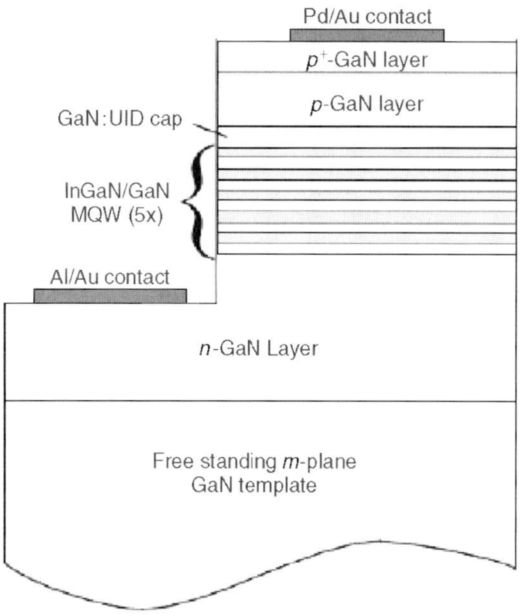

Figure 12.21 Schematic cross-section of the *m*-plane InGaN/GaN light-emitting diode structure.

measured by on-wafer probing of the devices. The $I-V$ measurements were performed with a Hewlett-Packard 4155B semiconductor parameter analyzer. Relative optical power measurements under direct current (dc) conditions were obtained from backside emission through the GaN substrate onto a calibrated broad area Si photodiode. The optical power and emission spectra for the packaged lamps were measured using an instrument system MAS40 mini-array-spectrometer with a built-in integrating sphere. The emission spectra and the optical power emission of the LEDs were measured as a function of DC drive currents. All measurements were carried out at room temperature.

12.3.3.2 **Results**

Figure 12.22 shows the current–voltage $(I-V)$ characteristic of LED1. The diode showed rectifying behavior with a sharp turn-on between 3 and 4 V with a dynamic series resistance of 16 Ω. The reverse current was ~300 μA at −15 V bias. The ideality factor, determined from the $(I-V)$ characteristic in the intermediate forward bias region, was ~4. The high ideality factor is most likely owing to the leakage current associated with the high extended defect density in the material.

Electroluminescence spectra of the devices were measured for DC drive currents ranging from 10 to 200 mA (Figures 12.23a and 12.24a). The devices emitted in the blue spectral range. LED1 had an emission peak

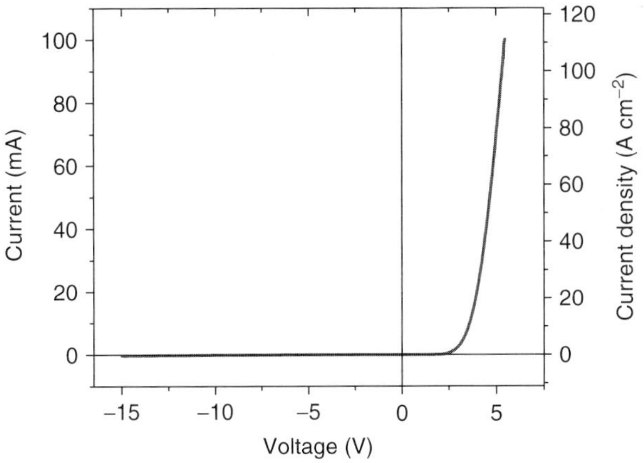

Figure 12.22 Current–voltage (I–V) characteristic of the m-plane LED (LED1).

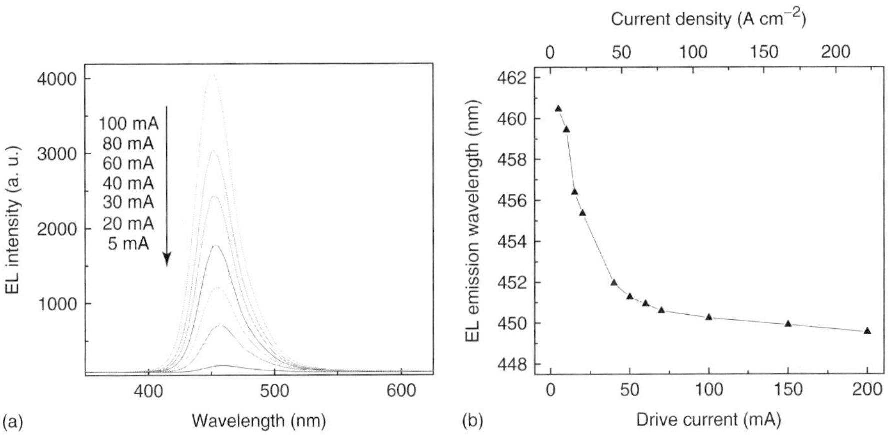

Figure 12.23 (a) EL spectra for different drive currents (LED1) and (b) the emission wavelength as a function of the drive current (LED1).

ranging from 460 nm at 5 mA to 449.6 nm at 200 mA. LED2 had an emission peak ranging from 432 nm at 5 mA to 423 nm at 200 mA. There was a blue shift (shown in Figures 12.23b and 12.24b) in the emission peak for drive current between 5 and 50 mA. This initial blue shift, also observed in some a-plane InGaN/GaN LEDs [26], can be attributed to the band filling of the localized states induced by alloy fluctuation in InGaN quantum wells [34]. However, there was minimal blue shift at higher drive currents (in excess of 70 mA, corresponding to a current density of 77.7 A/cm^2), indicating the absence of polarization-induced electric fields in the m-plane quantum wells. The emission wavelength of ~450 nm for LED

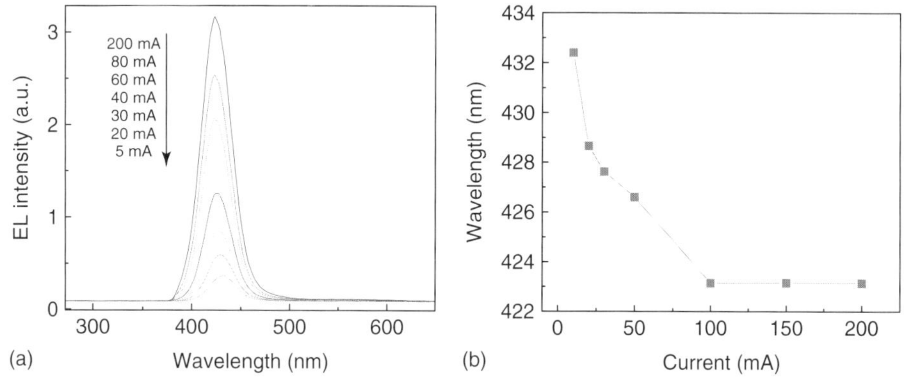

Figure 12.24 (a) EL spectra for different drive currents (LED2) and (b) the emission wavelength as a function of the drive current (LED2).

was the same as the PL emission peak. A linewidth of 38 and 35 nm was measured for the spectrum at 20 mA drive current for LED1 and LED2, respectively.

The output optical power increased almost linearly as the drive current was increased from 10 to 250 mA (Figure 12.25). There was no indication of 'roll-off' of the output power with the increase in drive current, as was observed in *a*-plane LEDs. The output power of LED1 at 20 mA forward current was 240 μW, corresponding to an EQE of 0.43%. cw output power as high as 2.95 mW was measured for a drive current of 300 mA, which represents the highest on-wafer output power reported for any nonpolar LED. The output power of LED2 at 20 and 250 mA forward current was 126 μW and 0.9 mW, respectively. The sharp drop in the EQE for drive currents in excess of 50 mA is probably a result of reduced radiative recombination efficiency at high currents owing to device heating.

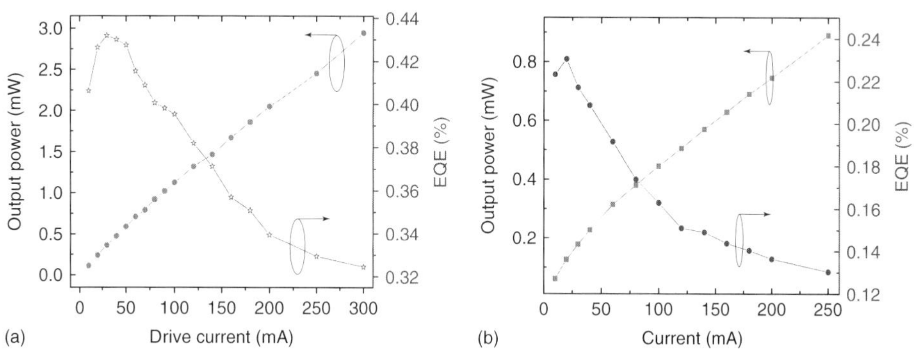

Figure 12.25 Variation of on-wafer cw output optical power and external-quantum efficiency with drive current for (a) LED1 and (b) LED2.

12.3.3.3 DC and Pulsed Performance of *m*-plane Packaged LED Lamps

The electrical and luminescence characteristics of the diodes were measured before and after packaging the LED1. The *I–V* characteristic of a typical *m*-plane LED lamp is shown in Figure 12.26. The diode showed a rectifying behavior with a turn-on voltage of ~3.2 V and a forward voltage of 3.9 V at 20 mA. The reverse current was <300 μA at a bias of −15 V and an ideality factor of ~4 was determined from the *I–V* characteristic in an intermediate forward bias region. The dynamic series resistance decreased from 16 to 11 Ω upon packaging owing to the improvement in the contacts.

Figure 12.27 shows the EL spectra of the LED lamp under pulsed operation. The pulse width and duty cycle were set at 20 μs and 10%, respectively, and the drive current was varied from 3 to 500 mA. The devices emitted in the blue spectral range with an emission peak of 452 nm for drive currents in excess of 50 mA. An initial blue shift from 460 to 452 nm, observed when the drive current was increased from 3 to 50 mA (not shown in the figure), was attributed to the band filling of localized states [34]. Wavelength shift was not observed for drive currents above 50 mA and up to 1 A, even for low duty cycles.

Figure 12.28a shows a comparison of the DC output optical powers of an LED before and after packaging. The variation in EQE with the drive current for the LED lamp is plotted in Figure 12.28b. The output power almost linearly increased with the drive current for unpackaged as well as packaged devices. This is probably because of improved heat dissipation owing to the conducting *m*-plane GaN substrate. The on-wafer output power at 20 mA was 240 μW, and a peak on-wafer output power of 2.95 mW was measured at 300 mA. The output power increased by a factor of 2.8 upon packaging. The

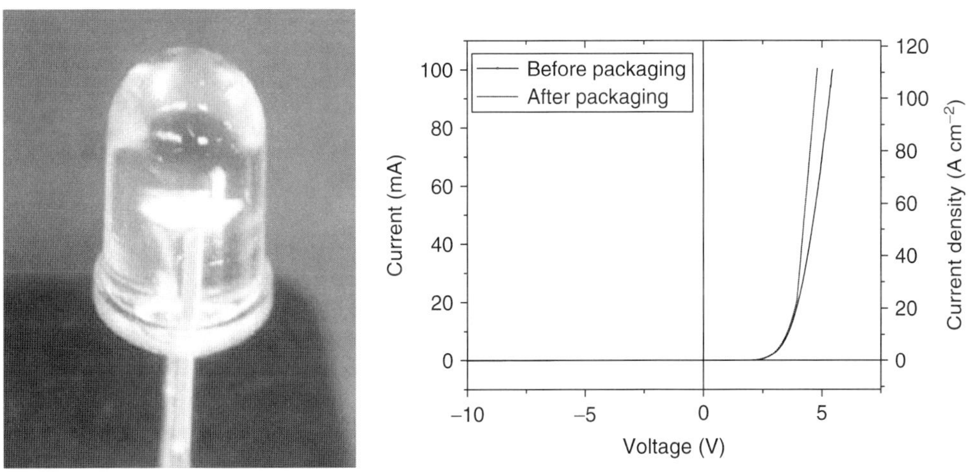

Figure 12.26 An image of the packaged *m*-plane lamp and comparison of the *I–V* characteristics of *m*-plane LED before and after packaging (cf. Color Plate XXXIV).

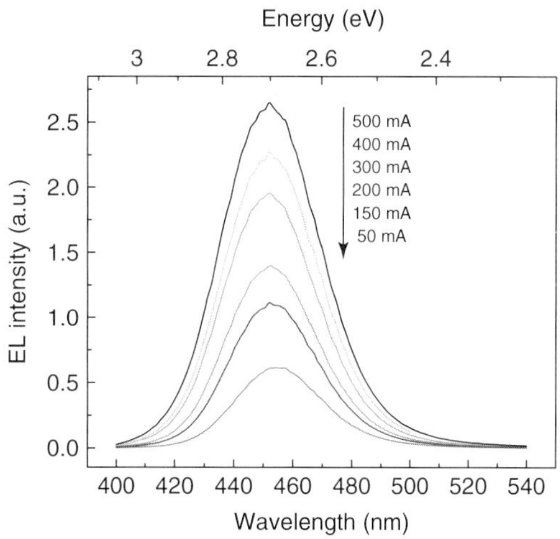

Figure 12.27 Emission spectra of LED lamp under pulsed operation.

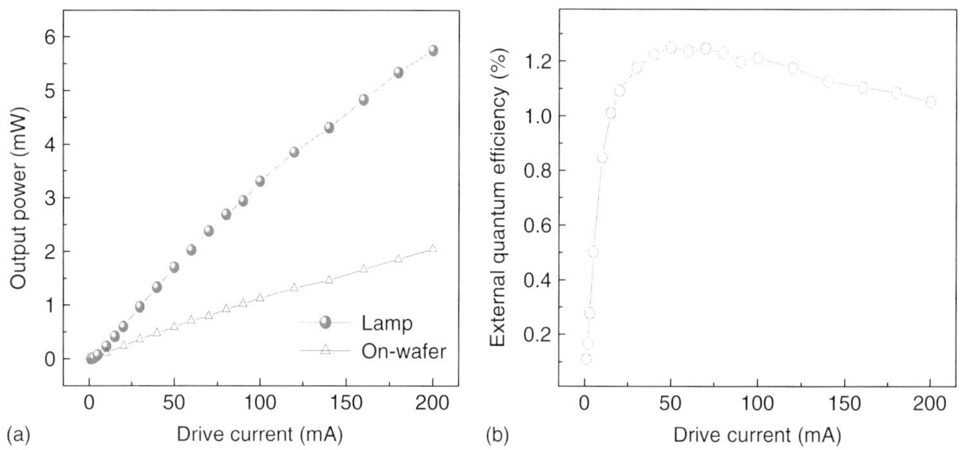

Figure 12.28 (a) DC output power versus drive current before and after packaging and (b) variation in external-quantum efficiency with drive current for LED lamp.

improvement can be attributed to the improved light extraction because of the epoxy encapsulation and light collection from the sides of the dies. The output power of the packaged LED lamp was 0.6 mW at 20 mA, corresponding to an EQE of 1.09%. An output power as high as 6 mW was measured at a drive current of 200 mA. A peak EQE of 1.25% was measured at 50 mA. Further improvement in output power is anticipated by employing transparent *p*-contact metals.

Figure 12.29 shows the pulsed-power performance of the packaged LED. The pulse width was kept constant at 20 μs, and the duty cycle was varied from 0.25 to 50%. The output power linearly increased with an increase in drive

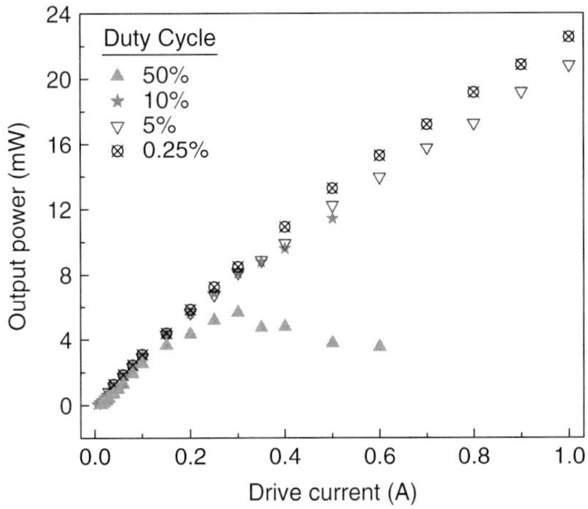

Figure 12.29 Variation in pulsed power with drive current for LED lamp for different duty cycles.

current for a duty cycle between 0.25 and 5%. Further increase in duty cycle resulted in a drop in output power for a drive current in excess of 300 mA. This drop can be attributed to a decrease in nonradiative recombination efficiency owing to self-heating. An output power of 23.5 mW was attained at a drive current of 1 A for a duty cycle of 0.25%.

12.3.3.4 **Transmission Through Free-standing *m*-plane Substrate**

Transmission through free-standing *m*-plane substrate was measured as a function of wavelength at different parts of the wafer (see Figure 12.30). It can be seen that there is a significant amount of absorption in the wafer, especially in the blue spectral range. To explain the increased absorption, secondary-ion-mass-spectroscopy (SIMS) was performed on the underside of

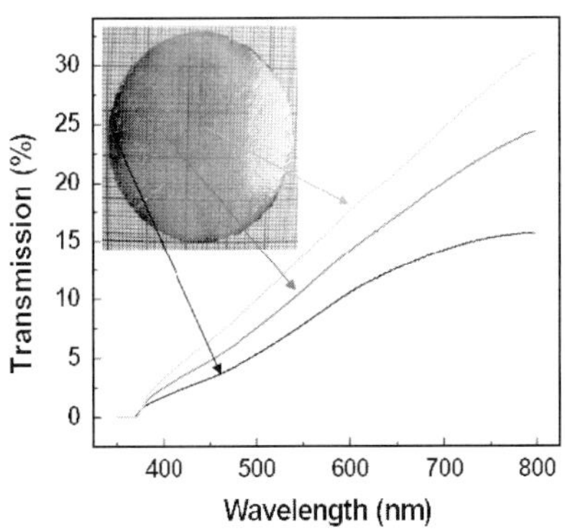

Figure 12.30 Transmission through *m*-plane substrate measured as a function of wavelength (cf. Color Plate XXXV).

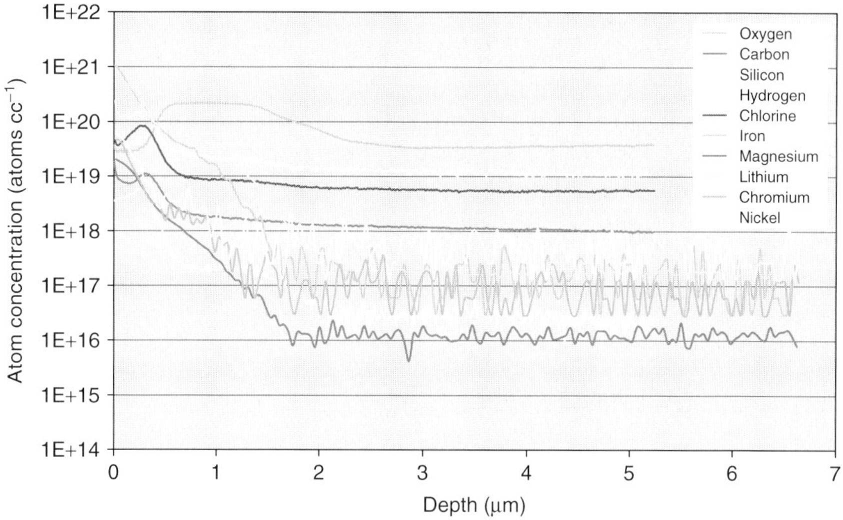

Figure 12.31 SIMS measurement performed at the back of the free-standing *m*-plane GaN substrate (courtesy of Haskell *et al.*) (cf. Color Plate XXXV).

the *m*-plane substrate (see Figure 12.31). SIMS revealed the incorporation of a significant amount of impurities like O, H, Cl, Fe, C, Li, Cr at the bottom 2-μm thickness of the wafer. The concentration of impurities dropped by three to four orders of magnitude after 10 μm of growth on $LiAlO_2$ substrate. The color of the wafer looked brown because of the presence of the impurities. However, the color was nonuniform along the radius of the wafer. This nonuniformity in the color resulted in nonuniform absorption during the transmission measurement.

12.3.4 Comparison of Power Performance of *a*-plane and *m*-plane LEDs

On comparing the L–I curves of the *a*-plane and *m*-plane LEDs (Figure 12.32), it was evident that the output power for the *a*-plane LED on *r*-plane sapphire substrate began to roll-off above 100 mA drive current. However, there was no sign of saturation for the *m*-plane LED on free-standing *m*-plane GaN substrate.

To explain the difference in the behavior of L–I curves for the two LEDs, thermal simulation was carried out using FEMLAB. The steady-state heat diffusion equation in 2D was solved and Joule heating in the LED was assumed to be the sole heat source. It was also assumed that the heat extraction was from the back of the wafer. In the simulation, it was assumed that the thermal conductivity of free-standing *m*-plane GaN was assumed as 1.3 W cm^{-1} K^{-1}

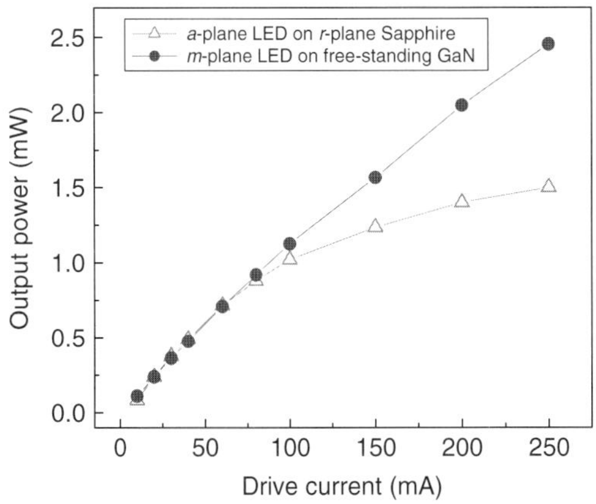

Figure 12.32 Comparison of DC *L*–*I* characteristic for *a*-plane and *m*-plane LEDs.

compared to 0.35 W cm^{-1} K^{-1} for sapphire. The results of the simulation are shown in Figure 12.33. It can be seen that the maximum temperature under 150 mA drive current for the *a*-plane LED is 105 °C compared to 39 °C for the *m*-plane LED. The difference in the temperature is even higher at higher operating currents. Thus, the difference in the temperature is owing to better heat sinking in free-standing GaN substrate because of its higher thermal conductivity.

Thus, the difference in the output power for *m*-plane and *a*-plane LED for drive currents in excess of 100 mA is owing to the difference in the operating temperature of the device. Higher temperature results in the decrease of radiative recombination efficiency, thereby resulting in the drop of the optical output power [35]. The efficiency of the nonpolar LED can be improved by adopting a flip-chip geometry.

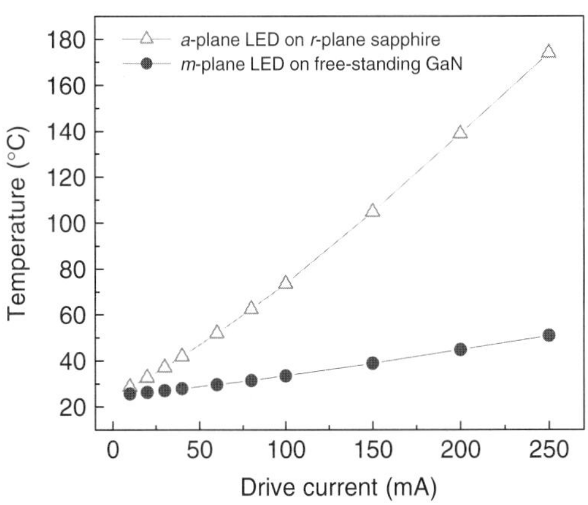

Figure 12.33 Comparision of the operating temperature for *a*-plane and *m*-plane LEDs under different drive currents.

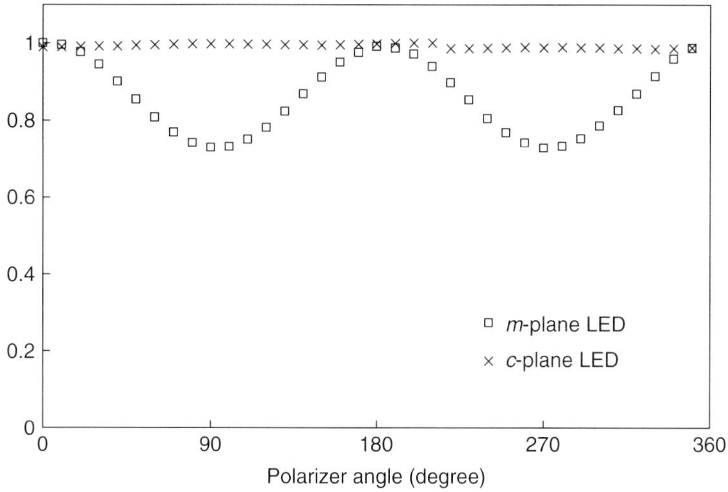

Figure 12.34 Spectrally integrated EL intensity of *m*-plane InGaN/GaN MQW LEDs as a function of polarization angle ($J_F = 400\ \mathrm{A\ cm^{-2}}$). The angle of 0° corresponds to a polarization parallel to the *a* axis. The same measurement on a *c*-plane LED is plotted for comparison (courtesy of Masui *et al.*).

12.4
Polarized Light Emission

Polarization of light emission from the top surface of *m*-plane LEDs was evaluated by Masui *et al.* [36] using a rotational polarizer between the LEDs and a Si photo detector. Figure 12.34 shows the dependence of EL intensity on polarizer angle, measured at 400 $\mathrm{Acm^{-2}}$ of LED forward current density (J_F). This result clearly shows the polarized nature of the LED emission. The EL intensity was higher in polarization along the $\langle 11\bar{2}0 \rangle$ direction (*a* axis) than along the *c* axis. The polarization ratio, ρ, has been defined as $\rho = |I_\perp - I_{//}|/(I_\perp + I_{//})$, where $I_\perp$ and $I_{//}$ represent EL intensities through the polarizer (proportional to the Si photodetector output current) for polarizations perpendicular and parallel to the *c* axis, respectively. The result in Figure 12.34 corresponds $\rho = 0.17$, and ρ showed a slight tendency to decrease as the LED drive current was increased.

12.5
Recent Nonpolar Optical Device Results Overview

There have been several recent reports on the performance of nonpolar devices that includes a report on the demonstration of first nonpolar-nitride-based visible laser diode. Okamoto *et al.* [37] reported *m*-plane LEDs on single crystalline *m*-plane GaN substrate emitting at 435 nm with output power of

1.79 mW at 20 mA, corresponding to an EQE of 3.1%. A polarization ratio of 0.77 was obtained for the LED at 1 mA drive current. Schmidt *et al.* [38] reported *m*-plane LEDs on *m*-plane bulk GaN substrate emitting at 407 nm with output power of 23.7 mW at 20 mA, corresponding to an EQE of 38.9%. The peak wavelength shift was less than 0.5 nm between 1 and 20 mA. More recently, Kim *et al.* [39] reported 402 nm *m*-plane LEDs on *m*-plane bulk GaN substrates with output power and EQE of 28 mW and 45.4%, respectively, at a pulsed drive current of 20 mA. With increasing current, the output power increased linearly; and a fairly flat EQE was observed with increasing drive current. At 200 mA, the power and EQE were 250 mW and 41%, respectively.

Demonstrations of *m*-plane InGaN/GaN laser diodes were reported by Okamoto *et al.* and Schmidt *et al.* [40, 41] Okamoto *et al.* reported continuous wave lasing at 400 nm with a threshold current of 36 mA (corresponding to a threshold current density of 4 kA cm^{-2}). Under pulsed operation, the threshold current was 28 mA (corresponding to a threshold current density of 3.1 kA cm^{-2}). The differential quantum efficiencies in continuous and pulsed mode were 16 and 30%, respectively. Schmidt *et al.* reported pulsed lasing at 405.5 nm with threshold current densities as low as 7.5 kA/cm^2 at 10% duty cycle. Most recently, Feezell *et al.* [42] reported *m*-plane InGaN/GaN laser diodes without any Al-containing waveguide cladding layers. The laser diodes utilized thick InGaN quantum wells to generate transverse optical mode confinement. Pulsed lasing operation was demonstrated, with threshold voltages and current densities of 6.7 V and 3.7 kA cm^{-2}, respectively.

12.6 Conclusions

Growth of light emitters along nonpolar directions is an effective means to circumvent the adverse effects of polarization. Growth of InGaN/GaN MQWs along nonpolar planes, both *a* and *m* plane, was investigated. The structural and optical properties of the MQWs were studied. The effect of defects on their emission properties has been addressed. Excitation dependent photoluminescence measurements revealed the absence of polarization in the MQWs. Internal quantum efficiency (η_{int}) of approximately 25 and 36% at 300 K was demonstrated for nonpolar *a*-plane and *m*-plane $In_xGa_{1-x}N/GaN$ MQWs, respectively. Although the η_{int} values are yet lower than that of conventional *c*-plane blue MQWs, the results encouraged one to realize high performance green, amber, and red LEDs by reducing the concentration of nonradiative defects. On the basis of the MQW optimization, light-emitting diodes, emitting in the violet-blue-green spectral range, were grown on nonpolar templates and their electrical and optical properties were studied. EL measurement confirmed the absence of built-in electric fields in the quantum well. The influence of defects on the emission characteristics of the light emitters was also studied. The electric field component of the blue surface emission in *m*-plane LED was polarized perpendicular to the *c* axis with the in-plane polarization ratio of 0.58 at 300 K. Nonpolar light-emitting diodes with milliwatt-range output power were demonstrated. DC output power as high as 0.6 mW at 20 mA and pulsed output power as high as 23.5 mW at 1 A were measured.

Acknowledgements

The authors are greatly indebted to the following collaborators who have made major contributions to this work: Dr Ben Haskell, Dr Takeyoshi Onuma, Dr Feng Wu, Hisashi Masui, Dr Cedrik Meier, Salka Keller, Dr Takahiro Koyama, Dr Mel McLaurin, Dr Kwang-choong Kim, Dr Stacia Keller, Prof. James Speck, Prof. Steven DenBaars, and Prof. Shuji Nakamura.

References

1. Waltereit, P., Brandt, O., Grahn, H.T. *et al.* (2000) *Nature*, **406**, 865.
2. Bhattacharyya, A., Friel, I., Iyer, S. *et al.*(2003) *Journal of Crystal Growth*, **251**, 487.
3. Kuokstis, E., Chen, C.Q., Gaevski, M.E. *et al.* (2002) *Applied Physics Letters*, **81**, 4130.
4. Sun, Y.J., Brandt, O., Cronenberg, S. *et al.* (2003) *Physical Review B*, **67**, 041306.
5. Sun, Y.J., Brandt, O., Ramsteiner, M. *et al.* (2003) *Applied Physics Letters*, **82**, 3850.
6. Sun, Y.J., Brandt, O., and Ploog, K.H. (2003) *Journal of Vacuum Science and Technology B*, **21**, 1350.
7. Sun, Y.J., Brandt, O., Jenichen, B., and Ploog, K.H. (2003) *Applied Physics Letters*, **83**, 5178.
8. Ng, H.M. (2002) *Applied Physics Letters*, **80**, 4369.
9. Craven, M.D., Waltereit, P., Wu, F. *et al.* (2003) *Japanese Journal of Applied Physics*, **42**, L235.
10. Craven, M.D., Wu, F., Chakraborty, A. *et al.* (2004) *Applied Physics Letters*, **84**, 1281.
11. Haskell, B.A., Wu, F., Craven, M.D. *et al.* (2003) *Applied Physics Letters*, **83**, 644.
12. Craven, M.D., Lim, S.H., Wu, F. *et al.* (2002) *Applied Physics Letters*, **81**, 469.
13. Wu, X.H., Elsass, C.R., Abare, A. *et al.* (1998) *Applied Physics Letters*, **72**, 692.
14. Haskell, B.A., Baker, T.J., McLaurin, M.B. *et al.* (2005) *Applied Physics Letters*, **86**, 111917.
15. Chakraborty, A., Wu, F., Keller, S. *et al.* (unpublished).
16. Chichibu, S., Wada, K., and Nakamura, S. (1997) *Applied Physics Letters*, **71**, 2346.
17. Chichibu, S.F., Uedono, A., Onuma, T. *et al.* (2006) *Nature Materials*, **5**, 10.
18. Chakraborty, A., Haskell, B.A., Wu, F. *et al.* (2007) *Japanese Journal of Applied Physics*, **46**, 542.
19. Chakraborty, A., Meier, C., Keller, S. *et al.* (2005) *Applied Physics Letters*, **86**, 031901.
20. Haskell, B.A., McLaurin, M.B., Hirai, A. *et al.* (unpublished).
21. Haskell, B.A., Chakraborty, A., Wu, F. *et al.* (2005) *Journal of Electronic Materials*, **34**, 357.
22. Off, J., Kniest, A., Vorbeck, C. *et al.* (1998) *Journal of Crystal Growth*, **195**, 286.
23. Bertram, F., Srinivasan, S., Geng, L. *et al.* (2002) *Applied Physics Letters*, **80**, 3524.
24. Tang, H., Haffouz, S., Powell, A. *et al.* (2005) *Applied Physics Letters*, **86**, 121110.

25. Chichibu, S.F., Uedono, A., Onuma, T. *et al.* (2005) *Applied Physics Letters*, **86**, 021914.

26. Chitnis, A., Chen, C., Adivarahan, V. *et al.* (2004) *Applied Physics Letters*, **84**, 3663.

27. Chakraborty, A., Haskell, B.A., Keller, S. *et al.* (2004) *Applied Physics Letters*, **85**, 5143.

28. Chakraborty, A., Haskell, B.A., Keller, S. *et al.* (2005) *Japanese Journal of Applied Physics*, **44**, L173.

29. Chakraborty, A., Haskell, B.A., Masui, H. *et al.* (2006) *Japanese Journal of Applied Physics*, **45**, 739.

30. Onuma, T., Chakraborty, A.,Haskell, B.A. *et al.* (2005) *Applied Physics Letters*, **86**, 151918.

31. Chakraborty, A., Xing, H., Craven, M.D. *et al.* (2004) *Journal of Applied Physics*, **96**, 4494.

32. Kozodoy, P., Abare, A., Sink, R.K. *et al.* (1997) *Materials Research Society Symposium Proceedings*, **468**, 481.

33. Mukai, T. and Nakamura, S. (1999) *Japanese Journal of Applied Physics*, **38**, 5735.

34. Chichibu, S., Sota, T., Wada, K., and Nakamura, S. (1998) *Journal of Vacuum Science and Technology*, **16**, 2204.

35. Zheng, R. and Taguchi, T. (2003) *Proceedings of the SPIE–The International Society for Optical Engineering*, **4996**, 105.

36. Masui, H.,Chakraborty, A.,Haskell, B.A. *et al.* (2005) *Japanese Journal of Applied Physics*, **44**, L1329.

37. Okamoto, K., Ohta, H., Nakagawa, D. *et al.* (2006) *Japanese Journal of Applied Physics*, **45**, L1197.

38. Schmidt, M.C., Kim, K.C., Sato, H. *et al.* (2007) *Japanese Journal of Applied Physics*, **46**, L126.

39. Kim, K.C., Schmidt, M.C., Sato, H. *et al.* (2007) *Phys Status Solidi (RRL)*, **1**, 125.

40. Okamoto, K., Ohta, H., Chichibu, S.F. *et al.* (2007) *Japanese Journal of Applied Physics*, **46**, L187.

41. Schmidt, M.C., Kim, K.C., Farrell, R.M. *et al.* (2007) *Japanese Journal of Applied Physics*, **46**, L190.

42. Feezell, D.F., Schmidt, M.C., Farrell, R.M. *et al.* (2007) *Japanese Journal of Applied Physics*, **46**, L284.

13
Growth, Structural, and Optical Properties of *a*-plane GaN Quantum Dots in AlN

Sebastien Founta, Fabian Rol, Bruno Gayral, and Bruno Daudin

13.1
Introduction

It is now a generally accepted fact that the current huge interest in III-nitride semiconductors is motivated by their potential applications as light emitters/detectors in a wavelength range extending from infrared to ultraviolet.

It is almost as readily agreed that the high efficiency of InGaN-based light emitting diodes (LEDs) and laser diodes (LDs) is related to the presence of localization centers that make them, to a large extent, insensitive to the high density of dislocations found in nitride material, which is a consequence of a lack of adapted substrates as well as of marked lattice mismatch between the various members of the nitride family.

Although the nature of localization centers is still controversial, their existence has been suspected very easily to be a determinant reason for the high luminescence efficiency observed in InGaN-based LEDs [1]. Among the various hypotheses invoked to account for these localization centers, the existence of In-rich regions in InGaN quantum wells (QWs) acting as quantum dots (QDs) has been postulated [2, 3]. Whatever the exact nature of localization centers may eventually be, the above speculation has concomitantly increased the interest in the intentional growth of GaN and InGaN QDs.

Indeed, restricting ourselves to the field of GaN QDs, it has been found that these nanostructures exhibit almost no quenching of photoluminescence (PL) intensity as a function of temperature; this is a clue that confinement of carriers in these virtually perfect objects makes them insensitive to the high dislocation density in the surrounding matrix [4].

However, for polar heterostructures in the III-N system (i.e. for QWs or QDs grown along the [0001] direction of the noncentrosymmetric wurtzite structure), optical properties are usually dominated by the quantum confined Stark effect resulting from the internal field [5]. The confined electrons and

Nitrides with Nonpolar Surfaces. Growth, Properties, and Devices. Edited by Tanya Paskova

ISBN: 978-3-527-40768-2

holes are spatially separated along the growth axis, leading to a red shift of the luminescence as well as to a reduced oscillator strength and eventually to a reduced radiative recombination rate that may turn out to be detrimental for LED applications.

The reduction/suppression of the built-in internal electric field in the MV cm^{-1} range which governs optical properties of polar heterostructures is the main motivation for growing and studying nonpolar heterostructures. For nonpolar QWs, the *c* axis does not cross any polarization discontinuity between the well and the barrier materials (in other words, the polarization in the materials is parallel to all interfaces). As a consequence, there is no internal electric field in these QWs as was, for instance, demonstrated in Refs. 6–8.

For nonpolar QDs, the situation is more complex. Indeed, the *c* axis does cross GaN/AlN interfaces at two of the QD lateral facets so that a charge sheet density appears on these facets, possibly leading to an internal electric field inside the QD. However, the GaN/AlN spontaneous polarization difference being opposite to the piezoelectric one, contrary to the case of polar heterostructures, the internal electric field is expected to be strongly reduced in nonpolar QDs.

Of course, the final value of the internal electric field will depend on the geometry of dots, that is, on the detailed geometry of polarization discontinuity. This issue, namely, the growth of (11-20) QDs and the study of the parameters governing their morphology, is still an almost unexplored field. It is our purpose to address it here and to correlate structural and optical properties of these new heterostructures.

13.2
Epitaxial Growth of Nonpolar Nitride Layers

13.2.1
Substrates and Growth Conditions

The lack of bulk nitride substrates, which has often been invoked as a serious limitation for the structural quality of the *c*-plane epilayers, is a particularly relevant issue for the case of the *a*-plane ones. To date, very few publications report on successful growth of bulk *a*-plane nitride layers exhibiting reasonably good structural quality. In a recent article, however, Paskova *et al.* reported on low-defect-density bulk *a*-plane GaN substrates, exhibiting a dislocation density in the 10^5 cm^{-2} range. These substrates, originating from Kyma Technologies, were obtained by slicing a *c*-plane hydride vapor phase epitaxy (HVPE)-grown GaN boule parallel to the (11-20) plane [9]. However, substrates of this type, although already commercially available, are not yet widely used by the GaN community, and the vast majority of groups are still using pseudosubstrates, i.e., substrates with a nature different from that of the epitaxial layer, to perform growth.

Two types of pseudosubstrates are currently used for the heteroepitaxial growth of *a*-plane GaN or AlN layers: (1-102) *r*-plane sapphire [10–16] and

(11-20) *a*-plane SiC [17–20]. This type of growth results in a high density of structural defects such as stacking faults ($\sim 10^5$ cm^{-2}) and dislocations ($\sim 10^9$ cm^{-2}) [9, 10, 20], an issue that has been partially solved by lateral epitaxial growth techniques employing SiO_2 stripe patterns [21]. To date, *a*-plane GaN QDs have been achieved only on *a*-plane SiC; this is why we will focus our study on the works carried out on this type of substrates.

Prior to the study of the structural and optical properties of self-assembled GaN/AlN QDs on *a*-plane SiC, it is worth recalling that nitride growth mechanisms are strongly dependent on experimental conditions. As a matter of fact, on the *c* plane, two different growth regimes have been identified as a function of the Ga flux (with the N flux kept constant), namely the N-rich regime (with an excess of N with respect to Ga) and the Ga-rich regime (with an excess of Ga with respect to N) [22]. In the latter, an autosurfactant effect of Ga (a segregation of Ga on the growing GaN surface) has been put in evidence [23], and a coverage diagram, describing the Ga surface coverage during growth as a function of Ga flux and growth temperature, has been established [24]. This growth diagram proved very useful to understand the suitable growth conditions required to achieve the Stranski–Krastanow (SK) growth mode of GaN on AlN giving rise to the formation of QDs, and how to control the growth mode by adjusting of the excess of Ga.

Further, it should be emphasized that the nature of the active N itself, that is, ammonia or N plasma, drastically influences growth kinetics. Actually, the above considerations on the growth diagram as well as all experimental results reported in this chapter refer to plasma-assisted molecular-beam epitaxy (MBE), active N being obtained by radio frequency cracking of N_2 in a plasma cell.

As regards the the *a* plane, it has been established that growth also strongly depends on Ga/N flux ratio value. More precisely, it has been found that GaN growth in Ga-rich conditions results in a rough surface, whereas growth in N-rich conditions leads to a smooth surface. Accordingly, SK growth mode of GaN deposited on AlN is observed in Ga-rich conditions. It is worth emphasizing that this situation is opposite to what is observed for polar nitrides. This suggests that surface diffusion barriers are lower in N-rich conditions than in Ga-rich conditions for (11-20) GaN, which again is the opposite of what is theoretically predicted [25] and experimentally observed [26] in the case of (0001) GaN growth.

13.2.2
Anisotropy of MBE-grown AlN *a*-plane Surface

Figure 13.1 shows an atomic force microscopy (AFM) image of an AlN buffer layer, about 25-nm thick, grown in N-rich conditions (Al/N $\approx$ 0.7) on a (11-20) SiC oriented substrate. AFM reveals a highly anisotropic morphology, consisting of ripples oriented along the [1-100] direction. A surface profile

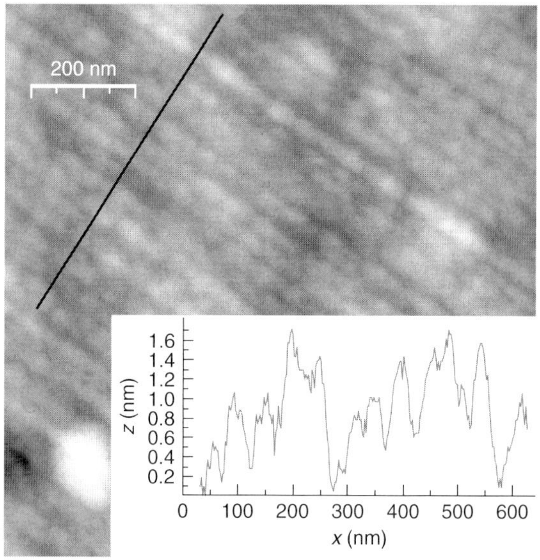

Figure 13.1 (1 × 1) μm^2 AFM image of an *a*-plane AlN layer. Inset: morphology profile taken along the [0001] axis (black line).

taken along the [0001] axis is presented in the inset, indicating peak to valley values of a few monolayers and a mean period of the features of about 30 nm.

These results are consistent with those reported by Onojima *et al.* [27]. Two possible causes may account for the anisotropy of the *a*-plane nitride layers, namely (i) an anisotropy of the strain relaxation in the buffer coming from an in-plane anisotropy of the lattice mismatch between AlN and SiC (+1.0% along [1-100] and −1.2% along [0001]) and (ii) the presence of *c*-plane stacking faults propagating through the nitride layer and emerging at the surface of the buffer, which results in the observed ripples [20, 27].

It should be noted that anisotropic growth resulting from anisotropic lattice mismatch has been invoked to account for the spontaneous growth of silicide nanowires on Si [28] and of Ge islands on high-index Si substrates [29]. It is then expected that growth proceeds preferentially along the direction of lesser lattice mismatch, as observed for AlN growth on (11-20) SiC.

13.3 Growth of Nonpolar GaN QDs

13.3.1 Impact of Growth Parameters: Nominal GaN Quantity

The QD samples have been grown using the following procedure: a 28-nm AlN buffer layer was first grown on the SiC substrate at 730 °C in N-rich conditions. On top of this buffer, a first GaN layer was grown at 700 °C in the Ga-rich regime. After deposition of about 2–3 GaN monolayer (MLs), the reflection high energy electron diffraction (RHEED) pattern abruptly changed from streaky to spotty, revealing the formation of GaN QDs following the SK growth mode. The GaN growth time (and hence the amount of GaN deposited) was varied in order to vary the mean size and density of the resulting QDs [19]. After suppression of Ga flux followed by a consumption of the Ga excess under N flux, the GaN dot plane was covered by a 20-nm AlN layer, and a second GaN dot plane was deposited following the same procedure as for the first one. The four samples which have been grown contain respectively 3.7, 6.4, 11.8, and 25.3 GaN MLs in each QD plane. The transition was observed after a growth of 10 s, corresponding to ≈2.7 MLs. The surface QD plane was analyzed by AFM, and the buried QD plane by PL [19].

In the case of a GaN growth stopped as soon as the transition is seen on the RHEED pattern, the corresponding AFM image (Figure 13.2a) shows GaN QDs with an average density of about 1.2×10^{11} cm^{-2} and a mean height and a diameter estimated to be 2.2 and 21 nm, respectively. An alignment of the dots in rows along [1-100] is observed. We note that the coverage of the surface with QDs is not homogeneous: zones with a high QD density coexist with other zones nearly free of QDs. Advantage was taken from this inhomogeneity to study the optical properties of single dots, which will be further illustrated in Section 13.4.2. Here again, the linear alignment of the QDs is possibly driven by the anisotropic mismatch of GaN with respect to AlN (+2.5% and +4.1% along [1-100] and [0001] respectively). Alternatively, ripples on AlN surface may act as nucleation centers.

In the case of a GaN growth stopped 10 s after the 2D–3D transition (Figure 13.2b), the surface is entirely and homogeneously covered by GaN QDs. Both the average size and the average density (2.9 nm × 22 nm and 1.7×10^{11} nm^{-2}, respectively) are larger than the ones measured for the first sample. Note that the dots exhibit a monomodal size distribution, in contrast to what is observed in the case of (0001) QDs [30]. The dots are about to coalesce along the lines and between the lines. As a consequence, the alignment direction is not so clearly seen in the Fourier transform image as was the case for the first sample, but it is still visible.

In Figure 13.2c, the growth was maintained up to 30 s after the transition. The QDs are in the process of coalescing along the two in-plane perpendicular directions ([1-100] and [0001]), which results both in a decrease of their density (down to 5.9×10^{10} cm^{-2}) and in a huge damping of the preferential direction in the Fourier transform image. Note that the density given above has to be considered with caution since an accurate counting of the dots is difficult owing to the coalescence process.

Finally, in Figure 13.2d, corresponding to a growth maintained 80 s after the transition, the QDs have fully coalesced, and a further decrease of the density of the objects is measured (5.9×10^{10} cm^{-2}).

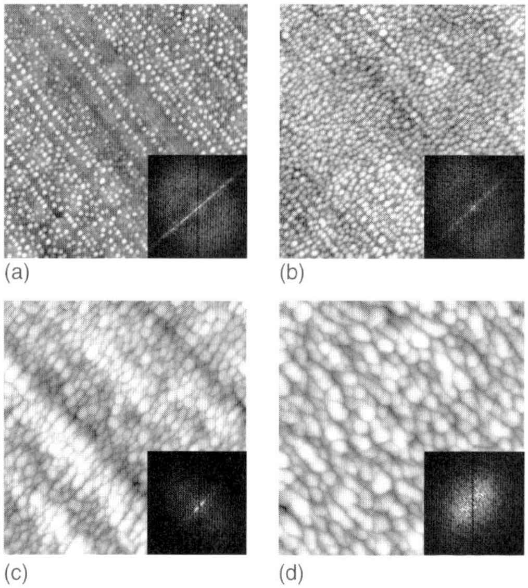

Figure 13.2 $(1 \times 1)\ \mu m^2$ AFM images of GaN QDs grown on *a*-plane AlN. The growth of GaN was interrupted just at the 2D/3D transition (a) at 10 s (b), 30 s (c), 80 s (d) later, which corresponds to a total of 3.7, 6.4, 11.8, and 25.3 GaN MLs, respectively. Note the Fourier transform image in the lower right corner of the images.

In conclusion, increasing the nominal amount of GaN deposited increases both the average density and the average size of the dots as long as they are well separated. Following this, the average density decreases as a consequence of the coalescence process. On the *c* plane, the coalescence was proved to decrease the structural and hence the optical quality of the samples by introducing dislocations at the junction between adjacent QDs. To avoid such a situation on the *a* plane, it may be necessary to avoid large GaN amounts, and the quantity corresponding to Figure 13.2b (6.4 MLs) may be regarded as a maximum value for this parameter. For the same reason, the only way to use this parameter to grow structures with a low GaN QD density, a prerequisite for the optical study of the single GaN QD, seems to be a decrease of the post-transition GaN amount down to the lowest value possible (cf. Section 13.4.2).

13.3.2
Morphology of *a*-plane GaN QDs

The AFM micrographs presented in Figure 13.2 were taken using a 'conventional' AFM tip, with a typical radius of curvature between 5 and 10 nm. This value is of the order of the typical diameter of *a*-plane QDs (about

20 nm) and thus is too large to enable an accurate study of their morphology. To overcome this limitation, 'super sharp tips' with a radius of curvature inferior to 5 nm may be used to image the QDs with a better lateral and height resolution.

Two QD samples, consisting of one buried plane and one plane on the surface, were grown specifically for AFM analysis using such high-resolution tips [31]. These structures were grown at 750 °C using the following process. A 7-nm AlN layer was first grown on the SiC substrate, followed by a GaN/AlN QW superlattice consisting of five periods of (1 ML GaN/13 MLs AlN) bilayers. A second 7-nm AlN layer was then deposited on top of this superlattice. The aim of this QW-based buffer is to prevent possible structural defects or chemical impurities originating at the substrate surface from propagating through the subsequent layers and thus altering their structural and optical quality [32, 33]. An enhancement of the luminescence intensity by a factor up to 5 was indeed observed on *a*-plane QD samples, thanks to this technique. On top of the buffer, a first GaN QD layer was grown in the Ga-rich regime. The growth was stopped by a suppression of both Ga and N fluxes, followed by a desorption of the Ga excess under vacuum. This technique, enabled by the high growth temperature (750 °C), was preferred to the one depicted in Section 13.3.1 because it enables a precise control of the GaN amount deposited by preventing the growth of an extra monolayer of GaN when consuming the Ga excess present on the surface under N flux alone. The GaN dot plane was then covered by a 30-nm AlN layer, and a second GaN dot plane was deposited following the same procedure as for the first one. The first sample (sample A′) contains 3.0 GaN MLs and the second sample (sample C′) 4.5 GaN MLs. These GaN amounts were chosen lower than the maximum value of 6.4 MLs corresponding to the second sample of the series studied in Figure 13.2 to ensure that sufficient separation between the QDs allowed for a reliable AFM analysis. The increase of the QD density from 8.4×10^{10} cm^{-2} in sample A′ to 1.6×10^{11} cm^{-2} in sample C′ is consistent with the evolution that was observed between the first and the second AFM images presented in Figure 13.2.

AFM images of the dot plane at the surface of samples A′ and C′ are shown in Figure 13.3. On sample A′ (Figure 13.3a), the edges of the QDs are just visible, and their shapes mostly appear square or rectangular. On sample C′ (Figure 13.3b), these edges are more pronounced, and the population of dots has a wide variety of distinct shapes: square, rectangular (with the biggest side along [0001] or along [1-100]) or trapezoidal.

To compare quantitatively the average shape of the dots in the two images, a statistical analysis was carried out on a population of 130 dots on each sample. For each dot, we measured its in-plane dimension along [1-100] (referenced as the *parallel dimension*), its in-plane dimension along [0001] (referenced as the *perpendicular dimension*), and its height. We found an average value of 20.8 nm × 18.7 nm × 1.6 nm for sample A′, and 25.3 nm × 24.6 nm × 2.6 nm for sample C′. On the basis of these figures, we note that the mean

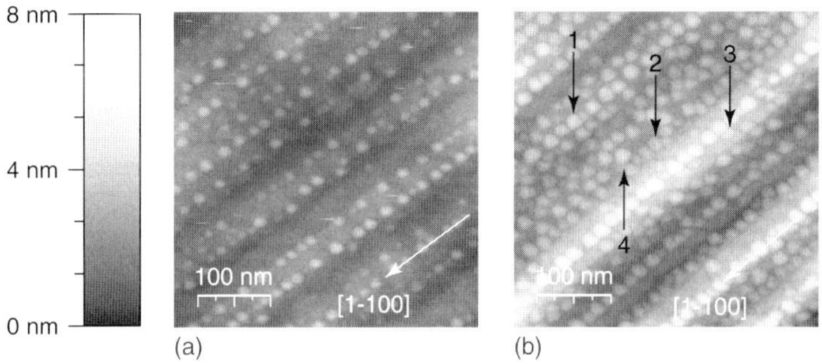

Figure 13.3 (1 × 1) μm² AFM micrographs of *a*-plane GaN QDs at the surface of sample A′ (a) and sample C′ (b). On sample C′ we can distinguish various shapes: square, rectangular (along [0001] or [1-100]) or trapezoidal, shown respectively by arrows 1 to 4.

perpendicular dimension is smaller than the mean parallel one, revealing an in-plane anisotropy that is more pronounced for the 'smaller dots' (sample A′) than for the 'larger dots' (sample C′).

The distributions of the in-plane dimension ratios for both samples are presented in Figure 13.4.

Figure 13.4a, corresponding to sample A′, shows a Gaussian distribution of the in-plane dimension ratios. It is centered on a value lower than unity (0.89), with a full width at half maximum (FWHM) of 0.29, and 80% of the dots exhibit a perpendicular dimension smaller than the parallel one. On the other hand, the distribution in Figure 13.4b, corresponding to sample C′, is wider than the one of the first sample (with an FWHM of 0.45), and is not significantly shifted with respect to unity (1.00). Even though the center of the distribution shifts toward unity with increasing GaN coverage, the increase of

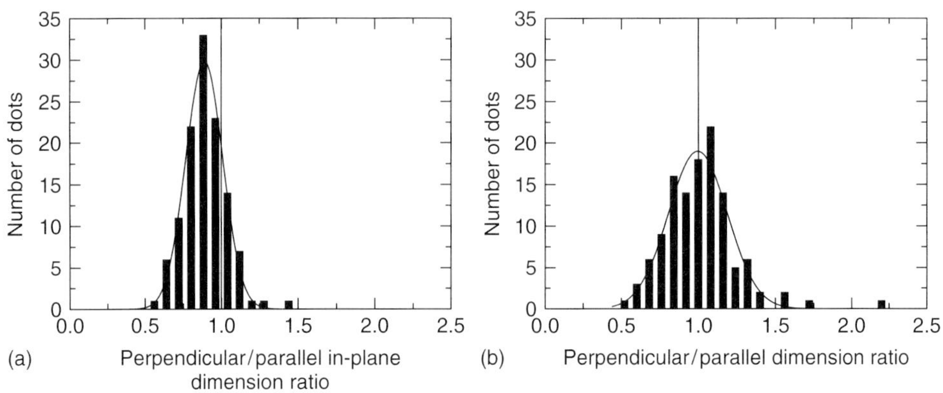

Figure 13.4 Distributions of the in-plane dimension ratios for samples A′ (a) and C′ (b).

the width of the distribution in sample C′ directly reflects the increased variety in shape with respect to sample A′ (Figure 13.4a) that is observed in the AFM image, meaning that the proportion of dots having a nonsquare shape (i.e. an in-plane dimension ratio different from unity) is larger. Please note that the absolute values given here have to be considered with caution since they may be affected by an unavoidable temporal drift of the AFM signal depending on the slow scan direction (from bottom to top or from top to bottom of the scanned area). To enable comparison, all the images were taken along the same slow scan direction, and to our minds only the relative difference between both statistics is relevant.

As regards the average aspect ratio of the dots (the height divided by one of the two in-plane dimensions), we can define two of them for each QD population: a 'parallel aspect ratio' and a 'perpendicular aspect ratio', depending on the in-plane dimension considered. We found these ratios for sample A′ to be 0.077 and 0.086 respectively, and for sample C′, 0.102 and 0.105 respectively. From these results, we can conclude that the aspect ratios of the QDs in sample C′ are systematically greater than those of the QDs in sample A′. This indicates that the increase of the volume of the dots during the growth is not isotropic. A dot nucleates with a low aspect ratio, and this aspect ratio increases as the amount of GaN deposited is increased.

13.3.3 Strain State of *a*-plane GaN QDs

It has been mentioned that AlN buffer layers grown on *a*-plane 6H-SiC substrates develop a highly anisotropic surface morphology during growth. This anisotropy, which may be correlated to an anisotropy of the lattice mismatch with respect to SiC combined with the generation of stacking faults emerging at the surface, influences the lateral ordering of GaN QDs by giving rise to rows of QDs oriented along [1-100]. Subsequent observations revealed an in-plane anisotropy of the dot morphology and an evolution of the aspect ratio that are dependent on the amount of GaN deposited. Since the relaxation rate and the dot morphology are strongly related, the next step of the study consists in measuring the strain profile inside a dot to see whether the relaxation process is also anisotropic or not.

Several techniques are usually employed to study the strain in QD layers. Raman spectroscopy and high-resolution X-ray diffraction, on the one hand, can be routinely used to address this issue, but the data obtained are averaged over the entire volume of the dots. High-resolution transmission electron microscopy (HRTEM), on the other hand, provides information at the monolayer scale, but the technique requires extensive sample preparation that may induce an extra relaxation owing to thin foil effects; the data gathered are local and may not reflect the structure of the whole sample. Medium energy ion scattering (MEIS) appears as an alternative: no preparation is needed, as in the case of Raman spectroscopy and X-ray diffraction, and it provides information over an appreciable part of the sample (typically a few mm^2) with a depth resolution of the order of one monolayer. Thus, we have chosen to employ this technique to study the strain profile from bottom to top of the *a*-plane GaN QDs.

MEIS is a near-surface analysis technique using a beam of light ions accelerated between 50 and 500 keV. The beam is collimated on the surface of the sample along a predefined crystallographic direction, and the backscattered ions are analyzed in energy and exit angle by a 2D detector. The backscattered spectrum exhibits a blocking dip at a precise angle corresponding to shadowing by atoms aligned along a crystallographic direction. The blocking angle can be expressed as the ratio of an out-of-plane and an in-plane lattice parameter and is thus related to a component of the strain tensor. A correspondence is made between the energy loss of a backscattered ion and the depth reached from the surface through the relation $\Delta E = \mathrm{d}E/\mathrm{d}z \times \Delta z$, $\mathrm{d}E/\mathrm{d}z$ being the stopping power of H^+ in GaN given by the SRIM 2003 software [34] (Note that the stopping power is determined in the case of a bulk material. Nevertheless, even though the GaN layers under study are only a few nanometers thick, the collection time – and thus the total number of scattering events – is high enough to make the use of statistically determined stopping power valid). Next, the variation of the blocking angle as a function of the energy of the backscattered ions helps to trace a depth profile of the strain component chosen. Within the energies considered, the penetrating depth ranges between a few tens and a few hundreds of nanometers, and the resolution is of the order of one monolayer.

The experiments were carried out at the Daresbury MEIS [35] facility using 102.8 keV protons on two samples (samples A and C) similar to samples A′ and C′, as well as on a third sample having an intermediate amount of GaN (sample B). The incident angle was chosen to be 7° off the [11-20] direction. The blocking dips considered were [01-10] and [11-21], and the backscattered protons were collected in a 2D detector [36] (with a relative energy and angular resolutions of about 2.4×10^{-3} and 0.3°, respectively) within a 27° range around these two directions, successively. For the analysis, it was chosen to describe the wurtzite unit cell with three lattice parameters, namely $a_{1\bar{1}00}$, c (along the two in-plane directions [1-100] and [0001], respectively), and $a_{11\bar{2}0}$ (along the out-of-plane direction [11-20]). For bulk material, $a_{11\bar{2}0} = a_{1\bar{1}00}/\sqrt{3}$. A scheme of the scattering geometry and an example of backscattered spectrum collected on sample C around [01-10] are presented in Figure 13.5.

As a reference for the exit angle scale, we measured the blocking dip from the bulk SiC substrate, corresponding, in the case of the [01-10] direction, to an angle of $\theta_{\mathrm{SiC}} = 30°$ with respect to the (11-20) plane (shown in Figure 13.5c). The proton energy loss in matter increases as a function of the depth in the sample. Therefore, the highest energy of 97.5 keV corresponds to the Ga at the GaN surface, that is, to the top of the dots and wetting layer, and the lower energies correspond to protons backscattered deeper in the dots. For a given

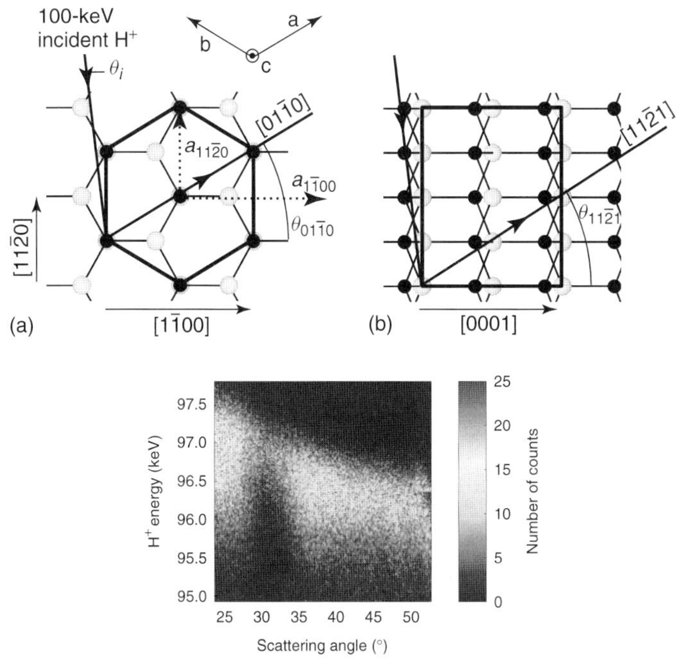

Figure 13.5 Scheme of the scattering geometry in the (a) (0001) and (b) (1-100) planes for medium energy ion scattering, and (c) energy and angular distributions of the backscattered protons for sample C, close to the [01-10] blocking direction. An angle of 30° corresponds to the SiC substrate, which is used as a reference (cf. Color Plate XXXVI).

depth, the Ga blocking angle can be related to the ratio $a_{11\bar{2}0}/a_{1\bar{1}00}$ (for [01-10]) and $a_{11\bar{2}0}/c$ (for [11-21]) through the relations:

$$\theta_{\mathrm{Ga},01\bar{1}0} = \arctan\left(\frac{a_{11\bar{2}0}}{a_{1\bar{1}00}}\right) \qquad \theta_{\mathrm{Ga},11\bar{2}1} = \arctan\left(\frac{a_{11\bar{2}0}}{c}\right) \tag{1}$$

However, the measured kinetic energy of an ion scattered from a given depth is distributed over a certain energy range, since the number of scattering events may vary for the same examined depth. This 'straggling effect' may play an important role in our experiments owing to the relatively small thickness of the samples. Taking this effect into account in addition to the finite-energy resolution of the detector, we estimate the depth resolution to be of the order of 1 ML.

The evolution of the $\sqrt{3} \times a_{11\bar{2}0}/a_{1\bar{1}00}$ and $a_{11\bar{2}0}/c$ ratios as a function of depth for samples A, B, and C are presented in Figure 13.6. The data are compared to the case of a GaN layer completely strained in both in-plane directions on a relaxed AlN [37] (label GaN/AlN in Figure 13.6), and to the case of

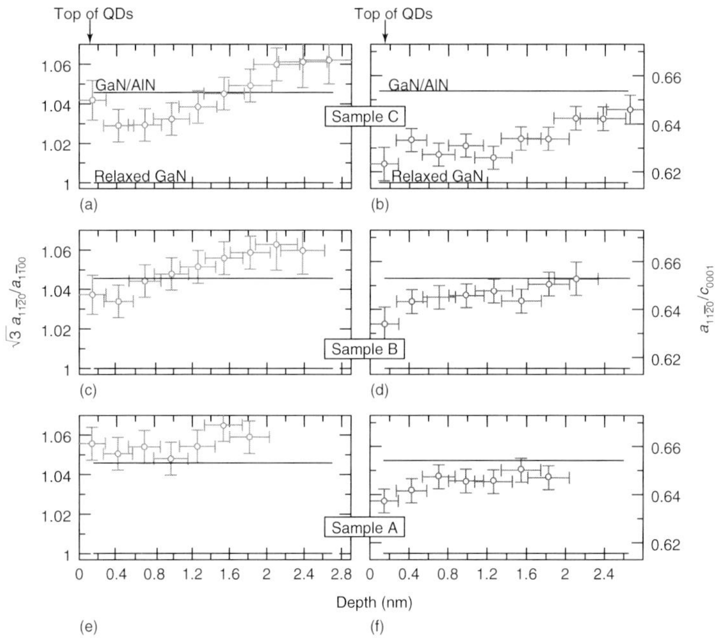

Figure 13.6 Depth profile of the $\sqrt{3} \times a_{11\bar{2}0}/a_{1\bar{1}00}$ (a, c, and e) and $a_{11\bar{2}0}/c$ (b, d, and f) ratios for samples A, B, and C. The GaN/AlN label refers to a GaN layer completely strained in both in-plane directions on a relaxed AlN (after [31]).

relaxed GaN (corresponding to $\sqrt{3} \times a_{11\bar{2}0}/a_{1\bar{1}00} = 1$ in Figure 13.6a, c, and e and $a_{11\bar{2}0}/c = 0.615$ in Figure 13.6b, d, and f).

Let us begin by commenting on the evolution of the $\sqrt{3} \times a_{11\bar{2}0}/a_{1\bar{1}00}$ ratio, corresponding to the strain profile along [1-100] (Figure 13.6a, c, and e). We observe for sample C a decrease of the strain from the GaN/AlN interface to the top of the dots and wetting layer: while the initial value at the GaN/AlN interface is found to be larger than the value of a completely strained GaN, the strain at the top of the dots is intermediate between the completely strained and the relaxed values. Similar variation is observed for the two other samples, but the amplitude of the strain profile is larger for sample C than for sample B and for sample B than for sample A. The difference revealed by the MEIS data between sample C and sample A is consistent with the perpendicular aspect ratio values deduced from the AFM measurements on samples C′ and A′. Indeed, the larger parallel aspect ratio measured on sample C′ than in sample A′ (0.102 and 0.077, respectively) may be correlated to a larger amplitude of the strain profile according to the Kern and Müller model of the stress relaxation in QDs, stating that relaxation is proportional in first approximation to the aspect ratio [38]. Please note that the data are calibrated with respect to the SiC substrate, assuming that this layer is completely relaxed. However, the corresponding angle was measured at a depth only a few nanometers below the SiC/AlN interface, that is, at an energy for which the signal corresponding to the Si in SiC could be best isolated from that of the other elements. At such a depth, the SiC may not be completely relaxed, but rather slightly strained by the AlN buffer. As a consequence, all the strain profiles presented may be vertically shifted with respect to the actual values, which could explain why the strain state measured in the underlying wetting layer was found higher than the expected maximal value corresponding to a completely strained GaN layer.

We now turn to the evolution of the $a_{11\bar{2}0}/c$ ratio, corresponding to the strain profile along [0001] (Figure 13.6b, d, and f). For each sample, the initial strain value at the GaN/AlN interface is lower than the corresponding initial value in the [11-20] direction. The amplitude of the strain profile is again larger in sample C than in sample A, consistent with the AFM analysis which revealed a larger perpendicular aspect ratio in sample C′ than in sample A′ (0.105 and 0.086, respectively).

From the experiments described above, it appears that the strain profiles along the two nonequivalent [1-100] and [0001] directions are remarkably different. The systematic enhanced strain relaxation at the base of the dots along the [0001] direction may be interpreted as a plastic process. Indeed, in the case of a GaN layer deposited on a relaxed AlN, the in-plane GaN strain at the GaN/AlN interface amounts to −2.4% along [1-100] and −3.9% along [0001]. The greater in-plane strain along [0001] is believed to favor plastic strain relaxation through the formation of misfit dislocations at the GaN/AlN interface along the [0001] direction. These defects would partly account for the 3.9% total strain, while additional elastic relaxation would accommodate part of the residual lattice mismatch. Then, the combination of these elastic and plastic relaxation mechanisms could account for the smaller residual strain measured along [0001] on the three samples.

To validate this hypothesis, HRTEM experiments were performed on a sample containing 19 GaN/AlN QD planes grown following the same procedure as the one used for the AFM and MEIS samples described above (Figure 13.7).

The experiments were performed in high-resolution mode using a Jeol 4000EX microscope operated at 400 kV and equipped with a slow scan charge-coupled device (CCD) camera. The sample was prepared in cross sections, with two sides oriented at 90° to each other to investigate both [1-100] and [0001] zone axes. To reduce the noise, a Wiener filter was applied to the HRTEM images using the strain determination software [39]. The image shown in Figure 13.7a was taken along the [1-100] zone axis. It gives a general view of a part of the sample containing two of the 19 QD planes. Note that we only see GaN wetting layers in their AlN matrix: the GaN QDs are not visible in this image, owing to their low density with respect to the scale. A careful examination of the image reveals the presence of dislocations at the GaN/AlN interface at different places such as the positions highlighted by circles in areas 1, 2, and 3. Figure 13.7b gives an enlargement of area 1, where the dislocation core has been indicated

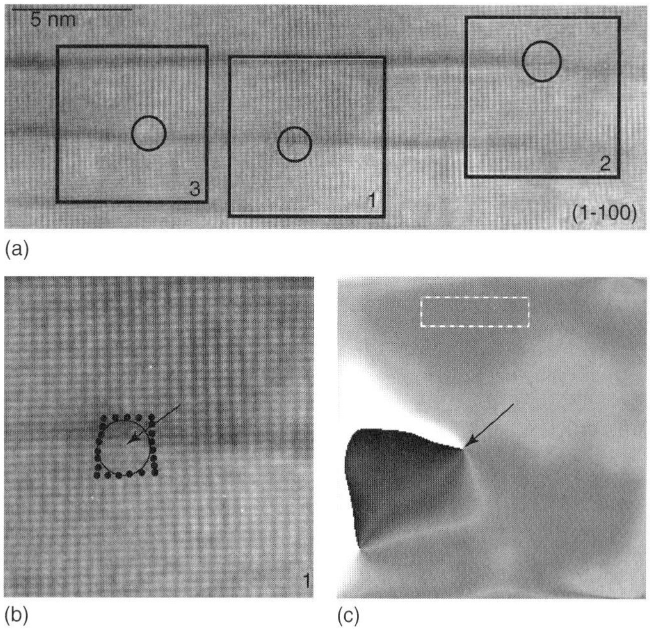

Figure 13.7 HRTEM cross-sectional image taken along the [1-100] zone axis of part of a sample containing 19 *a*-plane GaN QD planes showing two GaN wetting layers in between AlN spacers (a). Circles in the areas marked 1, 2, and 3 indicate the position of misfit dislocations. An enlargement of area 1 is presented in (b), where a Burgers circuit has been drawn around the dislocation (indicated by an arrow). The phase image of (0002) lattice fringes is shown in (c). The selection on the upper part of the image corresponds to the area chosen for phase reference. The 2π-rad variation highlighted by the arrow is characteristic of the presence of a dislocation (cf. Color Plate XXXVII).

by an arrow. A Burgers circuit around the dislocation has been drawn, leading to the Burgers vector $\mathbf{b} = 1/2$ [0001]. The image was further analyzed with the geometrical phase method, to obtain information about the local displacements and to detect the possible presence of dislocations [40]. This approach is based on combining real space and Fourier space information. Displacements are measured by calculating the 'local' Fourier components of the lattice fringes in an image, the amplitude describing the local contrast of the fringes and the phase their position. In the case of a dislocation, discontinuities are observed in the phase. A 2π discontinuity corresponds to the introduction of an extra plane. The resulting phase image of (0002) lattice fringes is shown in Figure 13.7c, where the 2π variation (full color range) pointed by the arrow indicates the presence of a dislocation [41].

We therefore have evidenced that misfit dislocations are generated to accommodate the stress in this orientation. Very roughly, the lattice mismatch accommodated through the formation of these dislocations can be estimated at 1–2%. Note that this value is consistent with the negative offset of the strain profiles displayed in Figure 13.6b, d, and f with respect to the ones displayed in Figure 13.6a, c, and e.

On the other hand, all HRTEM images taken along the [0001] zone axis reveal a very regular arrangement of the atomic columns, and no dislocations were detected, further supporting the evidence for anisotropic strain relaxation in (11-20) GaN QDs.

13.4
Optical Properties of (11-20) GaN QDs

As noted in introduction, the case of (11-20) GaN QDs with respect to internal electric field is quite different from the situation of polar heterostructures for which the internal field is due to two opposite charge planes facing each other in a planar capacitor type of structure. In the case of nonpolar QDs, the two surfaces that carry the surface charge are at an angle with respect to the *c* axis, are of relatively small area (typically 2 nm × 20 nm) and are well separated (typically by the dot diameter: 20 nm). The resulting electric field depends on the precise geometry of the facets and on the strain distribution inside the QD. This complex problem was first addressed by A. Cros *et al.* [42] who show that the value of the electric field in (11-20) QDs is much reduced compared to the polar-axis situation. They also show that the electric field is very sensitive to the dot geometry. Let us also point out that owing to the small charge carried at each lateral facet of a QD (on the order of 2e), the field can be easily screened by residual doping of the QDs.

The precise modeling of the electric field in such QDs is thus not a straightforward task as it requires a precise knowledge of the strain profile in the QDs, of the QD geometry, and in particular of the facets geometry, and of the residual doping. It is, however, possible to study the optical properties of such QDs by means of several spectroscopic techniques, and to deduce properties of the wavefunctions of the confined carriers from these studies.

13.4.1
Optical Spectroscopy on Ensembles of QDs

We probed by continuous wave (CW) PL a sample containing one (11-20) QD array (Figure 13.8). The experiment was made by exciting the sample at 244 nm with a doubled Argon laser on a ~50-μm diameter spot, the sample being placed in a He-flow cryostat.

The PL has a rather broad distribution (200 meV) which is typical of self-assembled QD systems, reflecting the distribution in QD size. The first striking feature is that the luminescence is above the GaN band gap [19, 43], which is the normal situation for a confined system. This is different from polar GaN/AlN QDs of similar sizes for which the luminescence is below the

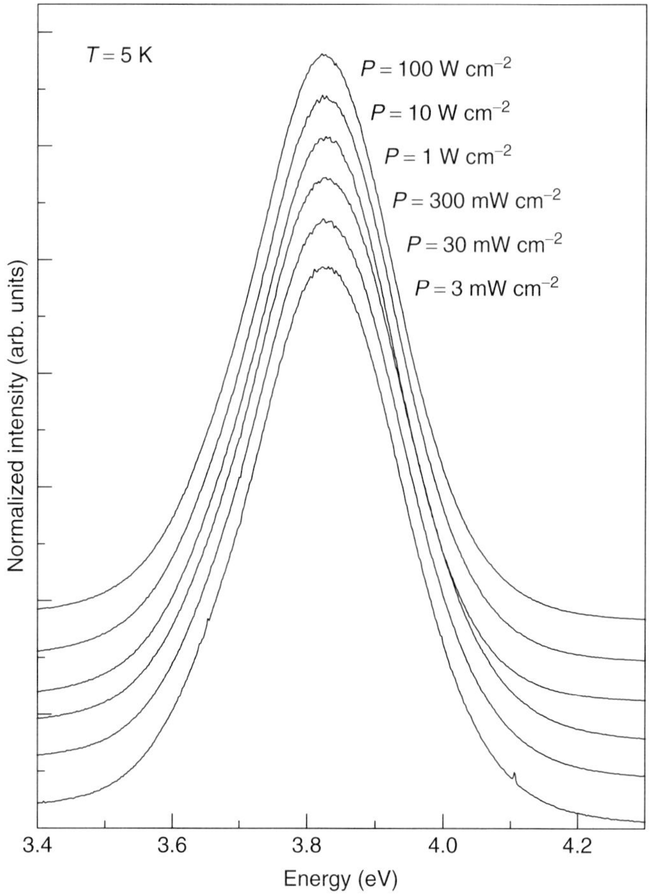

Figure 13.8 Low-temperature photoluminescence spectra as a function of excitation power density at $\lambda = 244$ nm. The emission is above the GaN band gap and no screening is observed.

GaN band gap owing to the quantum confined Stark effect [4]. This is a first indication that the effects of the internal electric field are reduced compared to polar QDs [19, 43]. The PL spectrum does not evolve as a function of excitation power density over 6 orders of magnitude, which shows that no screening or state filling effects are observed in our CW experiments.

A more sensitive way of probing the effect of the internal electric field is to study the luminescence decay time. Indeed for polar QDs, the reduction in spatial overlap of the electron and hole wavefunctions induced by the internal electric field has a dramatic effect on the radiative recombination rate [4, 44]. For three-dimensionally confined wavefunctions with negligible excitonic effect, the radiative decay rate is proportional to the electron-hole wavefunction overlap. This leads to very long decay times ranging from 5 ns to

200 μs for polar GaN/AlN QDs with sizes ranging from 1.5 to 3.5 nm [44]. We performed time-resolved luminescence experiments on a sample containing a single array of nonpolar GaN/AlN QDs, by exciting the sample with a tripled Ti:Sapphire laser emitting at 250 nm, with a repetition rate of 76 MHz and an average power of 5 mW. The signal was detected with a Hamamatsu streak camera with a time resolution of 4 ps. Figure 13.9 shows the luminescence decay curves for various detection wavelengths. We checked that the decay curves do not vary as a function of temperature in the 4–100 K range, thus showing that nonradiative processes do not play any important role at low enough temperature.

The first striking feature is that the luminescence decay is rather fast, the intensity reaching I_0/e after 150 ps and being less than $I_0/100$ after 1.5 ns. This is clearly a much faster decay than for polar GaN/AlN QDs of similar size for which the intensity decays to $I_0/100$ after about 500 ns [44]. This again is a strong signature of a much reduced quantum confined Stark effect in nonpolar QDs compared to polar QDs. A very important feature shown by Figure 13.9 is that the luminescence decay curves do not depend on the emission wavelength within the entire emission spectrum. The radiative decay properties thus do not depend on QD size in the nonpolar system. This again is very different from polar QDs for which the radiative decay time varies roughly exponentially as a function of QD height [4, 44]. The fact that the radiative decay curves do not depend at all on QD size is the most salient experimental evidence that we see no evidence of the effect of an internal electric field on the optical properties of an ensemble of nonpolar QDs.

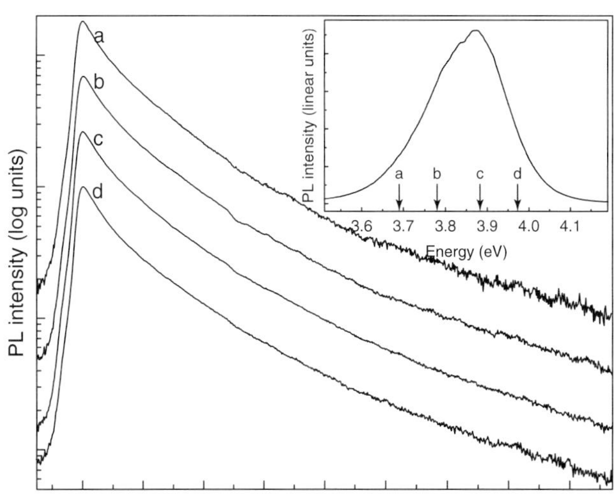

Figure 13.9 Time-resolved photoluminescence spectra, detected at various emission wavelengths. The decay curves are the same for all wavelengths.

13.4.2
Single QD Spectroscopy

While the optical properties on an ensemble of nonpolar QDs do not show any evidence of quantum confined Stark effect, it is interesting to get rid of the inhomogeneous broadening by studying a single QD emission in PL. In other QD systems, single QDs are studied by reducing the number of dots for which luminescence is collected either by etching mesas [45] or by opening holes in an opaque metallic mask [46]. In our system, the experimental difficulty is twofold. The first difficulty is that the density of QDs is rather high, around 10^{11} cm^{-2} so that even in a 100-nm diameter mesa, 10 QDs are still present. It is, however, to be noted that for a low GaN coverage, the density is inhomogeneous on a micrometer scale (see Figure 13.2a) so that some mesas might contain less QDs. The second difficulty is to design a microphotoluminescence set-up with good performances in the UV spectral range. For this we use a UV-dedicated microscope objective with numerical aperture (NA) = 0.4 and working distance 8 mm made by optics for research (OFR). The exciting laser beam at 244 nm is decollimated by a lens doublet before being focused on the sample so as to compensate for the chromaticity of the microscope objective between 244 nm and the collected PL around 300 nm. The excitation laser is focused onto a ~1-μm diameter on the sample, and the luminescence is collected through the same objective and refocused onto a 55-cm Jobin–Yvon monochromator with a UV doublet, the signal being then detected by a LN_2-cooled CCD camera. The spectral resolution around 4 eV is 300 μeV.

The first difficulty mentioned above was overcome by using samples with a reduced GaN coverage. Such a sample is shown in Figure 13.10. The nominal coverage was 2.5 MLs, leading to the formation of a few lines of dots and, as a whole, to an *average* dot density of about 1.8×10^{10} cm^{-2}. Such samples represent the state of the art in matter of low density (11-20) GaN QDs and make possible single dot spectroscopy experiments relatively easily. However, note that experiments presented in this section were performed using sample of the type shown in Figure 13.2a.

The samples with mesas are prepared by defining a 50-nm thick Ni mask by e-beam lithography and a lift-off step, and then by etching the unmasked semiconductor with a $SiCl_4$ inductively coupled plasma (ICP) etching. The mesa sizes range from 200 nm to 5 μm. Figure 13.11 shows a scanning electron micrograph of such a mesa.

The surface QDs are visible as small brighter dots on the micrograph, showing that in such a mesa there are still many QDs. Samples with openings in a metallic mask were prepared in the following way: polystyrene nanobeads with diameters of 300 nm were dispersed on the sample, on which 100-nm aluminum were then deposited. The removal of the beads thus leaves circular openings in the aluminum mask [47].

The PL spectra of the smallest mesas display sharp lines that are typical of single QD lines. Single lines are better isolated on the high-energy side of the

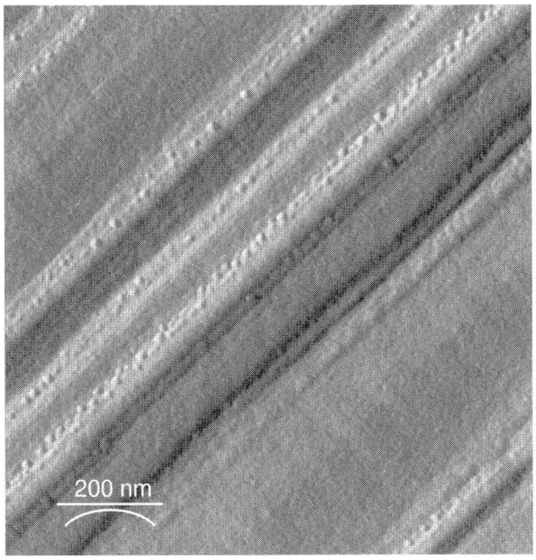

Figure 13.10 AFM image of low density (11-20) GaN QDs. Nominal GaN coverage is 2.5 MLs.

QD distribution owing to the lower spectral density of QD emission. However, no sharp lines could be isolated on the low energy side of the QD emission spectrum, for reasons that are so far not clear. We first checked that the intensity of these lines varies linearly with exciting power, in the low power regime; this shows that the observed sharp lines correspond to the recombination of a single electron-hole pair and not to excited states recombination.

The first information extracted from this microphotoluminescence study is the spectral widths of single dot lines. The measured line widths range from 0.5 to 2 meV. These are rather large line widths for single QD emission lines. The broadening is attributed here to spectral diffusion [48, 49]: loosely bound charges around a given dot are trapped and detrapped as a function of time, leading to a time-dependent potential variation in the QD that shifts the discrete transitions of the QD. As the spectra are time-integrated on timescales that are longer than the typical spectral diffusion processes, this leads to a broadening of the observed lines.

This phenomenon can actually be observed on the typical integration time (1 s) used in these experiments, as shown in Figure 13.12. A single line is witnessed to change line width and peak position as a function of time with typical peak position fluctuation amplitudes that are comparable to the observed line width.

Spectral diffusion was also observed in the study of polar GaN/AlN QDs [50] leading to line widths between 5 and 20 meV, that is one order of magnitude larger than for the nonpolar QDs we studied. Although spectral diffusion depends on the sample quality and excitation conditions, this result is also consistent with the fact that the internal field is much lower in nonpolar QDs, as the presence of an internal field enhances the spectral diffusion. Most importantly, the line widths that we observed were narrow enough for us to study single QD lines as a function of temperature and observe a temperature-dependent broadening of the lines (see Figure 13.13).

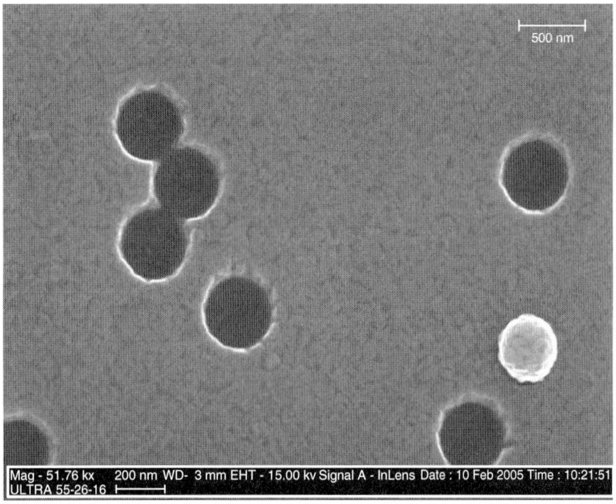

(a)

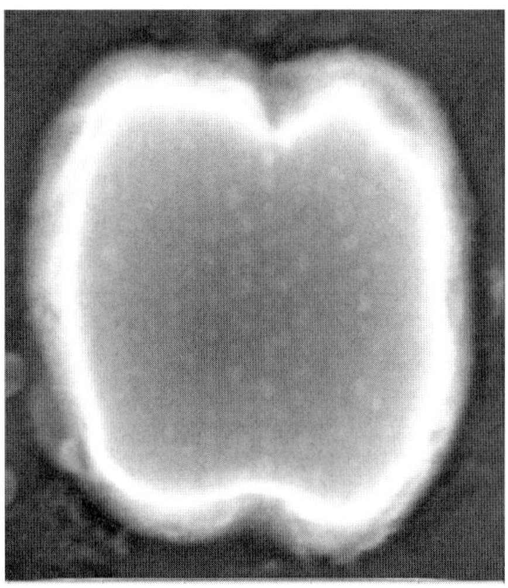

(b)

Figure 13.11 Various ways to isolate single QDs: (a) openings in an Al mask obtained by using nanobeads (one remaining nanobead is visible in the lower right corner) and (b) an etched mesa, on which the surface QDs are visible.

When the temperature is raised, the lineshape becomes asymmetric (see Figure 13.14), which is characteristic of the appearance of acoustic phonon emission and absorption events [51–54]. With an increase in the temperature, the acoustic phonon population builds up, thus increasing the probability of absorbance or emission of acoustic phonons when an electron-hole pair

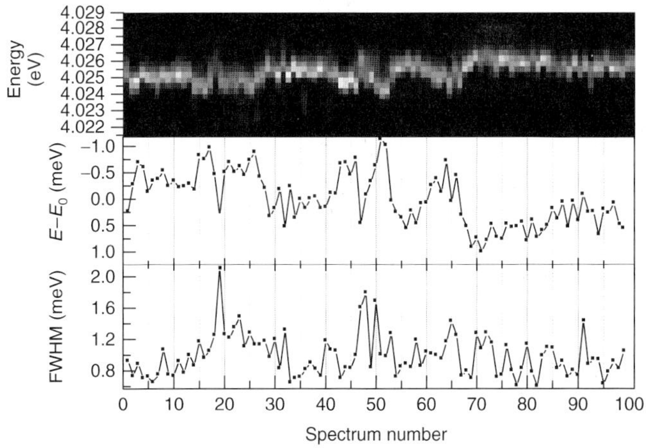

Figure 13.12 Energy position and full width at half maximum of a single dot emission (one spectrum is integrated each second).

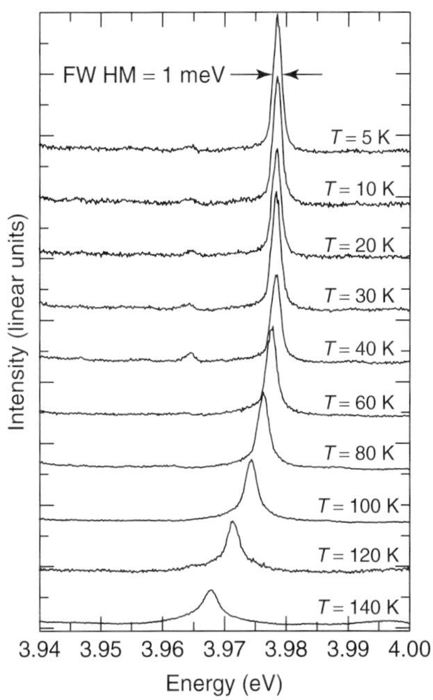

Figure 13.13 Evolution of the emission lineshape of a single QD. The broadening of the line at high temperature is characteristic of coupling to acoustic phonons.

radiatively recombines. The larger 'phonon wing' on the red side of the single dot line is owing to the fact that emission events are more probable than absorption events. The experimental observation of this coupling to acoustic phonons allows to probe the spatial extent of the confined wavefunctions [51].

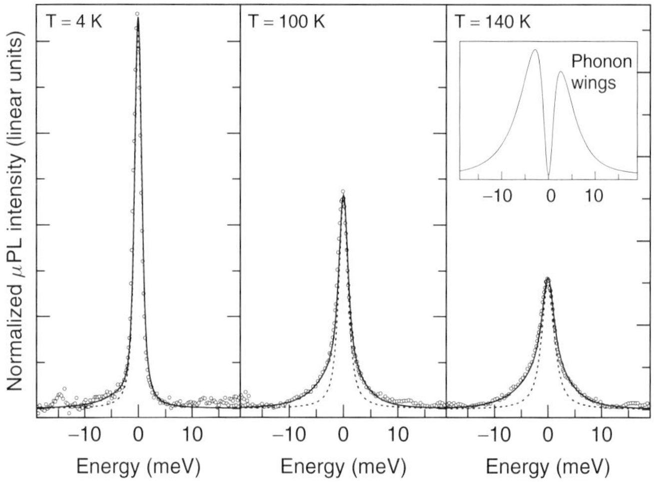

Figure 13.14 Evolution of a single QD emission line as a function of temperature (empty dots: same experimental data as Figure 13.13). The dotted line represents the zero-phonon line (Voigt profile) and the continuous line is the fit obtained with our model. The inset shows the difference between the zero-phonon line and the full model, which corresponds to the acoustic phonon emission and absorption events.

The modeling of the particular experimental situation described in this work is done in detail in reference [55] and we shall present the main ideas and conclusions here.

The coupling of localized electronic states to phonons was first described by Huang and Rhys [56]: in their framework, the acoustic phonon modes are discretized and each individual mode forms a new eigenstate (an acoustic polaron) with each confined exciton state. Radiative transitions then occur between these polaron states, possibly leading to photon emission with emission or absorption of acoustic phonons. The probability of emission of an acoustic phonon depends on the coupling strength between the confined electron-hole pair state and the acoustic phonon mode, as well as on the phonon mode population (i.e. on the temperature).

To model the coupling of a confined electron-hole pair to the acoustic phonons, we follow the model described in Ref. 51. We consider only the deformation potential coupling to longitudinal acoustic phonons and neglect the piezoelectric coupling. While this approximation is perfectly justified in zinc-blende semiconductors [57], it needs some more justification for III-N wurtzite semiconductors for which the piezoelectric coupling coefficients are much larger [58]. The deformation potential coupling depends on the overall extent of the electron-hole wavefunction, while the piezoelectric acoustic phonon coupling is sensitive to the polarity of the electron-hole wavefunction. In particular, the piezoelectric coupling vanishes if the electron and hole wavefunctions perfectly overlap [57, 59]; on the contrary, it dominates over the deformation potential coupling in polar heterostructures for which the electrons and holes are strongly spatially separated [59, 60]. In the case of the nonpolar GaN QDs considered here, the spectroscopy experiments on ensembles of dots, especially the time-resolved PL characterization, do not show any evidence of quantum confined Stark effect, so that it can be assumed that the confined electron-hole pair permanent dipole is small enough to neglect the piezoelectric coupling to acoustic phonons.

To model the exciton–phonon coupling, it is necessary to make an assumption concerning the confined electronic wavefunctions in the nonpolar QDs considered here. The Bohr radius in GaN is 2.8 nm so that the potential well along the growth direction is smaller than the Bohr radius and is therefore strongly confining. On the other hand, the typical lateral extent of the QDs is much larger than the Bohr radius so that the carriers are laterally weakly confined. The wavefunction can thus be assumed to be quasi–two dimensional and formally described as under:

$$\psi_X(r_e, r_h) = \left(\frac{1}{\xi\sqrt{\pi}} e^{-R_p^2/2\xi^2}\right) \times \left(\sqrt{\frac{2}{\pi a_{2D}^2}} e^{-r/a_{2D}}\right)\left(\frac{2}{L}\cos(\pi z_e/L_Z)\cos(\pi z_h/L_Z)\right) \qquad (2)$$

The third part of this wavefunction is the solution of an infinite-barrier QW with height L_z, which is a good approximation as band offsets between GaN and AlN are large enough for AlN to behave as an infinite barrier. The in-plane electron-hole correlation is described by a 2D hydrogenoid wavefunction, r being the relative position and a_{2D} the 2D Bohr radius of the infinite QW. The lateral extension of the center of mass position R_p is described by the localization parameter ξ. It can be shown that deformation potential coupling is a maximum for acoustic phonons with wavevector $q \sim 1/\xi$, so that the study of the phonon coupling as a function of temperature allows us to deduce a value for the typical center of mass extension ξ. Within this framework, the emission spectra for each temperature can be fitted in the following way: the probability to emit or absorb one or more phonons for each discretized phonon mode is computed following the Huang–Rhys procedure. Each of these probabilities is then convoluted by the zero-phonon line (i.e. the central part of the single QD emission spectrum for which no phonon emission or absorption event occurs) to obtain the final emission spectrum. For each temperature, the zero-phonon line is fitted by a Voigt function. Indeed, as spectral diffusion dominates the broadening of the central line at low temperature, the zero-phonon line cannot be solely fitted by a Lorentzian, and it is, for instance, well fitted at 4 K by a Gaussian with line width $w_g = 1.1$ meV. The ξ parameter that describes the electron-hole center of mass lateral extent is the only temperature-independent fitting variable.

The best fit as indicated on Figure 13.14 is obtained for a localization parameter $\xi = 2.1$ nm, corresponding to a FWHM extent for the center of mass of $2\xi = 4.2$ nm. This value is much smaller than the typical QD diameter, which indicates that a supplemental localization inside the QDs occurs. It is possible to further probe this localization hypothesis by analyzing the decay time of the PL, and this is discussed in the next subsection.

13.4.3
Radiative Decay Time Analysis

As described previously, the geometry of the nonpolar QDs studied here is such that the height of the dots is smaller than the GaN excitonic Bohr radius, while the diameter is much larger than the Bohr radius. For such a QD in the case of a quasi–two-dimensional wavefunction described by Equation 2, the confined carriers are expected to be in the 'giant oscillator strength regime' for which the oscillator strength is proportional to the coherence surface of the exciton center of mass wavefunction [61, 62]. Let us recall that for localized wavefunctions, the oscillator strength is related to the radiative lifetime by:

$$\tau_R = \frac{3m_e c^3 4\pi\varepsilon_0}{2ne^2\,\omega^2 f} \tag{3}$$

From the study of the radiative lifetime, it is thus possible to extract information on the localization parameter ξ. More precisely, in the framework of the wavefunction described by Equation 2 and granted that $2\xi > a_{2D}$, the oscillator strength is given by [61, 62]:

$$f = \frac{8E_p}{E} \cdot \left(\frac{\xi}{a_{2D}}\right)^2 \tag{4}$$

where E is the transition energy and E_p the interband matrix element (Kane energy). The conduction band to valence band matrix element for the 'heavy hole' valence band was calculated by Chuang and Chang to be 15.7 eV [63].

Let us first assume that the center of mass of the exciton is only laterally confined by the AlN barrier (in other words, no supplemental localization inside the QD occurs). In that case, we need to estimate the exciton center of mass extension parameter ξ. In a simple approach, we can consider that the exciton is laterally confined by infinite barriers in a cylinder that has the height of the QD and a diameter corresponding to the diameter of the top facet of the QD. Let us note that this model tends to underestimate the lateral extent of the confined wavefunction. For the center of the QD size distribution (height 1.6 nm, diameter 19 nm), the parameter ξ deduced from this simple approach is 3.4 nm. Equations 3 and 4 thus yield $f = 85$ and $\tau_R = 24$ ps. This is clearly a much faster decay than the experimentally measured decay, and thus shows that the expected giant oscillator strength regime is not reached in our structures.

The other limiting case is the lateral strong confinement for which the carriers are laterally confined on a dimension smaller than the two-dimensional Bohr radius. In that case, lateral confinement effects dominate over the Coulomb attraction, and as a consequence, the excitonic effects are negligible in the oscillator strength estimation. The oscillator strength is then proportional to the electron-hole wavefunctions overlap, and in the case of a perfect overlap is given by $f = \frac{E_p}{E}$ [62]. This yields $f = 4$ and $\tau_R = 520$ ps, which is a longer decay time than what is experimentally measured. The experimental situation is thus intermediate between an exciton coherent over the entire QD surface and a strongly laterally localized exciton. Let us now again consider a quasi–two-dimensional wavefunction, but with a localization parameter $\xi = 2.1$ nm as was determined in the acoustic phonon coupling study. In that case, Equation 3 gives $f = 33$ and $\tau_R = 62$ ps. This decay value is still smaller than the experimentally measured one. There are two possible explanations for this discrepancy: (i) the Kane energy we used ($E_p = 15.7$ eV) stems from a calculation [63], while Ref. 64 for instance, determines experimentally a value $E_p = 7.7$ eV; (ii) most importantly, for a localization parameter $\xi = 2.1$ nm, the lateral extent of the exciton center of mass is comparable with the Bohr radius, in which case, the excitonic enhancement of the oscillator strength that appears in Equation 3 should be reduced.

This analysis is thus consistent with the fact that excitons in the nonpolar QDs studied here are laterally localized inside the QDs, the decay time of each QD being related to the lateral localization of the exciton inside the QD and not to the QD size. This is confirmed by the experimental data presented in Figure 13.9. The luminescence decay curves are not monoexponential.

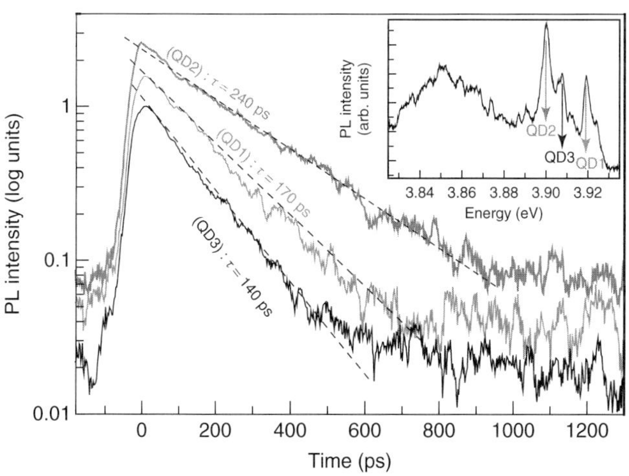

Figure 13.15 PL decay curves at 5 K for various single QD lines (indicated in the inset). The decay curves are monoexponential with various decay times.

They are better described by a collection of independent emitters with decay times varying between 70 and 600 ps. To test this hypothesis, we performed time-resolved microphotoluminescence to probe the decay of single QDs.

The results are shown in Figure 13.15; in striking contrast with the decay curves for ensembles of QDs, the PL decay for single dots is well fitted by a single exponential. Various isolated QDs have different decay times, and these decay times are not correlated to the emission wavelength. These various decay times can be attributed to excitons that have different lateral localization surfaces inside each QD, this localization parameter being independent of QD size.

13.5 Conclusion

In conclusion, it appears that the morphology of (11-20) AlN layers grown on (11-20) SiC is dominated to a large extent by lattice anisotropy which results in a rippled surface. As a remarkable secondary consequence of this anisotropy, it has been found that (11-20) GaN QDs grown on such an AlN buffer layer by plasma-assisted MBE are aligned along the [1-100] direction. The anisotropic strain state experienced by GaN deposited on (11-20) AlN leads to an anisotropic strain relaxation of (11-20) GaN QDs, purely elastic along [1-100] and elastic and plastic along [0001]. The optical properties of the dots are found to be unaffected by the plastic relaxation contribution which may mostly affect the wetting layer. Indeed, the QD PL is intense and is observed at higher energy than GaN band gap, consistent with the strong carrier confinement and with the strong reduction in internal electrical field that is expected for nonpolar heterostructures.

Analyzing the temperature dependence of the PL line width in the framework of a quasi-2D wavefunction model and combining it with time-resolved PL experiments, it is concluded that excitons are laterally localized at a length scale smaller than the dot size. No direct correlation was found between monoexponential decay time constants and dot size. As a consequence, it appears that the localization of exciton itself does not simply depend on dot size, making the nature of localization in these nonpolar QDs an issue still open till date.

Acknowledgments

We acknowledge the technical assistance of J. Curé and Y. Genuist for epitaxial growth. We thank M. Terrier and Dr D. Peyrade for their help in processing the samples for single dot spectroscopy. The additional help of Dr J. Bleuse for time-resolved spectroscopy measurements is gratefully acknowledged.

References

1. Nakamura, S. (1998) *Science*, **281**, 957.
2. Gerthsen, D., Neubauer, B., Rosenauer, A. *et al.* (2001) *Applied Physics Letters*, **79**, 2552.
3. Musikhin, Y.G., Gerthsen, D., Bedarev, D.A. *et al.* (2002) *Applied Physics Letters*, **80**, 2099.
4. Simon, J., Pelekanos, N.T., Adelmann, C. *et al.* (2003) *Physical Review B*, **68**, 035312.
5. Bernardini, F., and Fiorentini, V. (1998) *Physical Review B*, **57**, R9427.
6. Waltereit, P., Brandt, O., Trampert, A. *et al.* (2000) *Nature*, **406**, 865.
7. Craven, M.D., Waltereit, P., Speck, J.S., and DenBaars, S.P. (2004) *Applied Physics Letters*, **84**, 496.
8. Akopian, N., Bahir, G., Gershoni, D. *et al.* (2005) *Applied Physics Letters*, **86**, 202104.
9. Paskova, T., Kroeger, R., Figge, S. *et al.* (2006) *Applied Physics Letters*, **89**, 051914.
10. Craven, M.D., Lim, S.H., Wu, F. *et al.* (2002) *Applied Physics Letters*, **81**, 469.
11. Ng, H.M. (2002) *Applied Physics Letters*, **80**, 4369.
12. Chen, C., Yang, J., Wang, H. *et al.* (2003) *Japanese Journal of Applied Physics*, **42**, L640.
13. Paskov, P.P., Schifano, R., Monemar, B. *et al.* (2005) *Journal of Applied Physics*, **98**, 093519.
14. Kusakabe, K. and Ohkawa, K. (2005) *Japanese Journal of Applied Physics*, **44**, 7931.
15. Tsuchiya, Y., Okadome, Y., Honshio, A. *et al.* (2005) *Japanese Journal of Applied Physics*, **44**, L1516.
16. Chevtchenko, S., Ni, X., Fan, Q. *et al.* (2006) *Applied Physics Letters*, **88**, 122104.
17. Onojima, N., Suda, J., Kimoto, T., and Matsunami, H. (2003) *Applied Physics Letters*, **83**, 5208.
18. Craven, M.D., Wu, F., Chakraborty, A. *et al.* (2004) *Applied Physics Letters*, **84**, 1281.
19. Founta, S., Rol, F., Bellet-Amalric, E. *et al.* (2005) *Applied Physics Letters*, **86**, 171901.
20. Zakharov, D.N., Liliental-Weber, Z., Wagner, B. *et al.* (2005) *Physical Review B*, **71**, 235334.
21. Craven, M.D., Lim, S.H., Wu, F. *et al.* (2002) *Applied Physics Letters*, **81**, 1201.
22. Tarsa, E.J., Heying, B., Wu, X.H. *et al.* (1997) *Journal of Applied Physics*, **82**, 5472.
23. Mula, G., Adelmann, C., Moehl, S. *et al.* (2001) *Physical Review B*, **64**, 195406.
24. Adelmann, C., Brault, J., Jalabert, D. *et al.* (2002) *Journal of Applied Physics*, **91**, 9638.
25. Neugebauer, J., Zywietz, T.K., Scheffler, M. *et al.* (2003) *Physical Review Letters*, **90**, 056101.
26. Daudin, B., Feuillet, G., Mula, G. *et al.* (1999) *Physica Status Solidi A*, **176**, 621.
27. Onojima, N., Suda, J., and Matsunami, H. (2002) *Japanese Journal of Applied Physics*, **41**, L1348.
28. Chen, Y., Ohlberg, D.A.A., and Williams, R.S. (2002) *Journal of Applied Physics*, **91**, 3213.
29. Omi, H. and Ogino, T. (1999) *Physical Review B*, **59**, 7521.
30. Adelmann, C., Daudin, B., Oliver, R.A. *et al.* (2004) *Physical Review B*, **70**, 125427.
31. Founta, S., Coraux, J., Jalabert, D. *et al.* (2007) *Journal of Applied Physics*, **101**, 063541.
32. Meynadier, M.H., Brum, J.A., Delalande, C. *et al.* (1985) *Journal of Applied Physics*, **58**, 4307.
33. Gérard, J.M., Cabrol, O., Marzin, J.Y. *et al.* (1996) *Materials Science and Engineering*, **B37**, 8.
34. Ziegler, J.F. *SRIM Program*, (2007) http://srim.org/.
35. Bailey, P., Noakes, T.C.Q., and Woodruff, D.P. (1999) *Surface Science*, **426**, 358.
36. Tromp, R.M., Copel, M., Reuter, M.C. *et al.* (1991) *Review of Scientific Instruments*, **62**, 2679.
37. Craven, M.D., Waltereit, P., Wu, F. *et al.* (2003) *Japanese Journal of Applied Physics*, **42**, L235.

38. Kern, R. and Muller, P. (1996) *Surface Science*, **392**, 103.
39. Galindo, P.L. *et al.* (2005) *Proceedings of the XIV Microscopy of Semiconducting Materials Conference*, Oxford.
40. Hÿtch, M.J., Snoeck, E., and Kilaas, R. (1998) *Ultramicroscopy*, **74**, 131.
41. Hÿtch, M.J., Putaux, J.L., and Pénisson, J.M. (2003) *Nature*, **423**, 270.
42. Cros, A., Budagosky, J.A., García-Cristóbal, A. *et al.* (2006) *Physica Status Solidi B*, **243**, 1499.
43. Garro, N., Cros, A., Budagosky, J.A. *et al.* (2005) *Applied Physics Letters*, **87**, 011101.
44. Bretagnon, T., Lefebvre, P., Valvin, P. *et al.* (2006) *Physical Review B*, **73**, 113304.
45. Marzin, J.-Y., Gérard, J.M., Izraël, A. *et al.* (1994) *Physical Review Letters*, **73**, 716.
46. Gammon, D., Snow, E.S., Shanabrook, B.V. *et al.* (1996) *Physical Review Letters*, **76**, 3005.
47. Fischer, U.Ch. (1985) *Journal of Vacuum Science & Technology B*, **3**, 386.
48. Rice, J.H., Robinson, J.W., Jarjour, A. *et al.* (2004) *Applied Physics Letters*, **84**, 4110.
49. Moriwaki, O., Someya, T., Tachibana, K. *et al.* (2000) *Applied Physics Letters*, **76**, 2361.
50. Kako, S., Hoshino, K., Iwamoto, S. *et al.* (2004) *Applied Physics Letters*, **85**, 64.
51. Besombes, L., Kheng, K., Marsal, L., and Mariette, H. (2001) *Physical Review B*, **63**, 155307.
52. Bayer, M. and Forchel, A. (2002) *Physical Review B*, **65**, 041308.
53. Favero, I., Cassabois, G., Ferreira, R. *et al.* (2003) *Physical Review B*, **68**, 233301.
54. Peter, E., Hours, J., Senellart, P. *et al.* (2004) *Physical Review B*, **69**, 041307.
55. Rol, F. Founta, S. Mariette, H. *et al.* (2007) *Physical Review B*, **75**, 125306.
56. Huang, K. and Rhys, A. (1950) *Proceedings of the Royal Society of London-Series A*, **204**, 404.
57. Takagahara, T. (1999) *Physical Review B*, **60**, 2638.
58. Vurgaftman, I., Meyer, J.R., and Ram-Mohan, L.R. (2001) *Japanese Journal of Applied Physics*, **89**, 5815.
59. Krummheuer, B., Axt, V.M., and Kuhn, T. (2002) *Physical Review B*, **65**, 195313.
60. Krummheuer, B., Axt, V.M., Kuhn, T. *et al.* (2005) *Physical Review B*, **71**, 235329.
61. Kavokin, A.V. (1994) *Physical Review B*, **50**, 8000.
62. Andreani, L.C., Panzarini, G., and Gérard, J.M. (1999) *Physical Review B*, **60**, 13276.
63. Chuang, S.L. and Chang, C.S. (1996) *Physical Review B*, **54**, 2491.
64. Im, J.S., Moritz, A., Steuber, F. *et al.*(1996) *Applied Physics Letters*, **70**, 631.

14
Semipolar InGaN/GaN Quantum Wells for Highly Functional Light Emitters

Mitsuru Funato, Yoichi Kawakami, Yukio Narukawa, and Takashi Mukai

14.1
Introduction

Light emitters based on nitride semiconductors typically consist of [0001]-oriented quantum wells (QWs) [1] where the quantum confinement Stark effect (QCSE) is caused by piezoelectric and spontaneous polarizations, which lower the optical transition probability [2, 3]. To circumvent this issue, several groups have tried to fabricate InGaN/GaN and GaN/AlGaN QWs on nonpolar planes such as the $\{1\bar{1}00\}$ plane (*m*-plane) [4] and the $\{11\bar{2}0\}$ plane (*a*-plane) [5–7]. However, owing to the difficulty in growing high-quality crystals in nonpolar directions, the layers contain numerous nonradiative recombination centers.

Alternatively, there is increasing interest in semipolar planes, which are tilted with respect to the (0001) plane [8–23] because a reduced or even negligible electric field is theoretically predicted [24, 25]. Figure 14.1 shows example calculations for InGaN/GaN, which include the internal electric field as a result of the polarization effect and the consequent square overlap integral of the electron and hole wavefunctions that are proportional to the optical transition probability. The assumed In composition and InGaN well thickness are 10% and 3 nm, respectively. It should be noted that the internal electric field becomes zero at a given angle between 0 and 90°, but this angle strongly depends on the physical parameters used in the calculation. In fact, the crossover was calculated to be 35° or 48° in Figure 14.1a, depending on the parameter set used. The precise angle that provides a zero electric field is controversial. Hence, future studies should experimentally estimate the magnitude of the electric fields. Nevertheless, semipolar planes are quite attractive because a weakened electric field is predicted without serious difficulty in crystal growth due to the presence of the polarity in those planes [26], leading to a better optical transition probability (Figure 14.1b) while suppressing nonradiative transitions.

Nitrides with Nonpolar Surfaces. Growth, Properties, and Devices. Edited by Tanya Paskova

ISBN: 978-3-527-40768-2

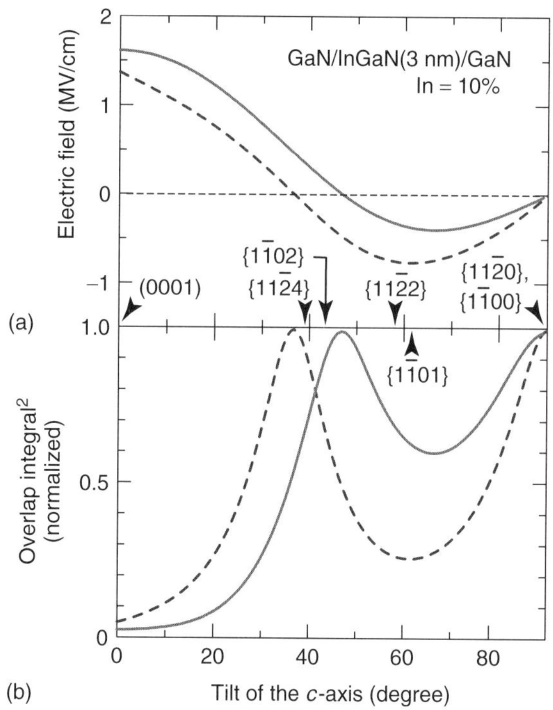

Figure 14.1 Example calculations of (a) the internal electric field and (b) the consequent square overlap integral of the electron and hole wavefunctions for $In_{0.10}$ $Ga_{0.90}$ N/GaN. Solid and broken lines are the calculated results for different parameter sets.

In addition to a higher optical transition probability, another interesting characteristic of the semipolar planes is optical anisotropy caused by the reduced crystal symmetry. When InGaN is unstrained or symmetrically strained in the (0001) plane, one expects to see two dominant transitions for **E** $\perp$ [0001] and one dominant transition for **E** $\parallel$ [0001] [27], where **E** is the electric field vector of light. Because the lowest energy transition (A exciton) is polarized perpendicular to the [0001] direction, (0001) surface emitters such as conventional light emitting diodes (LEDs) exhibit random polarization, that is, they do not display polarization. On the contrary, semipolar and nonpolar planes, in principle, involve in-plane optical anisotropy. Furthermore, the anisotropic in-plane compressive strain in an InGaN well layer strongly affects the polarization properties. These effects have already been investigated for {1$\bar{1}$00} GaN [28, 29] and (0001) GaN [30], but not for semipolar QWs. Thus, it is important to experimentally and theoretically investigate the polarization properties in semipolar planes. In terms of practical applications, the optical anisotropy can be applied to optical switching [31] and polarization-sensitive photodetectors [32], for example.

However, a major hurdle to realize the above-mentioned properties is how to create high-quality semipolar planes. A preliminary experiment using a regrowth technique has been reported in Ref. 8 where InGaN QWs were grown on {1$\bar{1}$01} facets. However, the QWs unexpectedly showed very weak luminescence because of the presence of numerous stacking faults. On the contrary, we have found that inclined semipolar GaN can be grown on GaAs (11*n*) substrates ($n = -8 - -3$) [9, 10], but high-quality QWs could not be obtained on GaAs. Recently, we have confirmed that regrowth by metalorganic vapor phase epitaxy (MOVPE) is still a promising technique for creating semipolar and nonpolar planes [11–13]. It has been demonstrated that singular planes such as polar (0001), semipolar {11$\bar{2}$2} and {1$\bar{1}$01}, and nonpolar {11$\bar{2}$0} and {1$\bar{1}$00} naturally appear as microfacets and form three-dimensional multifaceted structures, and that InGaN/GaN QWs grown on semipolar {11$\bar{2}$2} and {1$\bar{1}$01} microfacets exhibit high internal quantum efficiencies (IQEs), mainly due to weakened internal electric fields. The multifaceted structures add another novel characteristic, that is, multicolor emissions suitable for devices requiring sophisticated syntheses of color, including white and/or pastel [12, 13]. Around the same time, another research group began working on similar microfacet structures [14] and has recently fabricated semipolar {1$\bar{1}$01} microfacet LEDs [15]. The observed high IQW on the {11$\bar{2}$2} and {1$\bar{1}$01} planes has motivated us to fabricate planar structures with a semipolar orientation because current device processes have been optimized for planar structures. We have prepared semipolar {11$\bar{2}$2} GaN substrates, performed MOVPE of GaN and InGaN/GaN QWs [17], and then fabricated InGaN/GaN LEDs that emit blue, green, or amber [18].

This chapter is structured as follows: The crystalline structure of the {11$\bar{2}$2} plane is briefly described in Section 14.2 and the results for the microfacet QWs are described in Section 14.3. Finally, Section 14.4 reviews planar {11$\bar{2}$2} QWs and LEDs.

14.2 Semipolar {11$\bar{2}$2} Planes

In our study, semipolar {11$\bar{2}$2} planes played a key role. Figure 14.2 illustrates the crystal structure of GaN containing a (11$\bar{2}$2) plane. The (11$\bar{2}$2) and ($\bar{1}\bar{1}$2$\bar{2}$) surfaces possess different bond configurations and, thus, have different chemical natures. However, the influence of the bond configuration on, for example, the growth characteristics and optical properties has yet to be clarified. In terms of the polarization anisotropy in the optical properties, the key in-plane directions are the [1$\bar{1}$00] direction, which is perpendicular to the [0001] direction, and the [$\bar{1}\bar{1}$23] direction, which is parallel to the projection of the [0001]-axis on the (11$\bar{2}$2) plane and perpendicular to [1$\bar{1}$00].

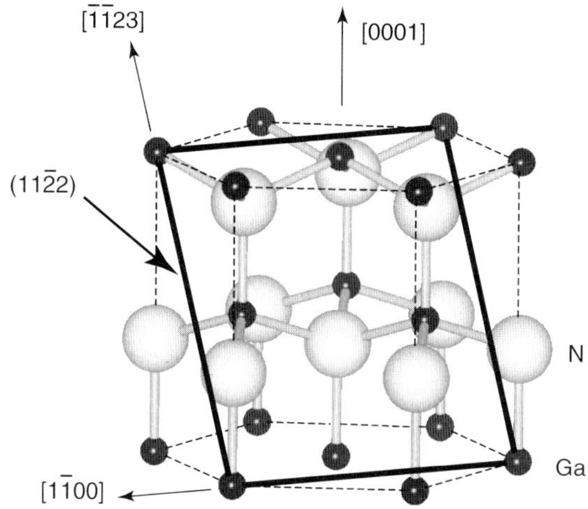

Figure 14.2 Schematic crystal structure of GaN and a (11$\bar{2}$2) plane.

14.3
Microfacet QWs

14.3.1
Fabrication and Fundamental Properties of Microfacet QWs

The samples were grown by MOVPE. Here, the formation processes of InGaN/GaN QWs on faceted microstructures were introduced. Two different methods, which produced essentially identical QWs, were established, as illustrated in Figure 14.3 by (a) → (b) → (c) and (d) → (e) → (f). In both cases, the starting template was well-established, high-quality (0001) GaN grown on (0001) sapphire (Figure 14.3a and d). In Figure 14.3b a stripe pattern was formed along the [1$\bar{1}$00] direction by reactive ion etching (RIE) where the GaN template was deeply etched into the sapphire substrate so that the sapphire substrate worked as a mask for regrowth. On the contrary, in Figure 14.3e conventional photolithography was used to form SiO_2 mask stripes along the same direction. The subsequent regrowth of GaN created microstructures composed of (0001), {11$\bar{2}$2}, and {11$\bar{2}$0} facets along the [1$\bar{1}$00] direction, similar to Refs. 33 and 34. (For the process using RIE, this GaN layer recovers surface damages.) Finally, InGaN/GaN QWs were fabricated on these microfacets (Figure 14.3c and f).

The width of the striped GaN seed region ranged from 4 to 15 μm with a period of 10 to 100 μm. The shape of the microfacet structure varied with the mask geometry; multifaceted structures tended to appear on masks with a large occupancy of the seed region within a period, while {11$\bar{2}$2} facets were dominant on masks with a small occupancy. On the contrary,

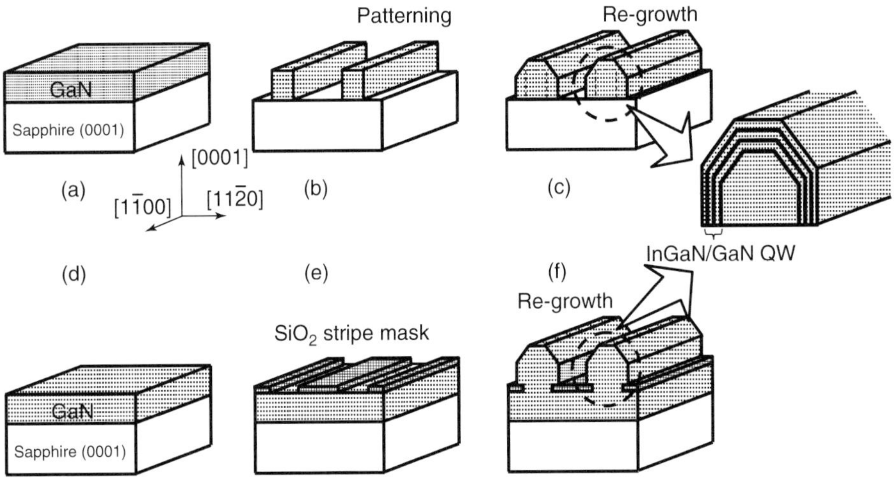

Figure 14.3 Schematic of the two different formation processes of microfacet QWs. (a) GaN template → (b) patterning by RIE → (c) fabrication of QWs by regrowth, and (d) GaN template → (e) SiO_2 mask patterning by photolithography → (f) fabrication of QWs by regrowth.

microfacet structures formed on mask stripes along the perpendicular [11$\bar{2}$0] direction mostly were composed of only semipolar {1$\bar{1}$01} planes. It was confirmed that {1$\bar{1}$01} microfacet QWs possess high IQEs that are comparable to {11$\bar{2}$2} microfacet QWs and, therefore, are promising for highly efficient light emitters. However, this chapter concentrates on mask stripes along the [1$\bar{1}$00] direction with a relatively large occupancy of the seed region and demonstrates high IQEs and multicolor emissions.

Figure 14.4 shows a cross-sectional scanning transmission electron microscopy (STEM) bright field image of a fabricated structure viewed along the [1$\bar{1}$00] direction. The specimen was prepared by conventional Ar^+ milling and the acceleration voltage for the STEM observation was 200 kV. The sample was grown by the procedure shown in Figure 14.3(a) → (b) → (c) with a seed width of 4 μm and a period of 20 μm. Figure 14.4a confirms that the (0001), {11$\bar{2}$2}, and {11$\bar{2}$0} microfacets appear on the top, slopes, and sidewalls, respectively. Almost all of the threading dislocations were propagated toward the [0001] direction and some were bent by 90° toward the ⟨11$\bar{2}$0⟩ direction near the GaN/sapphire interface. Consequently, dislocations were rarely observed on the {11$\bar{2}$2} facets, which can be advantageous for high emission efficiency because dislocations work as nonradiative recombination centers. In order to estimate dislocation densities, cathodoluminescence (CL) measurements were performed. Although it is impossible to identify the dislocation type, dark spots observed in panchromatic CL images corresponded to dislocations. Thus, the dislocation density on each plane was estimated to be 1×10^{10} cm^{-2} for (0001), 4×10^7 cm^{-2} for {11$\bar{2}$2}, and 1×10^8 cm^{-2} for

$\{11\bar{2}0\}$ by counting the number of the dark spots, which is consistent with the STEM observations. Figure 14.4b shows a magnified image of the (0001) and $(11\bar{2}2)$ facets. The clear contrast indicated that InGaN QWs were successfully formed on the semipolar $(11\bar{2}2)$ plane as well as the (0001) plane. (An image for the $(11\bar{2}0)$ facet is not shown because the contrast was faint due to the low In composition.) In addition, it is apparent that well thickness is facet dependent.

Detailed observations of many samples revealed that the InGaN thickness was nearly uniform within a facet, but varied vastly among the (0001), $\{11\bar{2}2\}$, and $\{11\bar{2}0\}$ microfacets. Energy dispersive X-ray spectroscopy (EDS) showed that the In composition also depends on the facets and, furthermore, varies within a facet. Figure 14.5 shows an example of the inter- and intra-facet variations of the well thickness, which is proportional to the growth rate, and composition. Although such spatial distributions of the well thickness and composition are controlled by the growth conditions, the trend shown in Figure 14.5 is typical, particularly for the In composition. The In composition increased from the $\{11\bar{2}0\}$ facet to $\{11\bar{2}2\}$ facet, reached a maximum at the border between the $\{11\bar{2}2\}$ and (0001) facets (position 2), and then decreased on the (0001) facet as it approached the center (from position 2 to 1). Planar (0001) QWs grown under the same growth conditions had a growth rate of 0.137 nm sec^{-1} and an In composition of 0.13. Both the growth rate and In composition on the planar (0001) QWs were lower than those on the (0001) facet, but higher than those on the other facets [8, 14]. These findings suggest that In and Ga adatoms migrate toward the (0001) facet because the diffusivity is higher on the $\{11\bar{2}2\}$ facet than on the (0001) facet and that In possesses a higher diffusivity than Ga due to its higher vapor pressure. The highest In composition, which was on the border between the (0001) and $\{11\bar{2}2\}$ facets, can be understood by considering that In atoms migrating from $\{11\bar{2}2\}$

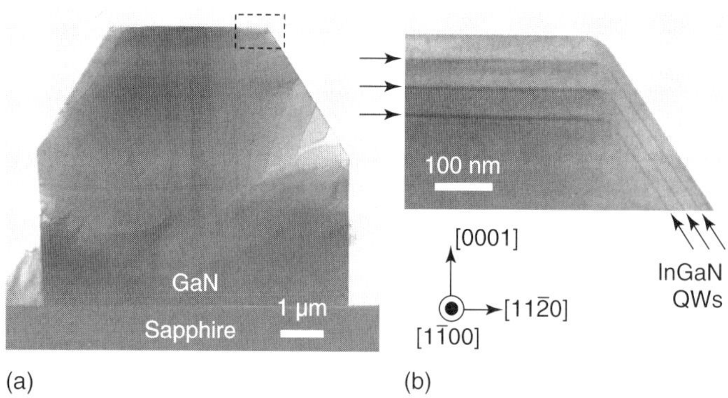

Figure 14.4 Cross-sectional STEM bright field images viewed along the $[1\bar{1}00]$ direction. (a) Entire region and (b) magnified image of the area designated by the dotted rectangle in (a), which includes (0001) and $(11\bar{2}2)$ microfacets.

to (0001) remain on the border due to the lower diffusivity on the (0001) facet.

The spatial variations of the InGaN thickness and composition may lead to multicolor emissions. Figure 14.6 shows microscopic photoluminescence (PL) spectra acquired at room temperature (RT) using a confocal microscopy system. The spatial resolution of the system is less than 500 nm. Hence, each

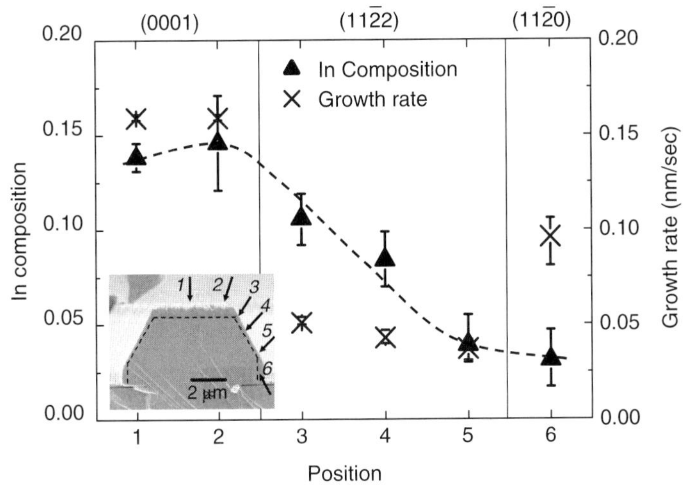

Figure 14.5 Local InGaN growth rate and In composition. Inset is a cross-sectional scanning electron microscopy (SEM) image viewed along the [1$\bar{1}$00] direction. Numbers in the image designate the positions where the growth rate and In composition are estimated. Broken line drawn for the In composition is to guide the eye.

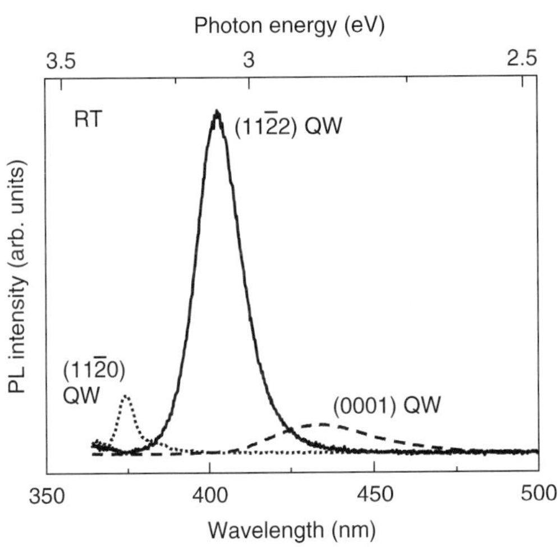

Figure 14.6 Microscopic PL spectra from the (11$\bar{2}$0), (11$\bar{2}$2), and (0001) microfacet QWs at RT.

facet was excited and monitored separately. In order to compare the emission intensities among the microfacet QWs, the same number of carriers were generated by exciting the GaN barriers with a frequency-doubled Ti:sapphire laser with a lasing wavelength of 353 nm. In this particular case, the respective InGaN well thickness and composition were 5.8 nm and 13.5% for the (0001) QW, 2.8 nm and 7.5% for the $\{11\bar{2}2\}$ QW, and 2.2 nm and 2.4% for the $\{11\bar{2}0\}$ QW. Considering the polarization field effect, the calculated transition energies of the (0001), $\{11\bar{2}2\}$, and $\{11\bar{2}0\}$ QWs at RT were 2.22, 3.15, and 3.39 eV, respectively. The PL peaks from (0001), $(11\bar{2}2)$, and $(11\bar{2}0)$ microfacet QWs were observed at 2.86, 3.07, and 3.31 eV, respectively. Small deviations from the calculated transition energies may be due to errors in estimating the internal electric fields. It is noteworthy that the PL intensity from the $(11\bar{2}2)$ QW was much stronger than the (0001) QW, which is reasonable because the $(11\bar{2}2)$ QW may involve a weaker internal electric field and has a narrower well width. The weaker PL of the $(11\bar{2}0)$ QW is attributed to a lower In composition of 2.4%, which provides a weak confinement of the carriers/excitons in the wells. The carriers/excitons can thermally escape from the wells to the barriers where they nonradiatively recombine. We also compared the $(11\bar{2}2)$ microfacet QW with another $(11\bar{2}0)$ microfacet QW, which emits a similar wavelength, and found that the IQEs at RT are comparable (data not shown).

14.3.2
Emission Properties of $\{11\bar{2}2\}$ Microfacet QWs

In order to assess the IQE in the $\{11\bar{2}2\}$ microfacet QWs, the temperature dependence of the PL was measured by selectively exciting the InGaN wells. The sample is the same as that in Figure 14.6, that is, a 2.8-nm thick $In_{0.075}Ga_{0.925}N/GaN$ microfacet QW. Figure 14.7 shows the variation of the integrated PL intensity in addition to the result for a conventional planar (0001) InGaN QW for comparison. The reference (0001) QW has nearly the same well width and emits at a similar wavelength of about 400 nm. The PL intensity is temperature-insensitive below 40 K, suggesting that nonradiative recombination processes are negligible in this low temperature range. Therefore, the IQE can be regarded as 100% at low temperatures and can be evaluated from the PL intensity ratio at elevated temperatures. This assumption was validated using thermal lens spectroscopy [35]. Thus, the estimated IQE of a $\{11\bar{2}2\}$ microfacet QW at RT was approximately 40%, which is about three times higher than that of the reference (0001) QW (Figure 14.7). A similar comparison was performed for green emitters. It was found that the IQE of the green-emitting $\{11\bar{2}2\}$ microfacet QW at RT was 33%, while that of the reference (0001) QW was 17%. As noted above, these observations can partially be attributed to the low threading dislocation densities in the $\{11\bar{2}2\}$ microfacet QWs. However, these observations are mainly due to weakened internal electric fields, which heighten the electron-hole recombination probability, as discussed below in terms of the radiative lifetime.

Typically, when the well width is thin, the internal electric field in the well does not affect the PL peak positions. In fact, our blue-emitting {11$\bar{2}$2} microfacet QW with a well width as thin as 2.8 nm and the reference (0001) QW showed almost identical blue shifts of the PL peaks when the excitation power was changed from 0.27 to 19.0 μJ cm^{-2}, in spite of the vastly different strengths of the internal electric fields in these two samples. (The blue shift can be attributed to the band-filling effect of the localization states.) On the contrary, our previous theoretical study [10] has shown that the exciton spontaneous-emission lifetime is strongly affected by the presence of electric fields even in a 1-nm-thick InGaN well. Therefore, if the electric fields in the {11$\bar{2}$2} QWs are weakened, then a short radiative recombination lifetime should be observed. To demonstrate this, the temperature dependence of the time-resolved photoluminescence (TRPL) was measured for the same blue-emitting {11$\bar{2}$2} microfacet QW shown in Figures 14.6 and 14.7. The obtained PL decay times (τ_{PL}) were analyzed using Figure 14.7 and the following equations:

$$\tau_{PL}^{-1} = \tau_{RAD}^{-1} + \tau_{NON\text{-}RAD}^{-1} \quad (1)$$

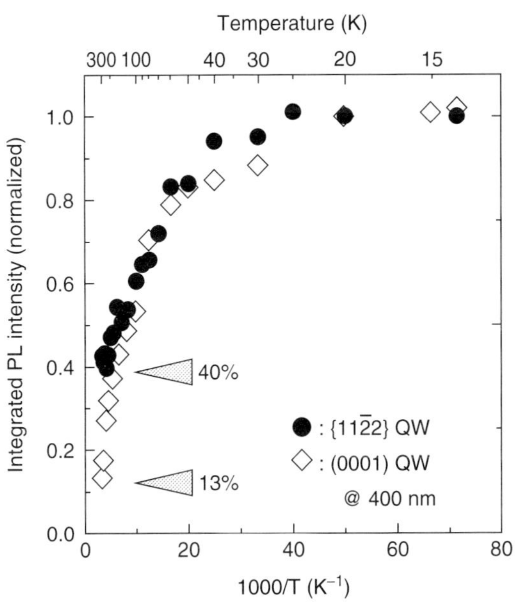

Figure 14.7 Temperature dependence of the integrated PL intensities of a {11$\bar{2}$2} microfacet QW and a conventional (0001) InGaN QW that emit at ~400 nm. Intensities are normalized by those at 13 K so that they are proportional to the IQE.

and

$$IQE = \frac{\tau_{RAD}^{-1}}{\tau_{RAD}^{-1} + \tau_{NON\text{-}RAD}^{-1}} \tag{2}$$

Thus, the temperature dependences of radiative (τ_{RAD}) and nonradiative ($\tau_{NON\text{-}RAD}$) lifetimes were extracted and are summarized in Figure 14.8. The inset in Figure 14.8 shows the PL decay curves and the evaluated lifetimes at 14 K where the nonradiative processes are negligible and the obtained lifetimes reflect radiative processes. The evaluated lifetime in the {11$\bar{2}$2} microfacet QW was 0.39 ns, while that in the reference (0001) QW was 1.43 ns. This much faster decay in the {11$\bar{2}$2} QW strongly suggests much shorter radiative recombination lifetimes due to weakened internal electric fields. This assertion is supported by a theoretical calculation; the spontaneous-emission lifetimes in the {11$\bar{2}$2} and (0001) QWs were calculated to be 0.1 and 2 ns, respectively, when the effects of the polarization fields and carrier localization were considered [10]. With Figure, 14.8 indicates that the radiative lifetime of the {11$\bar{2}$2} QW changes only slightly from 0.38 to 0.65 ns upon increasing the measurement temperature. The radiative lifetime at RT (0.65 ns) is still shorter than that of the reference (0001) QW at low temperatures (1.43 ns).

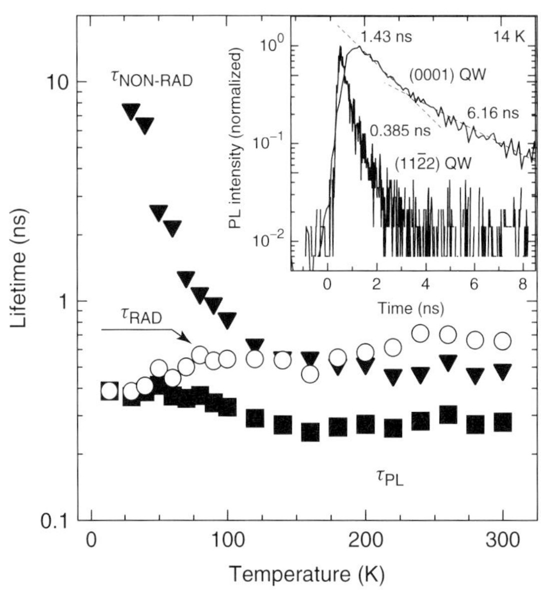

Figure 14.8 Temperature dependence of τ_{PL}, τ_{RAD}, and $\tau_{NON\text{-}RAD}$ of a {11$\bar{2}$2} microfacet QW that emits at ~400 nm. Inset shows PL decay at 14 K acquired at the emission peaks of the {11$\bar{2}$2} microfacet QW as well as a reference planar (0001) QW for comparison.

This small temperature dependence of the radiative lifetime suggests that the carriers are strongly localized, which prevents the carriers from being captured by nonradiative recombination centers. These factors contribute to a high IQE in the $\{11\bar{2}2\}$ QWs.

14.3.3 Multicolor Emission Toward Tailor-made Solid-state Lighting

Semiconductor optical devices emitting arbitrary colors, including pastels and white, impact various fields such as nano-biotechnology and solid-state lighting (SSL) [36]. In fact, white LEDs based on InGaN/GaN QWs have achieved a great success in SSL. However, because semiconductor emitters inherently provide nearly monochromatic output light, the current white LEDs consist of a blue LED pumping a yellow phosphor. A drawback to using phosphors is that the absorption energy of the phosphor must agree with the emission energy of the LED, which can limit the material and, consequently, the spectral design. Furthermore, Stokes loss due to the down conversion from blue to yellow is inevitable.

We believe that our microfacet QWs may solve the problems, because if the intra- and inter-facet variation in the In composition and InGaN well width (see Figure 14.5) are controlled, then InGaN/GaN microfacet QWs can be single-chip multiwavelength light emitters, which can control the color. In this section, we demonstrate that changing the growth conditions and mask patterns for regrowth can synthesize emission light from InGaN/GaN microfacet QWs to produce pastel and white colors.

14.3.3.1 Multicolor Emission Based on Intra-facet Variations of the In Composition

When the intra-facet variation of the In composition is emphasized, a broadband emission should be realized. Herein, a multiwavelength (rainbow color) luminescence from a $\{11\bar{2}2\}$ microfacet QW is demonstrated. The sample was a microfacet single quantum well (SQW). A STEM observation confirmed that an InGaN well with a thickness of 2 ± 0.2 nm was successfully and uniformly formed within the $(11\bar{2}2)$ facet. On the contrary, the In composition estimated by EDS equipped with the STEM system monotonously increased from 25% on the $(11\bar{2}0)$ side to 40% on the (0001) side. Considering a 2 nm uniform well width and an internal electric field due to the polarization effects, the estimated QW transition energy ranged from 2.43 (510 nm) to 2.79 eV (444 nm).

The influence of this surprisingly large In spatial distribution in the $\{11\bar{2}2\}$ microfacet QW on the optical properties was microscopically investigated by CL at RT. The lower inset of Figure 14.9 shows representative monochromatic CL images acquired at wavelengths of 420 nm (blue), 500 nm (green), and

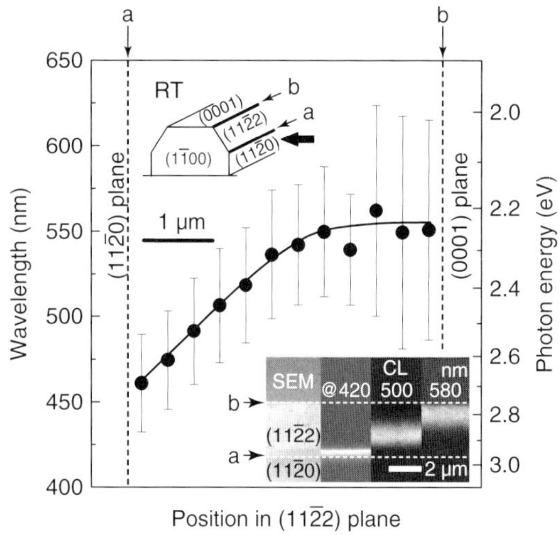

Figure 14.9 Position dependence of the CL peak wavelength of a (11$\bar{2}$2) microfacet QW. Vertical bars represent the emission line widths. Lower inset is a SEM image and monochromatic CL images monitored at 420 nm (blue), 500 nm (green), and 580 nm (amber), while the upper inset shows a schematic sample structure where the arrow indicates the measurement direction.

580 nm (amber) together with a scanning electron microscopy (SEM) image observed along the [11$\bar{2}$0] direction. As shown, the emission band formed along the [1$\bar{1}$00] direction and moved toward the top (0001) facet, while broadening its width with a longer monitored wavelength. It is noteworthy that blue, green, and amber were clearly observed from a single facet, suggesting that the luminescence from the {11$\bar{2}$2} microfacet QW covers the full-visible range. Figure 14.9 also plots a more detailed position dependence of the emission properties. The vertical bars represent the full width at half maximum (FWHM). The emission peak wavelength ranged from 460 nm (2.70 eV) to 551 nm (2.25 eV) from the (11$\bar{2}$0) side to the (0001) side, which is consistent with the above calculations based on STEM and EDS observations and suggests that the position dependence of the emission wavelength is due to the In spatial distribution. Furthermore, the rather broad line width, which is more than 60 nm (more than 200 meV), also supports the presence of a significant In spatial distribution.

The macroscopic optical properties of the {11$\bar{2}$2} microfacet QW were assessed by selectively exciting the InGaN wells with a frequency-doubled Ti:sapphire laser with a wavelength of 400 nm. The repetition rate, pulse width, and excitation density were 80 MHz, 1.5 ps, and 1.31 μJ cm^{-2}, respectively. Figure 14.10 shows a PL spectrum acquired at RT. The PL peaked at 535 nm (2.32 eV) and encompassed wavelengths from 450 to 650 nm, which correspond to almost the entire visible range. The estimated FWHM was

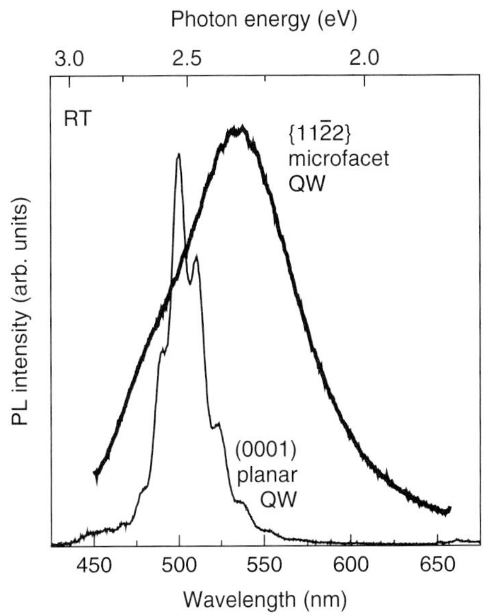

Figure 14.10 PL spectra of a {11$\bar{2}$2} microfacet QW and, for comparison, a planar (0001) QW measured at RT.

approximately 450 meV, which is about three times larger than the reference planar (0001) QW. This anomalously broad emission is due to the distribution of the In composition and characterized the luminescence color of the present {11$\bar{2}$2} microfacet QW. In terms of the Commission Internationale de I'Eclairage (CIE) chromaticity chart, the luminescence color is classified as whitish emerald green, far from monochromatic light. Because changing the growth conditions controls the In distribution within a facet, various apparent colors, including white, can be realized using only the structure proposed in this study. This is strikingly different from conventional technologies in which the light color is designed by, for example, using several emitters with different output colors or by using phosphors as a color converter.

It is worth investigating the emission efficiency and relevant properties. In the same manner as in Figure 14.7, the IQE was evaluated to be 33% at RT. Because the integrated PL intensity was used for this estimate, 33% means the efficiency averaged over the entire spectral range. Although the maximum IQE achieved in the red spectral range by InGaN-based LEDs to date is a few percent, our {11$\bar{2}$2} microfacet QWs with a red component show a high IQE. In addition, steep thermal quenching has not been observed on the long wavelength side of the PL spectrum, which suggests that the high IQE is nearly independent of the emitting wavelength.

To investigate the reason for this wavelength-independent high IQE in {11$\bar{2}$2} microfacet QWs, TRPL measurements were performed at 13 K using the same excitation conditions as above. Figure 14.11 shows the PL decay curves at 460 nm (blue), 530 nm (green), and 575 nm (amber) within the broad {11$\bar{2}$2} microfacet QW emission. Each curve is composed of fast and slow decays.

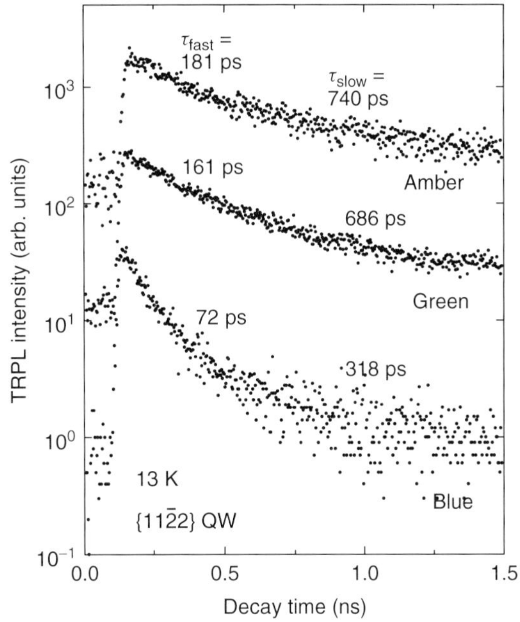

Figure 14.11 PL decay curves of a {11$\bar{2}$2} microfacet QW at 13 K. Decay curves are monitored at 460 nm (blue), 530 nm (green), and 575 nm (amber).

Because carrier/exciton transfer due to the In distribution may occur in the present microfacet QWs, the interpretation of these two decay components is rather complicated. Therefore, we have yet to draw a conclusion. However, it is interesting to compare the obtained lifetimes with those for (0001) QWs. Typical lifetimes for 3-nm-thick (0001) InGaN QWs are 14 ns at 454 nm, 85 ns at 521 nm, and 1 μs at 630 nm. Hence, it is clear that all the decay components in the {11$\bar{2}$2} microfacet QWs possess much shorter lifetimes and much less wavelength dependence owing to the weakened internal field than that in the reference (0001) QWs. Because the radiative lifetime is inversely proportional to the optical transition probability, the observed wavelength-independent high IQE in the {11$\bar{2}$2} microfacet QW can be attributed to the short radiative lifetimes over the wide spectral range. Again, we like to emphasize that even the amber component possesses quite short lifetimes compared to a (0001) QW, which suggests that {11$\bar{2}$2} QWs are promising for long wavelength emitters.

14.3.3.2 **Multicolor Emission based on Inter-facet Variations of the QW Structures**

The inter-facet variations of the In composition and well width may cause different emission colors from different facets. Various apparent emission colors can be realized by an additive color mixture of such emissions.

In this section, we demonstrate that simple growth processes achieve this controllability.

Two microfacet QWs that emitted different colors were designed. One emitted 500 nm from the (0001) facet and 400 nm from the $\{11\bar{2}2\}$ facet (sample **A**), whereas the other emitted 580 nm from the (0001) facet and 430 nm from the $\{11\bar{2}2\}$ facet (sample **B**). The In composition in the $\{11\bar{2}0\}$ facet was insufficient to emit visible light. Hence, the contribution from this facet was ignored. The emission color from each facet of the fabricated microfacet QWs was microscopically confirmed by CL at RT and was consistent with the design.

The macroscopic optical properties of the microfacet QWs were assessed by PL measurements at RT. Figure 14.12a and b show the PL spectra of the samples. The excitation source was a He–Cd laser with an excitation density of 10 W cm^{-2} and an excited spot of 300 μm, which included many (0001) and $\{11\bar{2}2\}$ facets. The origins of the emissions, which were indicated in Figure 14.12a and b, were identified using the CL spectra. The PL spectra were

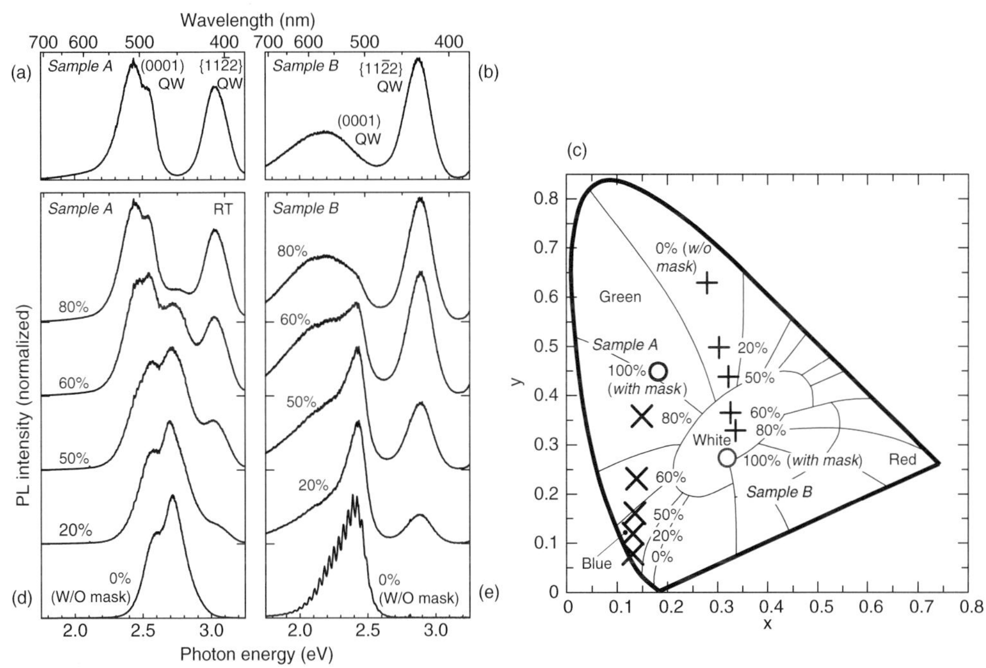

Figure 14.12 PL spectrum of (a) sample **A** and (b) sample **B** acquired at RT. (c) CIE chromaticity diagram where the open circles indicate the emissions from samples **A** (Figure 14.12a) and **B** (Figure 14.12b), while the crosses indicate emissions from samples with designed mask patterns. See text for more details. Variation of PL spectra due to mask geometries for (d) sample **A** and (e) sample **B** (cf. Color Plate XXXVIII).

characterized by multi-band emissions from the (0001) and $\{11\bar{2}2\}$ facets and rather broad line width caused by the intra-facet In distribution. This broad line width can improve the color rendering properties.

The two emission bands observed in the PL spectra were mixed and the apparent colors were greenish white for **A** and purplish white for **B**, indicating color syntheses of nearly monochromatic light output from each facet QW. In terms of the CIE chromaticity diagram, the emission of **A** was at (0.179, 0.460), while that of **B** was at (0.318, 0.277) as plotted as the open circles in Figure 14.12c. Monochromatic light from conventional LEDs should be on the boundary in the CIE chromaticity diagram. In contrast, the emissions of our microfacet QWs remained far from the boundary and were classified as green for **A** and white for **B**, indicating that pastel and white colors can be realized by nitride semiconductors without phosphors.

Altering the growth conditions controlled the output colors in the above example, while the mask geometry also changed the InGaN thickness and In composition, which consequently changed the emission color. In a preliminary experiment, masks, which were composed of regions with and without stripes, were examined because it is predicted that the emission shown in Figure 14.12a and b was synthesized with the emission from conventional planar QWs grown in a stripeless region. Without the mask stripes, the PL peaked at 460 nm for **A** and 520 nm for **B**, both of which were between the emissions from the (0001) and $\{11\bar{2}2\}$ microfacet QWs, as shown at the bottom of Figure 14.12d and e. Figure 14.12d and e shows the changes in the PL spectra at RT as a proportion of the masked region varied from 0 to 80% where the percentages in italics indicate the proportions. The spectral synthesis by changing the mask geometries is easily recognized. The crosses in Figure 14.12c plot the results in terms of the CIE chromaticity diagram. The output light color was easily controlled from blue to green through a bluish green and greenish blue for **A** and from yellowish green to white for **B**.

14.4 Planar QWs and LEDs

In the previous section, we demonstrated that semipolar $\{11\bar{2}2\}$ planes are promising for low internal electric fields when these planes naturally appear as microfacets via a regrowth process. However, for mass production, conventional planar structures have been investigated [19–23] for various orientations such as $(10\bar{1}\bar{3})$ [19–22], $(10\bar{1}\bar{1})$ [20, 21], and $(11\bar{2}\bar{2})$ [22]. In fact, LEDs on $(10\bar{1}\bar{3})$ [19, 21] and $(10\bar{1}\bar{1})$ planes [21] have already been demonstrated. However, because the films were grown on foreign substrates such as sapphire $(1\bar{1}00)$, spinel ($MgAl_2O_4$), and SiC $(30\bar{3}8)$ without optimizing the growth processes, a high density (10^9–10^{10} cm^{-2}) of threading dislocations remains and the surfaces are not flat on the atomic level [20, 22], both of which are detrimental to device performance [19, 21].

The superior optical properties of our microfacet QWs and reports on planar heteroepitaxial films lead us to believe that if planar structures are grown on high-quality GaN substrates, then device performances could be drastically improved. In this section, we describe the MOVPE of GaN and InGaN/GaN QWs on semipolar $(11\bar{2}2)$ GaN substrates, their fundamental optical properties, and the fabrication of LEDs.

14.4.1 MOVPE Growth and Fundamental Properties of GaN and InGaN/GaN QWs

Semipolar $\{11\bar{2}2\}$ GaN substrates were prepared as follows. A (0001) GaN bulk crystal was initially grown on (0001) sapphire by hydride vapor phase epitaxy and then sliced so that semipolar $\{11\bar{2}2\}$ planes appear. Both the $(11\bar{2}2)$ and $(\bar{1}\bar{1}2\bar{2})$ surfaces were planarized by chemical mechanical polishing (CMP) and were ready for growth. Unless stated, the $(11\bar{2}2)$ plane was used in this study. The substrates were intentionally undoped, but possessed n-type conductivity with a carrier density of 3.3×10^{17} cm^{-3} at RT. Consequently, the PL is dominated by donor-bound excitons at 10 K and by free excitons at RT. The transition energy determined by PL and photo-reflectance measurements indicated that the substrates are almost strain free, which was also confirmed by X-ray diffraction (XRD) radial scans. The line widths of the XRD omega-scan of the (0001), $(11\bar{2}2)$, and $(11\bar{2}0)$ planes were 15.1, 26.9, and 32.4 arcsec, respectively, from which an edge dislocation density of 3.0×10^6 cm^{-2} and a screw dislocation density of $2.6 \times 10^5/cm^{-2}$ were estimated [37]. These quantities are about 3 orders of magnitude lower than that in GaN epitaxial layers on sapphire substrates.

We first discuss GaN homoepitaxy. Two GaN layers were directly grown on the substrates (without buffer layers) at ~1050 and 975 °C. The optimal growth temperature for conventional (0001) GaN on sapphire is 1050 °C, which creates GaN surfaces with step and terrace structures. The surface morphologies of these GaN layers as well as an as-received bulk GaN substrate were observed by atomic force microscopy (AFM) and are shown in Figure 14.13a to c. The surface of the GaN substrate was covered with numerous dot-like structures (Figure 14.13a) and the root-mean-square (RMS) roughness was 2.6 nm. Judging from the density, the dot-like structures are not due to dislocations in GaN but are due to CMP. After growing 3-μm-thick GaN at 1050 °C, numerous dots remained with a RMS roughness of 1.8 nm (Figure 14.13b), which differs from (0001) GaN on sapphire. A magnified AFM image (not shown) suggested that facets are formed around the dots. Because the facet structure generally depends on the growth temperature, other temperatures were examined. Consequently, we found that a lower temperature of 975 °C provides atomically flat surfaces for $(11\bar{2}2)$ GaN and an example is shown

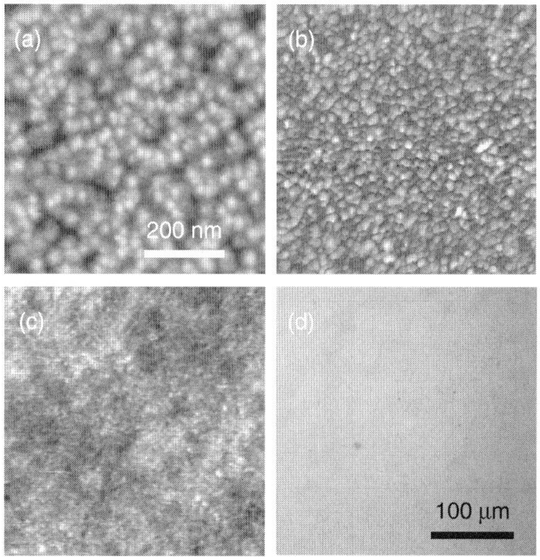

Figure 14.13 Surface AFM images of (a) (11$\bar{2}$2) GaN substrate and GaN homoepitaxial layers grown at (b) 1050 and (c) 975 °C where the scale bar is drawn in (a). Peak-to-valley distances are (a) 13.9 nm, (b) 9.9 nm, and (c) 1.0 nm. (d) is a surface Normarski microscope image of the GaN homoepitaxial layer grown at 975 °C.

in Figure 14.13c. As shown, the dots were replaced by fine structures and the RMS roughness was as low as 0.16 nm. Cross-sectional analysis indicated that the step height of the fine structures is close to the (11$\bar{2}$2) layer spacing (0.136 nm). Figure 14.13d shows a wide area of this GaN layer observed by Normarski microscopy. It should be noted that the area is comparable to a typical LED size (300 μm square). The surface is almost hillock- and pit-free, which is favorable for LED fabrication. A clue to understanding why a lower temperature is suitable for growth on the (11$\bar{2}$2) plane can be found in conventional regrowth techniques on patterned [0001]-oriented GaN where the {11$\bar{2}$2} facets tend to appear at low temperatures [33, 34].

In order to assess the fundamental optical properties of the GaN homoepitaxial layer grown at the optimal temperature (975 °C), the PL and reflectance were measured at 10 K. For the PL measurement, a He–Cd laser was used to excite the sample (325 nm). The reflectance measurements were performed using a Xe lamp at a normal incidence on the sample surface. Both the PL and reflectance were detected through a 1-m monochromator coupled with a photomultiplier tube. The resolution is ~0.08 meV at 350 nm. To view the optical anisotropy effect, a polarizer was set between the sample and detector. Figure 14.14 shows the PL and reflectance spectra where the reflectance was measured for **E** ∥ [1$\bar{1}$00] (⊥ [0001]) or **E** ∥ [$\bar{1}\bar{1}$23]. In the reflectance spectra, transitions, which involved ground state A, B, and C excitons, were well resolved at 3.4802, 3.4848, and 3.503 eV (denoted $X_A^{n=1}$, $X_B^{n=1}$, and $X_C^{n=1}$), respectively. These quantities closely match those reported for bulk GaN [38–41], suggesting a negligible strain in our homoepitaxial layers. A weak signal denoted as $X_A^{n=2}$ at 3.4977 eV was derived from $n = 2$ free A excitons. Comparing the two observed reflectance spectra, it was concluded that the $X_A^{n=1}$ signal is stronger for the [1$\bar{1}$00] direction and $X_C^{n=1}$ is stronger

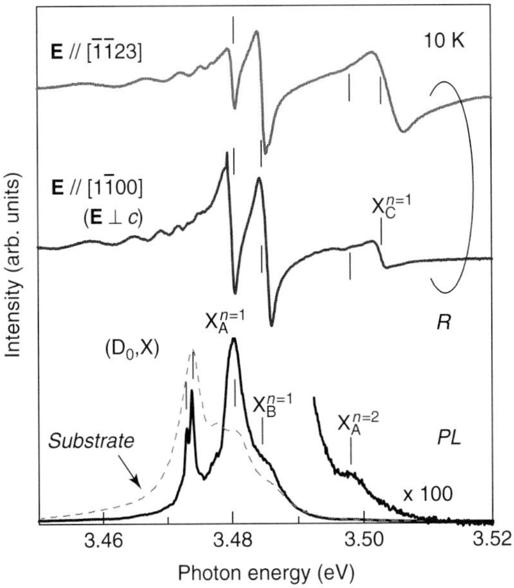

Figure 14.14 PL and reflectance spectra of the GaN homoepitaxial layer acquired at 10 K. PL indicated by the broken curve pertains to the GaN (11$\bar{2}$2) substrate.

for the [$\bar{1}\bar{1}23$] direction, both of which are well accounted for by the crystal symmetry of the (11$\bar{2}$2) plane. For PL, emissions from $X_A^{n=2}$, $X_B^{n=1}$, $X_A^{n=1}$, and the neutral donor-bound exciton (D_0, X) were observed. Unlike the substrate (broken curve in Figure 14.14), $X_A^{n=1}$ is dominant, which indicates high optical qualities in the epitaxial layer. The (D_0, X) emission was clearly split into a doublet at 3.4738 and 3.4729 eV; the relative intensities depended on the growth conditions, and therefore, the doublet is attributed to different donor species. A fit using two Lorentzian curves provides an identical line width of 0.6 meV, which is comparable with those of most state-of-the-art homoepitaxial or bulk GaN [38–41].

A five-period InGaN/GaN multiple quantum well (MQW) was fabricated on this high-quality GaN homoepitaxial layer. The XRD profile of the MQW consisted of satellite peaks, where the evaluated well and barrier layer thicknesses were 3.4 and 11.3 nm, respectively. It is noteworthy that the growth rates of InGaN and GaN derived from the well and barrier thicknesses are comparable with those on the (0001) plane. As for the optical properties, we were interested in if the short radiative lifetime observed for the microfacet QWs was preserved. Therefore, TRPL measurements were conducted at 10 K. The excitation pulses were from a frequency-doubled Ti:sapphire laser with a wavelength of 380 nm to selectively excite InGaN wells and a power density as low as 470 nJ cm^{-2}. The PL was detected by a streak camera. Figure 14.15

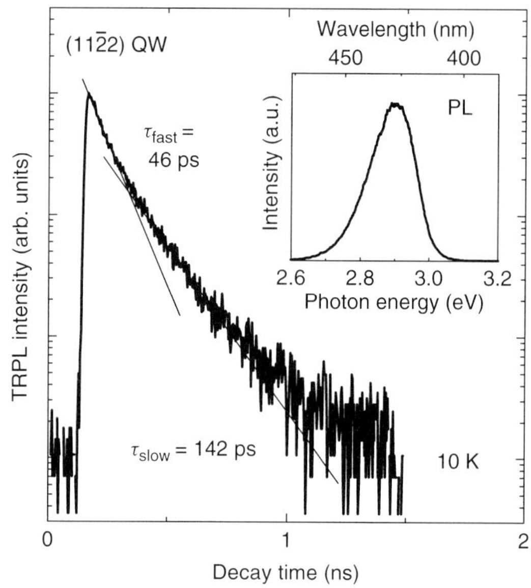

Figure 14.15 PL decay for the (11$\bar{2}$2) MQW at 10 K. Inset shows the time-integrated PL spectrum.

shows a PL decay curve monitored at the 428 nm emission peak and the time-integrated PL spectrum in the inset. The decay curve had fast and slow lifetimes, and was fitted by a double exponential curve. The fast and slow lifetimes were estimated to be 46 and 142 ps, respectively, and were nearly independent of the monitoring wavelength within the PL spectrum. This observation differs from conventional InGaN/GaN QWs because recombination dynamics in conventional InGaN/GaN QWs are well explained by weakly localized excitons and the PL lifetimes decrease as the wavelength decreases in the short-wavelength tail of the spectra. Analysis of this wavelength dependence is currently in progress. On the contrary, typical (0001) QWs involved lifetimes that were almost 2 orders of magnitude longer than those in the (11$\bar{2}$2) MQW. (For typical values, refer Figures 14.8 and 14.11.) Because the PL line width of the (11$\bar{2}$2) MQW was wider than those of (0001) QWs, potential fluctuations, which make the radiative recombination process slow, might be more remarkable in the (11$\bar{2}$2) MQW. Nevertheless, the lifetimes for the (11$\bar{2}$2) MQW were much shorter, indicating a higher transition probability in the semipolar MQW due to a significantly weakened internal electric field.

Owing to the reduced crystalline symmetry of the (11$\bar{2}$2) plane, the PL should exhibit polarization anisotropy. Our calculation based on the **k**·**p** theory predicts the following: (i) PL is polarized along the [1$\bar{1}$00] direction ($\perp$ [0001]), (ii) unstrained GaN has a polarization degree of 0.56, and (iii) strained InGaN on unstrained GaN shows a larger polarization degree where the polarization degree is defined as $(I_{[1\bar{1}00]} - I_{[\bar{1}\bar{1}23]})/(I_{[1\bar{1}00]} + I_{[\bar{1}\bar{1}23]})$. The polarizations were

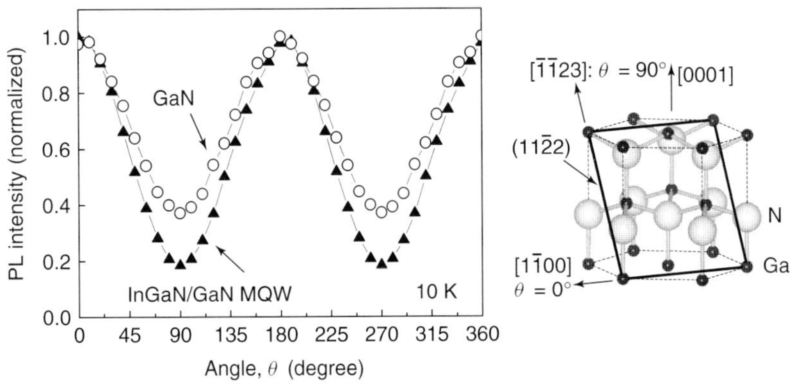

Figure 14.16 In-plane PL polarization property of a (11$\bar{2}$2) GaN homoepitaxial layer (open circles) and an InGaN/GaN MQW (closed squares) at 10 K. $\theta = 0^\circ$ corresponds to the [1$\bar{1}$00] direction, whereas $\theta = 90^\circ$ corresponds to the [$\bar{1}\bar{1}$23] direction.

measured for (11$\bar{2}$2) GaN homoepitaxial layer and InGaN/GaN MQW at 10 K and Figure 14.16 shows the results. To avoid overlap with the $X_B^{n=1}$ emission, the polarization for the GaN layer was monitored on the lower energy side of the $X_A^{n=1}$ emission (3.4792 eV). It is clear that both the GaN layer and InGaN/GaN MQW emit strongly polarized light. The estimated polarization degrees were 0.46 for GaN and 0.69 for MQW, which agree well with the prediction and indicate the high optical quality of our samples. The larger polarization degree for the MQW is caused by anisotropic compressive strain in InGaN.

14.4.2
Semipolar LEDs

Blue and green LEDs were fabricated on the (11$\bar{2}$2) GaN substrates and the amber LED was fabricated on the ($\bar{1}\bar{1}$2$\bar{2}$) GaN substrate. (However, clear differences in the LED properties due to the opposite crystalline orientations have yet to be found.) Figure 14.17 schematically illustrates the device structure, which consists of undoped GaN (1.5 μm), n-type GaN:Si (4.5 μm), undoped GaN (25 nm), InGaN (3 nm), undoped GaN (25 nm), p-type AlGaN:Mg (10 nm), and p-type GaN:Mg (150 nm). The expected carrier densities in n- and p-type GaN were 5×10^{18} cm^{-3} and 4×10^{17} cm^{-3}, respectively. The active region was a SQW embedded in undoped 25-nm thick GaN. After defining 320 μm × 320 μm mesas by RIE, electrodes for n and p-GaN were taken from the surface like in conventional LEDs on sapphire. The fabricated devices were then packaged in an epoxide resin. All the measurements were conducted under the dc operation at RT without temperature control.

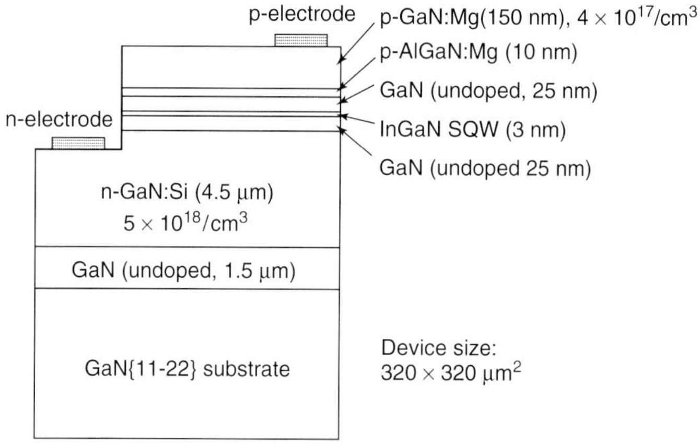

Figure 14.17 Schematic structure of $\{11\bar{2}2\}$ semipolar LEDs.

The current–voltage curves for all the LEDs displayed clear rectifying properties. A driving current of 20 mA was attained at 3.4, 3.8, and 3.0 V in the blue, green, and amber LEDs, respectively. Furthermore, a current of 200 mA was obtained at 4.7 V (blue), 5.2 V (green), and 4.1 V (amber). The scattering of the bias voltage to obtain a certain current was not due to an intrinsic property, but was due to a reproducibility problem.

The variation in the electroluminescence (EL) properties due to injection currents, which ranged from 0.1 to 200 mA, was assessed. Figure 14.18a to c show typical EL spectra at a driving current of 20 mA. The line width became broader in the order of blue < green < amber LEDs, indicating that a higher In composition leads to more significant potential fluctuations. Figure 14.18d to f summarizes the FWHM and peak wavelength of the EL. Although there are many factors that can contribute to the observed properties, we can exclude factors that appear with a high carrier density such as screening of the polarization fields by carrier injection [42] and band gap renormalization because the maximum estimated carrier density under the present experimental condition is 1×10^{17} cm^{-3} assuming the carrier lifetime to be 200 ps, which is a typical PL lifetime at RT. Our self-consistent calculation of the Poisson and Schrödinger equations indicated that such a carrier density does not bring a drastic screening. Currently, we consider that the observed properties shown in Figure 14.18 can be interpreted in terms of the state-filling effect.

We first discuss the green LED (Figure 14.18e), which exhibits typical characteristics. The peak position shows a significant blueshift in the low-current range (less than 25 mA) followed by a moderate blueshift in the high-current range (above 25 mA). As previously mentioned, our LEDs involve potential fluctuations, which work as localization sites for carriers. In the low-current range, injected carriers occupy such localization states and easily

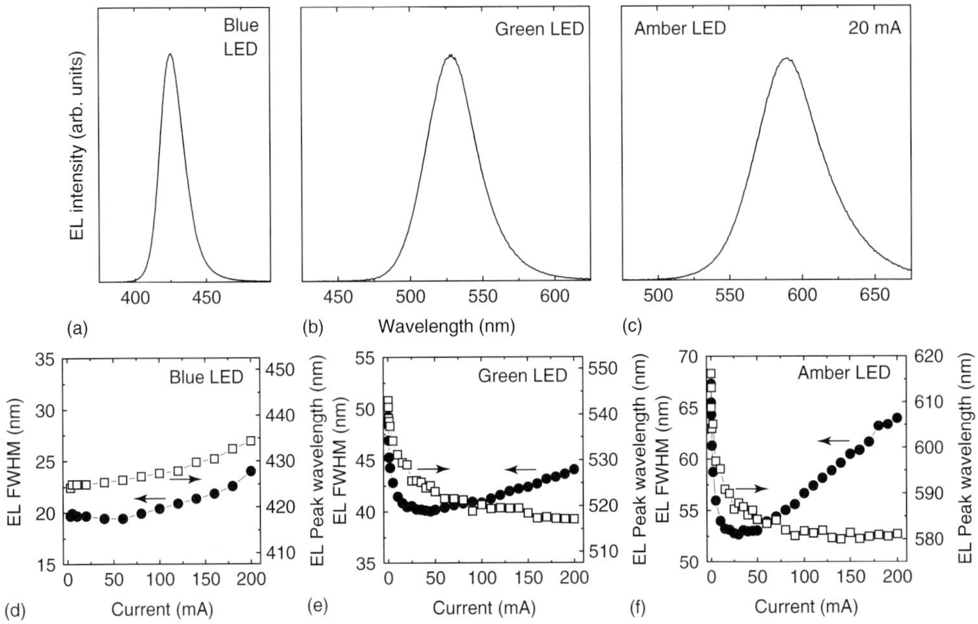

Figure 14.18 EL spectra of (a) blue, (b) green, and (c) amber LEDs at a driving current of 20 mA. (d) to (f) show variation of the EL FWHM (closed circle) and peak wavelength (open square) of the blue, green, and amber LEDs, respectively, as a function of the injection current.

move the quasi-Fermi level toward a higher energy, which is observed as a drastic blueshift, because the localization states have a small density of status (DOS). As the injection carriers increase, the pseudo-Fermi level approaches less localized levels with a larger DOS near the QW level and then the blueshift becomes moderate. The energy dependence of DOS emerges from the variation of the line width due to the injection carrier density. The narrowing of the line width in the low-current range due to the state-filling effect suggests that the higher energy states in the localization states have a greater DOS and dominate the EL spectra. On the contrary, the broadening observed above 25 mA suggests smaller energy dependence of the DOS in the less localized levels.

This interpretation, which is based on the state-filling of the potential fluctuation, is supported by the amber LED (Figure 14.18f) because both the blueshift and V-shape behavior of the line width are more pronounced in amber LEDs with larger potential fluctuations.

An additional factor to be considered is the QCSE, which is unavoidable before the built-in potential due to the pn junction is compensated by the applied forward bias. Because the internal electric field is weakened in $\{11\bar{2}2\}$ LEDs, the QCSE may cause a blueshift with forward bias, which is different from conventional (0001) LEDs. As previously described, the applied voltage exceeds the built-in potential (~3 V) with a current of ~20 mA. Thus, we attribute the significant blueshift observed below 25 mA to both the state-filling effect and the QCSE, while the following weak blueshift seen above 25 mA is only attributed to the state-filling effect.

It is interesting to note that the EL peak is redshifted in the blue LED (Figure 14.18d), although the line width shows a similar, but weaker V-shape behavior. The cause of the redshift is probably Joule heat generation. The V-shape behavior suggests the presence of the state-filling effect, but heat generation may hinder it to some extent because the blue LED involves the weakest localization effect due to the lowest In composition. The absence of a redshift in the green and amber LEDs (Figure 14.18e and f) is a result of the competition between the blueshift due to the state-filling effect and the redshift due to heat generation. Also, heat generation causes line broadening, which may influence the observed broadening in the high-current range in all the LEDs.

The optical output power and external quantum efficiency (EQE) were assessed as a function of the forward current and the results are shown in Figure 14.19. For the blue LED, the output power was 1.76 mW at a driving current of 20 mA, reached 19.0 mW at 200 mA, and displayed a weak saturation tendency. The EQE was evaluated as 3.0% at 20 mA, increased as the driving current increased, and reached a maximum of 4.0% at 140 mA. Then it decreased, probably due to heat generation. For the green LED, the output power was 1.91 mW at 20 mA when the EQE was 4.1%, and increased almost linearly to 13.4 mW at 200 mA. The EQE reached a maximum of 4.9% with a driving current of 0.2 mA and monotonously decreased to 2.8% at 200 mA. The amber LED (Figure 14.19c) had a measured output power of 0.54 mW at a driving current of 20 mA and 1.9 mW at 200 mA. The EQE increased sharply as the driving current increased and reached a maximum of 1.6% at 1 mA, but then it decreased to 0.4% at 200 mA via 1.3% at 20 mA. From Figure 14.19a to c, it is clear that the decrease in the EQE as the injection current increases becomes remarkable in a LED emitting a longer wavelength,

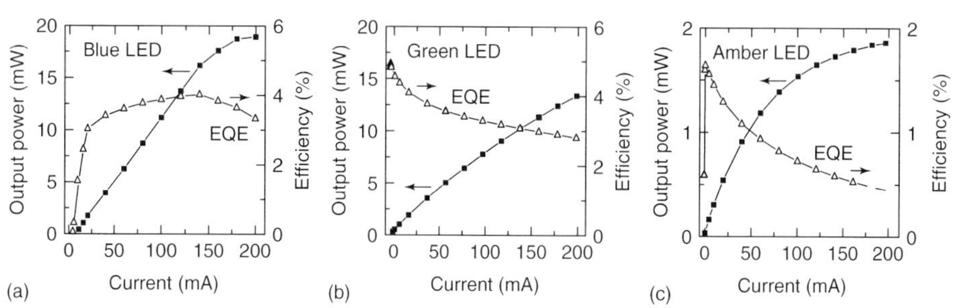

Figure 14.19 Output power and EQE of (a) blue, (b) green (cf. Color Plate XXXIX), and (c) amber LEDs.

that is, in a LED with a higher In composition. Therefore, this degraded device performance can be ascribed mainly to the inferior crystalline qualities of the InGaN SQW with a higher In composition and partially to a stronger internal electric field.

Similar to PL shown in Figure 14.16, the EL from the $\{11\bar{2}2\}$ LEDs should be polarized. Because InGaN is compressively strained, the transition between the conduction band and the heavy hole valence band is predominant in the EL, especially at a low driving current. Therefore, the EL should polarize perpendicular to the [0001] direction, that is, along the $[1\bar{1}00]$ direction in the $\{11\bar{2}2\}$ plane. To demonstrate this, polarization EL measurements on our LEDs were conducted with a driving current of 20 mA. To avoid the contribution of light scattered within the LED dies, a confocal configuration was adopted. It was found that the EL is polarized along the expected direction ($[1\bar{1}00]$) with a polarization degree of 0.60 for the green LED when defined in the same manner as described in Section 14.4.1. This strong polarization can be advantageous for devices such as liquid crystal displays that require optical switching.

14.5 Summary

In summary, we have demonstrated recent progress on the microfacet QWs and planar, semipolar QWs and LEDs. The $\{11\bar{2}2\}$ plane plays a key role in both structures. The major achievements are as follows: (i) the $\{11\bar{2}2\}$ microfacet QWs with a high IQE due to weak internal electric fields, (ii) white and pastel luminescence based on facet-dependent multicolor emission, (iii) high-quality $\{11\bar{2}2\}$ GaN and InGaN/GaN QWs on GaN bulk substrates, and (iv) high-quality $\{11\bar{2}2\}$ LEDs with optical anisotropy on GaN bulk substrates. Although subjects such as fabrication of the multicolor LEDs, investigation of the material properties in semipolar directions, and improvement of the device performances remain to be elucidated, our results lead us to believe that there is a bright future for semipolar planes for high performance light emitters.

Acknowledgments

The author would like to acknowledge K. Nishizuka, M. Ueda, T. Kondou, K. Hayashi, K. Kojima, and A. Kaneta at Kyoto University and T. Kosugi and M. Takahashi at Nichia Corporation for their support. This work was partly performed using the facility at the Venture Business Laboratory of Kyoto University (KU-VBL). This work was partially supported by the 21st Century COE Program (No. 14213201) and Grants for Regional Science and Technology Promotion from the Ministry of Education, Culture, Sports, Science and Technology.

References

1. Nakamura, S., Pearton, S., and Fasol, G. (2000) *The Blue Laser Diode*, 2nd ed. Springer-Verlag, Heidelberg.
2. Bernardini, F. and Fiorentini, V. (1999) *Physica Status Solidi B*, **216**, 391.
3. Hangleiter, A., Im, J.S., Kolmer, H. *et al.* (1998) *MRS Internet Journal of Nitride Semiconductor Research*, **3**, 15.
4. Waltereit, P., Brandt, O., Trampert, A. *et al.* (2000) *Nature (London)*, **406**, 865.
5. Chitnis, A., Chen, C., Adivarahan, V. *et al.* (2004) *Applied Physics Letters*, **84**, 3663.
6. Haskell, B.A., Wu, F., Matsuda, S. *et al.* (2003) *Applied Physics Letters*, **83**, 1554.
7. Ng, H.M., Bell, A., Ponce, F.A., and Chu, S.N.G. (2003) *Applied Physics Letters*, **83**, 653.
8. Takeuchi, T., Lester, S., Basile, D. *et al.* (2000) *IPAP Conference Series*, **1**, 137.
9. Funato, M., Yamamoto, S., Kaisei, K. *et al.* (2001) *Applied Physics Letters*, **79**, 4133.
10. Funato, M., Kawaguchi, Y., and Fujita, Sg. (2004) *Material Research Society Symposium Proceedings*, **798**, 347.
11. Nishizuka, K., Funato, M., Kawakami, Y. *et al.* (2004) *Applied Physics Letters*, **85**, 3122.
12. Nishizuka, K., Funato, M., Kawakami, Y. *et al.* (2005) *Applied Physics Letters*, **87**, 231901.
13. Funato, M., Kotani, T., Kondou, T. *et al.* (2006) *Applied Physics Letters*, **88**, 261920.
14. Neubert, B., Brückner, P., Habel, F. *et al.* (2005) *Applied Physics Letters*, **87**, 182111.
15. Wunderer, T., Brückner, P., Neubert, B. *et al.* (2006) *Applied Physics Letters*, **89**, 041121.
16. Srinivasan, S., Stevens, M., Ponce, F.A., and Mukai, T. (2005) *Applied Physics Letters*, **87**, 131911.
17. Ueda, M., Kojima, K., Funato, M. *et al.* (2006) *Applied Physics Letters*, **89**, 211907.
18. Funato, M., Ueda, M., Kawakami, Y. *et al.* (2006) *Japanese Journal of Applied Physics Part 2*, **45**, L659.
19. Sharma, R., Pattison, P.M., Masui, H. *et al.* (2005) *Applied Physics Letters*, **87**, 231110.
20. Baker, T.J., Haskell, B.A., Wu, F. *et al.* (2005) *Japanese Journal of Applied Physics Part 2*, **44**, L920.
21. Chakraborty, A., Baker, T.J., Haskell, B.A. *et al.* (2005) *Japanese Journal of Applied Physics Part 2*, **44**, L945.
22. Baker, T.J., Haskell, B.A., Wu, F. *et al.* (2006) *Japanese Journal of Applied Physics Part 2*, **45**, L154.
23. Kamiyama, S., Honshio, A., Kitano, T. *et al.* (2005) *Physica Status Solidi C*, **2**, 2121.
24. Takeuchi, T., Amano, H., and Akasaki, I. (2000) *Japanese Journal of Applied Physics Part 1*, **39**, 413.
25. Park, S.-H. (2002) *Journal of Applied Physics*, **91**, 9904.
26. Pfeiffer, L., Störmer, H.L., West, K., and Baldwin, K.W. (1991) *Journal of Crystal Growth*, **111**, 333; Ko, H.-C., Yamaguchi, S. Kurusu, H. *et al.* (1996) *Japanese Journal of Applied Physics* **35**, L366, It has often been reported for conventional semiconductors such as GaAs and ZnSe that the growth in the nonpolar directions is difficult.
27. Gil, B., Hamdani, F., and Morkoç, H. (1996) *Physical Review B*, **54**, 7678.
28. Gil, B. and Alemu, A. (1997) *Physical Review B*, **56**, 12446.
29. Ghosh, S., Waltereit, P., Brandt, O. *et al.* (2002) *Physical Review B*, **65**, 075202.
30. Alemu, A., Gil, B., Julier, M., and Nakamura, S. (1998) *Physical Review B*, **57**, 3761.
31. Flissikowski, T., Omae, K., Misra, P. *et al.* (2006) *Physical Review B*, **74**, 085323.
32. Ghosh, S., Brandt, O., Grahn, H.T., and Ploog, K.H. (2002) *Applied Physics Letters*, **81**, 3380.

33. Marchand, H., Ibbetson, J.P., Fini, P.T. *et al.* (1998) *Journal of Crystal Growth*, **195**, 328.

34. Hiramatsu, K., Nishiyama, K., Onishi, M. *et al.* (2000) *Journal of Crystal Growth*, **221**, 316.

35. Okamoto, K., Inoue, K., Kawakami, Y. *et al.* (2003) *The Review of Scientific Instruments*, **74**, 575.

36. Tsao, J. (2004) *IEEE Circuits & Devices Magazine*, **20**(3), 28.

37. Chierchia, R., Böttcher, T., Heinke, H. *et al.* (2003) *Journal of Applied Physics*, **93**, 8918.

38. Korona, K.P., Wysmołek, A., Pakuła, K. *et al.* (1996) *Applied Physics Letters*, **69**, 788.

39. Kornitzer, K., Ebner, T., Grehi, M. *et al.* (1999) *Physica Status Solidi B*, **216**, 5.

40. Torii, K., Deguchi, T., Sota, T. *et al.* (1999) *Physical Review B*, **60**, 4723.

41. Skromme, B.J., Palle, K.C., Poweleit, C.D. *et al.* (2002) *Applied Physics Letters*, **81**, 3765.

42. Sala, F.D., Carlo, A.D., Lugli, P. *et al.* (1999) *Applied Physics Letters*, **74**, 2002.

Index

Nitrides with Nonpolar Surfaces. Growth, Properties, and Devices. Edited by Tanya Paskova

ISBN: 978-3-527-40768-2